IT'S WARMER DOWN BELOW

THE AUTOBIOGRAPHY OF

SIR HAROLD HARDING, 1900-1986

Sponsored by

THE BRITISH TUNNELLING SOCIETY

Edited by AMANDA DAVEY

TILIA PUBLISHING UK

First published 2015
by Tilia Publishing UK
Sussex, UK

ISBN 978-0-9933965-0-2

© Amanda Davey

All Rights Reserved

Designed by Tilia Services
Printed by CPI Group, Croydon

Cover photograph: Tapestry chair cover 'Puffing Billy' designed and stitched by Sir Harold Harding

DEDICATION

to

Sophie
Caroline
Edmund
and
Robert

CONTENTS

FOREWORD by Douglas Parkes	i
PREFACE by Amanda Davey	iv
ACKNOWLEDGEMENTS	vi
INTRODUCTION	vii

PART 1

1:	RUN UP TO THE START	1
2:	TUBE TUNNELS	7
3:	DOWN TO EARTH IN THE EAST END	18
4:	GO WEST YOUNG MAN	23
5:	OPEN AIR – THEN BACK AGAIN 1929-1930	37
6:	CRISIS AT FORDS	46
7:	'THE SPECIALIST'	57
8:	BACK TO THE BIG TIME	73
9:	THE WAR AND THE BLITZ	76

PART 2

10:	WIDENING VISTAS – THE RISE OF SOIL MECHANICS LTD	89
11:	DOWN TO THE SEA IN SHIPS	103
12:	COMPANY DIRECTOR	107
13:	THE PERSIAN EXPEDITION 1954-56	124

PART 3

14:	CHANGE OF LIFE	134
15:	CHANNEL TUNNEL 1958-1972	137
16:	TRAVELLING BROADENS THE MIND	163
17:	ARBITRATION UNDER ENGLISH LAW	188
18:	THE ABERFAN TRIBUNAL	192
19:	ARBITRATION UNDER INTERNATIONAL LAW	204
20:	SERVICES TO CIVIL ENGINEERING	224

PART 4

21:	THE BRITISH TUNNELLING SOCIETY	228
22:	LAST YEARS AND A SIGNIFICANT LEGACY	235
Portrait of the President by Hugh Golder		241
PEOPLE INVOLVED		242
TABLE OF CONVERSIONS		253
Selected Bibliography		254
INDEX		258

ILLUSTRATIONS

1. XXI Club 5th annual dinner 1932 — vii
   Family tree — viii
2. Letter sent to the two-year-old Harold by his father, from the ship on which he was travelling in his attempt to improve his health, shortly before his return and untimely death in December 1902 — 1
3. Life in South Africa — 2
4. Wapener, site of the siege, c. 1900 — 2
5. HJBH in the role of Sherlock Holmes for the play *The Speckled Band* (1918) while in the Officer Cadet Battalion — 4
6. Uncle Jack Harding - stage name Rudge Harding - as the Prince of Wales in *The Scarlet Pimpernel*, Orcszy/Barstow 1905 — 6
7. Greathead Shield at Mornington Crescent showing clay brought down from the face — 8
8. Surveying and setting out London's tube - HJBH annotated sketch — 11
9. The engineer lines the instrument with the wires and sets out the angle up the tunnel — 12
10. Miners working in the shield — 13
11. Sketch of a timber heading in bad ground by 'Bumper' Harris — 15
12. Timber heading — 15
13. Impression of the intertwined tubes at Camden Town — 16
14. Sketch by Segrave showing the conductor and the vestibules — 17
15. Driving the supply tunnel showing the horizontal bars supporting the timbering in preparation for bricklaying — 20
16. Brickwork arch under construction — 21
17. The head of the service shaft at Piccadilly Circus, showing the hut arrangement on the Island — 24
18. The *Morning Post* photograph celebrating the inauguration of gyratory traffic in Piccadilly Circus, July 27th, 1926 — 25
19. Angels Guard Thee…Creed Vowler (l.) and HJBH (r.) Civil Constabulary Reserve during the General Strike of 1926, as drawn by SHBL — 28
20. The completed 'stomach' model of Piccadilly Circus station and connecting tunnels, made by HJBH and Sophie Blair Leighton. The model is now in the London Transport Museum — 30
21. Thank you note from H. Dalrymple-Hay for the setting out at Piccadilly — 36
22. Construction of Fords Power House, Dagenham showing the unusual siting for such a heavy structure — 47
23. Timber jetty for cooling water intake shafts at Fords Power House — 52

24. Stripping piles for pile caps	52
25. Standard diving suits have connected air tubes, which although cumbersome can be life-saving in complicated conditions	56
26. Consolidation of the ground below Bentall's Department Store	65
27. Poster for the film *The Tunnel*'s release in America as *The Transatlantic Tunnel*	66
28. Caisson ready for lowering, Pigeon House Outfall, Dublin 1935	67
29. Chemically solidified ground between tubes in compressed air, Dublin 1935	68
30. Sewer diversion Bow to Leyton extension, jetting down a well point after the first 4ft of trench was excavated	75
31. Map showing the intercepting, outfall and main storm relief sewers of the London drainage system from *Engineering Wonders of the World, 1909*	79
32. Ground-water lowering by well-points for bomb recovery at the National Physical Laboratory, Teddington	82
33. Thank you letter for recovery of the unexploded bomb (UXB) at the National Physical Laboratory	82
34. Barge assembly line at West India Dock	86
35. Concrete caissons forming main deep water breakwater for the Mulberry Harbours at Dieppe	86
36. Petrol barge ready for concreting the cross walls	87
37. *M.V. Grosvenor* driving box piles for an anti-submarine boom, Spithead, 1951	105
38. *M.V. Ebury* off Milford Haven	105
39. Robert Harding and friend as footmen to the Marquis of Bath at the Coronation	115
40. Bridge at Jammersburg Drift	123
41. Road building in Spain	126
42. Canyon country in Iran/Persia	127
43. Investigation work for the Channel Tunnel. Sparker plot	140
44. M. Malcor *et al* down the 1880 Sangatte tunnel	141
45. Original stark report produced by the Channel Tunnel Study Group	148
46. Shipping lanes in the English Channel, published by the Dock and Harbour Authority, 1961 - part of HJBH's scheme review	150
47. Artist's impression of the finished tunnels as proposed in the 1961 version, published by the Dock and Harbour Authority	150
48. HJBH (centre) on site at the Channel Tunnel stage 3 at Bruay	162
49. HJBH making notes on the Litani Project	164
50. The headline in Sweden caused some international exhanges of pronunciation	170
51. The Temple of Quetzalcoatl (the Feathered Serpent) at Teotihuacan	173
52. Twenty-three mile mouth of Chesapeake Bay bridge-tunnel, rough sea	179
53. DeLong platform at work during construction of the Chesapeake Bay bridge-tunnel	179

54.	Courtyard of the hotel on Paradise Island	180
55.	The cloister 'ruins' at Paradise Island Hotel	181
56.	Pipeline and power station, Norton Dam	183
57.	Pipe tunnel	183
58.	Arch over the road to Columbo	184
59.	The Report of the Aberfan Tribunal and its accompanying documents were widely read for the insights on slope stability	200
60.	Fish River intake, Hendrik Verwoerd (HV) Dam	212
61.	View across the top of the HV Dam	212
62.	Sophie, Mrs H and Mrs B at the HV Dam site	214
63.	Mac Mac Bridge	217
64.	HJBH as President of the Institution of Civil Engineers	225
65.	Poster for a 1973 lecture given by HJBH	227
66.	A partnership that lasted for nearly 60 years. HJBH and Sophie after repairs completed on the model of Piccadilly Circus that had cemented a pattern of collaboration	237
67.	The year that the Channel Tunnel was commissioned, the *New Civil Engineer* celebrated a key figure in the history of the project with a place on the cover using this photograph	239
	Portrait of the President	241

FOREWORD
Douglas Parkes

In civil engineering and especially in the close world of tunnelling, many people know each other through many routes and paths. So it was with my knowledge of Sir Harold Harding (HJBH). A close friend of his son, Edmund, from our shared time at Brand's, I was also privileged to spend much time with him in connection with the British Tunnelling Society from its earliest days and we became firm friends as a result. This book has taken a slow path towards publication but is not the less for that. In it HJBH describes starting his training in the twenties with John Mowlem under one William Rowell, a contemporary of Greathead. His first job was a step-plate at Euston: he caught the tunnelling bug at an early stage. However, he coined one of his famous 'Hardingisms' in connection with his work as Mowlem's agent on Ford's Power House at Dagenham – 'Experience should be measured by intensity'. That intensity was in some measure due to inadequacy in the initial risk strategy for the work. He had many and varied tasks during that first, post world-war reconstruction period, amongst them the construction of four hand-dug deep pile/pad foundations for the second phase of Lutyen's Britannic House, as I was fascinated to find out when my firm were involved there. He was the site engineer responsible for the setting out of Piccadilly Circus Underground Station and he and his future wife made a model of the work, partly to help understand what he was building: that model is now in the London Transport Museum.

He must have stood out, because he was, after some ten years with Mowlem, chosen as the firm's expert, to learn, promote and exploit the two new geotechnical processes of deep well dewatering and chemical injection of the ground. He ploughed a lonely furrow at this for some time until he was joined by Glossop, and later by Dr Golder who was then seeking general experience in the industry. He rejoiced in the value of being able to discuss the development of theory and practice with men who could speak his language. He had been involved with Golder, together with Cooling and Skempton when all three were at the Building Research Station. The three had produced their theory for analysis of the slip failure of the reservoir embankment at Chingford during its construction at the time by Mowlem. Robert Wynne-Edwards had been agent there and Glossop was one of his engineering staff. At that time too HJBH became friends with Professor, then Doctor Terzaghi who was called in and confirmed the BRE theory for the collapse. The investigative work at the time saw the establishment of the first, non-research laboratory on the site to enable the bank to be redesigned. Out of this was born, one night in Glossop's site caravan during the air raids of 1941, the blue print for Soil Mechanics Limited.

Sir Harold was later made first Works Director of Mowlem, and subsequently the first non-family Director of the firm, with Glossop. He continued with his pioneering work and enjoyed building up the firm's fleet of sea-going craft for marine geo-technical investigation and foundation work on a profitable basis. In 1956-7 circumstances dictated that he was to take his friend Terzaghi's advice and became an independent individual consultant. He took note of Terzaghi's adjunct that a consultant is a person who is supposed to know more about the subject under consideration than his clients, and yet recognised that one's personal practical training never ceases. He set great store by the value both of experience and the interaction of minds in problem solving. In another Hardingism he stated that the best engineers are those who treat the forces of nature with humility until they are sufficiently sure how they can outwit them and convert them unsuspectingly to the use and convenience of man. I need hardly comment further on the aptness of that one.

His recorded prescription for one over-confident individual was: 'Send him to me and I will put him down a great deep hole and teach him humility', and yet he recounted that he never failed to learn something that he did not know before from his many sittings as judge at student and graduate papers. He was keen that engineers should have both a sound academic grounding and a broad practical training and that they should learn how to express themselves well. He was intensely proud of his chosen profession, comparing the ubiquitous skill of the engineer favourably with that of a surgeon – and yet he was at pains to point out that there was no such thing as a 'British Standard Engineer' and that variety in skills, aptitudes and abilities was essential to fill the needs of our varied calling. He was instinctively sceptical of too much reliance on mathematics, without the leavening of the understanding of engineering principles, and was disposed to a little gentle teasing of the mathematically inclined. He published many papers, but used to boast that none of them contained 'π'. He applied his 'Coefficient of So-Whatness' to test the validity of any paper or formula in danger of offering overly theoretical solutions that bore too little resemblance to reality.

More than this, he deplored attempts to raise the academic standards for engineers beyond what was necessary, simply because this could exclude from the profession those of sound engineering instinct who could make useful contribution within the 'party of the extreme centre (of the engineering profession)' to which he modestly assigned himself, 'made up of those who conceive, give birth to and maintain civil engineering works'. He was not decrying the search for the best brains, but warning against any tendency to create an intellectual apartheid.

He was blessed with an extraordinary memory. Perhaps this was one reason that he set so much store by experience, for he was able to store facts and recall them in great detail later when a situation arose to which they were relevant. He recounts going to the aid of a bomb disposal squad who were in some difficulty with wet Thames ballast and telling them precisely how many wellpoints[1] they needed and how they should be sited. He would remember the plant and equipment necessary for large and complex operations, no doubt a useful estimating asset. But he was careful, when providing plant for use at a distance, to see that there was sufficient. He was, reputedly, taken to task for sending two piling hammers to an overseas project when only one was needed. He responded with Harding's First Law (apologies to Murphy) 'If I only send one hammer they are bound to drop it in the drink, but if I send two, then neither will be lost'.

He had views on the status of the engineer in society which are just as relevant today, but I will leave you with one prophesy which he did not live to see fulfilled, but which may startle those who have not recently re-read his presidential address of November 1963. He was referring to the future effects of the population explosion: 'But that explosion may produce a greater danger that humanity might find itself being choked by its own waste products, atomic, industrial and human. So in the future the work to prevent this may become our major task ---'.

---

1 Wellpoint: small diameter tube with slots, inserted into wet ground - water pumped out by vacuum pump.

Sir Harold's final major construction task, and the one for which he is more generally remembered, was the investigatory preparation for the Channel Tunnel. He was appointed, together with a French counterpart, M. René Malcor, joint Engineer to the Channel Tunnel Study Group – a title he later said was misguided, because it gave grounds for accusations that his work was biased and gave little credence to other types of fixed crossing. He argued that it should have been the Channel Crossing Study Group, since his studies were directed to determining the best form of crossing in an iterative process that sought to establish what form the crossing should take in the light of all known facts and conditions. The study brought into use many items of marine plant for investigative work which had given him almost schoolboy pleasure in his Mowlem days, although by that time many of the fleet employed belonged to George Wimpey and not to his old firm. His work was still the basis for the route the second time round, although there were some relatively minor changes, particularly on the UK side where the route was modified to avoid an area of suspected poorer ground. I say 'suspected' advisedly, for nothing is certain in tunnelling except the uncertainty of the ground, and here I will quote perhaps the best known Hardingism of all: 'The only borehole that can be relied upon to describe the ground conditions precisely would be one the length of the tunnel and several feet larger in diameter'.

He is quoted as ascribing to his friend Terzaghi investigative skills that would have made him a detective in the same class as Holmes, but he was no mean sleuth himself and shortly after completing his work on the Channel was called upon to be the third member of the tribunal charged with investigating the tragedy of Aberfan, when so many children were engulfed by the collapse of the coal waste tip above their school. There followed recommendations for the management of such tips that have protected those potentially at risk from them ever since.

His last years were spent still involved with much that was going on in civil engineering and in his world. Founding Chairman of the British Tunnelling Society, I had the unexpected request from HJBH to act as a character witness in an unlikely situation, which Amanda describes in her chapter tying up the stories from his last decade. It is quite remarkable and characteristic that right up to his last weeks he was lobbying for the Channel Tunnel and lived to hear that a twin-bored tunnel had again been commissioned.

PREFACE

Amanda Davey

Sir Harold John Boyer Harding (HJBH) was my grandfather. There were six of us: two children to each of his three children. Civil engineering plays little part in a child's world and we knew him as our grandfather rather than as anything else. Many years ago I watched a recording of a 1960's TV interview with Carl Jung. Jung was asked what his grandchildren felt, to be related to such a great man. Jung looked back through his wise gentle eyes and said simply, 'But I am their grandfather to them!'. This is what it was like for us. We knew that he had presence and stature, but I don't know that any of us had an inkling of the real contribution that he had made to civil engineering and our modern world.

My father had hoped to be able to pick up where his father finished with his writing, but unfortunately events got the better of him and the files lived in a cupboard after his death until I brought them home. This exercise has been one that has been a joy, a privilege and a revelation. HJBH's previous book, *Tunnelling History and my own Involvement* is a profoundly worthy work, but all who knew him and valued him were a little disappointed because it was less obvious that he had inserted the flourishes of humour that enlivened all of our lives with him. This autobiography is not only about a great deal more of his experiences beyond the tunnels, but it is also full of quirks and foibles and plentiful in humour and it has been very good to be part of bringing it into publication.

HJBH was a civil engineer of distinction and a grandfather to be inordinately proud of and to appreciate in his humanity, heart and kindness. He helped me with my undergraduate dissertation with funding, love, encouragement and some very sound advice about how to write in short comprehensible sentences. My contribution to this, his autobiography, is in a way a form of repayment for the training and help that he gave me at the very same time that he was writing it.

Although he had been requested by several publishers in the 1960s and 1970s to write his autobiography, by the time he was ready to start submitting it, publishing was in the doldrums and his life was considered to contain insufficient 'racy' material. While the contents of this book are not especially salacious in the terms favoured by publishers of the 1980's, thankfully publishing has developed more diverse taste. Now his stories are part history and part an instruction in how to look more carefully under the bonnet at how things have happened and continue to take shape.

It is a curious thing to be the daughter of a civil engineer and the grand-daughter of a civil engineer. It is not a world renowned for equal opportunities, although things have become much better in recent years. My genetic instincts were in tune when school geography suddenly diverted itself away from coals to Newcastle and brought in the joys of rivers, glaciers and volcanoes in Physical Geography. This in turn led to a Geography degree at the University of Exeter, coincidentally just up the road from my grandparents (not part of the plan, but a great source of joy). I am not too sure that HJBH was aware that my passion for geomorphology existed. I had yet to realise just how important slopes were to him! How self-centred we are as youngsters. We shared a great deal of interest in maps and in history. There was virtually no subject upon which he didn't take an interest and he would frequently reach for his trusted *Encyclopedia Britannica 1911* to check his remembered facts

- an action continued throughout this book.

Initially, I followed a map career, getting the post of Map Curator at Sussex University in the weeks before his death. Later, I retrained as a Landscape Architect. Quite a bit of Civil Engineering and Landscape Architecture overlap. This has meant that going through this text, much of which is fairly technical, I have understood more than I expected to be able to. I have also had the pleasure and luck to have Douglas Parkes, another important tunnelling engineer, as advisor and mentor for this book, for which I am very grateful indeed. He has been a friend to our family for many years and is also my brother's godfather.

It is the measure of the man HJBH was that in the last couple of months of his life, although he was very ill he still fought for the future of the Channel Tunnel. He did not find his illness easy, so it was no mean feat that he was typing letters, sending off photocopies of letters published in *The Times* and of articles written following interviews with him about the options being discussed. Mrs Thatcher, as Prime Minister, was keen to promote a bridge and immersed tunnel and not the far more straightforward sensible and safe twin bored tube tunnel as was in the end constructed. Storm events and loss of coastal cliffs of the early weeks of 2014 were an indication of just how unwise the bridge idea would have been and how important this lobbying was on behalf of the tunnel option. The English Channel is a dynamic and active place ravaged frequently by the powers of nature as well as by the wayward thrusts of distracted sea-going vessels.

My grandmother was a remarkable woman, married to a remarkable man and it is not chivalry that makes HJBH report much of what she contributed to their life together. Behind every great man is a great woman is oft said and little understood. I think she knew, very early in their relationship, that he was very talented and very driven. She was bright and intelligent enough herself to see that he would need her undivided support.

HJBH had a great sense of humour, which was professionally invaluable, but it did sometimes put him in interesting situations. One family anecdote has him at a social gathering where he was introduced to a man proudly sporting his school tie, a tie for a significant school that many people would recognise. Following the characteristic quip, 'Oh what year were you sent down?' the poor man blanched and shrank into the woodwork, whimpering 'Who told you? How did you know?' at which HJBH was seriously non-plussed.

ACKNOWLEDGEMENTS

This book has been through several phases in its existence and much help has been given in that time. While I am not wholly sure I can name all of those who helped in its first phases, some names need to be mentioned. The early drafts were sent out to a good gang of advisers, many of whom reported back with changes, corrections and suggestions. These advisors included the family; Caroline (whose instruction was to avoid being pompous), Edmund (who was happy to support whatever his father had in mind) and Robert (who believed that the entertaining stories were important). Professional support came from Hugh Golder and Roger Lloyd-Thomas, who had been the civil servant whose help had been appreciated by the Aberfan Tribunal and remained a staunch and detailed contributor.

In this recent phase, my own thanks are to be given to the great support from the British Tunnelling Society, who have been significantly helpful in bringing the book to publication. Particular mention must go to Damian McGirr (winner of the Harding prize in 2005 and Chairman of the BTS 2012-14), Roger Bridge (Chairman of the BTS 2014-16) and Shani Wallis (publisher of the journal *TunnelTalk*). Debra Francis, from the Institution of Civil Engineers Library has been a joy to work with and has become a valued friend as well. Greg James has also been of valuable assistance. Douglas Parkes has been nagging at me for quite some time to get going and his help has been hugely appreciated, in particular on the engineering and emotional support front, as is his Foreword. My uncle Robert has supplied missing bits of text and my sister-in-law Deborah Harding has been helpful in going through the text and resolving some of the bits and pieces that I got too close to. My cousin Claire and her father Philippe Oboussier have been supportive in so many ways and I thank them for it, including some very helpful proof-reading. John Bartlett, a co-founding member of the British Tunnelling Society and past Chairman, given his first job at Mowlem by HJBH, has also been very helpful with providing background and lending me his copy of the history of the Smeatonian Society. As word has spread about this book coming toward publication there have been many messages of support, all of which have been very gratefully received and appreciated. Gerald Fleuss has been of great help, particularly with regard to Edward Johnston and the London Underground typography.

I also need to thank members of my immediate family, for their support and encouragement, in particular Simon. All who have helped, thank you!

A proportion of the cover price for this book is to go to the Institution of Civil Engineer's Benevolent Fund. The construction industry is not always an easy ride and the Benevolent Fund supports casualties of the economic cycles and was an important early committee to which HJBH belonged.

The images on pages 225, 239 and 241 are reproduced courtesy of the Institution of Civil Engineers. The image on page 237 is reproduced with the kind permission of Chris Baker.

Every effort has been made to obtain permission from the relevant copyright holders and to ensure that all credits are correct. I have acted in good faith and on the best information available at the time of publication. I apologize for any inadvertent omissions which will be corrected in future editions if notification is given to the publisher in writing.

Contains public sector information licensed under the Open Government Licence v3.0.

INTRODUCTION

Harold Harding, KB

A reviewer once wrote in the Sunday Times, 'Autobiography is the most squalid of all literary genes'. The labour pangs which gave birth to this collection of genes came from the seduction by a publisher and the urging of many friends to bring a long period of gestation to fulfilment. Since the interest lies in the civil engineering, my focus has been on this aspect of my life in preference to the private, which consequently crops up only from time to time.

The period from 1900 to 1978 saw a third and fourth industrial and social revolution. At the start, 'the art of directing the great sources of power in nature for the use and convenience of man', was carried out only by man power, horse power and steam power. Through the twentieth-century, our work expanded enormously, due to the fresh sources of power by invention. Electric power, the internal combustion engine in all its proliferation by road and sea, air flight, the revolution in plant, the advent of atomic power, and not forgetting the telephone, radio, television and electronics, have been directed, not only for the convenience but also the destruction of man.

The story has been marked by much good fortune which nullifies the inevitable occurrence of bad so there are no real tears for the reader to shed. While I have no remaining early diaries, there were 116 letters written to the 'Twenty One Club' (XXI Club),[2] twice or more every year from 1923 to 1977. I have quoted extracts from these letters since they give an immediate colour to some experiences that would be hard to reproduce at a distance of time.

1. *XXI Club 5th annual dinner 1932.*

Standing, left to right: K. E Hyatt, K. C. Wilson, R. McC. Briggs, W. H. G. Roach, G. F. Hyams, B. J. Kirchner, R. D. Knight, J. C. G. Vowler, A. E. Coulson, J. Duvivier, J. F. Pain, A. C. Vivian.

Sitting left to right: H. J. B. Harding, J. A. Van Heel, R. N. R. Briggs, W. R. Grigor-Taylor, A. M. Holbein, A. McVie, R. T. James

---

2 The 'Twenty One Club' was founded in 1922 and formed of 28 original members from the City and Guilds Engineering College, who mostly graduated in 1921. Half were ex-service students and, although they covered several 'years' and different faculties, they had become friends by being leading activists in the social and athletic sides of the College Union.

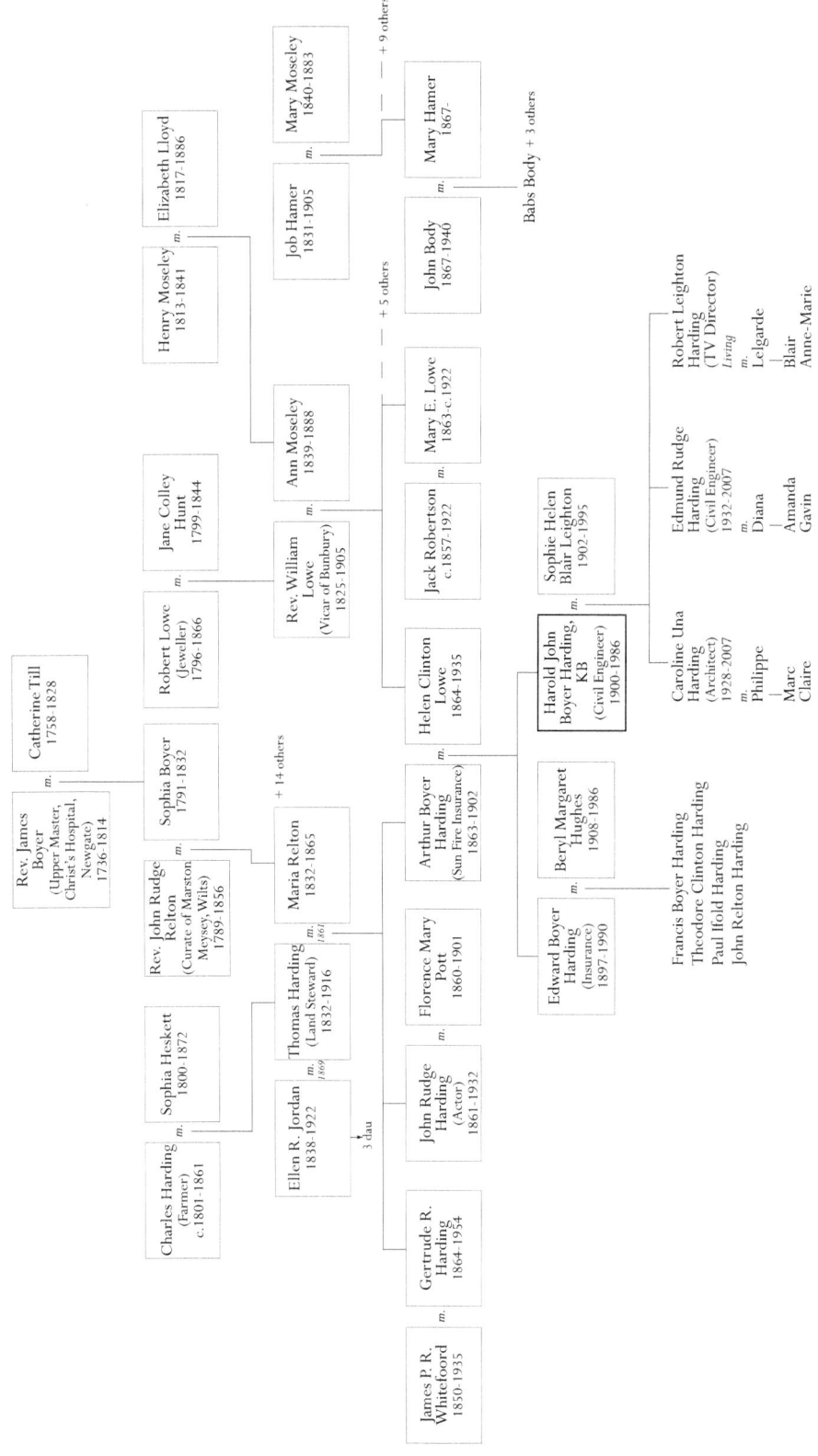

PART I

1

RUN UP TO THE START

My first escapade as a civil engineer came at the age of ten after my elder brother and I (at our request) were given three bound volumes of *Engineering Wonders of the World,* as we had been excited by an occasional fortnightly part. I used to browse over them during every holiday. At the age of eleven on Exmouth sands I tried to drive a shield tunnel from a hole in the sand using a Bath Oliver biscuit tin with the end cut out as a shield. This was inspired by an article in Volume One entitled 'The Tube Railways of London', by Harley H. Dalrymple-Hay. This was to be unexpectedly influential on my future career.

I had been born on January 6th 1900 (so always knew my age), on the fringes of Clapham Common. My father, Arthur Boyer Harding, an official in the Sun Fire Office, was the son of Thomas Harding, land steward to Lord Calthorp at Elvetham, Hartley Wintney, Hampshire. He came from a very long line of alternating yeomen, squires and gentlemen farmers largely in Wiltshire and Somerset.

Thomas Harding had married the daughter of the Reverend John Rudge Relton who had married the daughter of the Reverend James Boyer. Boyer was the grandson of a Huguenot refugee who came to Britain after the repeal of the Edict of Nantes. He has a small niche in fame as, after being educated at Christ's Hospital, he became Upper Master and taught Coleridge, Charles Lamb and Leigh Hunt and is immortalised in their essays. When he died, Coleridge wrote, 'How fortunate that the cherubs who carried James Boyer up to heaven had no bottoms or he would have assuredly flagellated them on the way'. This and later experience led me on an occasion to express a dislike of the old-fashioned insistence on parson headmasters. On

2. Letter sent to the two-year-old Harold by his father, from the ship on which he was travelling in his attempt to improve his health, shortly before his return and untimely death in December 1902

Sunday they would preach 'Suffer little children to come unto me' but on Monday morning it would be 'Come unto me little children and suffer'.

My father married my mother, Helen Clinton Lowe, in 1894 in London. She was the second of six daughters of the Reverend William Lowe, a country parson at Bunbury, Cheshire. The Rev William Lowe's father Robert had been a goldsmith in Chester. My family tree has parsons nesting in every branch, shows no connection with the 'unacceptable face of capitalism' and only a distant one with civil engineering.

3. *Life in South Africa*

4. *Wapener, site of the siege, c. 1900*

My father died in December 1902 when my brother Edward was just six and I was verging on three, leaving us with negligible income. My mother's elder sister had married Jack Robertson, one of a family of Scottish pioneers from Blair Atholl. He and his brothers made farms, built their own dams, built mills and also opened up Basutoland (the Kingdom of Lesotho from 1966). Uncle Jack promptly had us out to live with them at Jammersburg Drift on the Caledon River in South Africa on the Basutoland border. Two years before, his farm and the adjacent kopjes (small hills) had been the site of 'The Siege of Wapener' in the Boer War and the memories were fresh.

My aunt and uncle had no children due to the results of a miscarriage triggered by an accident to a Cape Cart when crossing a spruit in flood and Uncle Jack and I became deeply devoted to each other. My mother, brother and I returned to England in 1906 after four years clearly remembered. Uncle Jack had fed and clothed us and supported us completely. He became a Governor of Christ's Hospital in Horsham, Sussex to get me in there and financed us at the City and Guilds College. Sadly, he was to die just after hearing that I had completed my course in 1922.

I spent eight years between 1909-1917 at that wonderful but bleak and tough school, Christ's Hospital, where we made our own beds, cleaned our own shoes, waited on each other and marched into Hall every lunchtime to the school band. We learnt discipline and self-reliance. I struck a very bad patch for a year from September 1912 when I came under a classical master of the old sadistic style who, when he was not telling me that he saw no hope for me in future life at all, kept repeating 'the fact that you are always top in History and bottom in Latin proves that you are idle'.

The basic problem of specialist education was solved by a man to whom I owe a great deal. In that hot-bed of the classical education, T. S. Usherwood was allowed to form an engineering side and I was almost the first to sign on. Usherwood then told us that we were to be relieved of many periods of compulsory Latin and Greek, but, as engineers are bad at expressing themselves, we were to do extra periods of English literature, English composition and English grammar. As well as languages, mathematics, chemistry and physics a few periods were devoted to sound grounding in solid geometry and mechanical drawing which was interesting, educational and eventually valuable and gave us a sense of vocation.

Usherwood had a London degree in mechanical engineering and was a fine mathematician but he falsified the idea that there are two cultures, as he showed that he knew more about art, literature, music, ballet and drama than any of the Arts degree masters from older universities. Most of his boys obtained scholarships, which in those days were few and far between. When he found that I was slipping back in mathematics under an old master retained through war-time conditions, he gave up two evenings a week to give me free coaching so that I could scrape into the City and Guilds College. I want to use this opportunity to express my profound gratitude to his memory as well as to point a moral: the best brains and boffins can look after themselves. I decided to become a civil engineer at the age of 10 and throughout my adolescence this statement was usually met with surprise and sometimes derision, as there were no marked signs of mechanical aptitude or mathematical genius. But there is much to be said for following the example of Sir Charles Inglis and Usherwood in helping along those who know what they want to do, rather than in concentrating on gaining credit for the intellectual high fliers.

I duly went to the City and Guilds College[3] in October 1917 and was promptly enrolled in the London University Officers Training Corps (LUOTC). After three months of lectures followed by evening parades with the LUOTC, I became 18 in January 1918 so was automatically given leave from the college and started full time training with the LUOTC. The HQ was under the long timber examination hall behind the old Imperial Institute, which was the centre of London University Administration. Behind the Imperial Institute were two quadrangles left over from the 1870 exhibition surrounded by long-vanished exhibition galleries. The west quad was a depot for the Office of Works and was filled with horse drawn wagons with an odd name carved on the dash boards: 'John Mowlem and Co.', who had built the Institute. In the east quad we were drilled, except when doing company drill, with great precision, in the Imperial Institute Road, under a regimental sergeant major from the Guards. Company drill seemed to be the vital training for trench warfare, but I found later that it also helped with public speaking. We each in turn had to drill the company, bellowing out commands: 'Details for a company halted in line advancing in column of platoons from the right – the command will be' etc etc.

Naturally I put in for the Royal Engineers. At my medical at Duke of York's Headquarters I was reading off the carefully learned bottom line when the medical officer said that he had turned the card around, so I was passed Grade B.1 Garrison Duty Abroad. The reason for the learned bottom line was that I was put into glasses at the age of eight for being short-sighted.

Then came the German breakthrough in March 1918 so, 'at a stroke' as the saying is, I was turned into A.1 Infantry. In June I was posted to No 2 Officer Cadet Battalion after enrolment as a private in the Middlesex Regiment at Hounslow Barracks and found myself billeted in Queens College, Cambridge instead of huts in wild neighbourhoods. In spite of

---

3 Part of Imperial College, Exhibition Road, London.

5. *HJBH in the role of Sherlock Holmes for the play* The Speckled Band *(1918) while in the Officer Cadet Battalion*

my short sight I passed as a first class shot both in the Army and later in the territorials. This included having to fire fifteen rounds rapid from the left shoulder, for which the rifle was not designed, owing to the fact that my 'lazy' (right) eye could not see the bull.

The army intention was that after three months training we should find ourselves in France, but as things over there were getting better, our course was extended, after 40 per cent had been returned to their units as not coming up to standard. There were three Online Combat Battalions billeted in the Colleges, most of them being made up of Australians and New Zealanders. We assumed that they were sent there for a civilising influence. Half of my company were callow youths, the other half were back from overseas service, including two with Victoria Crosses from Passchendale. This seemed an odd mixture to me.

Our pay had been increased from the historic shilling a day to ten shillings a week. The Aussies seemed to be much better paid and we envied seeing them dining in the University Arms Hotel which we could not afford to enter. On Armistice Night[4] the Aussie element burned all the dummies in the bayonet fighting school and then climbed the walls of the sacred precinct of Newnham College to the alarm of the lady dons but not of the undergraduates!

The course was extended into January 1919, when one day a number of us younger men were told to report next day to the Orderly Room and the buzz went round that we were to be asked to volunteer to join General Ironside at Archangel to fight the Bolshies. We felt that it was our duty to accept as we had missed out in the war. How stupid can one get. On arrival we were quickly disillusioned:

> You are students so you are to be the first out, report tomorrow at the Crystal Palace where you will receive ration books, unemployment cards and a gratuity of four pounds. You will be commissioned for one day without pay and transferred to Z, Reserve of Officers.

So I found myself back at the City and Guilds College at exactly where, in the course, I had left. I was a Second Lieutenant Royal Fusilier unpaid Z reserve. I wonder if they ever knew. In due course, His Majesty King George V sent me a kind letter thanking me for having 'done the utmost in my power to bring the War to a successful conclusion'. Considering the comfortable way in which I had done so, it was very decent of him.

Then followed three happy but anxious years. The anxiety was the need to obtain a 50% mark every half session to avoid repeating or rejection, but I was making many good friends as more and more older men were demobilised. All had to pay their own fees except the

---

4 November 11th 1918.

few with scholarships and there were usually about four cars outside in Exhibition Road belonging to the few who had fathers with money. Fifty years later one could not find parking space in the whole Imperial College area as every space was filled by cars belonging to poverty stricken students starving on their grants.

In those days firms hated the sight of students so it was difficult to get vacation experience except through introductions. At the end of the first year, my contemporary Grigor-Taylor managed to find a month's work for us both at Robey and Co's works in Lincoln. They made heavy machinery, colliery winding engines, traction engines and so on. A speciality was the Uniflow Steam Engine which had central exhaust ports like a gas engine. Back at College, Professor Mitchell urged us to write papers for the College Engineering Society, he said that they might be of little value to our fellows but of great value to the author.

So to fill the gap in my mechanical deficiencies I wrote a paper on the Uniflow Steam Engine which was to be the first of many. This was valuable experience as it led me on a path of detection. Most useful of all, it led to the discovery of the little known Patent Office Library. Ten years later this became useful to my then firm as well as to me (see Chapter Seven).

In those days the college awarded its own Diploma, the ACGI (Associate of the City and Guilds Institute) and this was accepted by the learned institutions towards an Associate Membership. If a degree was desired one had to have taken the old London Matriculation and Intermediate and then sit for a separate exam three weeks after the College finals to a different syllabus with outside examiners. It was all a bit complicated.

In 1921 a mass meeting was held in the Central Hall, Westminster, of old and present students, chaired by Lord Rutherford. H. G. Wells also spoke, as an old student of the Royal College of Science. The meeting (with one dissentient) demanded UDI[5] for the Imperial College and the result was that some years later, London University allowed the college to grant BSc on the results of the college finals. After getting my AGCI, which was all that I needed, I sat three weeks later for the BSc. In Structures there was a question on a rolling load on a bridge carrying 50%. This takes a long time with cunning diagrams and tricky calculations. After two hours I had successfully rolled my load onto the bridge and then found that under pressure I had misread the question. The load should have rolled OFF the bridge. In the eyes of examiners this is unforgivable.

I found that as an ex-service candidate, I need only repeat in that one subject, so I did a fourth year post-graduate course in Structures with lectures from Sir Ralph Freeman and Oscar Faber. This gave valuable time to digest what had been hurried through in the first three years. I emerged not only with a BSc, but also a DIC (Diploma of the Imperial College) for a post-graduate course in Theory of Structures. In those days, no-one had heard of the Hardy Cross relaxation method or plastic theory and our computers were slide rules and seven figure logarithms. In spite of my diploma, I detected no spark of structural genius in myself, but luckily, as life unfolded, it was seldom called into play and gradually eroded away from lack of exercise.

Then came the search for work in time of slump. My father's brother and my uncle, John Rudge Harding, had been a well-known character actor with Sir John Hare, Dame Madge Kendal, the Terries and other actor managers. In 1914, at the start of the war, he had been invited to work for the British Red Cross Society and eventually was appointed the first Secretary of the Star and Garter Home at Richmond which by 1922 was being built by John

---

5 Unilateral Declaration of Independence, a concept made famous in the 1960s with UDI for Rhodesia.

6. Uncle Jack Harding - stage name Rudge Harding - as the Prince of Wales in The Scarlet Pimpernel, *Orcszy / Barstow 1905*

Mowlem and Co. In one of his meetings with Mowlem's secretary, Alfred Marsden, my uncle asked him if he knew any civil engineering firms who might be likely to want someone. The result was that I received two letters by the same post: one from Sir Robert McAlpine offering me ten shillings a week to hold the end of a tape on a road in Essex and the other from my uncle who said that if I called on Alfred Marsden, I should be offered a job at three pounds per week. The choice was not difficult.

In the slump, you took a job first and asked what it was afterwards. I found that my first job was the extension of the City and South London Railway from Euston to Camden Town to join up with the Hampstead Line and the engineer was Harley H. Dalrymple-Hay. So after living on family charity all my life so far, I became independent and found work which exactly suited my particular talents.

2

TUBE TUNNELS

In October 1922 I made the journey by underground from Ealing Common to my destination of 158 Hampstead Road in a mixed mood: unease as to what sort of people I should be working with, their personalities and their pecking order; and elation at starting a career on a magnitude which could prove a battle-honour in a future job reference. With me I carried an open memoranda handed to me by the secretary which is the only document that I ever saw to prove that John Mowlem and Co. would pay me for the next thirty-four years. Addressed to a Mr Rowell, it merely said 'This is H. J. B. Harding of whom we spoke, he is to be paid £3 per week, monthly'.

While I speculated on the future, I was fortified by knowing something of the contract, as a fellow student had given a talk to the College Engineering Society a year before on the plans of the London Electric Railway Co. The plan was to separate the journeys to the City and to the West End by means of the Camden Town flying junctions to connect the Hampstead Line with the City and South London.

158 Hampstead Road turned out to be the end house of a Victorian terrace, facing south over the open space of a front garden and the wide railway cutting where London and North Western trains began to enter Euston Station. Mr Rowell, in his late sixties, gave a non-committal grunt and handed me over to James Segrave, a cheerful character with enormous George Robey[6] eyebrows. It quickly turned out that he had been at the City and Guilds College twenty years before, so little interrogation was needed. We were to work in a large airy room with wide outlook, the last such comfort for the next ten years. On bare boards stood an enormous kitchen table, a high drawing bench with high wooden stools across the window, three hard chairs and a plan chest. It had none of your modern West End and City comforts and no Pirelli calendars. It took several days to find out who else was involved and the history and running of the firm.

Segrave first bade me bring my oldest clothes and heavy boots next day and handed me a dust coat, saying that he must now go below and would take me with him. The dust coat was for visitors and no protective clothing was ever provided, we had to buy our own. The obligatory hard helmet had not been invented but some foremen and old engineers wore bowlers as protection. Civil engineers on working sites are in effect on active service, whether on the Zambezi or in the centre of London. Guards officers and city gents might well feel that the bowler hat and rolled umbrella were indispensable walking-out gear, but we would never dream of appearing on the site in such accoutrements. It was years before I used either.

The first working site was outside Mornington Crescent Station, where a sort of palisaded

---

6 Robey was a music hall star.

7. Greathead Shield at Mornington Crescent showing clay brought down from the face

village had been built on a wide island. Electric cranes moved on heavy timber gantries connected to 'muck-hoppers' which fed the excavated material down to the new-fangled motor lorry below, or at times of stress a horse-drawn tip-cart. What first caught the eye was the large statue of Cobden[7] in his baggy trousers and frock coat emerging through the roof of the compressor shed, apparently unceasingly haranguing the passing stream of open-top buses and horse drawn drays.

There seemed to be much work to be done before the main shield drives could start. Two shafts lined with cast iron segments had been sunk. Why two shafts? Although most of the drives were twin tunnels side by side, at this point the future southbound tunnel had to be driven under the Hampstead Line tubes to junction with them from the east and so were at a lower level than the northbound, which came in at the same level as the existing tube. We climbed into a large bucket or skip and were swung out over surprised pedestrians and then disappeared down a hundred-foot shaft. Then we went along a short access tunnel to where strange noises could be heard, this was the grunting noise made by miners to release the stomach tensions as they pecked away at the London Clay. As an example, the same sound could be heard when tennis star Jimmy Connors served at Wimbledon. The greatest surprise to me was the hardness of the London Clay when found deep below the surface.

It gradually became clear that the gang of miners, miner's labourers and general labourers were starting the 'break-ups' for the running tunnels at right angles to the access heading. This heading was a complex affair to make a chamber of narrow width in which to build the first two rings of cast-iron lining. The clay seemed so hard and stood up so well that I asked why there had to be such heavy timber support to the top and sides. I learned my first important lesson, that London Clay can be treacherous, large blocks can slide out along greasy 'backs' as it is a fissured clay and, on large faces, has been known to kill. Also when exposed, the clay gradually exerts a remorseless pressure unless discouraged by early restraint. So, much work had to go on to give enough room to build chambers (two in each) in which to erect four shields for driving away in both directions. I was introduced to our two chainmen; nimble cockney Curly Bennett and elderly silent Scotty. They were the engineer's attendants who carried the instruments, held the end of the tape and carried chainmens' bags full of plumb bobs, chalk, hammers, chisel, nails and much else beside. Their title dates back to the 66 foot steel chain which preceded steel tapes and fixed the length of a cricket pitch.

A few days later, I was sent down with Curly Bennett to check the work, which involved hanging down plumb lines tucked around bolts to be out of the way. In order to tease a newcomer, the miners had tampered with them, when the first was swung into place a used contraceptive had been added to the plumb bob. The miners told Curly to take it home

---

7 Richard Cobden, MP, member of the Anti-Corn Law League and peace campaigner is commemorated by a statue funded by public donations (1868) outside Mornington Crescent Underground Station.

for his Missus. He replied, 'My Old Woman says she ain't goin' to be artificial done – the real thing or nuthin' at all'. This was my first encounter with the almost extinct London tunnel miner, with whom I made many friends of a sort over the years. There was not an Irishman in sight. There were Cornishmen from abandoned tin mines, Cumberland Joe, Jimmy Devon, with many Londoners, one called Tall Faced Charley.

As days went by, the set-up and the history of the firm began to unfold. Mowlem had been founded in 1822 when the original stonemason, John Mowlem, took a ship from Swanage to London having been employed in the Tilly Whim cave-quarries. After working, like Telford, on Somerset House, he developed his business into a firm of integrity, starting with Billingsgate Fish Market. John Mowlem was joined by his nephew, George Burt (I) who continued the name out of gratitude, he was succeeded by Sir John Burt and his brother George (II). Sir John got his knighthood for work which, as a child, I had observed on the route while being driven by four wheeler coach from St Pancras to Victoria when visiting my brother at Christ's Hospital. The construction was the widening of the Mall, the building of the Admiralty Arch and the erection of the Victoria Memorial, before rotary traffic had been thought of, and the refacing of the old stucco of Buckingham Palace with its present noble facade, circa 1908.

Sir John had a son Edwin Burt, whose sons were later to become my own close friends. Sir John's brother, George (II), produced Sir George Burt (III) KBE, and his much younger brother, Eric. So in 1922, these three were the directors of what was a very private company. It was divided into three parts: Civil Engineering under Willie Rowell, responsible for deep foundations and tunnels as their speciality; the Building side, responsible for monumental buildings of fine craftsmanship such as the Port of London Authority building, Lloyds and others, built on the foundations constructed by us; and a rougher department who did road works, drainage and such less-qualified work. Beyond this the firm had such a reputation for integrity that they were entrusted with maintenance contracts for the Office of Works, the London County Council (LCC) and similar bodies.

The civil engineering staff in those days consisted of only six engineers, of whom I was the lowest form of life, so it was several years before the directors became 'Harding-conscious'. Willie Rowell had worked for the firm on the original City and South London tube, from which so much has sprung all over the world. His reputation was made by driving the Waterloo and City Railway (the 'Drain') in compressed air under and along the Thames in 1894-98. Harley H. Dalrymple-Hay was the resident engineer and devised the revolutionary hooded-shield, clay-pocketing method of tunnelling. Between these two a close friendship developed and both became established as leaders in this field. So I can claim to have been taught by the 'Founding Fathers'.

Mr Rowell's habit was to rent an empty house close to his largest contract and install in it his head timekeeper so that the wife of the latter could cook his daily lunch. He did not hob-nob with the next tier below him and also, very sensibly, it made him independent of the head office. There were three middle-aged engineers whom he kept at arms length, perhaps because one might eventually be his successor and there was one other, a little older than me, who was his personal assistant. Rowell spared no pains to instruct us younger men.

The hierarchy in most contractors' organisations is for a chief engineer to be the equivalent of an Admiral and each separate contract would be a ship individually captained by a man known as the agent. This is derived from the traditional contract conditions dating from Thomas Telford where 'The contractor shall appoint a responsible agent to carry out the work'. Today some prefer to be called Project Manager to improve their image.

On the Camden Town to Euston contract the agent was a Scotsman who had lost an arm in the War in a Royal Engineer Tunnelling Company. As the Camden Town to Euston tube was to have eight separate tunnel drives, as well as all the junction chambers to be built when the trains were stopped for the night, there was a second in command, namely Segrave, who had to control all the work below ground, while I was the rest of the 'officers mess'.

Each of the other two were in command of smaller parts of the work, such as sinking shafts in the basement of the Nash building of Peter Robinson, to carry the new foundations below the Central Line tunnels at Oxford Circus before rebuilding the shop and escalator tunnels at Tottenham Court Road. Curly Bennett had a slight contretemps when helping out at Peter Robinson. The 'muck' from the small six foot diameter shafts was wound up by a primitive jack-roll, which is a timber drum with a handle at each end as used on country wells. A miner asked Curly to let him down this deep shaft by putting one foot in the bucket and holding the rope. Curly started to lower him away, but the weight began to take charge. Curly had his back against the wall. He lost control of the handle, which hit him under the chin and knocked him back to the wall where he rebounded in time to be hit by the next revolution of the handle. This rhythm went on until the miner reached the bottom while Curly took the count of ten.

A contract worth £1 million in those early days, would in more recent years cost nearly 20 times as much, due to Parkinson's Law and the growth of paperwork, two or three engineers doing the work of one. In modern large firms, the central control is so distant that reams of paper, returns, cost analysis and constant reports are needed so that directors and accountants in their executive suites can create the illusion that they are in control. The clerical staff have likewise expanded. In a paper on practical training I once wrote that ours was almost the only industry in which duplicate staff are engaged to annoy and frustrate each other, this was rather sweeping, to enforce a lesson.

At Hampstead Road the office work was minimal. Time sheets had only a few columns for Name, Hours, Rate and Wage. Modern time sheets are immense with many columns in addition for wet time, holidays with pay, PAYE, allowances, travelling time and a lot more besides, so one timekeeper can only handle a dozen men. In those days Dickensian thrift prevailed. The shield gangs would only work on piecework per ring, but the day rate of pay had just been established by the Civil Engineering Conciliation Board which for years set an example seldom followed. The labourer's rate per hour was x, the miner x plus 6 pence and a miner's labourer x plus 4 pence - x varied with the cost of living index. From 1922 to 1938 x stayed steady at 1'/3 (one shilling and three pence in old currency). The agent, like a film producer, is responsible for organisation, the supply of material, general discipline, the paying of wages and making sure of the monthly payments to the contractor. The setting out engineer is responsible, like a film director, for the performance of the actors below and above ground, to see that the work is done as specified and that the tunnels are in the right place and that they eventually join up.

After a few weeks, I was made responsible for the day-to-day setting out – that is to establish lines and levels for the general foreman and the pit bosses to work to, and keep extending them as needed to suit progress. Incidentally, at the start there was a general foreman with the look of a Regimental Sergeant-Major (which was rather oppressive), but he suddenly left and I was told 'George was a first class foreman until he bought himself a set of drawing instruments'. The setting-out engineer can lose the firm much money if he

2. TUBE TUNNELS

makes bad errors, but has little chance to make money, so he is not highly regarded in head offices. The agent, on the other hand, can make or lose a profit by his own competence, so directors begin to take more interest.

People often ask how tunnels can join up so accurately. The simple answer is intense care, constant checking and in my day sound training in trigonometry and seven figure logarithms. Before the days of metrication, steel tapes had to be divided in feet and decimals of a foot, in order to make such calculations possible. I have drawn this little cartoon in

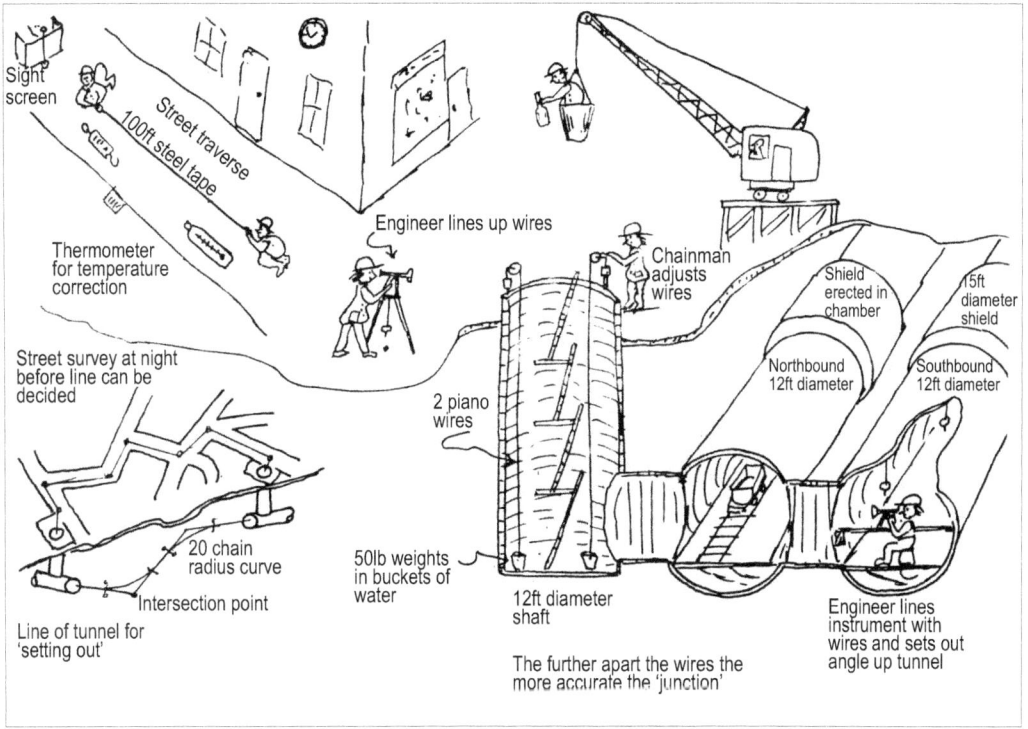

8. Surveying and setting out London's tube - HJBH annotated sketch

order to save many words (see Fig. 8). In a city a street survey is carried out, of necessity at dead of night, and this establishes the position and distance apart of the two sections to be joined. One suitable line is chosen to be north-south regardless of the compass. (In the past, time was needlessly spent in trying to establish true north by sights on Polaris). From this, all points can be given co-ordinates, as when reading motoring or Ordnance Survey maps. The engineer can then decide on the curves and gradients to join two stations. The tunnels used to be compelled to follow under streets except by special dispensation, which accounts for the extraordinary contortions which trains perform near South Kensington Station. But first, I was to test the vital steel 100ft tape against the British Standard 100 feet. To my surprise this turned out to be on brass plates set in the long stone seat in Trafalgar Square below the National Gallery. The tape was read while being pulled by spring balance at 10lbs and the temperature taken. Then, by looking up the coefficient of expansion of steel, a little brain work told at what temperature it should have read the 'true' 100 feet. The result was stamped on a brass label. When surveying or doing the opposite, setting out, a correction needed to be added or subtracted and I leave the reader to work out which.

Back to Camden Town; once a sufficient length of tunnel had been driven by simple approximate means there came the Ceremony of the Wires. This had to be done at night in dry and still conditions and, in my life, always on the coldest nights in winter. The chainmen fixed timber beams across the shaft as far apart as possible, which carried adjustable drums wound with fine piano wire. The wires suspended 50lb weights in buckets of water at the bottom of the shaft, to damp the swing. The further apart the wires, the closer could be the junction. The senior engineer then set up his ancient theodolite with immense care over the survey point and gazed up the road to the far station. Once, there was an obstruction to the line of sight, which Curly Bennett investigated, finding some drunks who had been thrown out of a club and so shovelled them into the gutter. The proper angle was then set out and the far wire lined up with timeless care. The angle was then measured many times on both faces and after transiting the telescope, if the errors were within reason the second wire was laboriously lined in. The junior engineer stood shivering by, to take down notes and the mind wandered to a school anthem, 'Watchman – will the night soon pass?'. Next, a climb 100ft down the ladders, carefully avoiding the wires, and the theodolite was then set up where the running tunnel crossed the access heading, where much time was taken up carefully lining the instrument with the wires. Then the chainman, in search of a cup of tea, climbed the shaft and accidentally knocked the wire which meant that everyone had to go back up to the top and start all over again. Then thankfully down to the tunnel again which was about 40 degrees warmer.

9. *The engineer lines the instrument with the wires and sets out the angle up the tunnel*

When truly lined up, the appropriate angle was set out and a point marked on a steel 'dog' driven into the joint between two rings of lining and checked by repeated turning of angles. Permanent marks were established which carried plumb bobs for going ahead. Grade was kept by a sighting along boning rods like inverted T-squares, suspended from adjustable hangers. That is a basic description of how work was done in those days. Now there are hydrodists, laser beams and other novelties, as well as much smaller compact instruments.

Figures 7 and 10 show a type of shield, six of which were driven simultaneously; a shield is not a digger, it is a protection. The clay was dug by hand, the cutting edge trimmed it to shape, the 'tail' held up the clay while the miners built the rings – six segments each of six hundredweight lifted by six men - which were then wedged tight by a small key segment. The jacks shoved and steered the shield in the direction which the junior engineer had set out by his lines and levels. Any ring one inch out of position had to be rebuilt at the contractor's expense. As the shield was long, once it faced wrong it took a long time to get back on line so the routine was not monotonous as there were daily crises. The senior engineer followed behind some weeks afterwards and checked the main survey lines at leisure. By the end of nine months one had had as good a training in shields in clay as ever likely to be needed, but there was much more to learn.

Much has been written elsewhere about shield tunnels but only absorbed by those who

2. TUBE TUNNELS

10. *Miners working in the shield*

are interested. Sir Marc Isambard Brunel invented the very first shield and laid down the principles governing his invention, which has never been challenged. His was a very large rectangular one, used for his amazing Thames Tunnel in 1825-1842, which now carries the trains from Whitechapel to New Cross. This amazing effort can be painlessly absorbed by reading the adventures of his son Isambard Kingdom Brunel in the biography by L. T. C. Rolt.

In 1869, Barlow produced his scheme for the tiny Tower Subway, using a circular shield designed by Greathead. This bore a strong resemblance to Brunel's patent and cast iron rings were used for the first time. Brunel had used brick lining but said that he would prefer cast iron for a circular shield. From this small beginning in Britain all such tunnels all over the world have sprung. As I had to repair the Tower Subway during the Blitz, it reappears in Chapter Nine.

The ground was not all London Clay. Two shields driven south from Ampthill Square ran into water bearing sand, sooner than expected. This was the lower Woolwich and Reading Beds, as variable as the London Clay was uniform. When the mottled clay appeared in the face, it was a surprise as it was like a brawn of many patches of blue, red, yellow and green and often called Shepherd's Plaid. It was much harder than London Clay. Simple air locks were built with two concrete walls twelve feet apart, five feet thick to take the pressure, and steel doors built into them. Large pipes were for the low pressure air and small ones for water. High pressure air electric cables were incorporated into the walls, not forgetting some spare ones. No one suggested telephones for safety. Medical examination before going in was perfunctory – a cough into a stethoscope and an attempt to provide 'a sample' into an empty medicine bottle. This valuable early experience was short as the sand petered out, so any thoughts on compressed air work will be left until later when there will be plenty of it.

In due course, electric locos, veterans of the Waterloo and City of 25 years before were installed on 18 inch track. They were powered by 600 volt bare cables hung from the roof, so it was advisable not to touch. The idea of transforming down the 230 volt lighting to 60 volts occurred to no-one, so instead of electric leads, the miners preferred to use dozens of candles in dark corners, as had been the immemorial custom.

The London Tube railway system was the first and is still the most comprehensive in the world. Deep tube tunnels started in 1886 with the City and South London Railway and coincided luckily with the invention of electric traction which made it possible. Its initial success set off an outburst of tunnels in the USA and elsewhere. In June 1975 I was able to organise an exhibition at the Institution of Civil Engineers (ICE) to celebrate the 150th anniversary of the Thames Tunnel and the development of British shield tunnelling ever since. This was later transferred to the Science Museum.[8]

The City and South London brought in developments almost unchanged until 1922: hydraulic power for the jacks; high pressure air to drive the hydraulic pumps and to blow the grout to fill the space behind the rings; the first use of compressed air in waterbearing ground; and the use of electric light and power. Then in 1922 salesmen appeared peddling new-fangled compressed air tools called spaders. The miners were loathe to abandon their short handled picks as the spaders were heavy to hold, noisy and the blades kept breaking, but both tools and miners improved until they became universal. At that time, welding and oxy-acetylene burning had not penetrated to the tunneller's consciousness.

Today the media assume that we all work in a disputatious mass on 'the factory floor'. In civil engineering there are no factory floors, which accounts for a strike-free record. Each tunnel face was isolated, so contact with each gang of less than a dozen men was easy and friendly. Once a rate per ring had been agreed, work went on day after day with the gang going home as soon as 4 rings had been built. The shifts were 12 hours day and night, nominally, with 8 hours on Saturdays and Sunday nights. I was at first mystified why some men changed their names every six months, until I realised that they made their own income tax returns. PAYE was an invention of the second war.

By July 1923, most of the six shield drives had joined up and I was abruptly summoned by Mr Rowell who said 'I want you to go on night shift to look after the big junction chambers. You will be paid an extra £1 per week'. This was in top of my rate of £3.

The Hampstead and Highgate was the last of the five original tube lines, opened in 1907, and was the first to branch out to two termini. South of Camden Town Station, two junction chambers were shield-driven like a station tunnel and provided the points of junction, which gave room for two running tunnels at their north ends. The Up line to Highgate had to fly over the Down line from Hampstead, so that is why the four station tunnels at Camden Town are at different levels. Dalrymple-Hay had devised a cunning scheme to allow our new City and South London tunnels to join the existing ones with no disturbance to traffic. We had to build two step-plate junctions at night on the 'old' H&H line, one some hundreds of yards south of the existing chamber and the other between it and the station. We drove a long loop tunnel between the two, working from our completed shield drives, this would allow the Hampstead-bound trains to branch at the new southern chamber and dodge the old chamber. The south end of the old chamber was laboriously moved over in a series of steps so that the new City and South tunnel could come in beside the old one. Then the platelayers would come in on a single night shift and lay the new points for the City and South trains to branch in the old chamber while at the same time, merging with the Highgate-bound line. The same thing in reverse was carried out for the southbound trains. Our new shield-driven southbound tunnel had to be driven below the old one and then gradually come up to the same level.

Step-plate junctions are so called because the running tunnel is widened to one side by a series of steps until there is room for two tunnels to emerge. This is done by starting with 12 ft 6in, then 14ft and by stages of standard sizes until several rings of 24ft can accommodate the running tunnels' twin exits.

This work, needing much improvisation, made a fascinating change from routine shield

---

8 Subsequently it was added to the collection of the Brunel Museum housed in the Brunel Engine House, Rotherhithe (see p. 233).

driving. The miners were paid for a 12 hour shift to encourage them to take on the work and some earlier hours were spent in getting materials down the lifts and getting ready. The only trouble was that I never managed to sleep much during the day which meant that my digestion became disorganised. I had a little cabin built against the wall on one platform, with a bench, electric fan, wash stand and a cabin boy, the youngest son of the foreman, who made me tea in the night watches, of a comforting but dubious nature. Sometimes he would go to the surface to buy me sandwiches from a coffee stall.

When the last train had gone, the miners would sit along the platform edge while the LERy platelayer planted a box of light bulbs on one live rail with a cable to the other. As soon as the bulbs went out showing the rails were no longer live, we all leapt into action. Plate-layers trolleys would be lifted off the platforms and quickly assembled. Two heavy axles with their flanged wheels would be dropped onto the rails and then the heavy timber base, with its U-shaped bearings, would be dropped onto the axles. Tools and timber would be loaded on with side pieces to be added if spoil was to be brought back. These would be pushed to the scenes of action, of which the four nearest Camden Town were serviced from there under foreman Tom Veasey.

The two distant junction chambers were worked from Mornington Crescent Station under the famous figure of 'Bumper' Harris, who had lost a leg at Waterloo (Station), while building the Bakerloo Tube. He had returned from his farm to oblige his old boss, Willie Rowell. In 1977 a question in a BBC Mastermind programme produced the answer that Bumper Harris was the man with a wooden leg who kept going up and down the first-ever escalator at Earls Court, to encourage the nervous. This first escalator tunnel had been a Mowlem job before the first war to serve the Earls Court Exhibitions.

I learned a great deal by questioning Bumper on his experience. When I told him that I did not understand how to tunnel in running ground, sands and gravel, he presented me with an exercise book in which, especially for me, he had made a series of perspective drawings showing every step in the tricky work of driving a closely timbered heading. He was a good untutored artist.

11. Sketch of a timber heading in bad ground by 'Bumper' Harris

12. Timber heading

He liked to answer questions and seemed to like my admission of ignorance, as one night he said, 'You know? That young fellow before you, he was one of these bloody BSc's, but he didn't know nuffin', a shrewd remark to be noted by all would-be PhD's.

Fig 13 shows an artists impression of the lay-out of the junction chambers to be built

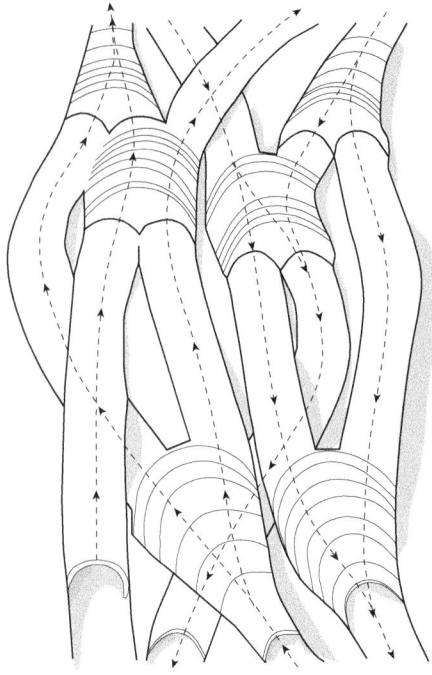

13. Impression of the intertwined tubes at Camden Town

at Camden Town. There was to be no curtailment of the normal train services so we worked from about 1am until 5am on the track with extra hours on Saturday and Sunday nights. Monday night was our 'day off'. Because of the weekend work there was none on Monday nights, so I would report to the office on Monday, and Tuesday became my 'Sunday'. I joined a cheap golf course at the bottom of Hanger Hill as it is the only game you can play by yourself when most people are at work. The need for fresh air was serious, as in those days the only tube ventilation was by the movement of the trains, i.e. when the trains stopped our supply of fresh air stopped as well. It was many years before ventilating works were built. Sometimes I got out in the early morning at Charing Cross and walked along the Embankment to Westminster to take the train to where I lived with my mother in a chilly flat overlooking Ealing Common Station. A poem came out in Punch at that time called the Lament of an Inner Circler:

> *From the mountains of Darjeeling*
> *To the Isles of Cocos Keeling*
> *Everybody finds in Ealing*
> *Their appointed Goal.*

On Sunday mornings the first train from Camden Town to Waterloo was filled with worthy citizens of all ages, sizes and sexes armed to the teeth with fishing rods, creels, umbrellas, chairs, food and drink, for an attack on the upper reaches of the Thames. Professional loneliness was prevented by Robert Ogle Barnes, who was an Engineer-Inspector for Dalrymple-Hay. When the dirt and lack of air became oppressive, Robert and I would go off at dawn and have a Turkish bath at the Imperial Hotel, Russell Square. The cooling off period on comfortable couches counted as a night's rest before breakfast in the hotel.

Down in the tunnel quite a lot stirred. In order to speed up the work, we drove a heading alongside the old northbound junction chamber from the end of our new tunnel. At night, timber stages were suspended from the bolts in the top of the junction chamber, just clear of the top of the trains. We drove small vertical shafts upwards and broke out a few iron segments to reach the stages so that men could work on the crown of the tunnel during the day.

In those days the only entry to the coach was at each end, where there were open vestibules and passengers would crowd into them. Between every pair of coaches there was a gateman, whose job it was to open pairs of light-weight, waist-high open work folding steel gates, balanced precariously over the couplings. There came a complaint that brick-bats kept dropping off our stages during the day onto the vestibules and the passengers did not like it. As Segrave and I seldom met since he went home long before I came to work, we wrote notes to each other, in passing this on he sent a drawing of two crowded vestibules with the Gateman calling out 'Next stop Camden Town. Mind your Nuts' (see Fig. 14).

There was some variety to my working life. In September, my kind cousins, the Body's,

2. TUBE TUNNELS

14. Sketch by Segrave showing the conductor and the vestibules

gave me my fourth consecutive holiday in Scotland. They had left things late and the only available place to rent was Beaufort Castle, from Lord Lovat. I had two weeks allowance which was generous for those days and Beaufort Castle made a contrast to my squalid life below ground and was not at all like the Ealing flat. A cheerful house party included my friend Sub Lt Billie Vereker, RN and we fished for salmon together on the River Glass. In the winter, I was Billie Vereker's guest to dinner at the Royal Naval College, Greenwich. This was an impressive affair followed by tough play in the Bowling Alley. I had changed into white tie and tails in his room, but had no time to change back before making for the last train at Charing Cross. I had not the nerve to parade before all the miners in my finery so got out at Mornington Crescent and changed in the Gents, before trekking back to work up the empty tunnel - not the usual end to a good dinner.

There were ten days of fresh air in October, when I was sent to Swanage to help a consulting engineer to set out a road through the wilds of Studland to the mouth of Poole Harbour. Edwin Burt was helping to promote the car ferry across this opening. We had seven men clearing a path through virgin forest so it became a course in jungle surveying. In the spring of 1924 the tunnel work petered out after the LERy had laid all their tracks through old and new tunnels. Once the tunnels were complete, the LERy's habit was to drive a train during the night with lead fingers about 4 inches long sticking up from the corners of the carriage to match the tight points in the structure gauge, roughly 2, 4, 8 and 10 o'clock. Luckily none of the fingers bent.

3

DOWN TO EARTH IN THE EAST END

After the Camden contract finished, a spell of reduced activity included tendering, unsuccessfully, for other work and tidying up old records. I was set to trace two tattered drawings which showed, in a series of sections on the lines of a comic strip, the steps to be taken when driving timbered tunnels in loose ground. One was for a tunnel of 30ft in diameter by what was called the English Method, by which almost all our railway tunnels had been driven when in loose ground. The other, more usefully, was for a sewer tunnel to be bricklined, leaving an inside diameter of 8ft. The miners called this 'tunnelling on bars' for reasons which will follow.

Following this, I had to look after the driving of a timbered heading along the lines that Bumper Harris had taught me. This was under a wide road junction in Shoreditch High Street and was to carry high tension cables. It ran from outside J. Lyons to the Unicorn Public House. Only a few weeks before, I had been reading the autobiography of the tunnelling engineer Sir Francis Fox and in it he described how he had encountered a spherical cavity in sand and gravel when driving such a heading. He had found that it was due to a 'run of ground' by careless tunnelling in a sewer which had been driven some feet below. Sir Francis Fox went on to write about the first railway tunnel under the River Mersey and how he used to visit the miners at three o'clock in the morning 'to cheer them on their lonely task'. My cynical youthful reaction was to think that the miners would regard him as a pious old snooper. Having read the piece on spherical cavities, we found just such a cavity half way across, about ten feet wide under the crossings of several tram lines. I was to meet similar cavities in the future – but always on the work of rival firms.

The method we used of driving a heading, with frames called side trees, with head tree and cill, to support timber boards as they were tapped ahead while the gravel is scraped gingerly away, is known in the USA as fore-poling. A Swedish journalist named Sandstrom wrote in his *History of Tunnelling* (1963) 'It should be noted that no tunnel would be driven by means of fore-poling today. For one thing there probably exists throughout the Western World no man capable of driving a heading in this manner. It is a lost skill'. Naturally I do not concur, as the judges say. No art or skill is ever completely lost even if practitioners become few in number. We read from time to time of holes suddenly appearing in streets here and there and in most cases they arise from broken pipes or sewer brick linings which allow the water flow to erode the ground outside. As many of these are 100+ years old there will be no shortage of maintenance work in the future, of a kind beyond the help of a silicon chip!

In October 1925 the firm sent me to the Isle of Dogs to complete an extension of the storm-water Pumping Station after the LCC had driven the previous contractor bankrupt. I was left on my own with only occasional visits from the agent, Charles Connor. This was my first view of the East End of London, where I was to spend many days in peace and even more in war. It was also my first, but by no means my last, encounter with the LCC Main Drainage Department. They were a self-contained unit who had their own methods and as their new works were almost continuous they disdained consulting engineers. Their ideology was quite different so an explanation on their function may not come amiss.

To many people, civil engineering conjures up visions of imposing bridges in romantic places. Those of us who burrow below ground leave little visible signs of our efforts, so words written by David Piper have bearing upon what will follow. He was reviewing for *The Times* a book called *London, 1808-1870; the Infernal Wen*:

> So London slowly groped....to a centralised administrative authority: grappled confusedly with education, from primary schools to the founding of London University: was consolidated and rent asunder by the railways into its heart: drained off (but only after cholera had struck repeatedly) the horrible challenge of such concentration of human effluvia, involving a sewage system that was one of the great engineering feats of the Century.

He was referring to the London Main Drainage produced between 1856 and 1874 by a hero of mine, Sir Joseph Bazalgette, on behalf of the Metropolitan Board of Works. They bequeathed it to the LCC and then to the Greater London Council and it has been washed up into the Thames Water Authority.

On both banks of the Thames, Sir Joseph planned High Level, Middle Level and Low Level main sewers which converge on large pumping stations at Crossness on the south bank, and Abbey Mills on the north. From Abbey Mills the contents flow by gravity down five 8ft brick tubes on the top of a high embankment made of concrete, for over four miles to the outfall at Beckton.[9] I know, as I had to mend all five later during the blitz. The water ran into the Thames and the sludge was pumped into ships, each aptly named after the chief engineer who had commissioned them. They then steamed out to the 'black deep' and dropped their offering which made it blacker still. It was not until a few years later that modern disposal works were started, which are now of enormous proportions. This helps the salmon come back to us.

Innumerable local sewers are cross-connected into the main ones. When rain falls on a heavily built-up area like London, the run-off is voluminous. In order to prevent flooding, the main sewers were designed with large weir chambers at intervals, so that the surplus rain water can tumble into storm-relief sewers. These eventually discharge their much diluted effluent into the Thames where other pumping stations are built so that, at high tides, they can lift and pump out the flow against the tide through tide-flaps. On the north bank these are at Hammersmith, Lots Road, Pimlico and at the Isle of Dogs. Bazalgette built the Victoria Embankment, which is the most imaginative improvement to London and in it he not only incorporated the District Railway and the first ever pipe-subway but also the Low Level Sewer.

The then existing Pumping Station at the Isle of Dogs had been built in what the humorist Osbert Lancaster might call Bazalgette-Victorian, with an ornate tall brick chimney. Inside, boilers were kept always under steam so that, when rain poured down and the tide rose, lovely old 1870 horizontal steam engines would start up and in majestic slow motion, pump

---

9 There had been a students' visit in 1920 and it was found that the effluent settled in some disused reservoirs under brick arches.

the storm water from unseen depths through the tidal flaps against the head of the river water outside. Innumerable Malthusian objects were also discharged.

The trip to work from Ealing to the Isle of Dogs took at least one and a half hours, so was apt to become tedious. On leaving Aldgate East Station, the centre of the wide Whitechapel High Street was lined with large hay-wains and horse drawn wagons, as this was an active haymarket until the second war. In the evening many jellied-eel stalls appeared. The next step was to board an open-top bus or LCC tram to bump along the granite setts of the Commercial Road for three miles, past shop fronts with strong Russo-Polish-Jewish flavour. Over that year I read all three volumes of Carlyle's *French Revolution* on these journeys.

The journey continued past the surprising Limehouse Church and signs of China Town, down the East India Dock Road to the entrance of the Blackwall Tunnel. This was then followed by a mile walk along the east side of the Isle which included crossing the lock gates of two entrances to the West India Docks. The bridges seemed to be almost always raised for ships to pass, meaning that the bus could only pass infrequently. When I first reached the site, I could see that all the ejected contractor had done was to build the concrete walls of the new Engine House extension and little else. The north side abutted on an abandoned ship repairing dock while on the east or river side, we had a small length of promenade above the chamber housing the tidal flaps, from which to admire the shipping. The rest was bounded by the Samuda Wharf scrap yard – not an aesthetic sight. The west side was bounded by an 8ft high wall beside a little street with a terrace of small workmen's houses.

The walls built by the previous contractor were in trenches and they had sunk two shafts by the centuries-old method of timber 'runners'. These are 3 inch thick planks held vertically by frames of heavy timber and driven down by 14lb hammers as the ground is dug away at their toes. All our work was in deep deposits of Thames sands and gravel, and pumps had to work day and night to lower the water in the ground by 18ft. Steel sheet piling, noisily driven, had not yet come on the market.

15. *Driving the supply tunnel showing the horizontal bars supporting the timbering in preparation for bricklaying*

The centre of the engine house, called the 'dumpling', had to be dug out and the floor concreted. We had to drive a supply tunnel to bring the future flow through holes in the wall to the pumps. This brick-lined sewer (6ft inside when built) had to be driven by the English Method, so it was a lucky break to have drawn out the method not long before. I could then conceal my ignorance while watching the work. The photograph in Figure 15 shows a completed length and the way in which the horizontal bars hold up the timbering, while awaiting the bricklayers. Larch bars are specified as they groan and creak for a time before breaking which naturally endears them to the miners. I had ordered seven-inch bars, but what were delivered were useless tree-trunks, full of knobs. When I came next day, the miners had started work in the night with beautiful smooth spars which had appeared by magic. There was a deserted ship repair dock over the boundary wall, which could explain their appearance.

3. DOWN TO EARTH IN THE EAST END

This form of close timbering in running ground is a fine art. Each miner has his miner's labourer behind him as he works along each bar. He never takes his mind off the face in tricky moments and just as a television surgeon calls for 'scalpel, forceps, clips', the miner calls for 'club hammer, wedges, litter'.

When I first got an urgent call from below for 'long litter', I was baffled. I was told to send for a barrow load of straw from the floor of a nearby stable. This contains a valuable malleable additive, which makes the litter ideal for stuffing into chinks and holes in the timbering to stop sand from running in. It took 48 hours to complete each 'length' as it is called, usually 10ft long to allow for 7ft 8inches of lining.

Then came my first encounter with the famed London sewer bricklayer, an artist in his own right, and in those days a different animal from those employed by the jerry builder. First a deal had to be made by haggling and then two bricklayers and a labourer would go below while their linesman on the top fed them with bricks (soaked in water) and cement mortar. They would work round the clock for twenty-four hours to 'turn the arch' and rub down the finished length, leaving it beautifully smooth.

16. *Brickwork arch under construction*

The arch of brickwork had to be built over a drum with expensively smoothed horizontal laggings, which were placed on semicircular ribs as the brickwork rose. To complete the arch, the bricklayer worked out backwards, pushing in bricks end-on over short blocks, adding one at a time. One night, on leaving for home, I saw one of our expensive laggings on the shoulders of 3 nine-year-old Giles-type Isle of Dogs kids disappearing at a smart trot down the main road. In those days one did not tangle with the natives. On another occasion a heap of timber from the diggings had been piled up near the street wall and much of it vanished in the night. Next day, the chimneys of most of the terrace houses opposite seemed to be on fire.

One day I was standing with my foreman gazing down a hole which was being dug, when I saw on the bald head of an enormous man an elaborate tattoo of the Crucifixion. The man looked up and revealed a moronic and villainous countenance. I found that he had been sent to us by the Poplar Labour Exchange on release from prison. At that time all general labour had to be recruited from the Poplar Labour Exchange. According to the story, he was about to decapitate his wife on a butcher's block in Crisp Street Market when the butcher intervened and knocked him out. Many policemen were needed to take him away. That night I took care to walk home in the middle of the road. The foreman eventually found him uncontrollable and paid him off. This marked the start of the dilution of the stock of old-time navvies. Also at that time all wore 'Yorks', the straps under the knee to loosen the trouser knees. A few years later Yorks vanished completely in a change of fashion.

The rest of the work had its amusing moments in between minor excitements. One of the lasting lessons covered problems of pumping, in those days no one seemed to have heard of self-priming pumps. We inherited six-inch petrol-driven centrifugal pumps which were placed down shafts just above normal water level. So, with suction pipes nearly 20ft long hung from them down to the sumps, every time a small object got under the clack valve at the bottom, the pump lost its water and the suction had to be hauled up and cleared and the whole re-primed with water. Once self-priming pumps appeared I used no other. This was an unsuspected introduction to my later efforts to introduce 'groundwater lowering methods'. At Camden Town I had learned only one form of tunnelling under engineers and miners with years of experience. I now had had to manage on my own with many different forms of work. I also had to deal with money, wages, ordering materials, programming the work, suppressing quarrels, wrestling with the plant department, coping with a head office and repelling attacks of the LCC engineers. The memory is still vivid. At the end of twelve months I had been reconditioned into a young man with varied experience and increased confidence.

In the autumn of 1925 the work on the Isle of Dogs was nearing its end and a small piece of road construction had to be faced. A cousin of my mother had married a British civil engineer, John Body, in Mexico (and I had holidayed in Scotland with their help). They were now living in Castelnau, Barnes as he was a director of the road firm, Constable Hart. On the way home one Saturday afternoon (we always worked Saturdays), I decided to drop off at Hammersmith to ask his advice and learn suitable professional catch-words to be one-up on the LCC. I found my cousin (his daughter) was in the middle of a tea party, in which I joined. A strange-to-me girl seemed cheerful and intelligent, with very nice legs, so I offered to see her home to Kensington. She spurned my offer but invited me to come to a small dance at her house a few weeks later. She cracked a joke, I capped it, she capped that and away we went, finding that we made each other laugh. Not obviously romantic, but two years later we were married. Was the Isle of Dogs the *Deus Ex Machina*? Her name was Sophie Blair Leighton.

4

GO WEST YOUNG MAN

Charing Cross and Piccadilly Circus Stations

Towards the end of 1925, I was lent to Dalrymple-Hay for a few days to survey the underside of the platforms of the District Railway Station at Charing Cross. This involved lifting manhole covers on the platforms and crawling about in different compartments below. I have a clear recollection of the way that, on emerging with eye level just above the platform, girls' skirts had, almost overnight, shot up from the ankle to the knee with the invention of artificial silk stockings. The survey was followed by the work of building passages under the District railway track and escalator tunnels to the new station below. A firm run by Sir John Norton-Griffiths,[10] had extended the Hampstead line under the Thames to join the City and South at Kennington. Until then the Hampstead line had ended in a loop, driven partly under the Thames, coming back to a single station beside the two Bakerloo Stations.

The District Railway took up their track on successive nights, laid steel joists between the platforms, filled them with concrete and put back the track. We then drove a centre heading under the joists and broke out on either side at intervals, so that bricklayers could pin up the steel joists off the brick invert of the District tunnel. We sank two sumps, one under each platform and pumped down the ground water so that we could drive the 12ft diameter escalator tunnels through the sands and gravel to the new station tunnel in the London Clay.

Our working site was in Embankment Gardens and we sank a shaft and burrowed through the headwall of the District Station to get to work (the ground conditions we encountered will keep recurring in this story). In the glacial era, the water level of the North Sea had sunk as water was held by the ice. The Thames 'rejuvenated' itself three times, by cutting into the London Clay hills on either bank. This left three terraces of sands and gravel, all containing water. The Boyn Hill Terrace is on the level of Pimlico, the Taplow Terrace is under Piccadilly and Holborn and finally the Flood Plain gravel is where the river flows at present.

Between 1921 and 1926 the Francois Cementation Company had successfully sealed fissures in rock in mine shafts by injecting a weak slurry of cement and water and it was hoped to turn our sands and gravel into concrete by this means. Injection pipes were driven 2ft apart above the arch of the sloping tunnels. Cement and water at a pressure of 100 lbs per square inch were pumped down the open-ended pipes as they were being pulled out. Unfortunately it was found that the cement would not enter sand unless it was very coarse, so the process only produced boulders of concrete with untreated sand beds between. This meant that if the boulders dropped when tunnelling, the sand above ran in. One day the pressure hose burst and

---

10 Grandfather of the one-time leader of the Liberal Party, Jeremy Thorpe.

hit me in the stomach which meant a dirty journey home for a fresh suit. Years later I was able to tell the ICE that I had been baptised into the injection process by total immersion.

When our small short tunnels reached the station, we checked the junction. The station tunnel was empty but planks had been left in place of one segment of lining for this purpose. At this point came my first encounter with the press. I had just got back to my hut when in bounced a cocky young man 'I am the Evening News. Have you junctioned yet?'. I told him that he must ask the railway company and also asked if he had ever been down a tunnel. He replied that he did not have to in order to write about one. His piece came out under banner headlines: 'Today the tunnellers at Charing Cross completed their difficult task. A mighty cheer went up and hand grasped hand through the hole'. So much for the media.

Early in 1926, while we were working away at Charing Cross, the firm was awarded the contract for the first section of the new Piccadilly Circus station. The Piccadilly and Bakerloo lines had both opened as rivals in 1906 but shared lift shafts at the top of the Haymarket, where the Piccadilly station had the trains on the outside and the platforms opened easily between them. Access from the lifts was by a single 10ft tunnel of interminable length and passengers entered and left through a single opening in the station headwall. The Bakerloo stations were at a higher level, with separate passages and the running tunnels just scraped over the top of the Piccadilly ones. The congestion had become intolerable and Dalrymple-Hay persuaded the LERy that he could not only drive the escalator tunnels but also build the whole of a new station concourse under the road concrete entirely in tunnel, without stopping road traffic. At that time there were two islands in the Circus with a direct route from Shaftesbury Avenue between them, the public 'comfort stations' were under the eastern island and on the other stood the statue of Eros on its fountain, erected to the memory of Good Lord Shaftesbury and described in a poem in Punch as:-

> *Practicing Toxophilitics in the Bare or the Buff*
> *Which would have shocked Lord Shaftesbury so much!*

The LCC dumped Eros and his fountain in one of their parks and handed his island over to us as our working site (the story goes that his later re-erection was delayed as the LCC had forgotten which park they had put him in). At the start, the department store, Swan and Edgars, and the County Fire Office stood in their original Nash design, but they subsequently disappeared, leaving large holes in which the modern versions appeared as we watched. Swan and Edgar were not allowed to use their top floor because the Fire Brigade could not reach them with their equipment. How unlike the more recent film of the *Towering Inferno*.

17. *The head of the service shaft at Piccadilly Circus, showing the hut arrangement on the Island*

The first thing that I had to do was to design a layout for this tiny island working site. A shaft of 18ft diameter destined to be an emergency stair-well had already been sunk, half under the road and half just inside the site. The heavy timber 'muck stage' with two electric locomotive cranes hugged the west side, so that lorries could drive in at one end

4. GO WEST YOUNG MAN

18. The Morning Post *photograph celebrating the inauguration of gyratory traffic in Piccadilly Circus, July 27th, 1926*

and out at the other. The sheds for mess-room, cement store and compressor shed took up much room, leaving little space for storing the many different sizes of cast iron segments. In those days the modern smart prefabricated units did not exist. Instead, we put two little box-huts on the top of the sheds for the foreman and the inspectors, with a larger one for the engineers, both ours and the consultants. In this we kept our old working clothes, drawings and a wash basin. A photograph was published on the 27th July 1926, the day when gyratory traffic started, while the miners strike was still in full swing, and it shows our island very clearly. Two days later, Lowe produced his cartoon of 'Political Piccadilly' with our site labelled as 'Underground Profits Station', with buses labelled Miners and Owners.

From our hut we gazed towards the London Pavilion and could observe the historic old flower-girls sitting beside their baskets on the other island with the red-coated shoe-black - they had been ousted from Eros Island for the duration. Beside them stood a tall man in slightly better dress than the normal news vendor, with three papers under his arm, he had speech at times with passers by but we never saw him sell a single paper. We thought him rather sinister.

Mr Rowell decided to move nearer to the scene of action, so moved the office from the airy house at Hampstead Road to a dingy little empty shop in Litchfield Street, opposite the Ivy Restaurant and close to St Martin's and the Ambassadors Theatres. He installed the head timekeeper and his wife, as was his custom, in a flat on the top floor so that she could cook for him and provide room for his siesta. Mr Rowell occupied the first floor as his office, which had the only view into the street. The senior agent previously mentioned had left to be a director of a rival firm, so that left Segrave and I to run both Charing Cross and Piccadilly Circus simultaneously. We occupied the shop but with no view, as the window had been blacked for much of its height for the privacy of the previous occupants. It was only on visits to the first floor that a glimpse could be had of theatrical celebrities entering the Ivy Restaurant. On matinée afternoons the buskers would line up outside to entertain the theatre queues. Our junior clerk received a torrent of abuse one day, when they saw him watching the blind musician as he removed his 'blind' notice, in order to read the racing news.

Litchfield Street added momentarily to the public gaiety as the only other shop was occupied by the now forgotten 'Colonel' Barker. When he appeared in court charged with many pretences, his wife declared that they had been happily married and that she had not known that the Colonel was a woman! The latter had got away with this imposture on many now-forgotten occasions. A friend of ours had, with other men, shared a changing room with her quite unknowingly.

Many a night had to be spent in the Circus in order to survey and to set out the work, this included carrying out the survey down the shaft by means of plumb wires as previously described. For the first two years my opposite number from Dalrymple-Hay's staff was John Leigh Hunt, who had been in my year at College and which added greatly to my pleasure. Being in Piccadilly Circus all night was a special case, with the traffic only easing up at about 1 am. In every doorway lurked a hopeful professional 'lady-in-waiting' – except on Cup Final night. We had to work briskly, as the Westminster Council workmen started to hose down the roadways about 5 am. On one occasion, Dalrymple-Hay sent out an old engineer to help us. This engineer had been a rigid resident engineer on the original Bakerloo tunnels under the Thames but had fallen on hard times. He was gazing through the theodolite when three prostitutes started teasing him and asking him to take their photographs. As he showed shocked disdain, they turned their backs, whipped up their skirts and said 'Take a look at that then!'. Like a number of engineers of his generation, he was strictly pious and straight-laced so he took a lot of consoling.

About 2am on one cold winter's night, we were kneeling in the middle of the road marking a vital survey plug by the light of acetylene hand lamps with the chainman's bag beside us. A Rolls Royce drew up alongside and in the back were two astonishingly beautiful girls in evening dress and expensive furs. The driver, in white tie and tails, leaned out and asked, 'What are you doing there? Ferreting?'. Their night seemed more enjoyable than ours.

The first months of 1926 coincided with my brother's six months leave from India. He had been out there for a five year spell in spite of a promise of only three. This meant that my social life became more active. Many times I worked all Friday, danced most of Friday night, worked as we always did in those days on Saturday morning and then spent all Saturday night in the Circus or down the tunnel when the men were not at work. Sometimes we broke off for a snack about 3am in the Criterion Restaurant, part of which was open all night. To compensate for this, Hunt and I would treat ourselves to a Sunday breakfast at the Regent Palace Hotel before going home to bed.

The Underground commissioned an artist's impression of the work, which gave some idea of its complexity, but those who use the station only see half of the work that we did. The last part was to tunnel out the circulating area under the road concrete, with the stairways to the street, not forgetting the direct entry into Swan and Edgar. First, the conglomeration of water, gas and hydraulic mains, gas mains, telephone cables and power cables which criss-crossed this area had to be picked up and diverted. To do this, we drove a pipe subway of the same size as the tube running tunnels all around the booking hall and under much else. This twisted and turned, with forty-five degree slopes up to the chambers in which the pipes and cables were intercepted and turned into the finishing tunnels. These chambers crossed each street, first by driving parallel timber headings in the Taplow Terrace ballast and then filling them with brick walls. The next step was to drive again over the top of them, by 'offering up' steel joists to hold the road concrete, supported by timber props until the bricklayers came back and underpinned them off the walls which had been built in each heading. Just to add to the fun, several live sewers had to be surveyed and then diverted to avoid the pipe subway. In our spare moments we drove four escalator tunnels and the six chambers, four on the middle landing and one for each station.

These large tunnels were enlarged in three stages in order to reduce the area of the clay face which had to be supported. First a pilot tunnel was driven all the way, made by using four of the six segments which make up a ring of 12ft diameter. They make a bulgy sort of square when they are bolted together through specially shaped wooden blocks and became known as 'novelty headings'. These were enlarged to 14ft tunnels and then to the final sizes – 22ft 4in for escalator tunnels and 25ft or 26ft for the chambers. Agility is called for when building the heavy segments on the thirty degree slope by means of hand winches and brute force. The ground for the first 20ft or so below the roadway was made of sand and gravel, mixed with old cellars at the upper level. So the first 22ft 4in rings had to be built in heavily timbered slots in the clay at 60 degrees to the horizontal. Then the final drive up the hill into the gravel was a tricky operation.

The cross passages between the four middle landing chambers were at first only crawl holes about 4ft wide. When the chambers had each been built, the linings were supported by heavy steel frames, known as horse-heads, while more segments were being taken out to make the permanent openings. These were made by erecting heavy steel jambs, lintols and cills to resist the heavy earth pressures when the horse-heads were released.

This was a particularly happy contract, as Dalrymple-Hay and Rowell had worked together so many times. Rowell had a faithful following of miners, the cream of the London tunnel world, with not a trade's union in sight. He paid them slightly 'over the odds' by some private arrangement and they preferred this varied work to repetition shield driving on the Malden extension of the City and South London line, which was to have gone to Wimbledon. The LSWRy had created so much resistance that it had to end up at Malden. At about 11am something would happen, which nowadays would seem strange. The yardman and the tea-boy would be seen advancing across the Circus from the pub in the Haymarket carrying a washing basket of Falstaffian size, full of quart bottles of beer. The crane would lower this down the shaft for distribution to the many faces.

The General Strike 1926

None of our miners belonged to trade unions in those days but when the General Strike started in June 1926 we still came to a halt. However, after two days the railways and buses started some sort of service run by volunteers. We stood by in case we were needed as platelayers, then the Government appealed to ex-servicemen or Territorials to join the Civil Constabulary Reserve as special constables, by reporting to local Territorial Headquarters. I came under both headings.

One of our XXI Club members was an officer in a unit whose drill hall was in St John's Wood and he recruited three fellow club members including me. We lunched with him in his Officers Mess and then were sworn in and issued with a steel helmet, a truncheon, a striped armlet and a first aid dressing. We were then put on fatigue to wash up in the Officers Mess. Officers became instant superintendents and inspectors. They ordered us to go on parade in the street so we ran out, fell in, numbered off, formed fours (as was then the drill) and proceeded to march up and down the street as a disciplined body. As most of us were complete strangers to each other it was a heartening example of the use of training and discipline.

We slept on the drill hall floor, ready to move as a body to quell any rioting which was feared might occur in Camden Town. In order to keep us busy, we were marched into Regent's Park every morning for 'physical jerks', sometimes at the double on the way there. We passed some workmen who called out, 'Anyhow, the buggers can run!'. After our exercise we were invited into the Zoo free of charge, which helped to pass the waiting time. When it was over, Sophie Blair Leighton drew a cartoon of two of us with helmet, truncheon and enormous wings sprouting from our shoulders while between us loomed a gloomy giraffe. The caption ran 'Angels Guard Thee'. No respect for our unheroic effort.

When the strike collapsed we were disbanded, so the four of us went off to see the young Fred and Adele Astaire in a matinée of *Lady be Good*, long before he became a film star.[11] We were allowed to keep our truncheons, so I then had a collection of three issued to special constables:

My own	1926	The General Strike
My father's	dated 1887	possibly for the Fenians (pre IRA)
My grandfather's	dated 1848	The Chartist Riots

Directly the General Strike started, John Mowlems had set up the emergency milk depot in Hyde Park under H. B. T. (Teddy) Wakelam, who had joined the firm as a war-time friend of Eric Burt and who was a considerable asset. He had been Captain of the Harlequins and a noted sportsman. Soon after the BBC had become firmly established, he wrote to them and offered to give a running commentary on a Rugby International. This had never before occurred to them, but the result was a great success.

19. *Angels Guard Thee…Creed Vowler (l.) and HJBH (r.) Civil Constabulary Reserve during the General Strike of 1926, as drawn by SHBL*

And so back to work. We discovered a curious relic when we were building the pipe interceptor across Coventry Street. We started with a 4ft high heading from a cellar in an empty shop. As we neared the London Pavilion, we ran into an empty 7ft diameter well, lined with brick. There we found an iron suction pipe which experts dated to 1750. There were peculiarities about the top, which was only 2ft below the road level and consisted of a brick dome capped by a York stone slab. There were pipes and cables all around it but it had never been encountered before. Many dozen lead-covered cables, each carrying several hundred pairs of telephone wires passed through the interceptors and fresh cables were laid round the pipe subway. GPO wiremen then spent months re-connecting each pair of wires, which involved much telephoning to each other at each end of the cable to be diverted.

My chainman, George Cash, whom I had inherited from a friend, had been an army gym Staff Sergeant-Instructor, so his agility came in handy when climbing about in the 26ft diameter chambers. We worked harmoniously together until I was promoted beyond setting out work and he then joined the LCC as a flusher, stationed at Old Ford Road. One night a while later, a friend was walking down St James Street when George Cash bobbed up from an open manhole:

'Why George, I thought you were stationed at Old Ford!'
'Yes but I am up west as a relief man'.
'Do you prefer it up here?'
'Oh indeed yes. The bath water does smell lovely!'

These sewer diversions were located in the Taplow Terrace sands and gravel, so had to be tunnelled with closely timbered headings by 'fore-poling'. These had to be driven from end

---

11 She was much more attractive than Ginger Rogers and a better dancer, but she got married to a Cavendish and left showbusiness.

to end and then the bricklayer would work back to the shaft as he bricked the lining. The setting out called for skill. First the points of junction had to be fixed by co-ordinates from the survey, then the main survey line had to be carried down the main shaft from the street, along an access passage into the pipe subway. Next, turn left for fifty feet, up the 45 degree slope, turn left again and on. At a selected point, two segments had to be pulled out of the top of the 12ft lining. Miners then had to drive a timbered shaft about 3ft square vertically upwards in the gravel and drive sideways for several feet. The main heading would then start, driving in each direction, northwards to Shaftesbury Avenue and southwards to the other sewer. After a few feet, the latter had to go down a thirty degree slope in a larger size as it was to contain a 'tumbling bay' with a small staircase at one side for the flushers. A few feet later the heading had to be driven round a 15ft radius curve, which luckily brought it alongside the brickwork of the Criterion sewer. The bricklayer would then lay this curved portion.

The next step was to occupy a part of the street and sink a shaft for a new manhole on the live sewer. After the top was broken into, a pipe would then be laid between half-dam headwalls to carry the dry-weather flow. Then a bell-shaped chamber was constructed to embrace the sewer and our finished brickwork. At this juncture, a miner knocked out one brick too many and found himself gazing into the bar of the Criterion theatre. Luckily it was on the night-shift and the hole was quickly bricked up. The brickwork was then completed to join the two as, by the Grace of God, we had done it right.

During 1926, with two hours a day less spent on travelling, my private life became more active. I was seeing more of Sophie Blair Leighton, who was living in Warwick Gardens, SW5 with her widowed mother - her father had died a few years earlier. Edmund Blair Leighton had been in the forefront of the late-Victorian-Edwardian artists and his pictures had hung 'on the line' of the Royal Academy for forty years. Sophie liked to give small dances at her house or join in small groups for a dinner dance at the old Hotel Cecil on The Strand, before it gave way to the Shell-Mex building. She was studying art at the Slade School, so had a supply of girl students. Apparently, at that time the girls were kept apart from the men by studio and meal times, but I could supply acceptable males. She once said that the only male at the Slade who looked clean, well turned out and nicely dressed was Rex Whistler, who was also far more accomplished than any of his fellows.

One night she put me up in the top room which she used as a studio. A number of large canvases had been turned to face the wall so, of course, I turned them round, to find they were all nudes, most skilfully painted but of appalling plainness. Her subsequent explanation was that tall willowy girls cannot 'stand' for long periods and constantly faint, so short dumpy utility models are more suitable for students to work upon.

My brother got in touch and described playing golf on the Calcutta Maidan, so I sent him a parody of one of Hilaire Belloc's Cautionary Tales. Sophie got hold of it, drew a surround of pen-and-ink drawings and sent it to the Tatler who paid us £10 and published it. This was our first collaboration. She was once described as 'that girl with a man's mind' and she encouraged us to talk about our work and showed interest. One day I said that I had once thought of making a model of the Camden Town complex, and she said 'Why don't we make a model of Piccadilly Circus Station?'. After work one Saturday afternoon, Leigh Hunt and I surreptitiously smuggled Sophie into the workings so that she could see them for herself. She climbed down the 30ft vertical wooden ladder in the shaft to the first landing and then down the sloping pilot tunnel on the slatted duck-boards which served instead of stairs. We had completed two of the large middle chambers and in order to impress her we said 'Look at that 26ft diameter headwall, it is over 5ft thick and the concrete was all mixed down here by hand, as there was no room above for a mixer.' She took one look and said 'Why! It is cracked from top to bottom!'. Luckily it was only a hair crack. The moral of the story is that young contractors should not take their girlfriends down tunnels with the assistant resident engineer. We then slid our way down the second escalator pilot tunnel to the Piccadilly Line level. Once there, she showed us how to play music on a saw, held between the knees. Then we climbed back to the surface up the four ladders fixed in the 100ft working shaft. Thus, she had a preview of what we were about to model. We worked on the model for six months starting in the autumn of 1926 and became engaged in the middle of it. Our target was to show it at the ICE Annual Conversazione in June 1927 so we pressed on. We had started off working at 8ft to one inch but, finding out that it wouldn't pass through the door, we halved the scale. Working together proved better training for marriage than the traditional period of mooning about.

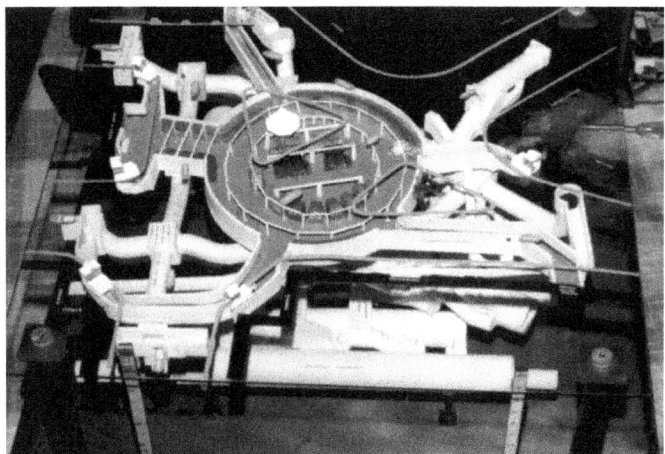

20. The completed 'stomach' model of Piccadilly Circus station and connecting tunnels, made by HJBH and Sophie Blair Leighton. The model is now in the London Transport Museum

I had been a keen supporter of the ICE and had registered as a student of it in 1919. Then, early in 1927, I was elected an Associate Member. The Institution likes to add dignity to its Conversazione by inviting firms to show models, which are almost always of above-ground structures or bridges, made by professional model-makers. Our model proved a centre of considerable interest, as it was made by young amateurs who were present to explain it. Also, at that time, the work it showed was much in the public eye. Dalrymple-Hay was naturally delighted, as he got considerable professional advertisement. He said, 'I have rung up Frank Pick and insisted that he should buy the model from you. It should be worth over £100.' This would certainly be acceptable to an engaged couple at 1927 rates. In due course I was sent for by Frank Pick, who kept me waiting for over two hours beyond the appointed time. He then merely said, as rudely as possible, 'I am not interested in your model'. I retorted that I had not suggested that he might be, but that it was Dalrymple-Hay and so the meeting terminated after half a minute. Frank Pick did much to improve the Underground practically and artistically, but he was a notorious bully, unlike his Chief, Lord Ashfield, who was courteous to all ranks. Winston Churchill made a widely reported outburst later during the second war which gave me the greatest pleasure – 'Never let that impeccable Busman into my sight again!'.

We lent the model to the Science Museum, who at once put it on display. A few weeks later the curator found two men sent by Frank Pick, trying to take photographs of it. Having found that they had not had my permission, he sent them away with fleas in their ears. I was then able to write an extremely terse letter complaining of this disgraceful behaviour and extracted a sum which covered the cost of the elaborate case and materials, for giving permission to use the photographs as aid to the artist's impression. This illustration became widely published, showing the work in a cross between a bird's eye and a worm's eye view.[12]

Towards the end of 1927 old Willie Rowell died quite suddenly, just as we were about to tender for the final contract, which was the construction of the booking hall. Segrave had to take over and promptly went down with mumps, so George Burt himself came every day to the Litchfield Street office. He re-called my only contemporary in the company, Raymond Head, who in the past had been Rowell's personal assistant and together the three of us flogged out a scheme for doing this complex work and put a price to it. Luckily we were successful as this was my first long-term connection with the head of the firm.

Segrave duly recovered and there was a sufficient lull for Sophie and I to get married and have our two weeks holiday. This was spent in Lenno, a little place on the north bank of the mountainous centre of Lake Como in Italy. The recommended hotel proved to be full of British, who love to keep together, and although the end of September was too cold for swimming, we spent much time on the lake. When we heard a bugle blow 'tootle a toot' and a voice cry 'A Lenno', it was a sign that the paddle steamer of *vascello a vapore* was coming in to the jetty and it was time to go aboard. These plied for the whole length of the lake and took us on several trips, where at different stops there came aboard; peasants, soldiers, priests and once a herd of pigs, who settled down all round us. We visited Mennagio and Bellagio, but that was the haunt of the truly affluent who only moved on expensive launches. The best spot was Bellona, an old castle high up on a cliff. We got much exercise by rowing across the lake or else round a photogenic peninsula to a little island, Isola Bella, which was a good place to paint (when the local Italians were not shooting sparrows with twelve-bore shot guns). There Sophie would knock off a highly accomplished water-colour in less than an hour while, for the first time, I tried to follow suit. In the attempt to be quite conventional my results were apt to turn out as poor Picassos. Still, they can always be torn up, and it does concentrate the eye upon what it is really seeing.

For the next fifty years, I faithfully tagged along with my wife in pursuit of landscape painting and ruined much paper, with only two or three passable results. My attempts to paint trees in full leaf led to the utterance of much blasphemy. After many years, I was able to pronounce that an artist, frustrated from painting when under the urge, resembles a hippy who is missing his fix. The true artist has sudden urges to paint which we earthbound, non-artists do not feel even if we succumb to other urges. So on holidays, I soon learned that to paint before playing golf prevents withdrawal symptoms. She revealed abilities, not only to paint portraits, but to sew, tailor, play golf and tennis, dance beautifully, paint and paper walls, plaster, carpenter, knit German fashion with extraordinary speed (the needles are held under the hands), drive cars, become a dedicated and 'certifiable' gardener, very musical, very well read, with fluent French, some Italian and snatches of any language to be picked up on foreign visits. Apart from being an experienced hostess, she was always a genius in handling children. As a result my best man, Creed Vowler, who was a Wodehouse fanatic, nicknamed her 'Jeeves' – so dashed efficient, so devilish competent. Talking of Creed, until my wedding I had been acting as secretary of the XXI Club with the help of others to copy the letters by various means, but long working hours made this difficult, so

---

12 Years later, during a pause in the bombing in 1943, the ICE staged an exhibition in the Great Hall on aspects of London. On HJBH's suggestion, Caroline's headmistress brought a group of girls to visit it as part of their education. A large poster of the artist's impression was on view and he described how they did the model and the artist's impression came about. The highly intelligent and decorative head girl said, 'That is interesting', she then went on, 'I have often used the station but it never occurred to me that anyone had ever built it. I always took it quite for granted'. There is a moral in this somewhere.

Creed Vowler took it on and ran it until 1977. He worked civilised hours in a normal office and had young secretaries who were interested and willing to copy and roneo the letters for a slight consideration.

After Sophie and I were married we took over her mother's house in Warwick Gardens for two years and during this time members of the Club began to come home from Assam, Rhodesia, USA, South America, Sarawak and elsewhere, so we were able to give welcoming parties for the exiles. Thus the work of the Club had been fully justified as it kept us together and continued to do so for fifty years. One member, on having left Malaya, came bounding up into my hut about 7pm crying out 'What a place to work, in the middle of Piccadilly Circus!'. The reply was 'What is the use if you have neither the time nor the money to exploit it?'.[13]

Back on our island in the middle of the Circus, the mystery of our encampment could hardly pass unnoticed, so the LERy (or TOT as it formed part of the Trams, Omnibuses and Tubes) allowed one pressman each month to visit the site. We tried to impress them with our work, but they all said that what they needed was a 'human story'. Polite ones were in short supply but eventually we gave the *Morning Post* a chance. Friday September 28th 1928 read:

> **Explosion in Piccadilly**
> People thrown into the air.
> Traffic chaos.
> Gas main pierced by electric drill.
> Shortly before 5pm yesterday, when traffic in Piccadilly Circus was at its height and hundreds of people were crossing to and from the various refuges, there was an ominous rumbling and the ground began to move as if at the approach of an earthquake: then came a loud report and stones and masonry were flung into the air from three different quarters of the Circus.
> The three upheavals were almost simultaneous and were followed by clouds of fumes and smoke.
> A woman crossing the street opposite the Criterion Theatre was flung into the air, a bootblack turned a complete somersault over his wooden box, a flower girl was thrown from her seat and her roses were scattered in all directions and women and children screamed with fear.
> A few minutes later fire engines rushed to the spot and ambulances from Charing Cross were soon on the scene.

In actual fact it was not as bad as all that. I was sitting in my little wooden hut above the mess hut, writing orders for the night shift, when a man scrambled up and said 'Get the gas company. We have a leak'. I rang the Gas Light and Coke Company, but some man replied that they had no emergency squad. At that moment there was a blast, not very noisy and the road split alongside the lavatory island, just at the point that the new passage to the Haymarket had reached. Certainly the flower girls turned base to apex and the Circus became packed solid with people gazing into the crack, smoking cigarettes. Some men came up from below with cuts and I rang for ambulances. Then the Fire Brigade all dashed down into the Ladies. As I was wearing working clothes and not looking 'official' it took some time to convince them that they should come down my tunnel, but by then someone had cut off the gas. It turned out that the pneumatic pick of a miner had punctured a two

---

13   This young man, R. T. James, eventually built up one of the most successful structural consulting firms in London.

inch gas main buried in the bottom of the road concrete. Another man went to try to find some clay to stop the hole but enough gas had mixed with the air to make an explosive mixture. We never knew what sparked it off as those working below denied that they were smoking.

The blast from this small pocket of gas travelled along the completed passage-way into the main booking hall and so blew some planks, which had been laid at three different corners over the unfinished street openings, off their seating. But only one claim reached the LERy; it was from a well-known local lady who claimed to have been knocked over on her usual beat outside Swan and Edgar. Certainly, some ladies who were below the Island had been unseated but no harm done beyond a fright. There was also a two-foot gas main suspended in chains from our booking hall roof and I had thought for the moment that this had been punctured, with me sitting on top of it. If this had been so, this memoir would never have been written. At least it made the gas men get on with diverting it, as gas mains were by then forbidden to run along pipe subways.

The change to married life coincided with changes in the cast of actors on the Piccadilly stage and foreshadowed changes in my firm. On the engineer's side, Leigh Hunt left to become an Admiralty engineer and was succeeded by a 60 year old 'stickit' engineer, as the Scots would say, so I had no-one of my age group to talk to. Then Dalrymple-Hay objected to dealing with Segrave, as a lower form of life than his old friend Mr Rowell and made trouble. Segrave confided to me that he himself was never happy unless he had a boss over him and on his advice, George Burt engaged a legendary figure from the Underground world named I. J. Jones, who must, perforce take his place in this story.

I. J. Jones had just finished widening the City and South London tube by direct labour for the consultants Mott, Hay and Anderson. This was done at night when the trains were not running (until the tunnel collapsed). 'I. J.' had earned the reputation of being a Victorian taskmaster, as he had spent all night down the tunnels, slept for three hours and spent all day in the office. As things turned out, I worked under him at various intervals for twenty years with never a cross word. He was a lonely man, living in lodgings and as he confided to me, the son of a hard Lancashire ironmaster. He had little interest outside work, which led to long hours for the rest of us. We got on very well together in spite of some peculiarities, which were diagnosed by a later colleague[14] who had a taste for psychology. He diagnosed I. J. as 'being pathological on the subject of money', which would later prove to be true. Also, he included a further diagnosis that he had 'the gift of enveloping jobs in an atmosphere of misery and gloom'. This is partly true, but in moments of disaster he would become very cheerful.

During 1928, unknown to us, John Mowlem and Co were in the process of changing from a very private company into a public company with quoted shares. This operation called for the appointment of an outside director, whom someone once described as representing the overdraft. This director was a Mr E. B. Beck who had been concerned with the iron and steel industry and had been at one time a member of a merchant bank and had two sons who were about to leave school for Cambridge. In the meantime, the two sons of Edwin Burt, Tommy and Kenneth, were already working in the firm, so there would be, in due course, two dynasties. The firm now ceased to be a fatherly feudal family firm and as the years went by, slowly carried on what Chairman Mao would have called continuous revolution.

---

14 Rudolf Glossop.

The whole of 1928 was spent on the Booking Hall and entry passages. The road concrete was to be supported from a framework of structural steel, which had to be bolted together. It was fabricated by Westwoods, whose works were upon the actual site in the Isle of Dogs where Brunel had built his Great Eastern steamer. The framework was trial-erected in the yard and left in position so that each member could be ordered forward when it was wanted. The whole operation was so complicated that it can only be summarised. First we drove a series of 4ft wide headings. Six inch square timber posts supported 6in x 3in steel joists from which the road concrete was held by packings. These frames were two feet apart. Shafts were sunk from these headings, into which heavy steel columns 25ft long were persuaded to sit on grillage foundations. Then at night, heavy steel plate girders 4ft deep were placed upon the columns by jacks, block and tackle and brute force. These formed a series of squares and slightly smaller plate girders were bolted to them to fill the gap. These all held the 6in x 3in joists which still held up the road. All this was done through the part of the working shaft inside Eros Island. This led me to describe tunnel work as trying to dress for dinner under the bed clothes.

After the deep area which was to contain the machinery chamber had been covered, there next came two concentric rings of beams which needed pits to take shallower columns to support them. As each beam was in place, it had to carry its share of the road concrete. In all, over 400 steel girders of one size or another were erected and each required its own separate little tunnel. Meanwhile, we sank four small shafts from inside the first headings and then drove a whole series of timbered headings in the sand and gravel, which had to follow the curved walls of the finished booking hall. Bricklayers then worked out backwards building the walls. As the outer boards could not be taken out but would leave a space of 1.5 inches if they rotted away, a new idea was inaugurated. Steel plates, ¼ inch thick were used on the outside, instead of boards. Then we drove another lot of headings upon the top of the first lot and again the bricklayers laid the next lift of wall, with the inner face built to the correct curve, while working out backwards. The last spread of joists from the outer ring beam were then erected, each in its separate little tunnel and packed up about three feet from the outer end. We then dug down and exhumed the curved brick wall and called back the bricklayers to build their third lift, in order to pin up the final joists. All this time the traffic could be heard a foot or less above us. In a quiet moment a bicycle could be heard, sounding like the peeling of a banana.

Just to complete the picture, the approach passages, some as long as 150ft, were built in much the same laborious manner. A heading was driven on both sides and brick walls built, five feet high. Then miners tunnelled over the top, 3ft at a time, erecting steel beams across the full width as they did so to hold the roof. When each pair of beams was securely propped, the bricklayers returned to pin them up by completing the short lengths of wall. The same was done in making the 50ft long pipe interceptor chambers across each of the six streets. Naturally the space around the beams was filled with concrete.

In a long Alpine tunnel the moment of junction is a single incident. On this complex station work, the drives are short but the junctions are innumerable. In these headings, the brickwork had to be set out, for better, for worse, without any visible sign that it was where it should be, until the drive over the top revealed the brickwork. As a mere matter of statistics which are so fashionable, I have calculated that, apart from the rest of the work, I was responsible for driving over 3,000ft of timbered headings from 30 different 'faces'.

One other item of news was published in the up-market Press, but it had to be paid for:

4. GO WEST YOUNG MAN

the birth of our daughter Caroline on July 11th 1928. Then, towards the end of the year, preparations had to be made for the Opening Ceremony. These were threatened by the possible demise of H. M. King George V, but he survived for the Nation by recovering at Bognor, which then became forever Regis. The flavour of the period and the proceedings can be gathered from my 11th letter to the XXI Club, Feb 3rd 1929:

> Piccadilly opened in a blaze of glory before Christmas. The opening ceremony was quite humorous. On reading of it in the papers one learned that it was a Distinguished Gathering — appearances are so deceptive.
>
> The place was beautifully decorated, all the show cases were full and bunting hid the unfinished portions. The ticket machines (temporary) were in position, so were the lamp posts (temporary) on the middle landing. The entrance at Dunn's Corner and the London Pavilion (temporary) were in position and so was the kerb (temporary) carrying the bronze rails at Swan and Edgars, also the signs (temporary).
>
> A neat hoarding hid the existing Gents Lavatory which sticks into the new Booking Hall and a neat handrail (temporary) led to the emergency stairs. Luckily it was a fine day so no buckets were needed to catch the drips. The ventilation arrangements (temporary) were in full blast and the escalators (temp-sorry permanent) were escalating.
>
> Lord Ashfield spoke for 45 minutes and congratulated and thanked for their co-operation the LCC, the Electric Light Companies, the GPO, the this, the that and the other, with at least five lines for the London Fire brigade and quite two lines for John Mowlem. It is of course more usual not to mention the contractor at all.
>
> The Lord Mayor of Westminster then made a speech as long as it was inept, carefully explaining all the difficulties that the Westminster Council had created, being apparently proud of them. He then announced 'Ladies and Gentlemen, I am proud to tell you that the accommodation in the Public Conveniences has been increased by 50%'.
>
> We then toured the station and were provided with free bubbly and sandwiches and all got quite jovial.
>
> On the previous Saturday the Underground had provided a lunch for our 250 men in the middle landings, all among the escalators. They installed 10 electric cookers and 8 coppers with 50 rough Nippies[15] to wait on us. We had fish, meat, sweet coffee, cigars and a pint of beer in an inscribed pewter pot with a cardboard box to take away the pot after the proceedings. It was quite amusing, some navvies arrived rather shot-up before the event. Dalrymple-Hay and foreman Johnnie Robinson made speeches which were drowned in a chorus of smothered raspberries.

In actual fact, after Dalrymple-Hay had congratulated himself, his resident engineer and I. J. Jones, a miner had shouted from the next chamber, 'What about young Harding? He did all the bloody work!'. This was an embarrassing moment but, as I was below with them almost every day, they saw more of me and the results than of anyone else. Dalrymple-Hay

---

15 Waitresses.

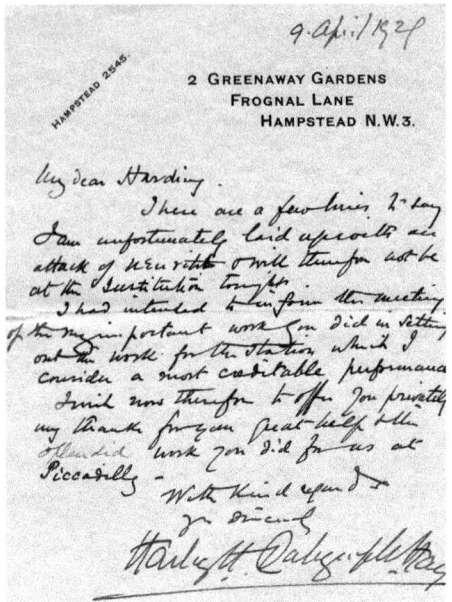

21. Thank you note from H. Dalrymple-Hay for the setting out at Piccadilly

had been so pleased at the way that all his schemes had worked with no errors to be corrected that he sometimes forgot that I was not on his staff. During the opening ceremony inspection, I descended the first escalator to find myself at the bottom, confronted with Dalrymple and my director, George Burt. Dalrymple said 'George, this is your best young man, you should double his salary'. He then turned to me and said 'There my boy, I can't do more than that for you, can I?'. George and I smiled wanly at each other. My salary remained the same.

When we were shaping Piccadilly Circus Station into its present form, we did not expect it to become the focus for demos, hippies, or drug fixes but the three years spent on it from 1926 to 1929 saw important changes in my professional and private life and I regard it as my favourite *tour de force*.

5

OPEN AIR – THEN BACK AGAIN   1929-1930

There were two short interludes before the next move. The firm's building department needed someone to set out the complicated curved lines of the east extension of Bush House for them and then to set out the main lines for a new Nurses Home for the old Seamen's Hospital at Greenwich. In Greenwich, the men who were digging out the new foundation holes were wearing rubber gloves and baskets lay beside each man. From time to time as they dug, they would toss bones into the baskets. The site had been the old hospital burial ground, so as each basket was filled it was carried to a wooden shed in which boxes, 6ft long, were laid out. An expert deftly sorted the bones and made them up into tricks of three in each box. Thus, each one carried three skulls, six limbs and three torsos. At dead of night, the boxes were transported to another suitable site, where a devoted Minister of Religion read the appropriate service over them as they were re-interred. The remains were victims of many a naval battle so it is possible that some were short of a limb or two but, no doubt, 'it will all come right on the Day'.

As things in the Holborn area are prone to be, the situation at Bush House was rather different, as the learning was done from a distance. I once inherited six volumes of *Old and New London*, by E. Walford and G. W. Thornbury which were published in 1878 and cast much light on London's past history. They record that the Pneumatic Dispatch Company was established in 1868 and laid a series of smooth-bored pipes in a trench from High Holborn, certainly along St Giles High Street and probably up Tottenham Court Road, as they eventually ended at a General Post Office at Euston. The diameter was about 5ft or so. Pneumatic apparatus was installed at their HQ in High Holborn, which blew a train of containers for mail and small parcels as far as Euston and then sucked them back by vacuum. This worked for 18 months and then lost money. The pipe was connected to the Metropolitan Railway tunnel instead and the pneumatic apparatus pumped in air to ventilate the smoky tunnel. *Old and New London* carries on thus: 'The tubes are still in-situ, doubtless only sleeping for a time, to be revived when London is ripe for its services'. The time became ripe with the invention of the telephone and this pneumatic tube made a subway to carry fifty main telephone cables.

The vital process of gaining experience can be indirect as well as direct and varied by observing the misfortunes of other people. The gas explosion in Holborn High Street in 1929 is one example as it took place only a few hundred yards from our office. Lack of oxygen is the risk to be met in such closed tunnels and the GPO very wisely gave their maintenance men electric blowers to freshen the air before patrolling the tube, but the socket into which the blower had to be plugged was placed at the bottom of a small manhole. On this winter morning, the first man descended to plug in the blower but it was too dark to see the plug, so he lit a match! That was the end of him, several others and the pipe. The tube split along the whole length, from the then Princes Theatre in High Holborn, around St Giles to a short distance from Tottenham Court Road Station. A long crevasse was formed for the whole distance, several fires were started, gas and water mains burst and manhole covers were thrown over nearby roofs. Several walking wounded were cared for. It seemed a mystery how gas could have accumulated in pipes which had originally depended on their air-tightness for their function, but joints may have rotted away near a leaking main. According to the newspaper, the gas company fought hard to prove that it was marsh gas and not town gas, but apparently without much success. The experience of seeing such damage and the efforts of the various services, who coped with it so well, was a useful preparation for future work in the Blitz.

In January 1929 another incident, which was none of our making, had a strong influence on my next assignment. John Mowlem's building department had just earned many laurels for completing the first of the two new headquarters for Lloyds, to which the famous Lutine Bell was transferred. The design by the architect Sir Edwin Cooper was in his best Classical tradition, in the highest standard of craftsmanship, with much carved detail and swags of wooden fruit. The same team were all set to start on a new headquarters for the National Provincial Bank opposite the Mansion House when events fundamentally changed the methods to be used.

These events were caused by problems incurred a few months previously, when another building contractor was making a deep excavation alongside the Commercial Union Building in Cornhill, opposite the Royal Exchange. In it was to be built a Midland Bank. This hole was being done in the old style to a considerable depth, but the sides below the Commercial Union footings and along the edge of Cornhill were supported by long raking shores that held up vertical timber 'soldiers' which held all the frames of timber. After heavy rain, the feet of the raking shores slipped and the old masonry Commercial Union Building cracked all over and the staff had to be evacuated. A day or two later, watched by crowds of excited spectators, the building collapsed onto the raking shores which supported Cornhill. The effect was most educational - Cornhill roadway began to fall into the hole, gas mains broke and caught fire, water mains burst, hydraulic mains, with their 600lbs/square inch pressure made their presence felt and a good time was had by all.

This proved to be the end of the raking shore era and led to reaction and at times over-reaction. The new National Provincial (or NP) Bank was to replace an existing one on the corner of Poultry and Princes Street, exactly opposite the Mansion House. It also touched on the edge of the Bank Underground Station and the other inside walls adjoined the Midland Bank Headquarters which were being built. Sir Edwin Cooper had a consulting structural engineer to design his steelwork but he was no match for the heavy-weight engineers deployed by the Underground and the Corporation of London, who insisted on a really safe method of digging down over 50ft to provide three basement levels. The decision was made to build the 50ft deep concrete wall, which had to hold up Princes Street and Poultry, in a series of alternate pits which were to extend for 16ft back from the building line. This matched the engineer Sir Benjamin Baker's dictum that the thickness of the base of a retaining wall should be one third of the height. The pits were only 12ft wide along the line of the road. The consulting engineers for the Underground, Messrs Mott, Hay and Anderson insisted on the work being carried out to their entire satisfaction, so it was entrusted to I. J. Jones (and me) and not to the building department. As there were many complications to be dealt with, we used our tunnel miners, who were trained in sophisticated methods beyond the skill of the simple timberman.

There had been only one basement in the old Bank, below this we had to dig through about 30ft of Roman tannery refuse before we reached the London Clay. This refuse deposit consisted of dry, black material which was like John Innes compost, riddled with old scraps of leather, pieces of sandal, bones and sheep's horn. The Guildhall Museum was in periodic attendance which interested the men. In one of the pits two curved sheep's horns had been nailed to the boards, one pointing upwards and the other down, a miner had written beside one 'Julius Caesar before meeting Boadicea' and the other 'Julius Caesar after meeting Boadicea'. It was good to see that education has not been wasted. Along the west of the site we unearthed the bed of the Walbrook. It was lined with 10ft long piles made from tree trunks, about 8 inches in diameter and with tapering points. In one of them a Roman chisel was embedded and the edge was still bright when it was pulled out. Then, when a heading was driven under Poultry towards the Mansion House to make a drain connection, we had to break out a 12 inch thick slab of Roman road, about 13ft below the road above.

The Bank of England was being greatly enlarged to become Britain's answer to Fort Knox. The famous outer wall had to be underpinned to help to make ever-deeper basements and vaults and this was carried out in alternate pits, 70ft deep. So, along Princes Street this cautious method was being used on both sides to support about 35ft width of mixed earth, riddled with cable, pipes and a 6ft main sewer.[16]

This was my first encounter with an eminent architect and was enlivened by watching his struggle with eminent engineers and it taught me a lesson. To such an architect, itching to reveal his visual masterpiece to his admiring public, foundations are frustrating and a pain in the neck. The civil engineer who works below ground regards such work in the same way as a surgeon regards a critical operation - the more complex the case, the greater the satisfaction in completing a challenging exercise with success.

The lower part of the mass concrete retaining wall that we were building in the pits was wide, but the top half was only 3ft thick. The eminent engineers who were defending the tube station and the Mansion House thought that this was much too thin. Briefly, the great men insisted that the top of this wall must be strutted across the whole site, before any of the inside earth or 'dumpling' was removed. This was to be done by using the steelwork of the upper basement floor. How could this be supported? Dalrymple-Hay came to the rescue with help from another site, where he was starting to sink a number of deep pits, lined with 25ft diameter cast iron rings. If we were to sink three pits, using his available rings, we could not be accused of risk. At the bottom of these we laid the large square grillages which held the main columns, which then held up the steelwork of the floor. Then, as we dug out the dumpling, the cast iron rings were released, one by one, and reused on the other permanent work. This is an over-simplified version of a comedy which I watched with the same interest as Shakespeare's *Much Ado*. Being a simple-minded youth, I suggested that it only needed some steel reinforcement up the back of the 3ft wall and was soundly snubbed. Three years later, we built much greater lengths of 3ft thick walls, reinforced as I had suggested and under much worse pressure, with complete success.

One last incident I recall was where I was rung up by the railway company to ask if we

---

16 When the Barbican rebuilding scheme was first mooted in the early 1950's, HJBH suggested that a large area could be dug out within one encircling retaining wall, in which several buildings could be built, instead of trying to support every narrow street with a wall on either side.

could take a five horse load of steel rods. Next morning the traffic in the main centre of the City was brought to a standstill while a long dray drawn by five horses, two pairs and a leader, slowly wound its way to our site. The five horses were then picketed around our site consuming their elevenses in the nose bags while the steel rods were unloaded.

In May, while the National Provincial Bank was in full swing, John Mowlem was given the contract to build Thames House, Millbank. I. J. Jones was put in charge of the foundations, so I found myself setting out all the work, which meant shuttling by District Railway between the two sites. The Millbank area, to the south of Westminster Abbey was familiar to me. It had become the property of the Dukes of Westminster when their ancestor, Sir Thomas Grosvenor married the heiress of the Manor of Ebury and all Belgravia. The area had little significance until the building of the Gas Works in 1812 (later suitably occupied by the Ministry of the Environment) and of Millbank Prison in 1821, which was erected 'for the punishment of offenders of secondary turpitude'. In the 1890's the prison was replaced by the Tate Gallery,[17] with the Royal Army Medical College and Hospital on either side. To the north of the hospital lay John Mowlem's depot and within its precincts were several massive, three storey buildings, thought by some to be remains of the prison. In one of these monumental buildings was the best joiner's shop in London, where works of the finest craftsmanship were produced to the satisfaction of Sir Edwin Cooper and other great architects.[18] On one side of the depot entrance stood the Speaker's Stables which housed his lumbering coach, which always needed half an hour's start in any Coronation Procession. There were no horses, they came from the local brewery. But John Mowlem's horses were the directors' pride and joy, as they took all the prizes at the Regents Park Easter shows. They lived in the depot, which we called the Yard, and were not paid off until 1932 (this hive of activity vanished in 1957 when Mowlem built the Millbank tower on the site).

Between the wars the previously rather seedy area between the River and Marsham Street was completely rebuilt, with flats for the working class, apartments for the gentry, Westminster Hospital and also some opulent offices. The land on either side of the old Lambeth Suspension Bridge (down-graded to foot passengers) was occupied by rather decrepit housing.

In 1928, after Sir Alfred Mond, later Lord Melchett, had brought off his combine of chemical firms, Mowlem built the new headquarters of Imperial Chemical Industries (ICI) to the north of the bridge. In January the River Thames had suffered one of its surges and broke the river wall along Millbank, swans swam in the Mowlem depot and a family were drowned in a basement nearer to the bridge. The whole of this area was then demolished and ICI decided to build 2 office blocks to be called Thames House on either side of Page Street. Grosvenor Wharf was the postal address of the depot and the firm had made much use of it in the past but the wharf was now to be replaced by lengthening the granite river wall to give an extension of the Victoria Embankment Gardens. Mowlem also did this work, under another engineer. To add to the local activity, Dorman Long and Co. started to build the new Lambeth Bridge for the London County Council. Thames House brought about changes in long established London foundation methods and other changes began to take

---

17 Now Tate Britain.

18 These can still be seen in Lloyds [main building demolished in 1979, but the entrance portal remains], the old Port of London HQ [10 Trinity Square] and many other fine buildings, not forgetting the Institution of Civil Engineers and the rebuilt chamber of the House of Commons. Coincidentally, his work can also be seen at the Royal Star and Garter Home, Richmond, built using plans from 1915 (see Chapter 1).

5. OPEN AIR – THEN BACK AGAIN 1929-1930

place in the firm. One novelty was almost the first use in London of the new material – steel sheet piling. The British Steel Piling Company (BSP) had just been formed and was peddling Universal Steel Piles. These universal piles were flat steel joists held together by shorter joists, so cunningly rolled that they clutched the piles together. BSP also hired out or sold steel piling frames and winches and distressed the inhabitants by the introduction of the noisy double-acting steam hammer. Universal Steel Piles later became extinct with the increasing popularity of the Larssen Pile.

Timber piles for carrying loads are among the oldest expedients but sheet piling was now to take over from this old-fashioned technique. The ground on the site consisted of 30ft or so of floodplain gravels and sands resting on the London Clay. There was a head of water of over 20ft in the gravels but the sheet piles, driven well into the clay, excluded the water. In the past, the water level would have had to be pumped down by masses of steam pumps. This time we had a clear open site, so the first step was to make a sort of dock wall all around. Two rows of sheet piles were driven 17ft apart, held by struts and walings while the ground was grabbed out between the piles. Then we built a retaining wall tapering towards the top which encircled the site, nearly a quarter of a mile in all. A Ruston steam dragline, the first seen in London, then roamed about on the surface as it gradually dragged all the ground out within the walls and loaded it away on steam road wagons. It also dug away the whole of Page Street so that there could be a continuous basement between the two buildings. Foundation pits were concreted following behind the drag line, so that Dorman Long could follow with the steel frame for the building. Page Street was eventually laid back on top of the basement steelwork.

In the winter of 1929 there was another panic over possible flooding by the Thames, so the populace lined the rebuilt parapet wall along Millbank in the morbid hope of seeing another inrush. Precautions were of little help to us, if the flood came our site would merely look like a swimming bath. But I. J. Jones posted the following notice:

> In the case of the Thames overflowing its banks and flooding occurring, three blasts will be given on the buzzer. On this signal, all men at the bottom will come to the surface (!!)

One useful lesson was learned - when the steel piles were driven down one could admire a beautiful straight line at ground level, but when the digging out of the ground had reached 40ft things were different. Flints in the gravel had affected the driving, so some piles were inwards at the toe and others further out. The lesson to be well remembered was the way that the occasional pile had parted from its clutch, letting in the water.

The architect of the ICI building and Thames House was the highly colourful Sir Frank Baines, who had earned the KBE for his drive and forceful work in the war as head of the Ministry of Works under Sir Alfred Mond as his Minister. He terrified all of his staff, especially the clerk of works on our site. To appease his boss, the latter wrote a foolscap letter nearly every day with any particular fault that he could think up, but this was hard for him as the work went so smoothly. I. J. merely stacked these letters, one above another and when the pile was several inches high I asked if he was going to answer any of them. He replied 'If I do, I should only say – "Your letter is before me and will shortly be behind me"'.

At this time new blood was beginning to make itself felt. When the firm had become a public company they had taken over another one in full going order which specialised in work in the Docks and the river. This continued to function as a separate unit known as 'Basin South' because its office and depot lay between the entrance to the King George V Dock and the Gallions Basin. This increase of strength enabled the firm to carry on the maintenance contract for the Port of London Authority, from its heyday to its decline. It also provided us with gangs of men who were used to pile driving and could be utilised for this current contract.

The amount of shuttering for the long length of retaining wall called for much fast and athletic work so I. J. managed to recruit a squad of four or five Scottish 'chippies' led by Jock Robertson. These agile and hard working young men had served their apprenticeship in the Glasgow shipyards as shipwrights' carpenters, before joining the tough firm Sir Robert McAlpine and Sons. They provided a fresh impetus and eventually each one became a foreman with the firm. Today there is much talk of business efficiency and man management. In those earlier days one had to work it out for oneself. There are strengths and weaknesses in feudal hereditary firms. One peculiarity was the lack of any hierarchy or official pecking order, each little unit reported direct to the top, so collaboration between units depended upon friendliness, persuasion, cajolery or, if all else failed, direct confrontation.

The Burts, like the Forsytes, had been 'Superior Dosset'.[19] They had a tradition that all their descendants should go through a course in the joiner's shop before working on sites to gain experience but some of their modern rivals sent their sons to universities. The strength of the firm at the top was Sir George Burt, who was a leader who attracted loyalty. On leaving Clifton College he had undergone this rigorous training. He had a strong sense of civic responsibility, was shrewd although in some ways simple, hard working and approachable. The weak link was his much younger brother, Eric Burt, a man of much charm and good nature, although inclined to be stubborn. He was only an echo of his brother. He had gone into the Army straight from school and when the war ended became an 'instant' director with no experience behind him. He revealed a slight lack of perception when, some years later he said that when he was in France, with 'signals', he always got back to the Mess on his motorbike in time for lunch and seemed oblivious that the Infantry did not. Eric had been an amateur racing motorist, so perhaps that was why the plant department was left with him. The man in charge of it was the son of the previous 'steam boss' and had learned what he could under him in the simple days before petrol and electricity. In contrast, Basin South brought in their own well-equipped repair shop, under a very efficient, shop-trained plant boss. The firm's electrical side was stronger under an ex-naval electrical artificer.

My increase in experience over six years had been under seniors who were masters of their particular art of specialised work, which required simple mechanised plant. At Thames House, plant problems increased, as we used ten pile frames with steam boilers and winches, three electric Scotch Derricks with 100ft jobs, five 5 ton steam locomotive cranes for grabbing, six 3 ton cranes, two old steam navvies and the newest steam dragline, as well as pumps and concrete mixers. Most of this plant was hired or bought for the work. I had suffered from the plant department in the Isle of Dogs campaign and I began to realise that the civil engineer does not have to be an expert on the mechanical interior of machines but must know their capabilities and how to deploy them. Later I was aware that Rommel and Montgomery could successfully deploy their armoured forces, but did not need to know how to repair them. For the next twenty years, I had to struggle to overcome reluctance to buy new plant or to scrap the old and to try to point my employers in the right direction. Some years later, I told my immediate assistants that the 'tactical handling of company directors is an advanced science'. When a director myself I still maintained the truth of this.

---

19 A Forsyte nickname.

5. OPEN AIR – THEN BACK AGAIN 1929-1930

Battersea Power Station – Enter Sir Henry Japp

During 1929 there were other portents of change, including the unannounced recruitment of Sir Henry Japp, KBE, MICE. Sir Henry had made a big name for himself while working for S. Pearson and Sons at the beginning of the century by driving the four railway tunnels under the East River, New York at very high air pressures. This followed the first successful tunnel under the Hudson River by his cousin, Sir Ernest Moir. We should not forget, in spite of the lapse of time, that it was the British who taught the USA engineers how to do it when they had failed. The British, and especially the Scots, provided the engineers and the mining gangs who established themselves in an ever growing number of such tunnels and formed the nucleus of those romanticised by the Americans as 'Sand Hogs'.

Japp was the first to pioneer scientific methods of decompression by stages. In spite of pressures of 40lbs no severe or fatal cases of the bends occurred on the East River tunnels and such cases as were treated were slight. During the 1914-18 war he remained in the United States as a member of the British Government Purchasing Mission and was made KBE for his services. After the war, Lord Cowdray[20] was winding up his firm of Pearsons to feed his cash into pastures new, so Sir Henry returned to Britain and became a Director of Kirk and Randall to build a new harbour at Workington, Cumbria. For this work he had recruited a small staff of young engineers who had trained in harbour and maritime work. Kirk and Randall went into liquidation, but Sir Henry and his staff stayed on to finish the work. Then E. B. Beck persuaded him to take on the building of a pumping station on the River Tees, to supply the works which ICI were building at Billingham.

What happened next is succinctly described in my XXI Club letter of May 1930:

> Early this year the foundation company were unfortunate enough to go into liquidation while engaged in the questionable undertaking of building the Battersea Power Station in face of much opposition. Out of evil cometh good (for other people) and while the London Power Company carried on the foundations by direct labour, John Mowlem took on the contract to build the coaling jetty in the river and also to drive a 10ft diameter tunnel under the river to carry the cables. This part fell to me under the auspices of Mr Isaac J. Jones.

Sir Henry Japp brought along his staff and took on the jetty, which involved trying out new forms of sheet steel piles which were coming onto the market. I learned a lot by watching without being involved. This was my first encounter with Sir Henry, who like the rest of us had his virtues and defects. The latter were shortness of temper and an aptness to bully, however when he tried this on me I fought back hard, so we got on well thereafter, as so often happens. I was also well briefed on the East River tunnels from my *Engineering Wonders*, while my mother's cousin-in-law, John Body, had been a director of Pearsons, so we had a source of conversation.

Sir Henry possibly thought that he was going to be top dog, but as I. J. Jones had already been taken on and was what the Americans called the 'Dean of the London Tunnelling World', he could not be supplanted. The head of the building department for many years would brook no one between himself and the Board and the same went for the Basin South

---

20 Born Weetman Pearson, he became Lord Cowdray in 1917.

manager. Thus the feudal chiefs jealously guarded their direct access to Sir George Burt, who was 'apt to want to push the barrow all by himself' as the saying goes. So Sir Henry had to be content with a smaller sphere of influence, but he persuaded the Board to adopt Lord Cowdray's invention of calling such tribal chiefs by the title of Works Director. Until then they were only known as 'Mr this' and 'Mr that'.

The cable tunnel, in compressed air, was a neat affair with few incidents. On the Grosvenor Road north bank we built a shed for the compressors, with the plane trees sticking through the roof. The top of the tunnel started with only 5ft below road level, from a square timbered shaft and the shield drove 50ft parallel to the river wall to give room for building the boiler lock. Then, with the pressure on, it started on a 325ft radius curve at a down grade of 1 in 10 and just scraped under the foot of the river wall. In the sand and gravel the classical method of clay pocketing was followed until the hooded shield had dived deep enough into the London Clay. The air pressure more or less adjusted itself to the rise and fall of the tide, although carefully monitored. No more than ten pounds pressure proved necessary but the bubbling on the foreshore mystified the locals. From Battersea the drive was straight all the way at the depth of 70ft below high tide level. There was sufficient cover of London Clay and so pressure was kept at 10lbs /sq inch as a precaution against running into a buried channel of gravel. A small sub-contractor had made two boreholes in the river bed from primitive plant balanced on a stage between two Thames barges, but they could easily have missed a sudden depression. The paper on Battersea Power Station revealed some extraordinary variations in the clay level over the site, which geologists found hard to explain. This was a recurrent theme.

My first of many dealings with the River Superintendent of the Port of London Authority involved extracting permission to build a companion-stairway down the face of the north embankment wall and to use a boat with outboard motor for travel between the shafts. Until this was done, the alternative was a walk of one and three quarter miles over Chelsea Bridge down to the Nine Elms Road, then up to the famous Dog's Home and a long walk over the site to the river bank. A car was seldom available and a short cut through the Goods Yard was forbidden. The survey called for careful triangulation across the river to establish distance as well as line. A base line was easily established along the embankment pavement. The distance was too far to read a level staff, so the answer was to make a plain staff with a sliding target of alternate black and white checks, which could be finely adjusted in reply to flag signals. The curvature of the earth shows a drop of 3/16 inch in the 1,000ft of level sight and it insisted upon being measured. Each time we established the opposite level value and crossed over to check it back again the accumulated error came out as 3/8 inch.

We were enlivened by other aspects of life on the Thames which were recorded at the time in another XXI Club letter:

> Doggetts Coat and Badge race for watermen, sculling in rum-tums or whatever you call them caused a stir. It was from London Bridge to Chelsea Pier and was a very tight race. The two leaders, Barry and another, led by 300 yards. The launches, steamers and tugs following kept close behind them and left four other competitors to try to scull in their wash. They very nearly sank. It seemed unsporting as either or both the leaders might have sunk or had a fit or been disqualified or something.
>
> On another cold wet morning we were surprised by a

5. OPEN AIR – THEN BACK AGAIN 1929-1930

> curious procession of skiffs being towed by a police launch who paused at our landing stage. Each skiff was manned by two oarsmen and a cox. The first flew a big ensign with GRV on it and the crew wore red rugger jerseys and straw boaters. The others wore a variety of rugger jerseys and yachting caps, all except one which wore trilbies. They had banners with swans on, fore and aft and were also labelled as the Vintners Company and some other company and were off Swan–Upping. Judging by the whisky which they were putting away from hip flasks as they left us, and also their complexions, the swans were in for a bibulous time.

One more extract from the next letter has a little historic significance:

> On August 14th 1930 at 8am my tunnel beneath the river junctioned with, as the papers said of Sydney Bridge, 'unparalleled exactitude' and 20 hours later my wife presented me with a son, which you will agree, was enough agitation for one week, however we are all three getting on very well, thank you.

There is always a strain before a junction, especially when no one else has checked one's work, echoes from the pneumatic spaders in the approaching face often sound as if they had passed by to one side. A short heading was driven in order to check the line so that the final drive of the shields could be adapted if necessary. The error turned out to be 1/8 inch for line, ¼ inch for level and 3/18 inch for distance. The shields were taken down, leaving their steel skins behind, as is the custom, so that the rings could be built under their protection. The last rings from either side bolted together with no sign of the junction. Thus ended, satisfactorily, my instrument work once and for all.

My private life proceeded with the usual ups and downs. We were living in a second floor converted flat in Cromwell Road, so as to be near each of our very elderly mothers in case of need. The flat overlooked the District Railway triangle where the British European Airlines 'Terminal' building (now Point West) stands and was incredibly noisy. I was working long hours and every Saturday, so rather than make my wife a golf-widow as well as a work-widow, I preferred to help her to push the pram into Kensington Gardens. She maintained that she was the only pusher who was not a nanny and she learned much of their curious customs. Today the foreign au-pair fills the gap left by this almost extinct species.

My work had all been in London, where, in those days, travel was quick and reasonably cheap as buses, tube trains and even taxis were plentiful, so with little money there was less need for a car, especially with so little time to use it. My first driving licence dates from 1925, with spasmodic attempts at occasional practice during holidays. Most of 1930 was spent on the Battersea Cable Tunnel followed by tendering unsuccessfully for future work. Then one evening in October, I. J. Jones, who had a firm's car and driver, came back from head office and said 'Can you come with me, right away, to Dagenham?' I said 'Yes, if I can spend a penny first', and that was the start of a new chapter.

CRISIS AT FORDS

In Dagenham did Henry Ford
A Stately Factory decree
Where Thames, polluted River ran
Through mud-flats measureless to Man
Down to the cold North Sea.

In the early 1920's, Henry Ford decided to build an English motor works so complete that iron ore would come in at one end and Tin Lizzies[21] would rattle out at the other. To do this he needed access to road, rail, water and labour, so he bought an area of land on the River Thames at Dagenham, fourteen miles downstream of London Bridge. Behind it ran the London and Tilbury railway, beyond which was the main road, east to west. Soon after the first war the LCC had built two vast working-class housing estates at Becontree and Dagenham, which meant that labour was available.

The site itself was a disused, flattened-out refuse tip, sitting on a deep bed of very soft clay, which meant that many thousands of concrete piles had to be driven before anything at all could be built. The only access road down to it was Chequers Lane, which ended on the west side at Dagenham Dock. This dock sits on Perrys Breach, where Captain Perry of the Royal Navy closed a wide breach in the river protection bank in the 18th century. The evidence was still there in the shape of a small inland lake named Dagenham Breach. The east boundary was the River Beam, which flowed into the Thames through tidal flaps at low tide and many acres behind were well below river level. On the other side of the River Beam, the active refuse heaps of the City of London towered 100ft above the ground.

The 1929 slump was beginning to bite when it was announced that Henry Ford in person was to visit the site with his pockets full of many potential contracts. On the day of his visit the area was thick with the cars of innumerable company directors to pay him tribute, but not a single car was a Ford.

The general layout was decided in Detroit, where Henry Ford was still an autocrat with no sympathy for trades unions. By October 1930, a formidable jetty was being built by Mowlem in the Thames, parallel with the land, which was to carry heavy plant to unload coal and iron ore for transfer to a very large Ore Yard, as well as to handle the export of finished cars overseas. The machine and assembly area was to be housed in a vast open steel frame building. I wrote at the time:

> ...the main sheds are 1,000ft along the river front and much more deep. One notices every 200 yards or so little

---

21 Ford Model T cars.

6. CRISIS AT FORDS

galleries of steel troughing about 20ft square with stairs at each end. It turns out that these are not for foremen or for tool stores but are for the 'convenience' of the workmen, to prevent them leaving the job during working hours. I do not know if a moving band will be installed and the whole organised on mass production lines. As well as this there is a blast furnace and its accessories with ore storage beds and coke beds and coke ovens, also sintering plants and dry quenchers (by the way the whole of the area is teetotal by decree) and wet quenchers and so on. There is also a large Power House which will run the factory and provide steam for various processes and which will sometimes consume London refuse which is at present being dumped beside our site. The City Corporation dump being about a mile square and 100ft high.

Whenever Ford's men would decide to go on strike the BBC always showed on the television a tall building with four steel chimneys, each bearing a letter F. O. R. D. This was the Power House. Some believed that it was at the last moment, after all the plant layout had been started upon, that Fords decided to generate their own electric power, so instead of being able to choose the site with the best conditions, someone in Detroit had to draw a rectangle on the only available space on the plan and mark it 'Power House'. This was in the extreme south east corner where half the area was on the mud flat outside the earthen river wall and the other half was out in the tidal part of the river. Sir Cyril Kirkpatrick and Partners then had to design it as best they could.

22. *Construction of Fords Power House, Dagenham showing the unusual siting for such a heavy structure*

No-one would visit Dagenham unless they had to. During that first hour-long drive in the dark, I. J. Jones explained that the Basin South outfit had completed the foundations for the blast furnace and were involved in the Power House contract. But this had got into such a mess that I. J. was to take over as works director and I was to be the agent, taking over the staff from the Basin South organisation. I also learned that Sir Henry Japp, with his own organisation, was building the jetty as a separate contract, with an office at the west entrance where the only access road ended. I also found out that the Power House was at the opposite extreme end of the site and that Fords had refused to build a road up to it as the area behind was full of the many other plants which have been mentioned and the railway sidings needed to feed them. Thus, everything for the Power Plant had to be brought in by rail or else water, after transhipment to barge at Basin South. The only access was by walking for quarter of a mile from the west gate, in the mud, in front of the Assembly Shed. Smaller items were carried along this path by porters.

On our first visit, we left the car by the jetty contract office and struggled in the dark and mud past the sheds. We sloshed our way through the unfinished Ore Yard and past the blast furnace until we reached the brightly lit area of the Power House site. We found a portentous meeting in progress in the usual conglomeration of wooden huts, called offices. Sir Cyril Kirkpatrick and his men were confronted by Sir George Burt and the Basin South hierarchy. After introductions all round, tactful words were spoken to complete the take-over. We then struggled out to view an alarming state of affairs.

The Turbine House was to be 350ft long and 70ft wide, with concrete walls 3ft thick. The bottom was to be founded on sand and gravel 40ft down. The area was surrounded by Larssen steel piling 59ft long and was divided up into two parallel series of compartments about 50ft long and 35ft wide, to be heavily supported by timber frames. The whole design depended upon the decision to get Francois Cementation Company to solidify about 6ft thickness of the gravel below the intended floor level, in order to keep out the considerable head of water, as forecast from experience at Charing Cross. This had been a failure. Work had only started on three of the fourteen compartments but, due to the failure of the cementation, one was flooded and old steam pumps were battling to keep down the water in another. The third one was smaller and the wall and floor had been built. Two heavy excavators on caterpillar tracks were derelict beside the compartments, as they had, not surprisingly, ground themselves 6ft into the soft clay where they had been used to grab out the spoil. We later found another 6ft of assorted timber buried under their tracks. This had been tossed in from time to time, to try to keep them on the surface.

A thousand cubic yards of material for concrete had been unloaded from sailing barges on to the river bank. This had caused the bank to slide gracefully into the river, pushing over the timber piles of the unloading jetty (unfortunately, soil mechanics was then an unknown phrase). This was my inheritance on instant promotion, at the age of thirty, to agent or project manager.

At the meeting there seemed some panic in case the second compartment was to 'blow' its bottom and Sir George Burt told me that he wanted me to come back to 'see over the high tide' at 3am. He would send a car at midnight to my Earls Court flat to collect me. So I was back home by 9pm to break the news to my wife, then managed a few hours imaginary sleep before going all the way back to Dagenham, carrying sandwiches and a torch. I made the driver struggle with me over the muddy approach so that he could sit in the warm while I wandered round making myself known to strangers like the night foreman and others. I had no idea what was expected to happen or what I was supposed to do if it did happen, but the tide passed with no effect, so I was driven home by 5am and back on the site by 9am the same day.

That day I examined my new command to find out what it was all about. The general foreman, Redwood, was a man of 50 who had been an agent on over-sea contracts for other firms. He had wide experience, education, was vigorous, slightly piratical and showed a good sense of humour, touched with mischief, once his '*amour propre*' had been tactfully restored when having to work under a much younger man. From experience in West Africa, he had built a proper cook-house protected by close wire mesh to keep out the hoards of flies from the nearby refuse dumps!

I took over three young engineers: one, 26, was a born engineer with 5 years experience in piling and concrete work, but the other two were only 21 with almost no experience. The only timekeeper and the materials clerk were even younger. Luckily there were good under-foremen and gangers. The Basin South branch had intended to use derricks to cover the site and steam locomotive-cranes to run on sleepers which should spread the load on the soft clay, but a director at head office had refused to spend money on new plant and had insisted on the use of the existing tracked excavators with the results described. This forms a parallel with many a British military expedition – an able commander given too few staff, inadequate reinforcements and the wrong equipment.

I found myself instantly involved in disputations over the driving of concrete piles which were meeting obstructions. The boiler house inland and the switch house on the river side of the turbine house were carried on over 2,000 concrete piles. These, over 75ft long, driven by six-ton drop hammers, called monkeys, handled by 10-ton steam winches on 90ft high steel piling frames, were all of them the largest used in Britain at that time. I found it wise to conceal my ignorance by remaining silent during acrimonious talk until I had learned the appropriate jargon and the real nature of the problem. The piles themselves dropped 20ft into the clay under their own weight and one day, when one monkey had been wound up to the top of the frame to make room for a pile, the bond snapped and we had to sink a pit 25ft into the clay to retrieve it. The driving of the piles into the gravel was so hard that the Pinkado timber used to soften the blow on the pile helmet constantly caught fire and had to be renewed. The piles were cast using high alumina cement, which had recently caused many building failures. Its virtue was the ability to set or harden so quickly that we could drive piles only three days old, but it needed relays of 'water boys' day and night, who had to spray the piles continuously with water to prevent them crumbling due to the heat which the cement generated. The cementation of the rest of the area was nearing completion, from a central mixing and pumping station through a maze of hoses to where pipes were being driven down at 3ft intervals. A tic-tac man made signals to the pumps as the pipes were pulled up (it was before the days of short wave radio).

I spent the early part of the winter in trying to bring order to chaos, not helped by a young resident engineer who had a liking for power and a sadistic streak and also liked to stir up jealousies among my gangers from a sense of mischief. One and a half hours travel each way from Earls Court did not help matters.

We set about digging out another compartment, but just as we reached the hardened bottom there was what is called a 'blow'. Water burst up through an area of sand where the cement had failed to penetrate and the hole rapidly flooded. We had to concrete the floor under water and use divers from the jetty contract to release the timber frames. This affair provoked an instant crisis and fears arose that if this happened again, the river might burst its bank by the River Beam and flood countless Essex acres. By coincidence, just at this time, one of Kirkpatrick's men stumbled across an article which described how the bank had burst at this very spot around 1600 and that the famous Dutch engineer, Vermuyden, who had been busy at the time draining our fens, was called in to close the breach. He did this by sinking timber boxes filled with chalk and we eventually dug up some remains of these.[22]

Fords, the engineers and Mowlem went into a deep huddle to find a safe solution. The contractor was blameless – it was a specified method which had failed, while the design of the work was also not in question. The problem was – how to reach the required depth safely. Although the two turbo-generators would be at ground level, there had to be much going on below them. They were to sit on complex concrete turbo-blocks, which also contained the condensing complex. The vital cooling water from the river needed to be controlled by intake and discharge shafts, and passed through the screen chamber into the pump house for

---

22 An echo of this method came years later, when the Dutch sea defences were closed after the 1953 floods by sinking surplus monoliths left over from the Mulberry Harbour.

circulation. The decision was made to complete the deep digging by working in compressed air – an art in which the engineers Sir Henry Japp and I. J. Jones had much experience. The intake, discharge shaft, screen chamber and pump house were to be sunk as compressed air caissons. The method of containing the air in the turbine house was more original. There were nine more compartments of different size and depth to be completed inside the steel sheet piling. First, we sank steel cylinders, 8ft 6in in diameter in each compartment in turn, a total of 56 in all, with much changing over of air locks. The lower parts were filled with concrete to a level and the muck was then dug in the compartment until about 6ft of soft clay was left above the gravel, which would hold back 12ft head of water. After this, a reinforced floor was cast in each compartment 4ft thick, resting on the concreted cyclinders like a billiard table. More reinforced concrete was placed above to make a thickness which varied from 8ft in one to 11 ft in another. Shafts to carry air locks were cast in each floor and the muck below was dug out in compressed air to depths carrying from 7ft in one to 17ft in the deepest (the one in which we found Vermuyden's boxes). These spaces were filled solid with concrete and the air pressure, which varied from 10 to 19 pounds per square inch depending on the depth, released.

While this decision was being reached we had other dicey moments. We found that the many concrete piles which we had driven were displacing ground, which took the line of least resistance by pushing the sheet piling towards the river. We had to mobilise our piling barges to drive clusters of raking timber piles into the river bed, on which we had to build large timber boxes full of rubble to resist the movement.

The decision over the choice of caissons had been made earlier, since we had already sub-let the sinking of a skip-hoist pit beside the blast furnace to Dorman Long and Co., who had just completed sinking the caissons for the piers of Lambeth Bridge. Their engineer, A. E. (Sandy) Reid became another of the good friends which I collected on my way. From the experience at Fords, I became an ardent advocate of compressed air caissons, in the right place. A caisson only needs one-tenth of the volume of air to be pumped in, compared with an equivalent area of a tunnel face.

The change in method of working required much new plant and materials, with a great deal of extra work and while the plant was accumulating I had to revise the entire layout of the site and much else. We were to have steam cranes to wind out of the air locks and three new 12 ton Scots derricks with jibs 120ft long. Two of them had to stand with their legs in the river, so we had to use our piling barges to drive clusters of timber piles to carry them and many other piles to carry a platform on which to work. The Port of London floating 50 ton crane had to be towed to the site to erect the derricks.

I began to have doubts on the complete wisdom of my elders. I kept finding it difficult to make I. J. Jones take needed decisions, especially where money was concerned, so began to adopt my own version of the Nelson Touch – when the signal is too silly, turn a blind eye. I took it on myself to build a large shed in tubular steel scaffolding, clad with galvanised iron sheets, in case of fire, for all the air compressors which were beginning to arrive. This was in spite of instructions to use up an old timber shed. An additional source of electric power was laid on, but I was prevented from doing the same to protect the distribution bus-bars as there was already a set in the wooden hut in which our electricians had pitched their camp. The results will appear shortly.

I noticed that the jetty contract was fully staffed with engineers who Sir Henry Japp had brought with him and they had a competent chief clerk from head office, on whom I kept

6. CRISIS AT FORDS

my eye for future work. I kept urging I. J. Jones to get me more staff but with no result. He developed the vice of calling out his driver to take him to Dagenham on Sundays to gloom over the idle job and worried if I did not come too. However, when, following my desperate demand for staff, he replied, 'If they cannot do the work in twelve hours, they must work twenty-four', I took courage - and my job - in my hands and went to head office to 'blow my top'.

The result was salutary. Once the powers realised the feeble staff with which I was expected to do twice as much work in half the time, changes were made. I. J. had picked up another tunnel contract, so he was tactfully detached. Sir Henry Japp, whose jetty contract was going well, took over. He had wider vision, from being trained in Lord Cowdray's firm, and I was given several more engineers, a mature timekeeper and clerks. In addition, I was asked if I minded one, McCracken, coming as joint agent. He had stayed on to complete the foundations of Battersea Power Station after his employer, the Foundation Company, had gone into liquidation. He was a slightly older, quiet Scotsman and we worked together most harmoniously. With two of us, one could be in the office answering all the phonecalls from head office and elsewhere and repel any attacks from the resident engineer, while the other was free to get on with the outside work.

I had rescued myself from an almost intolerable position and the entire picture changed. The contract forged ahead in style until the end of the work, except for one or two dramas which occurred. A short time after this revival, I found Sir Cyril Kirkpatrick's young engineers sitting in my drawing office looking miserable and being consoled with cups of tea. They and the inspectors had taken what is now called 'industrial action' by walking out and refusing to work anymore under the sadistic resident engineer already mentioned. They won, so he was replaced by an older man of much experience and good sense, harmony reined all round and progress increased noticeably.

At this time I was too pre-occupied with work, so my wife went house-hunting and rented a semi-detached villa on the outskirts of Ilford from a lady teacher. This had a garden and was on the edge of countryside, so my two small children at last had a garden to play in. Sir Henry provided the contract with a car and driver and the driver picked me up each day and then picked up others of the staff as we passed Barking Station. This ended the travelling and the very long hours.

We did have some lighter moments, our night foreman Fred Timpson, who could have understudied Tommy Trinder, was whisked off to hospital during one night with intense internal pains. Next evening we were mourning his possible demise and talking of a successor when in bounced Fred!

> 'Why Fred! We all thought that you were a gonner'.
> 'Gor blimey Guv'nor, you wouldn't ardly credit it, they gets me on the table, pulls darn me trousers and in comes a nurse wiv a bloody great tube. Cor Guv'nor – would yer believe it – right up me jacksie. They filled me up proper. Then they says 'Off yer go quick. It's at the far end of the corridor. It was the 'ell of a long one wiv all the other people sitting along it, waiting. So I ups wiv me trouser in each 'and and didn't 'alf ave ter run to get there in time. Gor blimey Guv'nor, you ought to 'ave 'eard them cheer as I went by!'

We were being much pressed by Fords to get on with the work and Sir Cyril held site meetings at intervals to discuss progress. At a later one, Fords local chief engineer, Mr Boyd, who was a mechanical engineer, attended. He was a very brisk little man with a high-pitched voice and a lot of pep. Sir Cyril was heavy and consequential, he was President of

23. Timber jetty for cooling water intake shafts at Fords Power House

the ICE that year and a churchwarden as well. Mr Boyd opened up:

'Well, Sir Cyril. We started one way and then everything happened quite differently'.

Sir Cyril replied, 'We never thought that was what would happen'.

'Ha!" says Mr Boyd, "that's what the girl told the sailor'.

Sir Cyril blushed deeply.

Mowlem then won the contract to complete the cooling water system. We built a timber piled jetty as an extension of the main concrete jetty and sank two steel shafts, 14ft 6in in diameter, under compressed air, one for the intake and the other for the discharge. In the bottom of each we built 7ft diameter shields and drove twin tunnels under the river to join up with the two land shafts which Dorman Long were sinking. We first drove a line of sheet piles so that the cold and hot water would be kept apart, this work had its fraught moments. It is amusing to remember that to hold up the cylinders until they were safely in the river bed, we used the same simple lowering gear the firm had used to sink the shafts in the Thames for the Waterloo and City tube in 1895.

24. Stripping piles for pile caps

The pump house caisson, 54ft long and 25ft wide was sunk to a depth of 55ft below ground level. This one had to be hung on lowering gear, although built on the ground. The 80ft long steel piles, which had been driven before the new layout, had refused all efforts to extract them, so had to be burned off from inside the working chamber at frequent intervals. Dormans' lowering gear was much more sophisticated, consisting of several 100 ton hydraulic jacks, which worked a complicated arrangement of hangars, with special safety devices to keep a jack extended if the hydraulic pressure failed. First the working chamber was built with 8ft of headroom under a strong steel air deck. The sides were built up as sinking proceeded, by building thin steel sheets outside supported by light horizontal steel trusses. These were supported by the same sort of trusses built vertically. Shutters were fixed to these on the inside and then concrete placed in the 3ft space at intervals as the caisson sank. This concrete provided the weight to overcome the friction and also became the permanent wall of the chamber. Access to the chamber was made by climbing down 'figure of eight' air shafts from the air lock bolted to the top. In old-fashioned locks, the shaft was circular with steel rungs on one side, and if anyone climbed up or down there was always the chance that a skip would be lowered down by mistake and wipe you off the ladder. In these more modern shafts, the skips went up one of the circles

6. CRISIS AT FORDS

and the men went up the other. The Dorman Long lock consisted of a chamber with a tiny man-lock welded to one side, with a cylindrical lock for the skips above. The skips were bottom-opening and, ingeniously, the crane bond passed through the centre of the top door. This was secured by pneumatic jacks for safety when in position but, when the bottom door was closed by the inside lock-keeper, the jacks were released and the crane swung away the skip, complete with its lid. When the smaller screen chamber caisson had sunk some way into the sand and gravel, we found that the whole of the bottom was paved over by a foot of hard, naturally cemented material, due to ferrous and ferric salts in the ground. This is a local Thames-side phenomenon called Blackwall Rock. This proved that it was the cause of our piling difficulties, so we were duly compensated.

While Dorman Long was sinking the cylindrical intake shaft, a bad tragedy occurred, involving a gang of Irish Sinkers. They were so called as they had to be capable of heavy manual digging in air but were not as experienced as tunnel miners. The shaft had been sunk about 30ft into the 'bungum' in free air, with about 20ft of shaft still above ground. Sandy Reid used to sleep in his hut on the site, as the work went on day and night. His Irish ganger came into my office, where Sandy and I were talking, to report that water with alot of air bubbles was seeping through the clay and was smelly. Sandy told him strictly to leave the work below until next day, when the air lock could be fixed. That night the ganger woke Sandy up at 2am, as was his custom, to report on progress and said that all was well. They had lowered a small air driven pump to the bottom to keep down any water and the exhaust provided fresh air. However, during the night the pump stopped, so they turned off the air supply. Then they thought that they would restart the pump, in spite of his instructions not to go down the shaft. The first Irishman climbed down the ladder into the half enclosed working chamber and promptly fell into the foot or so of water. Another man followed instantly to help him and did the same, then another and then another. The only witness had gone for help as the ganger too went down, so that when my night foreman climbed the stage, he was just in time to stop a sixth man from following the five who were already dead in the bottom. They had gone down with no thought for themselves and with no pause to find ropes or even turn on the air supply. One of the divers on the jetty contract was lowered by crane in his diving suit to get the men's bodies out.

A Home Office man instructed us how to fish for samples of the air by filling large Winchester bottles with water and, after enclosing them in a form of cat's cradle of string, lowering them to the bottom. By pulling another string, the bottle could be up-ended so that the water would run out and the bottle filled with the air or gas at the bottom. The resident engineer, Redwood, and I spent all the afternoon of that day collecting samples. The analysis showed 0.1% oxygen, 4% methane, 9% carbon dioxide and the rest nitrogen. The compressed air which passed through the ground met so much peat and vegetation in the soft ground that the oxygen content had been almost totally absorbed and the lack of oxygen had been the cause of the mens' collapse. Sandy Reid's instructions had been precise and given in our presence, so there was no blame attachable to him, especially as he was going beyond the call of duty by spending his nights in his office hut so as to be available in case of need.

Naturally we were bombarded from above by much conflicting advice on precautions to take in the chambers under air, such as canaries, Davy lamps and white mice. We chose the latter, although there could not be any danger inside the caissons when under air as many cubic feet of fresh air were being pumped into them. As the work under the compartments lasted several months, our stock of white mice multiplied rapidly, one family producing sextuplets. One day I went down under one of the air tight floors and made the routine question of another Irish ganger on the health of his mouse, he said:

> 'It was a funny thing. The little so and so kept running around and then fell down dead'.
> 'What did you do?'.
> 'Why sent up to the office for another one, of course.[23]

During the summer of 1931, the Labour Government was still in rapidly dwindling power and a XXI Club letter records a visit thus:

> One day we had to entertain 100 of our choicest legislators who came by a specially chartered pleasure steamer. They were mostly Labourites and if those are the flower of our National Legislators, we need a National Government of 10 and can afford to sack the remaining 603 members. The few non-Labour members were the opposite extreme – mostly Haw-Haws and middle-aged Bertie Woosters. The Labourites are aggressive in their questionings and some observed that if we had Nationalisation we should have works like that all over the Country. They did not seem to realise that there is nothing for the existing ones to do.

(This was in the deep slump).

> One Labour member, on learning all the various efforts to find a foundation with compressed air sinking and so on, remarked that 'God gave the land to the People, but he left us to do the rest'.
> Another body of distinguished visitors consisted of the shipload of Russians whom you may have read about in the papers. A rum crowd, mostly in blue slop-suits, but some with silk smocks and skull caps to give local colour. Our own down-trodden slaves of capitalism seemed highly amused at them and attempted to crack jests with some of the women teachers or whatever they were.

A few months after this the Labour Government was wiped out and a National Government took its place.

That summer, my wife and I abandoned our children for a short time and took advantage of an offer from the General Steam Navigation Co. For the sum of £6 each they took us to Holland, offering five days aboard while in Rotterdam including meals. We sailed at 2pm from Tower Pier in a small tramp steamer, with reasonable cabins and found the trip down the Thames, including the view of Dagenham, very instructive, reaching open sea by 8pm. Next day we woke up to find ourselves at the Hook of Holland about to enter the New Waterway up to Rotterdam. This wide cutting up to the River Maas is cunningly designed so that underwater groins make the tide scour the passage clean of silt. Variations of width with the flow of the River Maas diverted into it contrive to make a remarkably constant depth of water, so that no locks are needed. The largest vessels can get up to Rotterdam, 30 miles inland without hindrance. We decided to abandon ship and took the train to Amsterdam for three nights in a grand hotel, which we felt that we deserved. We explored the canals, the

---

[23] Sandy Reid had a different sort of adventure later when he was building a bridge in the Middle East. He had a workman who provided a good following of fellow workers and was very helpful. But he was being sought by the police for alleged murder. When the police arrived they sought for their man everywhere with no success. During the search, the call to prayer came from the Minaret of the local Mosque. Reid recognised the voice of his helpful workman! The police went away empty-handed.

6. CRISIS AT FORDS

Reichsmuseum with all its Rembrandts, went to the Hague for the pictures in the lovely little Mauritshuis and to the Frans Hals Gallery. On the voyage back, as we entered the Thames Estuary we found ourselves in the annual race of the famous Thames sailing barges, forty sail in all. Sadly, the commercial fleet of sailing barges has since been wiped out.

The work itself was going well in spite of several traumas professional and domestic. One day I was rung up by a city dentist who was removing an impacted wisdom tooth from my wife, to say could I come at once as they couldn't revive her from the anaesthetic. Happily they did so by the time that a harassed husband arrived. Some weeks later, she was very tired and tripped on the top stair tread, went headfirst down the stairs and cracked her skull across the iron radiator at the bottom. We had an old family retainer helping us at the time and luckily she was there, five minutes later and she would have been gone, leaving the two small children and going out for the day. My wife crawled into the dining room and got her to find a doctor. The retainer brought her a glass of brandy but as she looked so shaken, my wife made her drink it herself. Luckily the crack healed in time but it was a near thing.

One night in the winter of 1931 the fog was so thick that few men were on the site. Someone blundered into a store containing six oil drums and accidentally started a considerable fire. This soon spread to the electrician's wooden shed and destroyed the distribution board, which controlled the duplicate supplies of current, and put out all the lights and stopped the electric compressors which were in my fireproof steel compressor shed. This shed luckily protected the diesel driven compressor which was there as a safety margin, and this continued to hold the air in the chambers. There was a six-inch pump close by, but it was electric. The Scots derrick could have grabbed sand from the concrete plant and dumped it on the flames, but it too was electric. There was a steam crane which could have done the same, but the timber stage on which it ran was burning merrily. There was a three inch water main, but it ran under the shed which was burning, so only steam came out and the extinguisher in the shed could not be reached. The Dagenham Fire Engine could only travel as far as the west entrance, so the firemen had to struggle along in the dark in front of the main sheds, carrying whatever was portable. By the time they arrived the fire had burned itself out. The cooling water tunnels were served by a second electric supply along Fords jetty so were not endangered.

Fords jetty contract next to us was a separate job, but we were involved at times. The work consisted of sinking a great number of 14ft diameter steel cylinders deep into the river bed by grabbing under water. In these, concrete piers would be built up. In order to overcome the friction, 500 tons of 'kentledge', in the form of two ton cast iron blocks, was stacked up on steel frames balanced across the top of the cylinder. When the cylinder reached its depth, divers would clean up the bottom to receive the concrete which was placed under water.

One day, the diver was down below, with his relief on the top beside the air pump man, with the cylinder full of water. The tide went out sooner than they expected and the head of water blew the diver out under the cutting edge of the cylinder, so that he was buried under 30ft or so of river mud and gravel. Two divers spent all day in a boat using heavy water jets to jet themselves down through the mud to reach him. The divers got hold of his hand, while another diver at the bottom of the cylinder paid out his air line and they worked their way to the surface. By then, the tide had risen so high and his length of hose had become so short, that he had to be held upside down outside the boat for his helmet to be unscrewed. This was all done by hand signals as it was before the days of diver's telephones. We entertained the victim for some hours in our medical airlock and he seemed none the worse. The divers

25. *Standard diving suits have connected air tubes, which although cumbersome can be life-saving in complicated conditions*

all got gold watches and one got an award.

Work went well over the icy winter on the mudflats, although we had to buy our own protective clothing. By the time spring came, so did a sudden change come to me, as will follow in the next chapter. Experience is measured by intensity, not time, and Fords taught me more in two years than many learn in ten or more.

7

'THE SPECIALIST'

In my Institution of Civil Engineers Presidential address I used words that summed up how things were at this stage in my career and I quote them because they make more sense as part of a speech:

> Parkinson, in his Law, described the eight ages which a man will follow in his career, starting with the age of qualification and passing through the ages of discretion, achievement and wisdom, ending in the age of obstruction. I will divide my personal experience into roughly four decades: the age of obtaining experience, the age of exploiting it, the age of directing work and the age of advising.
>
> On the Charing Cross escalators and at Ford's power house, the attempt to inject cement into sands and gravels failed where the sand was too fine. At the same time my firm were engaged on the George V Graving Dock at Southampton, where that far-sighted engineer, Wentworth Shields, had forestalled possible disaster by boring deep enough to disclose water bearing Bracklesham sands under artesian pressure. He got Siemens Bau-Union to install the first deep-well groundwater-lowering system in this country, a method which had only recently been developed in Germany. These two things impressed Sir George Burt, who obtained the rights from Siemens to exploit several of their processes in this country.
>
> The change to the age of exploitation was sudden, for I received a telephone call telling me I was to be the firm's expert. The principal processes concerned were groundwater lowering by means of deep and shallow filter wells and the Joosten process of Boden-Verfestigung. For simplicity I translated this as 'chemical consolidation', but later on the pundits considered this an incorrect usage. This was the first time that a number of alternative expedients were grouped under one body and was a form of pioneering.
>
> The most unexpected result from carrying out such specialised work is the experience gained in listening to other people's troubles. More cases occur of people getting into trouble and then hoping to be rescued by some magic process than of engineers thinking out beforehand where such processes might be applied with advantage. One was also affected by the widespread lack of information on the nature and characteristics of the ground.
>
> In addition, one travelled about the country and met a great number of engineers, architects, contractors and others and learned their problems and the way in which they approached or organised their work. It also proved the truth of a remark by a former colleague, which I have [often] quoted: 'You have only to put up a brass plate and call yourself an expert and you at once begin to accumulate experience'.

Thus the ten years which reached their climax at Fords was the age of experience. The age of exploiting it started with an unexpected turn. The work over the next four years from April 1932 to October 1936 provided a small and unexpected foundation for a revolution in civil engineering and can be likened to the story of changes in surgery, from the use of anaesthetics, through sterile methods, X rays, new instrumentation and so-on, not forgetting injections!

During this time, Dr Karl Terzaghi was developing and preaching the gospel of soil mechanics, of which he was the founder and chief prophet in Europe and the USA. When he arrived in Britain in 1938, like Saint Augustine to convert the heathen, a few disciples were already working in the newly formed Building Research Station (BRS).

There are two useful short words in German:

Grundbau = Work carried out in or below the ground

Bau-grund = The ground in which the work is carried out.

Dr Terzaghi devised ways, hitherto unknown, of measuring and defining Bau-grund – the strength, particle size, properties and behaviour under various stresses. So it became possible to forecast and control the infinite malice of the ground beneath us instead of blaming 'Acts of God'.

I was now to undertake pioneer work in which the nature of the ground became vital. Little did one foresee in those slump-ridden times where these new paths were to lead. The expedients, which gradually multiplied, are now dignified as geotechnical processes and they form an integral part of the science of soil mechanics. In 1932, compressed air was much used, but sheet piles were a comparative novelty. Bored piles were just coming in, only 12 inches in diameter. Cementation proved useful in closing fissures in rock, but was a flop in most sands. Water was still pumped from sumps sunk the hard way and self-priming pumps were almost unknown. The caterpillar tractor was still to reach these shores.

Below ground 'water is the root of all evil' and there are several ways of dealing with it: exclude it by compressed air or sheet piling; remove it by pumping; change the nature of the ground by injection or freezing; or, best of all, go somewhere else. In the work which I have described so far, piling, pumping and cement injections had not been very successful, so the experience was there to be exploited.

Letter 22 to the XXI Club starts off this period with a description of ground consolidation and a young man's view of Germany in 1932, a year before Hitler came to power. The description of Germany is best given in the phrasing of the time and as I had been to see Boris Karloff in *The Mummy* this lends an Ancient Egyptian tone to the remarks.

> Consolidation of the ground where siliceous or sandy. This is an ingenious business of injecting chemical No 1 into the ground with a pump as the injection pipe enters and following up with chemical No 2 as the pipe is withdrawn. When No 1 and No 2 meet they combine and form a residual chemical which binds together the particles most successfully, much on the principle of the Seidlitz Powder.[24] This Siemens did for 100ft of tunnel for us at the Monument Station and while this was underway and proving successful, Mowlems came to arrangements with them to perform these exploits in England and one or two other sidelines into the bargain.
>
> All of a sudden I was yanked from the fast finishing Dagenham and informed that it was to be my business to look after this, so one had to orient one's mind to a totally fresh direction. I made the acquaintance of various German engineers, all very pleasant, and we were provided with a foreman, known as Hermann, to work the big medicine and train up two foremen of ours in the 'higher mysteries'. We

---

24 A stomach remedy.

7. 'THE SPECIALIST'

did a trial length of vertical consolidation through 8ft of water at Chancery Lane Station, which was 90% successful and would have been completely so but for the buffoonery of the alleged timberman provided by Cochrane's whose job it was. Needless to say, one of the difficulties to be overcome is the obstructiveness of other contractors if one is called in on their jobs.

We also did some most impressive waterproofing of leaking retaining walls at Dagenham by forcing the chemical into fissures where water was issuing. The first is pumped in and the second solidifies immediately it meets the first, and so seals the fissures.

Then behold the great ones informed me that I was to go to Berlin and visit the Siemens people in their own temple and see what was to be seen. One of our directors suggested that I should take my wife on a holiday (at my expense) and so there we were, having parked the brats in a country cottage with a Norland Nurse, preparing to march on Berlin towards the end of July.

We booked our passage via Harwich–Hook for Friday, July 21st. On the Wednesday, Herr von Papen debagged the Prussian Diet.[25] On Thursday, martial law was declared. On the Friday 'Berlin in a state of siege. Police to shoot at sight'. So we crept out of the Bahnhof Zoologischer Garten and peered furtively round the corner. However, everything was absolutely normal and much quieter and less dangerous than London.

We had armed ourselves with a phrase–book to eke out our scanty Deutsche, called All you want in Germany but it seemed entirely designed for picking quarrels, such as 'Disputes' 'your taximeter does not register correctly – don't try to cheat on us – drive us to the police station'.

We put up at the Grand Hotel at the west end of the Tiergarten. Berlin is on much the same general plan as London. The Tiergarten is the equivalent to Hyde Park but with no fences and roads running through it in all directions. East of this is the central portion and Government buildings, while to the south west is the main residential portion.

We found it very easy to get about – excellent Underground and state railway, electric trains, trams and buses. What a town to live in! At any rate at the west end all the streets are wide and planted with trees and all the windows and walls foggy with red geraniums. Tram lines on each side clear of road traffic and men appear with small mowers and mow the grass between the tram lines. Dozens of restaurants and cafés all with large open–air gardens in front of them full of flowers and plants. In ten days we had no meals indoors at all. And to think that all this was built about the same time that the horrors of Earls Court and the Cromwell Road were perpetrated.

To the south west and west are a chain of very large lakes surrounded by small hills and pine woods called

---

25 von Papen was instrumental in Hitler coming to power.

Wannsee and Nikolassee, which are on the River Havel. All are sandy bottomed and crammed with sailing boats and the lakes are white with the sails. Also multitudes of a new type of Rob Roy canoe for two with rudders steered by the feet, generally containing a man and his girl in bathing costumes, done to a nut brown colour. They have little lug sails and sometimes very miniature outboard motors on the side with long propeller shafts. As the weather is usually hotter and more settled than ours, they practically live on and in these lakes and a darned good life, too.

There are also dampfschiffs to take one for luftfahrts and rundfahrts, on which we penetrated to Potsdam, where we were struck by the sight of [comedy actor] Jack Hulbert proceeding in a motor car, which was of interest as we had seen him in 'Jack's The Boy' only a day or two before and by the way it is worth seeing.

The lakes extend for nearly 100 miles one hears and on its shores dwell the fabulous nudists, but it must have been the off-season as none were observed.

Well to resume, on the Monday I found my way to Siemensstadt, which is a large town entirely built by Siemens, and got lost, as although I asked for Siemens Bau Union, I did not realise that I should have asked for 'Hauptverwaltungsgebäude'. The offices for all the Siemens activities are about twice or three times the size of Harrods, or seem like it. I paid six visits to the place and got lost every time. The notable feature is or are the 'Paternoster' lifts, a form of vertical escalator. These are a series of boxes one above the other suspended from chains like the carriages on the Blackpool big wheel. There is one lot going up (and over) and one lot going down (and under) at a steady speed, so as the box approaches your floor you step in and are carried up, and you step out at the floor you require. If you fail to get off at the top floor you are carried over quite comfortably and after a long descent suddenly start to bob up again at the ground floor, still the right way up.

They have an engaging way of growing small plants and cacti in their office windows, which takes away any feeling of austerity and might well be introduced into lawyers' offices except that the atmosphere would be hostile for growth.

The Herr Doktor that I dealt with is a very pleasant and able fellow who hails from the Thuringian Mountains. We went to an 'After Dinner Cocktail Party' at his flat from 8.30pm till 12pm. There one sits around a round table and drinks many glasses from a bucket of Moselle with lemon, peaches, orange and other fruits in it, all surrounded in an ice cooler, while light conversation is carried on and curious one-story single-decked sandwiches consumed. They told us that sometimes they make 'cheerio' until 2 or 3am.

The main buildings, palaces and government offices, etc, are much what one would expect, but the whole place is littered with monuments to 1870, more than we have for the whole of our history, and one can see how their success

7. 'THE SPECIALIST'

ran to their heads and blew them up till they had to be pricked.

The political business was too mysterious to follow. There are 29 parties and they do not vote for candidates but for the party. Each party has a list and as every so many thousand votes come in, so another man is ticked off on the list as elected.

We saw a certain number of Nazis running about. They seemed mostly of the college student type, irresponsible variety. They have a very comfortable uniform with red armlets and blue swastikas on a white ground and it also looks very dashing. My own theory is that if they had hard hats and a row of brass buttons to polish and the uniform not quite so attractive to the female eyes, their numbers would diminish. Flags are hung out of the houses denoting one's political beliefs, all red, but the Nazis have a blue swastika on a white circle, the Reichbanner, three white arrows and the Communist, blue sickle and hammer in a white circle. This enlivens the place alot.

As far as one could gather the general public wanted someone to govern them firmly and did not care much who it was as long as it was firm, hence von Papen's success.

Our taxi man drove us round the sights on the way to the station coming away and passing one large building, he flung both arms in the air, just missing a dog, and shouted 'Das ist VALVAARK!' Expecting to see at least a Palace, we found a red front which said 'FW Woolworths – Nicht über 50 pfennings' or words to that effect.

We proceeded on the Friday evening to Frankfurt, past big brown coal mines which are large pits dug out with ladder dredgers, and spent Saturday there. A most pleasant spot with wonderful old fifteenth century buildings and narrow streets round the Cathedral with scenes painted on the plaster and all in cheerful colours. It was a cloudless sky and we spent 5 hours in one of the many floating baths on the [River] Main. These are supported on steel caissons like rafts and have big plank decks all round, and one can dine, have tea, swim, sun-bathe, etc and drink 'bier' in very comfortable circumstances, fenced off from the shore but open to the river. We went that night to Mainz and next day came down the Rhine in a Rhine steamer, arriving at Cologne at 5pm and thence to Brussels. This was their election day but no-one seemed to worry.

It was a very pleasant and interesting trip and the weather was perfect. We saw many steel arched bridges to remind us of our Sydney Expert, and they all seem to have travelling stages slung below as a permanent portion of the structure for painting and so on which seems a good idea and not unsightly.

There was upon the dampfschiff a posse of American tourists from Tallapoosa, Alabama, who had been doing Europe to some purpose, judging from the labels. One piece had such a high voice that another 3 notes would have been beyond the pitch of human hearing, unfortunately it wasn't. After several hours, the tourists began to flag visibly and the screecher voiced their general feelings by

> screaming, 'Oh Gee! I wanna GET somewhere!'
> Colossal paddle wheel tugs tow enormous Rhine barges upstream and all the barges are festooned with canoes and canoists, who paddle down and get a tow back. We passed several eights and a double sculling four of frauleins, a pretty sight indeed. Our rowing pundits will be outraged to hear that one eight seemed to be so far forgetting themselves as to be enjoying it. After giving us a cheer and rowing stoutly alongside us, stroke proceeded to wave lustily to a girl on the bank while rowing hard with the spare hand. He then stood up in the boat and proceeded to harangue his crew. I am sure the IC[26] Boat Club would not behave in this way.

The letter went on to say that our daughter Caroline, aged 4½, had nearly perished with double pneumonia before the days of wonder drugs and that her heart had been badly affected. She had to be kept on her back for a year, which was a period not easily forgotten but she fully recovered to later become a competent architect.

Kenneth Wolfe-Barry the head of a famous consulting firm, was engineer for Grimsby Fish Dock. At this time my wife and I, as was our custom, put on our best evening dress to attend the ICE Conversazione, which as usual was exceedingly hot in mid-summer. We were standing in the Council Room in the middle of a large crowd of distinguished seniors bedecked with medals and accompanied by their ladies. While we were talking to Wolfe-Barry, a tall and very large daughter of an elder member suddenly let out a large sigh and fell to the floor in a dead faint that shook the building. At once there was a crowd of agitated elder members bending down, fussing, with all their decorations jangling. My wife Sophie took three steps forward bent down and seized the ankles of the victim. With a most graceful gesture she raised the strapping legs high in the air, whereupon the victim promptly came round. Sophie quietly laid the legs back on the floor and stepped back beside us. Wolfe-Barry said 'That is the most remarkable and sensible thing that I have seen in such a case'. Sophie replied, 'That is what we used to do at the Slade when the models fainted.'

The next XXI Club letter was dated May 1933, as for a time it was agreed the members should write at closer intervals when they felt like it. It mentioned chemical injection work over Monument, Knightsbridge-East, Knightsbridge-West and Chancery Lane escalator tunnels as well as success in Grimsby with the water lowered 40ft. Then comes in September 1933 an important piece of news, the buying of my house in Putney in which, except during the war, we lived until 1970.

A second visit to Berlin in the summer of 1934 is described in another letter. It was after Hitler had only been in power for one year and still wore a shabby raincoat. The full nastiness made itself felt later after he had murdered his friends in the Rhöm purge. The visit was to make a joint tender with Siemens for injecting and water lowering work on the heightening of the Assiut Barrage on the Nile. It was ultimately unsuccessful, but it showed the ease of using decimals - until having to reconvert metres and reichsmarks into Egyptian pounds and tons.

> I had four very pleasant and interesting days in Berlin three weeks ago, not unconnected with the use of chemicals on a large barrage in the Near East. Whether one likes

---

26 Imperial College.

7. 'THE SPECIALIST'

the Nazis or not, one cannot deny that the general run of people seem happier and more settled than two years ago when I was last there and 43 [sic] parties were settling down to an election.

I have to report that in four days I saw six Nazis and one SS special police. Several of these were leading their children by the hand (though not pushing prams). On leaving I went to the station. There were no police, no Nazis, only the ordinary ticket collectors. One man inspected my passport at the frontier, the same man with the funny face who did so when I went there two years ago. This was not what I had expected after reading my 'Daily Telegraph'. I saw an excellent German film called 'Krach um Iolanthe', whatever that means, all about a prize schwein and a farming village. Very well done. There was a news film of Hitler addressing crowds at Hamburg. Half the audience clapped politely as we do when Ramsay Mac appears. I was quite prepared to leap up and screech 'Heil Hitler' with the best, if large stormtroopers looked like hitting me on the head from behind as one would expect from the papers. But the audience was comparatively unmoved.

Over the doors at Siemens is the motto in red 'Here only give the German greeting – Heil Hitler'. One could get very humorous on the subject, though to be candid and truthful what happens is that when people meet, they half raise a hand by bending the elbow and generally forget the rest. It is a little startling when a man opens a telephone conversation to order a ton of cement with 'Heil Hitler', but the rest is hardly noticeable. Also a policeman commenced 'Heil Hitler' before telling off my porter for crossing the road against the traffic lights.

I saw various bodies going to a meeting of Hitler youth. First about two or three hundred girls about 17 or 18 in blouses and blue shorts marched along singing. Then various bodies of boys in shorts and shirts and no hats marched by, carrying flags like Boy Scouts, with a Scout master to every detachment. They also sang. Actually the sight was really rather pleasing. Whether it is a hidden menace or not I cannot say but they looked healthy, pleasant and happy and were singing exactly as we did when pseudo-soldiers, cadets, territorials, college students and on similar occasions. This is not an expression of pro-Germanism but a statement of fact as observed.

I saw some very interesting jobs including the new Reichbank (though no one knows what they will put into it). They are lowering the ground water by 30 deep well pumps for a depth of 40ft.

****

What is the truth of the great RIBA scandal? I see that His Majesty had presented to him about two score of architects, right down to the one who designed the toilet fixtures and even the plaster foreman, but I could see no trace of R. T. James (or Partner) in the report. Presumably a mere engineer, whose only responsibility is to make it possible:

a) for a building to be created;
  b) for it not to fall down;
  c) to carry all the structural responsibility;
is as naught compared with the genius required to conceive illuminated glass banisters and pneumatic seated soundproof 'usual offices'.

  For oneself, we have performed our curious antics in the cellars of Messrs Bentalls Department Store, where we sank some 15 wells and pumped out most of them with one self-priming pump and lowered the water with remarkable success. For those seriously minded, we recommend the January number of Concrete and Structural Engineering as to how and why.

  The work we are doing there is to put new foundations in the cellar before demolishing the building above. Incidentally, one has seen some curious side lights on mammoth stores and how to run them. Mr Bentall, the uncrowned King of Kingston, local JP and similar initials, has created his rapidly growing emporium. His office itself is furnished with tasteful old world grace and he sits between portraits of Mussolini and Napoleon. The new building built two years ago by Mowlems is equipped with escalators and above them glows the motto of the 'House of Bentalls': 'to strive, to seek, to find and not to yield', which is a good slogan when searching for the soft goods department.

  Wireless booms continuously throughout the building and many and varied methods of attracting custom are diligently resorted to. One week the 'It' machine was on view for measuring your personal magnetism by means of voltmeters, animeters and a woman in nurse's uniform to keep it scientific. The next week 'free osteopathy' was practised. Last year Mr Bentall rang up Sir Aston Webb and Son to ask if they had designed his top floor strong enough to carry an elephant.

  This store shows where all the money has gone in the post war changeover. The place is surrounded with large vulgar cars from which emerge large vulgar people, who invade the Mock-Tudor Restaurant and other departments as if they owned the neighbourhood, which they probably do. Here they are the local aristocracy, but would be lost in the crowd in the West End and so this is one secret of the place's popularity. It is really a super Woolworths. Little of real Army and Navy Stores solid value can be obtained.

The reference to R. T. James was made because he and his partner Hausser were designing the structural work of the new headquarters of the Royal Institute of British Architects and I was injecting chemicals to solidify the loose ground under the next door building. This was a great step for so young a firm which is now one of the leaders. The letter started with the usual love-hate relationship between architect and engineer. It was only by the ingenuity of R. T. James and Partners' design of the structural part that the RIBA building could be squeezed into the space available, but little acknowledgement was given.

The work under Bentalls corner building had importance as it was the first shallow well job in this country. In order that the building could be kept open during the Christmas

7. 'THE SPECIALIST'

26. Consolidation of the ground below Bentall's Department Store

shopping period the consultant, a very practical man named B. L. Hurst, along with my help, came up with a scheme whereby the walls of the brick building could be underpinned by chemical injections in the basement. There was only 8ft headroom in the basement so with specially adapted tripods, seventeen wells of 2 inches diameter were bored for a depth of 30ft and equipped with 6inch filter tubes surrounded by filter gravel. The suctions were connected by a ring main to duplicate self-priming pumps and lowered the ground water by 15ft. The foundations for the new building were then dug out and concreted in the basement before the old building was pulled down. This was in contrast to the foundations of the previous main building built by John Mowlem building department where they tried pumping from very shallow sumps and made very heavy weather of it, even to placing concrete underwater within bags.

As my work was entirely new and took up much space in the basement while it was being installed, I met with obstruction and disbelief from my own side - the process work was on its own. One day B. L. Hurst turned up with an American engineer, a director of the Gow Company of Detroit which was part of the Raymond Company. I asked him if he was familiar with well-point pumping (where many small diameter pipes are put down by water jet in place of bored wells) and as he was familiar with that, and while installation was still going on, I showed him what I was trying to do and he was most encouraging.

We then asked him and his wife to dinner. He had us back a few days later to Grosvenor House and we got on splendidly. We went on writing to each other regularly until he died twenty-five years later.

At the Movies

> Behold, Gaumont British went to the Federation of British Industries and said 'who can help us with a film about a tunnel?'. Surprisingly the reply was 'Why Mowlems'. The film representatives then rolled up to our office and the Directors bade me find what was wanted and try to get back some of my salary. So in due course I entered the heavily guarded studios in Shepherd's Bush and after the productions manager had fixed a fee for six visits, unfolded his mystery and announced that they were working at a film on the construction of a tunnel in AD 1960 from

England to America! This caused a complete collapse of our technical expert.[27]

After efforts to persuade them to try something at least within the bounds of possibility, they explained that production had already started. So after obtaining written assurance that on no account should my name or that of the firm be mentioned in any connection with the film, I decided to get any fun out of it that was possible.

The art director is Herr Metzher and his job is to draw pictures of what he imagines, to suit the story and have the sets built, and clothes and details chosen. He explained at great length that it did not really matter what absurdity one put in the films as long as it looked reasonably coherent and that they wanted me to suggest one or two things for men to do in backgrounds and to look over some of the dialogue. Actually the whole thing was beyond engineering aid, although a very 'strong' story. Do they not tunnel through a volcano? For a disciple of Professor Brammall of the Royal School of Mines it was all very difficult. For instance they wanted the hero to say that the pressure varied as something or other, so I suggested that the 'pressure varied inversely as the reciprocal of the depth' and they thought that was fine!

As the studio is not big enough to take a 50ft diameter tunnel, the bottom half was built full-size and the top was a model one sixth full size, suspended in front of the camera including a fearful wonderful radium drill on caterpillars. When you see it you will be unable to detect the deception.

The fact that with ordinary progress, about 300 years would be required for the job and enough muck excavated to fill a large part of the Irish Sea, apart from geological and other minor points, does not disturb them in the least. The story is the thing and the careful avoidance of anything likely to raise what they technically call a belly-laugh.

The script is full of amusement. 'Camera pans back', 'Big moment for McAllan', 'Big showmanship angle' and so on. I had intended to give some extracts but that would take too long.

The actual filming is a wearying business. They keep on going over the same thing, while make-up men rush in and smother supers in cement dust to make them look dirty and so on. When the clapper-boy shouts 'Turn her over', he

27. Poster for the film The Tunnel's release in America as The Transatlantic Tunnel

---

27 This was quite a while before plate tectonics had been accepted.

refers to the camera and not the continuity girl.

On the first visit, I saw a studio where Sonny Hale and Jessie Matthews (dressed in man's evening dress) were doing a night club scene, with hungry supers (at 30/– per day and buy your own evening clothes) dancing in the background. The next time they were filming a cabaret scene. The lift stopped at that floor and I saw the room was full of dancing girls in Chinese hats, two saucers and triangles; but this time I was ruthlessly rushed aloft and not allowed in.

The question of what men should wear in the tunnel was settled by the producer. They are stripped to the waist and wear riding breeches and field boots! It was seriously explained that nowadays people are tired of girls in bust bodices and little else and that torsos are now the thing. Nothing looks better, they say, than a manly torso and in all the latest films, 'King of the Damned' etc, you will see acres of them (20/– per day and bring your own torso). As the majority of film fans are female, this probably is psychologically sound.

There are some very fine scenes taken with models, autogyros landing on skyscrapers and so on which are most realistic.

But please do not imagine that anything you see in the picture is any fault of mine.[28]

Back to work

Another excitement was an expedition to Dublin where there was a large Power Station at the Pigeon House, which was on a sandbank in the middle of a large breakwater that runs right out into the sea. New cooling water tunnels were being built and the contract fell to a combination of a German and Irish firm: Weiss and Fritag of Frankfurt and the Pioneer Road Construction Co of Dublin. The Germans' design was for sinking a number of reinforced concrete caissons about 80ft long, which contained two ready made 6ft tunnels above the working chamber. These were then joined together in a mysterious manner and the result was twins – tunnels running out into the Liffey or Dublin Bay or both. Well, it so befell that at one point while going under this colossal breakwater they had to go under the main sewer of Dublin. This runs inside the breakwater and decants into the sea at a special bathing

28. *Caisson ready for lowering, Pigeon House Outfall, Dublin 1935*

---

28 A review in the *Evening News* Nov 12 1935 (held in the British Film Institute archive) thought better of the film…. 'George Arliss plays his smallest film part …but the compelling thing about this most enterprising talkie is the brilliant direction by Maurice Elvey of the work of making the tunnel. You will shudder over some of the experiences of the brave fellows who dug into the earth.'

29. *Chemically solidified ground between tubes in compressed air, Dublin 1935*

station at the far end, so they got very cunning and arranged to sink adjacent caissons there, 12ft apart, while, built into each of the ready made tunnels were rings of timbered Larssen piling with a timber diaphragm to keep the sea out. The pleasant concept was that these piles should be jacked out until they met and the material enclosed dug out and the tunnels then united. However, the caissons arrived about 18ft apart and the piles would only jack out about 4ft and it is hardly surprising that this left them with a gap of over 10ft to get through. They could not keep enough air, they could not open the doors in the diaphragms and their penalty clause was on the point of commencing. At this point the Swiss engineer for the Frankfurters threw up his expressive palms and could do no more.

The resident engineer had been an assistant of Kirkpatricks on Fords Power House and he had the contractor's agent (the Irish portion) dash to London to spend a day trying to persuade me to go to their rescue. They did not know what the ground was like, but thought it varied from fine mud to cobbles about 12 inches in diameter and it had been disturbed in sinking the caissons. All work had to be done in a 6ft tunnel under 15lbs of compressed air. Finally, we took on the job and having first grouted any voids with lime and cement, we drove a ring of pipes from inside the tunnels radiating out and made a solid ring of solidified ground to join up the piling. My foreman then taught them how to tunnel through and we did one half of the tunnels for them and left them to complete the remainder. The solidification was very successful and we had to get them to make precast concrete rings to support the tunnel. The little Swiss engineer was a perfect curse the whole time, rushing about and screaming and swearing at the men, who shouted and swore back at him. When all was over, Herr Professor of his firm said that they had always intended to use the chemical process on this job, as they were the first people to use it in Germany etc etc. This project was very interesting and rather exciting as we had not performed anything like this from inside compressed air before and did not know how much the stuff would be upset by air fizzing through it.

I remember that one of the men in the tunnel, called Pheynix, was heavy-weight champion of the Navy, a stoker in the Lion at Jutland and sparring partner to Dempsey or someone in New York. He once knocked out Phil Scott with a shrewd one in the plexus, but Bombardier Wells was the ref and gave it a foul. 'Shure and it was a bloody scandal, sorr'.

I was very heartily entertained on my several visits. One of the pioneer directors during lunch spat on the floor with the greatest eclat and accuracy. Young Harty, the resident, discussed politics and said there was nothing so hopelessly diehard and mentally impossible as a Southern Loyalist. I, as an ex-reader of the *Morning Post* was shocked. He replied, 'Well I ought to know – my father is one'. I cannot expand on the situation as I quite failed to understand it. The walls were all painted with slogans 'Join the IRA', 'Down with Fascism' and other cries.

7. 'THE SPECIALIST'

I worked from the firm's head office for these four years which brought me into constant contact with the directors of two firms at once and their behaviour could be studied at close quarters – a curious form of wildlife. As I have said before and continued to tell my disciples 'the tactical handling of company directors is an advanced science'. Many a firm is saved by senior members persuading directors to point in the right direction.

At intervals one matter took up my time, this was an involvement in the attempts to save Old Waterloo Bridge. Some of what I said in the discussion on the paper on the new London Bridge[29] gives part of the story which may be unknown to later generations:

> In 1923 Waterloo Bridge, which like London Bridge had been settling, started to settle in earnest.... Directly this crisis became known, Sir Harley H. Dalrymple-Hay devised a scheme for underpinning and asked my chief in John Mowlem and Co to give him an estimate in which I was involved at a lowly level. Publicity led citizens who had never heard of Rennie, and who had hardly noticed the bridge, leap to the defence of Rennie's Masterpiece, and they held a conference of societies urging the preservation of Waterloo Bridge...
>
> The Royal Commission on Cross River Traffic was formed in 1926 and recommended that the bridge should be reconditioned and a new bridge built at Charing Cross. In 1932, the LCC ignored this and decided on a new bridge, but the battle still raged.
>
> In 1932 I was starting to introduce the Joosten process into the UK and it interested Sir Harley. It proved unsuitable in the conditions at Waterloo, but this did not deter him from making an even more ingenious scheme for underpinning.
>
> He suggested that I ask my directors for another estimate and they told me to do it myself. Sir Harley then briefed Sir William Davidson who told the House of Commons that John Mowlem and Co 'one of the most experienced bridge building firms' had made an estimate for underpinning of £500,000. Members of Parliament then voted to retain the bridge. I knew that the railway bridge across the Thames from Victoria was the last one Mowlems had built - in the 1890s!...
>
> Then Herbert Morrison, leader of the LCC, leaped on to the parapet and started to demolish the bridge with his own pick-axe. I was gratified. Eventually the Government provided money to build a fine new reinforced concrete bridge, now matched by the fine pre-stressed concrete London Bridge. Both of these would have given pleasure to the Rennies.
>
> In the discussion on Buckton's paper on the demolition of Waterloo Bridge it states 'Mr H. J. B. Harding observed that when the Royal Commission on Cross River Traffic was appointed in 1926, it issued a preliminary questionnaire to some of the principal witnesses. Question 3 read:
>
> If the present bridge should have to be removed and not rebuilt:
> a) How long would the operation take?
> b) How much would it cost?
> c) To what extent would waterborne traffic be interfered with?'
>
> Two witnesses answered 24-30 months, costing £275,000 to £300,000 and no interference. These two witnesses wanted to pull it down. Two other witnesses who were keen to keep the bridge said 3.5 – 4.5 years, costing £500,000 and interference to some extent! I went on to ask the authors which estimate was nearer the result but they did not reply. However, Sir Harley was rather cross. This proves that even among our great men, the wish can be father to the thought.

Another characteristic of this period was the oratory of Sir Henry Japp who had become a leader in the heretical sect of Christian Scientists formed by Mrs Eddy and had taken a course on public speaking for spreading the gospel. Public speaking is an art at its more abysmal in the learned professional institutions, as members deal in facts rather than ideas

---

29 ICE Proceedings 1971 'On New London Bridge'.

and the subjects are unavoidably dry and realistic. Sir Henry then became in great demand to enliven discussion at the 'Civils', he was also the first that I heard to inject a little pawky humour into the dull discussions and I sometimes fed him ideas. Again the Presidential address:

> Knowledge cannot be disseminated successfully without good speakers and this also affects the impact of the engineering profession on the community. Sir Henry Japp, under whom I worked in the early thirties was a trained public speaker and had strong opinions on the matter. He came up to my desk in 1933 and said 'Well, are you taking part in the discussion on my paper?'. I said with the deepest feeling, 'God forbid'. Sir Henry replied, 'Young man, you should never lose the opportunity of public speaking. Civil engineers are dumb, but if they can learn to speak, they can make themselves understood and not be over-ruled by different committees. You will take part in the discussion'.

As Sir Henry was in constant demand it became his custom to throw a paper on my desk and say 'Tell me what to say about this paper'. This was an apprenticeship in discussion and it was an education to listen to his delivery.

In Institution meetings one is talking to men sometimes older or more experienced, so it is always a nervous business as we are not by nature politicians. Fate led me into writing many papers and I have wasted many words over the years, but was encouraged to find that Mr Macmillan always felt the butterflies in the stomach when he had to speak in the House. So after a baptism of fire in public speaking, I was invited in this period to lecture in Manchester, Leeds, Hull and elsewhere on these new processes which were gaining ground.

In 1935 one particular little job was to grout up a small underground chamber at Marshgate Lane, Stratford East beyond Bow Bridge. After looking around I went home and declared that if there was to be another war, which God forbid, I trusted that I should not find myself in that desperate neighbourhood. God did not forbid and Marshgate Lane was to be my base all through the war.

The start of a Foursome

Many unexpected developments began to take place. During the specialist period I had no personal friend or possible collaborator in the firm. Then, three others entered the scene and we became close personal friends.

In 1935, while I was based in Head Office, Mowlem obtained the contract to build a large reservoir at Chingford for the Metropolitan Water Board (MWB). When I returned from some foray to the north of England, I found an interesting looking fellow sitting rather forlornly in our drawing office, so I nipped into Japp's secretary and was told that he had been taken on as deputy agent for Chingford Reservoir. I went back, put out my hand and said, 'My name is Harding'. He brightened up at once and said 'Mine is Wynne-Edwards'. I heard myself say, by sub-conscious reaction, 'You wrote that paper on piles in Vancouver' and so we went off to lunch together and he became my closest professional as well as personal friend. Later, as a couple of grass-widowers, we were to share a flat for most of the war. Some years later Wynne-Edwards told me that he had sat in that office for a week and none of the other four engineers in the office had even said good morning to him. He was about to tell Sir Henry Japp where he could put his (much needed) job when I took the trouble to be civil, so he decided to stay.

To assist when I was moved so suddenly to Dagenham, I had been able to introduce Ralph Glossop to the firm to take over my work of setting out tunnels. He was cousin to my particular college and XXI Club friend Grigor-Taylor and was our contemporary while at the Royal School of Mines. Glossop had practised as a mining engineer in several countries including Mexico but owing to the slump he was 'available'. After looking after all the below-ground work on Leicester Square Station with considerable success, he found that the directors did not seem to be aware of his existence, so he went back to mining. He spent four years developing new gold mines on the Gold Coast (now Ghana), then on his return I presented him to Sir Henry Japp as the best man to take over the special process work. This meant that for the first time there was a collaborator and a kindred spirit in whom to confide one's ideas. This made up three of our foursome - the final party is introduced next.

The Decisive Battle of Chingford

When I was writing a paper in 1947 on *The Choice of Expedients in Civil Engineering* I told Wynne-Edwards that I was stuck for an opening paragraph which would not bore the reader at the very start. He said without pause, 'Say "In most civil engineering works worthy of the name, the unexpected happens".'

The unexpected certainly happened at Chingford Reservoir and this became the catalyst for many changes in civil engineering, both at home and abroad. The reservoir was to consist of an earth bank encircling many acres. First, a cut-off trench was dug down through gravel to the London Clay and then filled in the traditional style. Then clay was dug from an enormous clay pit inside the reservoir and puddled by passing through pug mills. The clay was spread in layers in the trench and trodden under foot by large navvies in thigh boots. This puddle core wall was also brought up above ground as the bank was built, up to the top of the bank. In South Africa, when dams were built by farmers, they used sheep to tread in clay to make their dams watertight. Perhaps this is what inspired the inventor of the sheep's foot roller. The bank was formed by digging earth from the inner area and then placing it and compacting it by Caterpillar tractors and scrapers. Mowlem were the first British firm to use this newly developed American equipment and Chingford was the first occasion when it could be seen. The firm urged the removal of the bed of very soft clay which overlaid the gravel beds, but was rebuffed as it would have cost money.

In July 1937, a large part of the completed bank slid gracefully down into the reservoir area and this caused confusion, alarm and despondency. Several old-time Victoria Street consulting engineers were called in by the MWB. Their theory was that it was all the fault of the contractor. At this time, three young men, Cooling, Skempton and Golder, were studying the 'new-fangled' science of soil mechanics at the BRS and they asked to be allowed to carry out full-scale research by studying this slip. They reported that it was a complex circular slip of the classic 'Swedish' style and gave their reasons for its occurrence. They published the results in a paper to the ICE in 1942, with the conclusion:

> The analysis of the present problem points to the conclusion that the failure was largely caused by the presence of an existing layer of weak clay below the bank, the weakness due to the central puddle core being a contributory factor.

The Victoria Street consultants, who had not heard of soil mechanics, were contemptuous of their assessment, but Wynne-Edwards was more up-to-date. He had worked in Vancouver and on the Detroit tunnel at the time that Terzaghi was working and talking and writing in the USA. Wynne-Edwards was by now in charge of the Chingford contract. He told Sir George Burt that it was vital to the firm to get the opinion of Dr Terzaghi. But where was he? So they asked Siemens Bau-Union, who said that the great man was in Paris and gave a personal letter of introduction to him. Wynne-Edwards flew to Paris to see him and there unfolded the tale of the slip. Terzaghi said that he was very busy and if it had not been for Siemens' letter he would not have come, so our link-up with Siemens accidentally saved the firm. When 'Terz' arrived on the site and studied the slip, he said that the design was wrong, so were the consultants and that the young men from the BRS were right in their diagnosis. Terzaghi drew up a new design for the bank and insisted that it must depend upon borings and tests made on the spot as the work went on.

At this time Glossop was sent as deputy to Wynne-Edwards and handed back the 'processes' to me and my assistant. The BRS then ran a course covering laboratory work on the newly designed soil mechanics testing equipment. Glossop attended the course and so became intimately involved with Skempton and Golder, as well as their senior, Cooling. Glossop then returned to Chingford and set up the first 'commercial' soil mechanics laboratory in this country which was vital to the redesign of the dam.

Firms began to wake up and send men to the BRS for instruction, as did the Water Board and the railways, who have a vested interest in embankments. A great and well-deserved fuss was made of Terzaghi. He gave a course of lectures at the Imperial College and in 1939 he delivered the James Forrest Lecture to the Institution of Civil Engineers: *Soil Mechanics - a New Chapter in Engineering Science*. This had a dramatic effect on the thinking of the older members. Terz became very attached to Wynne-Edwards and whenever he came to England after the war he always stayed at his home. I became known to him from the start and so was enrolled as one of his early disciples. Terz had a very strong personality, great wit and exerted his effect in a quiet and subtle way.

During the war, Hugh Golder left the BRS and joined us. It is odd to look back and see that in 1938 only a few British engineers had taken degrees at Harvard in Soil Mechanics and their voice was inaudible for lack of an audience. The select few in the BRS and one or two in the new Roads Research Laboratory were the only ones trying to develop the subject.

For the next twenty years, we four can claim, with all due immodesty, to have led the field in putting theory into practice in actual work, both collectively and separately. We expanded the results of actual work by writing papers, articles and giving lectures. All four of us served on committees which drew up codes of practice in site investigation, foundations, earthworks and earth retaining structures. After the war, along with Glossop and Golder, we founded Soil Mechanics Ltd, the first of its kind in this country. How this came about will be unrolled in Chapter Ten.

8

BACK TO THE BIG TIME

Central Line Extension Bow – Leyton

In 1936, after the years of depression, the newly formed London Passenger Transport Board started a big tube railway extension scheme. One of these was to extend the Central Line from Liverpool Street through Whitechapel via Stratford to Leyton. It then ran on the Eastern Region tracks towards Wanstead and then in tunnels through Redbridge to Newbury Park. The trains then looped back on the Eastern Region tracks through Woodford. Trains could also go as far as Ongar. Mowlem won several of the contracts. One was to straighten out kinks in the old Central London tube and to lengthen the stations, all to be done at night.

In October 1936, I was offered the chance to take charge of a contract to drive 10 shields, all in compressed air, from the extension of the Central Line beyond Bow Road Station under the North Eastern Railway, all the way to Stratford Station, and then from the east of the station to join with the LNER tracks at Leyton Station. This route went through a mixture of mottled clay silt, shells, old boots and bone boiling factories. Six shields from 2 shafts were driven passing under the Lea River, the canal of City Mill River, Pudding Mill River and Waterworks River, at shallow depth while all the way under the railway bank, carrying 8 tracks. The other 4 drove 2 in each direction from a site in Stratford towards Leyton or towards Stratford Station.

Simultaneously with the change of livelihood, my son Robert was born. At the same time I progressed to the status of proud father, having been merely a harassed parent for many years as, at the age of 6, my eldest son Edmund started at a nearby prep school, where he became a bit of a heavy. He was given boxing lessons and aimed to deal me grievous bodily harm when in reach.

Bow to Leyton was a major contract by any standards. When tunnels have to be driven across a railway the 'four foot eight and a half' boys make a great song and dance, yet here we were driving six shields in compressed air for nearly a mile under and along the main line from Liverpool Street, luckily without causing an accident. As Captain of one's own ship, responsibility is easier to bear when it has been possible to recruit staff of reliable self-starters, as was the case for me here. My 'Admiral' for the last time was again I. J. Jones, luckily as all went well, he soon left me alone and concentrated on other work.

I. J., who haunted my life for so long, yet with no decline in mutual affection, was as previously mentioned described by Americans as the Dean of the London Tunnel World. His great experience failed him at times particularly on the subject of money. This was the failing of his generation and the lesson that we learned was to avoid 'reckless economy' and realise when it is necessary to spend money rather than to save it. As an example: One day several of us were sitting inside the boiler lock while the pressure was nearing 15lbs per square inch, when we noticed with some alarm that the seams were beginning to open. We decompressed with unusual celerity and sent down my blacksmith-welder, who rapidly welded up all the seams in time for the men to lock out. This lock had been I. J.'s pride and joy, as it had only cost £50 and he insisted on its being used. On inquiry it was found that it had only been designed by its previous owner for a pressure of 5lbs, as this was all that was needed for Tooting escalators. For many years after that I insisted that any steel air lock, old or new, must be tested (hydraulically) up to 50lbs per square inch, which is beyond working pressure. I had extended my strong belief in the Nelson Touch and had secreted my much-needed blacksmith behind the compressor shed, so that he would not be noticed as he had not been allowed for in the tender.

This was a happy job. Mott, Hay and Anderson were the engineers and the resident engineer, Symington, had been I. J.'s deputy on the widening of the City and South London tube. We agreed that whenever we had to argue or discuss the usual conflicting problems which are built into the contract method, we would always be alone without our seconds-in-command, thus we could agree or disagree impersonally. I always believe that those who have not the final responsibility like to find shots for the boss to fire and do not always understand the virtue of taking it quietly until success is reached.

Although we both had a staff of keen young men, we each had to find room for a much older man than ourselves. Sappers are said to be mad, married or Methodist; our two were all three. Mine, W. A. Child, was a recent convert to Christian Science. He was a good engineer but distinctly pyxilated (sic). One day Glossop went into the empty office at Litchfield Street and heard a constant noise. 'Bong-bong-bong-bong', he searched around and there under the drawing bench, crouched into the corner, was Child making thrusting gestures with a T-square. He explained that he was trying to work out what progress a man would make in a four foot heading in soft sandstone. The other man, Penfold, punched another form of the Bible and either of these would have upset the younger men, so I arranged with Symington that we should keep them away from the tunnels and let them both look after the rather complicated open-cut section where the tube emerged at Leyton. Together with a good but eccentric foreman this worked beautifully and the work was most successful, but carried out in an Alice-in-Wonderland atmosphere of dottiness.

Courses in management have become the fashion, although one wonders how many of the teachers have ever had to manage anything so volatile and constantly changing as the work described. Different people practice different methods. For instance, Symington so designed his long office hut with himself installed in state at one end, then came his clerk, then the washery, then the drawing office for the engineers. Finally his deputy, Jack Kell was allocated a room at the far end. In contrast, I planted myself in the centre of my office with my engineers next door on one side, on the other was my second-in-command but with a communicating door between us for instant palaver.

Child sought to advise me as to how, on one big contract, the agent insisted on all his staff and foremen lining up in his office at ten sharp to report their actions and he could not understand my methods. I never interrupted my foreman Joe Treleaven from his work, as he had three separate working sites to control, although I was instantly available to him at any time. I preferred to walk over to his hut in the evening, when the shifts were changing, to talk over details with him and the night boss, and to meet him in the tunnels. Site meetings, cost sheets and reports are no substitutes for close personal supervision by spending time going into all the faces as often as possible.

In 1938, I was involved in an unusual water-lowering work inside a vast gas-holder in the Old Kent Road, while Glossop was doing more and more special process work. This was linked to the introduction of American well-point pumping.

30. *Sewer diversion Bow to Leyton extension, jetting down a well point after the first 4ft of trench was excavated*

The tunnels cut through a large West Ham sewer under Carpenters Road Bridge which called for half a mile of diversion with 30 inch earthenware pipes in trench in waterlogged gravel. I was also trying out Moore-Trench type well-points and bought a trailer fire pump for jetting purposes and jetted down a trial point 24ft in ballast in under one minute. This pump delivered 350 gallons a minute at 150lbs pressure and had a 6 inch suction with two 2.5 inch discharges. One was connected to the well-point and the second sealed off. The usual thing when visitors were present was to leave a long suction hose attached to this on the ground and forget all about it. The Highland Ford engine was then started up and down went the well-point. As the pressure rose, the second suction gradually stiffened and raised itself up like an elephant's trunk, and, curling over, discharged itself full upon the jetting party – we did this several times and it never failed to raise a considerable belly laugh among the audience.

The well-points were jetted down at 6ft intervals and connected to a 'sow-pipe' of 6 inches diameter, so called because of the nipples which were welded to it every 3ft for this purpose. The Moore-Trench pumps had large piston vacuum pumps of a clumsy kind for priming. I asked Harry Sykes and Co. to design us British pumps with even more powerful vacuum pumps. After then, Sykes sold many hundreds of these for use with or without well-points. The vacuum prevented the centrifugal pumps from losing their water, which before self-priming had caused a great deal of trouble.

9

THE WAR AND THE BLITZ

The Second World War then came upon us. After the usual two week holiday of those days, I had arranged for my family to stay on for a week or so in the village of Woodbridge in Suffolk, while I returned to my tunnels. In the case of war we had an invitation to go to Hindhead, but when evacuation was officially ordered I could not get away to rescue them. This meant that my wife, with the three children, set out from Woodbridge and were diverted by the police to cross the Thames as far west as Slough, by which time it was dark. A local policeman took them to sleep on the floor of his house. They duly arrived at Hindhead the next day after good map-reading by Edmund, aged just nine.

As John Mowlem and Co. had been a tower of strength to governments and local authorities for 100 years or more, the firm had many duties expected of it. I became responsible for organising and maintaining in a state of readiness a squad of foremen and leading hands, who had been trained in the erection of Callender-Hamilton bridging material by a crash course at the Roads Research Laboratory. This preceded the Bailey Bridge and was made up of single pieces to be bolted together and launched out into space. With this squad I was to be responsible for rebuilding any Ministry of Transport bridge in Kent, Surrey and Sussex if destroyed. Luckily none were bombed, but in the height of the raids I carried out a reconnaissance of my biggest target, Rochester Bridge, considered of vital importance. We searched for the Callender-Hamilton material to no avail, until we found that the Ministry had stacked it under the north abutment – bang on target for bombing. Trust the men from the Ministry.

Another of my duties was to be ready with labour and plant to repair any tube tunnels which might be damaged in certain specified areas. These were divided up between several of us. I was to use my depot at Marshgate Lane, Stratford East as a base for repairing any damage to the tunnels of the LCC main drainage department north of the River Thames. In addition to these responsibilities, I had to press on to complete my new tube tunnels, as in the shallower sections they were to be used as air-raid shelters. It was a merciful thing that they were not hit, because at night they were filled with local humanity.

We were all in Reserved Occupations even though I was coming up to the age of 40 and therefore beyond re-call. After some weeks, my wife was able to share a house at Liss with an old friend, married to a Royal Navy Captain. Caroline was just 11, Edmund 9 and Robert exactly 3, so I was glad not to have them 'under my feet' in London. A letter to the XXI Club dated 16th October 1939 contains:

> Briefly. Still carrying on with the Bow–Leyton tube, now scheduled as a work of major national importance! Also shafts and compressed air tunnel at Cannon Street for the LCC to make a permanent pumping station for pumping

9. THE WAR AND THE BLITZ

> river water about London in the case of air raids in 1941.
> Family evacuated to Hindhead and shortly going to Liss. Eldest unmarried son at school at Eastbourne. Self living in my palatial office in Marshgate Lane, Stratford at the lower end of Bow goods yard all as described for the abortive war a year ago. Have installed electric cooker, already had hot shower, bath and two engineers living with me. Clubs – the Swan, Stratford. Recreations – occasional visit to Stratford Empire.
> Have built a reinforced concrete dug out for RE and selves alongside 2,000 gals of used oil in 40 gal drums.
> Our house derelict, deserted and dilapidated. Started living in office out of heroic call of duty and remain for comfort and economy, principally the latter. Heroic duty still can be claimed. Outlook – fed-up-ness steeped in determination.

Whenever my frugal-minded chief walked the tunnels, he would instruct me to take out every other light bulb to save money. As a firm believer in the Nelson Touch, I resolutely did nothing so stupid. This was just as well, as early in 1940, while the 'phoney war' was still on, an Irish man driving our 2ft gauge battery locomotive forgot to look in front of him and hit his head on the timber across the tunnel. This supported a staging for caulkers who were caulking the joints of the tunnel to keep out the water. Unfortunately, his legs remained in the loco, and he bumped his head on every sleeper until he reached the working gang at the face. Not surprisingly this led to a Coroner's inquest. The factory inspector came at once to see the scene of the fatality and uttered words which contain a moral: 'Mr Harding, I congratulate you upon having such a well lit tunnel. If it had not been, I should have had to take proceedings'. It was afterwards alleged that the victim's brother was an IRA man planting bombs in Underground left-luggage offices.

The letter of October 14th 1940 records when the war became real and earnest but with natural reticence, which was needed in those days.

> In reply to your request for a concise letter, let us say that on June 1st I was suddenly shot off to Seaford to dig up anti-plane landing trenches around Newhaven, but within a week was put on as supervisor of one of the Defence Lines from Eridge to Newhaven.

This was after Dunkirk but before the fall of France. At Seaford I was briefed by the War Office to find excavating equipment from local firms and river boards and to liaise with the military as to where to dig. The trenches were to trip aircraft that were expected to try to land with troops. One day a dragline was happily digging trenches about 100 yards apart on Seaford Golf Course, while spitfires and hurricanes flew over-head to try to rescue what was left in France and convoys huddled in Newhaven Bay. Up came two men in bowler hats who addressed my man: 'I am the Secretary of the Golf Club and this gentleman is the Borough Surveyor. Now if you would dig your next trench 150 yards away, it would not spoil the drive from the fourteenth tee'. To which my man replied 'OK Guvnor, no trouble at all' and it was done. There'll always be an England…

> This mighty work would merit a long letter, but to be brief consisted of organising local contractors to take on pill box construction as and when the Army made up its alleged mind as to what it wanted and also an anti-tank ditch with

road and rail blocks. We built 265 boxes mostly 3ft 6 inches thick in three months and spent £50,000 in the time. Brisk work.

The instruction was to use as many small firms as possible to keep them and their men occupied. Much can be done in such time. On the Monday I left Seaford, called on the County Surveyor at Lewes for names of firms, rang up some to meet me that evening at Crowborough, phoned to my office to send an engineer, my chief clerk and a gang of men and leading hands the next day, drove into Crowborough station and commandeered the goods yard, ordered a train load of beach concrete material, met builders and told them to send all available tarpaulins, scaffolding, timber, bricks and other oddments for next day with carpenters to make shuttering, took over an empty bungalow as an office and met the Army for drawings and instructions. All in one day with no paper of any sort to say that I was not an enemy agent or that I had any authority!

Next day all the excavating equipment from our Staines Reservoir contractor arrived on low-loaders with engines running. The plant drove off the loaders straight in the line which we had pegged out for them. After a week of effort we got an urgent message from the War Office 'Stop-stop! We've sent you the wrong drawings'.

We had two officers to site the pill boxes and to decide from where the enemy would come so that the entrance would be at the back. One was a Canadian with a kilt, he vaulted athletically over a wall with his kilt billowing around him – and landed in a thick bed of nettles. This quickly and luridly solved the question 'What does a Scotsman wear under his kilt'. The answer seemed to be negative.

We had a grand party with the C. in C. Home Forces and many and various generals came to test a tank obstacle and watch mine demonstrations by the Canadians. Glossop remarked 'Only one Heinkel and what a twinkling of gaitered legs!'. I remember a scene from the television programme *Dad's Army*, where generals attend a demonstration which gets out of hand and this reminds me vividly of this occasion. Sir Alan Brooke arrived in his stately Rolls with many a military police motor cycle outrider fore and aft. After much saluting, a tank of some lightness rolled up to test one of the road blocks which we were building. These consisted of blocks of concrete on each side, so that the troops would have free access between, for advancing or possibly retreating. Then, when the enemy tanks approached, devoted men would drop four railway rails (if no-one had forgotten them) into slots in the blocks. The tank charged the obstacle and bounced back off it, which audibly pleased the designers. The tank charged again and got through, which pleased the opponents, so everyone was happy.

The Canadians were part of a mining company and produced a light diamond drilling machine. With this they drilled in a flexible pipe horizontally under the surface for some 100ft. They then filled the tube with liquid explosive. When this went up, so did clouds and masses of Sussex Clay, leaving an instant anti-tank trench, but this was not all that went up. The split piping leapt out of the ground, described a graceful swoop skywards and then settled across the high-tension overhead wires of the nearby electricity pylon line. The fireworks were most impressive. I don't think that the script-writers of *Dad's Army* ever thought of that.

On the railways, we found the problem was how to make the blocks stable on embankments as they were likely to topple down the bank. I suggested that they should be anchored by bored piles. The boring rigs for the 12 inch diameter piles of those days were small and could be carried by hand and not stop trains. This was taken up with enthusiasm and I found myself being appointed to organise members of several rival firms to do this work. During this we had our attention firmly on the war, as we were in the direct line of the German Air Force on its daylight raids in the Battle of Britain. They would come over high up in perfect formation in the wonderful summer, several hundred in all. Later they would come back all over the place, no formation left and being harried down by our fighters, in considerable disarray.

```
    We got plenty of random and other bombs round
Crowborough as Jerry came over on his way in vast
numbers. On September 18th or so at 30 minutes notice
I had to return to London to rebuild the Northern Outfall
Sewer, completely broken in all five barrels in 9 places
including a 1500lb time bomb. All sewage flowing into
the River Lea. It is now mended but regularly 'near-missed'.
Also I have six other breaks to repair in and around the
Docks and the East End. Rather hot work, entailing too
much lying on the pavement to suit my temperament.
    Spent the first week sleeping in my brother-in-law's
basement in Earls Court while Jerry slung plenty of his best
into it. I went away for one night, during which a bomb
arrived in his basement – since then have 'lived' in head
office. To Trafalgar Square last night to observe and assist
as bomb blown in the top of 27ft diameter cast iron lining
at 50 ft below street level! So much for deep shelters. A
nasty business.
    Well the sooner Hitler gets tired of this the happier we
shall be. Family at Liss still and a 'near-miss' even there.
```

Sir Joseph Bazalgette, one of my personal heroes, built the great London Main Drainage scheme over 150 years ago. It fully withstood the bombing, which its designer cannot have foreseen. On both banks of the Thames, he planned the Low Level, Middle Level and High Level main sewers which all converged on Abbey Mills Pumping Station on the north bank and Crossness on the south bank. All the smaller sewers which crisscrossed the area led into them with overflows into the riverside pumping stations. Flushers could drop in dam boards at designed points, so that short lengths could be isolated for maintenance. This proved vital when the bomb damage occurred. The sewers, many of which had been driven in tunnel, varied from 3-15ft in diameter, and some were as much as 40ft deep. They were lined with brick or cast iron.

The original Abbey Mills Pumping Station is a cruciform rococo cathedral that would have

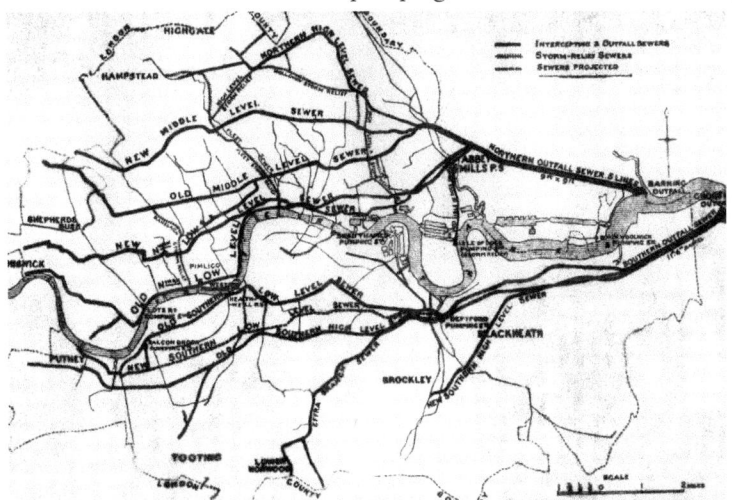

31. Map showing the intercepting, outfall and main storm relief sewers of the London drainage system from Engineering Wonders of the World, *1909*

brought joy to the poet Sir John Betjeman. In it, stately beam engines pumped the flow of rain water, industrial waste and homely sewage from the depths to which the sewers had reached and delivered this into the high level Northern Outfall Sewer in a high embankment, so that it could flow by gravity nearly three miles to Barking outfall at Beckton. There, as previously described, after treatment the liquid flowed into the Thames and the sludge was pumped into the ships appropriately named after distinguished LCC engineers and dumped in the Black Deep out at sea. More recent research showed that it then flowed back again, but that is another story.

The pumps had been changed from steam to oil and gas before the war. The locals claimed that the chimney attracted the bombers, so this was felled by steeple jacks. One bomb did drop through the central tower, but it never went off. Another blew off all the roofing and I had to take over the retiling, which was carried out by one of our building foremen. To repair other damage we had two refugee Belgians, who spoke no English, as scaffolders. One started to fall with his scaffold, so his mate ran to catch him, but unhappily it was the mate who got killed. This meant that in the heart of the blitz a solemn peace-time-like inquest was held by the local Coroner, with witnesses, while at the same moment victims of the previous night's raid were being buried in a mass grave alongside the Northern Outfall.

The Northern Outfall consists of five brick barrels 8ft in diameter, carried on a concrete embankment. It crosses railways and roads by bridges made of plate girders, between each of which is suspended a cast iron tube. The unexploded bomb mentioned above was in one of the brick barrels and was lifted out by my crane driver, who got a medal. Four of the barrels had been breached in several places, but more importantly one of the tubes crossing the London –Tilbury railway had been hit. It was mended by suspending a new tube made of bolted rings of cast iron lining, which happily were free from abandoned work at Charing Cross. When we sent transport for them, the storekeeper refused to part without written authority from County Hall, as they were 'strictly reserved for air raid damage'! The suspenders were made by Starkie Gardner, an architectural metal firm managed by G. Friese-Greene, who had been a buddy in the London Rifle Brigade.

Bombs in streets were apt to penetrate into the ground before exploding, so each repair gang had to dig down in timbered trenches through all the broken pipes and cables. Sometimes there were two sewers one above the other. As no two sewers were broken in the same place, the LCC sensibly had each repaired at leisure in a permanent form. This saved reopening if temporary repairs had been made. It took several months in each case before all the water, gas, hydraulic mains and cables had been restored.

The historic Tower Subway had been converted into a pipe subway. When a bomb had crushed 100ft of tunnel without penetrating, the maintenance of the water mains was declared of national importance. The solution was to drive 100 rings of 10ft diameter tunnel around the crushed length. The problem was how to do so. This was later described in the running-up contribution to the ICE ingenuity competition. The shaft was opposite the entrance to the Tower of London, so lights were strictly forbidden. I devised a complete gantry out of tubular scaffolding, covered in with galvanised sheeting so that work could go on day and night to keep the water mains in use. The shaft was flanked by several large Victorian warehouses when we started, however they no longer existed when we finished. Between the wars there had been an appeal to buy up such properties around the Tower to make open spaces, the Germans did it for us free of charge.

9. THE WAR AND THE BLITZ

April 11th, 1941

When I last wrote I had returned to our head office having been bombed out by proxy from my hideout in West Cromwell Road, as I told you, I went away for the night and a smallish bomb arrived in the basement. I then arranged to go to a Doctor in Putney but his house got half demolished. At the same time a very large missile dropped onto the newly finished station at Sloane Square at 'you know where' and I was bidden to take over this incident and clear it up. This was done in 12 days with the help of 12 RE's and 50 Indian AMPs but it was a dreary business until the 43rd body was removed after a week.

So I stayed on at head office being nearby and then managed to escape to some friends in Wimbledon, where I am now sharing with two men, one the owner of the house who has a staff and who works at Hawkers and the other who has been bombed out nearby. There are several very pleasant families nearby so one gets some social life for a change. My family are still in Liss when not at school and I have got down there pretty often.

I have spent all the time repairing bomb damage of one sort or another. On March 19th we had as unpleasant a night as any so far met and on driving up to my palatial range of offices next day I found nothing but ashes. My handsome edifice in which I had lived as comfortably if not happily for so long was entirely consumed and so was the building adjoining.

Happily we had the most important papers in a dugout, but I lost much that I would sooner have kept. Seven men were available but could not cope with it. While trying to restore order, messengers of ill-omen arrived to hurry me to Abbey Mills Pumping Station where a general shambles appeared to have taken place owing to a mine landing in the forecourt – de-slating the roof and cracking the trusses. However, the pumps were alright. At the same time another bomb had chopped right through ..yes you have guessed it, the dear old Northern Outfall Sewer where it ran in a bridge over Abbey Lane. A plate girder 68ft x 10ft had been blown right out and over and the cast iron flumes of two of the 4 barrels (9ft diameter) had gone. As they ran for 4 hours before being shut down, the road below was a remarkable sight being 2ft deep in road sweepings and Malthusian appliances. We wandered around carefully skirting an unfortunate dead man who lay in the road among the passers by. Eventually he was taken away about 3 in the afternoon.

I suggested to the Main Drainage Engineer LCC that it was the ideal place to use a Callender-Hamilton truss – but he was most hostile, said it would make them the laughing stock of London; and wanted to build brick piers (in the crater) and use lots of bits of joist (24ft x 7.5 inches) and other messes. However, I persisted and had my way and we built a truss 65ft long x 10ft deep, of their patent bridging material, in 2 shifts which is a great success. Now the LCC give themselves the credit.

I also have to mend the Tower Subway which is the first tunnel built by Greathead in which a shield with cast iron lining was used. It is 627.5ft diameter and contains 2 water mains and 2 hydraulic mains which have to be left in. We are building 10ft diameter iron around it to make room to work in.

I also had to go to Teddington to advise a Bomb Disposal Squad who were trying to use well-points in the effort to remove a time bomb which had dropped under a very valuable tank in an 'Institution of National Importance' where laborious work of a physical nature is undertaken. Eventually they had to go down 35ft, 20ft through ballast and 10ft of it waterlogged and then the rest in clay, in pit and later heading. We eventually found it 2ft 3 inches diameter, 9ft long and weighing 1.5 tons. I was taken down the heading to pat it while they were digging round it. Two very cheery BDS officers I dealt with kindly said they would never have got to it without my advice and it was the most difficult they had met. They then offered to make me an Honorary member of the BDS and to paint my mudguards red for me. I had great difficulty in dissuading them. So they dragged me off to get my views on another under a sewer in Battersea!

The last paragraph of the letter above referred to a bomb that fell at the testing tank for ship models at the National Physical Laboratory. This was the only one in the country of large size, so Sir Winston Churchill was alert to its importance. The RE Bomb Disposal Squads at the Duke of York's headquarters had been told by the firm that if any advice

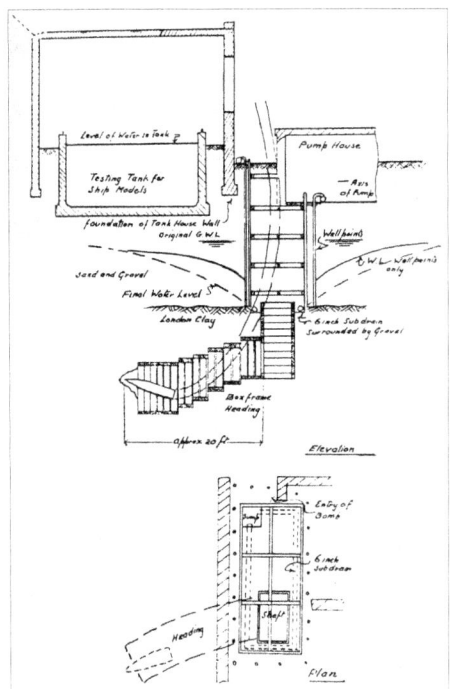

32. Ground-water lowering by well-points for bomb recovery at the National Physical Laboratory, Teddington

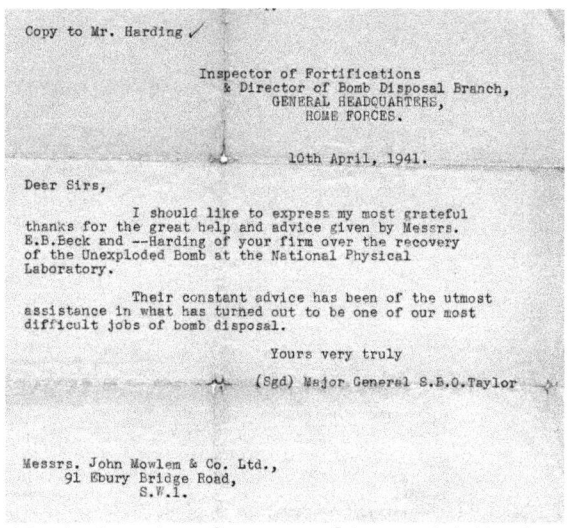

33. Thank you letter for recovery of the unexploded bomb (UXB) at the National Physical Laboratory

or help was needed in excavation, or plant was needed, I was to advise them free of charge, so I was involved in several cases.

Bombs do not drop vertically. The deepest unexploded bomb was under Mount Pleasant Post

Office, where our miners helped to drive a heading in London Clay down the bomb path for 70ft. We lent my well-point foreman, Birkett, to the Naval Mine Disposal to pump out for a mine at Yarmouth. The Admiral insisted on his being given a commission as a sub-lieutenant in the RNVR and he was eventually awarded a well-deserved BEM.

Sir George and Lady Burt slept in their flat in Victoria Street during this time. When asked why he did not sleep in his country house near Lingfield, he replied that he could not do so when so many of his men were doing repair work in London. When the bomb landed in Trafalgar Square in the middle of the night, I went with him to the site. The bomb had not penetrated but had cracked the heavy cast-iron concrete-filled segment of the 25ft diameter chamber at the foot of the escalators. This, with large lumps of London Clay, had dropped onto the unfortunate people sleeping there with fatal results. We were met by a Chief Constable who was much concerned for the many shelterers sleeping on the platforms, as water was flowing down the escalators. I was able to reassure him that it would drain away to Charing Cross where there were pumps. The vast cavity overhead was making the rescue men very jumpy. I assured the Chief Constable with great (but dubious) assurance that there was no fear of further collapse. It seemed the right answer and luckily proved right.

Except for these more dramatic interludes, the whole winter was spent in finding one's way to the 'incidents under repair', to guide and control the work. One wet Saturday afternoon comes back to mind. Everyone was rather miserable in the cold, while a German bomber roamed around overhead above the cloud, dropping the occasional random stick of bombs, while I pondered whether to withdraw my men from the trench under a London Docks warehouse before the remaining wall collapsed into it. Up dashed a telegraph boy on a bicycle, 'Say Guvnor, can you tell me where is St Dunstans-in-the-East?'. The reply wrung from me was 'Gone west by now I should imagine!'.

The nightly raids had reached a climax in March 1941 when my office and many papers went up in flames. Raids were then diverted to other towns, with large-scale returns to London in April and a final effort in May. Beside my office was a shelter for cars, with a roof and one side of galvanised iron. We rapidly built sides to make two rooms and were again in business in a couple of days, but in less comfort. The damage of the May raid was so extensive that I was diverted north to Old Street. Driving down Bunhill Row, on my right the remains of the London Rifle Brigade HQ was smouldering away and so were the surrounding buildings. I had a clear view now into the green lawns of the Honourable Artillery Company's ground. What should I see in the bright sunlight slightly clouded by smoke, but a cricket match in progress, all in immaculate white flannels. Adolf then turned his attention to Russia, so for a year or two we slept in comparative peace. The next job was to sink shafts outside Westminster Abbey and in Horseferry Road and to drive secret tunnels. Each night, a section of Coldstream or other Guards would bivuac in our Mess Hut. When we finished, we filled up the shaft with concrete and cleared away, after which along came the Guards to guard nothing at all, as no-one had cancelled them.

It may come as a surprise to learn that during the war the Institution of Civil Engineers became more vigorous than before, in spite of paper rationing. Meetings were held at lunchtime. Papers on Soil Mechanics became numerous, as the war brought new problems which were solved by the Building Research and Roads Research boffins. During this time I was appointed to my first ICE Committee, to advise the War Damage Commission on War Damage due to Earth Movement, in order to distinguish false claims from true. I also sat on the Benevolent Fund Committee and the start of the Code of Practice Committee on Site Investigation.

Another interlude was to thicken up the dam at the end of the Serpentine in Hyde Park, by tipping London Clay from the deep shelter tunnels into the water. It had been realised that the south east corner and the south overflow were well above ground and, if breached, would flood St George's Hospital and other buildings, as well as the Piccadilly tube. This involved a visit to the Park Authorities on Duck Island in St James' Park. There we learned that in the 1880s John Mowlem had driven a timber heading under the grass east of the Pelicans, in the water-logged gravel. There was a pumping station concealed on the island (as well as Princess Elizabeth's rabbits) and it collected water from this heading and pumped it up to the Serpentine when this needed topping up.

At this time alot of tunnelling was going on, forming deep shelters with 16ft diameter concrete segments. With unusual intelligence, these were driven in a line with transitional curves so that after the war they could be joined up to form express tube lines to relieve the existing ones and so have a second use. However, reverting to type, it was later decided not to use them, but have new lines in different areas.

While all this was going on my family obtained a tiny cottage at Liss with a derelict garden. My wife did much voluntary work sorting out the troubles of official evacuees and running clubs for children for the Red Cross Society. She also became Hon. Sec. of the rabbit club. Robert was given Benjy, the Buck and received a stud fee of 2/- a time from members. One day a retired General sidled sheepishly up the path with a basket, exuding embarrassment.

> 'Madame, I hear that you have a buck which could service my doe'.
> 'Yes, indeed. My son will see to that'.

Immense relief crossed the General's features, until Robert aged 7 emerged. He took the doe and put her in with Benjy and watched the proceedings with the eye of an expert.

> 'Now Sir, we will give him one more go to make sure'.
> 'That will be two shillings, Sir.

The General tottered away in a visible state of shock.

Mulberry Harbours and 'Phoenix' Houses

The work of building 47 concrete petrol barges for the coming invasion, in case Pluto did not work, was secret and 'unsuitable for description'. After the war several of these still floated for a few years outside County Hall at Lambeth as moorings for pleasure craft. They were built on the Export Quay of the West India Dock and each weighed 150 tons and were lifted into the water by the London Mammoth floating crane. During this, bombing started again and my office was blasted but, as usual, I was not there.

```
October 25th, 1943
My younger son, Robert, being now seven, has started
shaving. He has made himself a wooden razor blade which
he keeps in a matchbox next to my Gillette. He then
mounts the blade on the end of a penholder with a lump
of plasticine. On Sunday he seized my shaving brush after
I had used it and had a good lather-up and proceeded
to scrape it off with the wooden razor. He then uttered a
profound truth. He said: 'During the week while you are
away I use ordinary soap. It is just as good but not so
real'.
```

9. THE WAR AND THE BLITZ

I had just embarked upon making some of the concrete monoliths for the Mulberry code word for the invasion harbour at Arromanche. I was given the South Dock of the Surrey Docks to build in, one of the oldest in London but with new warehouses along one side. It had an entrance into the Thames which had been bombed, so that only one gate survived and naturally that was designed to keep the dock full of water, while we were trying to empty it. The work of getting the dock into an empty and usable condition took longer than the making of the monoliths and included making a massive concrete block wall with the help of divers to seal the river entrance. The other one into the Greenland Dock was simpler to deal with.

Directly after D-Day the flying bombs and V2 rockets arrived. It is interesting to note that they did much superficial damage to houses and people, but failed to do the damage to below-ground services, so made work for builders and not civil engineers in repair work. By various means I had kept my nucleus organisation going on my working site in Bow Goods yard, Marshgate Lane, so was able to turn it into a depot for the new firm of Soil Mechanics Ltd (of which more in the following chapter). In this way we were able to keep it free and independent from the deplorable Plan Department. This new firm of ours was the first of its kind in this and many other countries and led the field for decades, having developed beyond all expectation. This started almost a revolution in the profession and industry of civil engineering and was an example of free enterprise.

Late in 1944 our first large site investigation was for an extension of the Naval Depot at Portsmouth – HMS Vernon. I had the pleasure of being entertained in the Officers Mess by Sub-Lieutenant Birkett, BEM, my former foreman who was based there.

```
14th October, 1944
Today the beans have been spilled about the synthetic
harbours and all that, about which we have had to keep so
mum.
   The Twenty-One Club was well-represented in building
the caissons or Phoenixes, Holbein and Mac Briggs deciding
to do more construction and less demolition, having built
several of the largest down on the mud-flats, while I had
the task of making eight of them of the second size in the
Surrey Docks.
   This was a spectacular arrangement, but no photo could
be taken. However, one of our leading consulting engineers,
named Duvivier, guided our erring footsteps and made a
remarkable perspective or landscape drawing which would be
worth reproducing if our Secretary ran to illustrations.
   We had to dry out a wet dock with a ballast bottom, tip
in 70,000 cubic yards of bombed brickwork from buildings
and level them off to a benching. The sides of the entrance
lock turned out to be curved below water level and in the
end it was found that there was not much more than 6
inches clearance in towing out these affairs. Although they
are no longer secret, I am told that their dimensions are.
The Daily Express says that they were 'secretly prefabricated
by bewildered workers in Britain'. So that explains their
peculiar behaviour and habits – they were bewildered. Still
somebody must have worked between the tea intervals
or they would never have got built. Perhaps all the flap
and stress of trying to get them done made one take a
```

34. *Barge assembly line at West India Dock*

jaundiced view, but it was certainly an experience for all concerned.

Glossop built four more, close to DuV and me, in the Surrey Docks but in an excavated basin. The other basins were dug as deep as was safe in order to build a caisson 14ft deep and then float it out and finish it afloat elsewhere. But in his case we put in a very pretty system of shallow wells and lowered the water so successfully that he was able to build his completed in a much deeper basin. He has since reclaimed his wells and completed some more.

These had to be towed out by a Dock Pilot who is a civilian in a bowler hat, who seems to have a monopoly of handling ships in the Dock. In this case the pilot wasn't so hot and missed the opening by six feet. After much pulling and swearing, Glossop persuaded him to let two bulldozers pull it back for him from the bank which was done very successfully (Dock Pilots do not like advice). The pilot then blew his whistle and gesticulated. The tugs churned and puffed in the entrance and nothing happened for a long time. It was then found that the Pilot had forgotten to cast off the bulldozers.

35. *Concrete caissons forming main deep water breakwater for the Mulberry Harbours at Dieppe*

DuV and I took a trip down to Gravesend on one of these units which was very interesting. The remarkable thing is that one makes retaining walls in concrete, with asphalt and every convenience and they leak like sieves. One then makes concrete boxes with 12 inches of concrete and float them and they remain dry. Perhaps the secret is having them floating and not pushed around with earth movements. We managed to take some days off at the end of June and stayed near Blandford to visit our son Edmund who is now at Bryanston School. This has proved a very successful venture and we are thoroughly pleased with our decision. It would be a good idea if someone finished off this war as soon as possible. Things have certainly moved since June. During July and August

9. THE WAR AND THE BLITZ

36. *Petrol barge ready for concreting the cross walls*

the sky was thick with planes doing things in a big way and all flying south to Normandy. They now fly off east which shows how far we have gone.

April 30th, 1945

It is strange to think that we should not for the moment hear any more things that go bump in the night – or day. We can boast having received one of the final flying bombs on the roof of a warehouse next to my office in the Surrey Docks. As usual for me this was a daring escape by proxy – because I wasn't there – but it quite surprised my staff.

We turned the shed we had used for the welded landing craft into a mass production factory for what has been called a grim solution to a grim problem, to whit 'Phoenix' houses made with welded frames of tubular scaffolding.

This has developed into a sort of portal house made of tubular frames, asbestos and what have you (if any). It is an attempt by contractors to show what can be done by private enterprise – but of course in a controlled world it just does not work like that. After waiting for months, ready to start at any day, we got the word to make 100 instead of 1,000.

Making pre-fabricated houses of this sort is expensive and more trouble than a brick one as everything has to fit and there are so many of them. In brick you bung up a wall, knock holes where required and then smarm plaster on to smooth out the irregularities. Our houses are masterpieces of mass production. There are 28 internal panels to the exterior, of which 24 are different from each other. 18 partition panels every one different and so on, while the floor is so thin that when you sit down after breakfast to read the paper, the door flies open. However, they are all equipped with portal fittings, frigidaires and what not and will be far better than nothing for as long as the kids don't throw stones through them.

That gives me something to allocate my time to, while Glossop and I amuse ourselves running our subsidiary company 'Soil Mechanics Ltd'. This bouncing, if slightly illegitimate child with the unfortunate name – not of our choosing – not only carries out all the trials for chemical consolidation, ground water lowering by known and unknown methods, freezing and whatever you fancy, but also carries out trial borings and extracts samples in an exemplary manner. We have H. Q. Golder, one of the leading 'footnote boys', firmly in our clutches and hope to revolutionise this aspect of civil engineering [see Chapter Ten].

The family flourish after a startling performance in the Christmas holidays. My No 1 son (Edmund) broke out in chicken pox the day after Boxing Day. Ten days later, going down for the weekend, I found my wife, daughter and both sons with chicken pox. Edmund had staggered to his feet that day after having been very bad and was cooking a meal for the others. Bodies in every room in fact. So I had to take a few days off to cook, shop and feed as there was no help of any description. My wife had a really bad do but everything came right. I am now trying to resurrect and repair my Putney edifice but this will take time, hoping to be open to callers in the summer.

So, that was the end of the war and the start of a new and different chapter. One of the lessons which should be remembered is the dependence of city-dwellers for their health and survival, even more in war than in peace, on the efforts of the profession of Civil Engineering. Bodies need food to survive but the provision of water and the removal of the human residue is vital. The raids over Britain never reached the intensity of the Allied retaliation, where damage was concentrated on city centres. However deplorable this may have been, it certainly made the re-creation and rebuilding more comprehensive and plannable.

We were able by diligence and ingenuity to keep the water supplies and sewage disposal going in spite of the malice of the enemy, but it was a close-run affair. If Goering had continued focussing his daily attentions on London instead of diverting to other towns, we should have been hard pressed to keep up with the rate of damage in order to prevent epidemics and other risks which follow the breakdown of essential services. I was able to see the map in the Ministry of Works on which every flying bomb was pinpointed. Although they were widely spread, there was a thick curved swathe on the south bank of the Thames about ¾ mile from the City and Westminster. It looked like a slight range miscalculation by the enemy.

In London, the damage was so widely scattered that few areas called for overall concentrated new construction except for the Barbican – with deplorable results. When only a few houses are destroyed in a street the surviving ones are too valuable to replace. It is worth reflecting on how emergencies can be handled. In the repair work which I have described the Government paid money to the LCC so that they could pay their contractors, who would then have money to pay their men. These in turn could then buy food and supplies from the shopkeepers and the shopkeepers then had the means to replace their stocks and so on.

PART 2

10

WIDENING VISTAS – THE RISE OF SOIL MECHANICS LTD

One of the side effects of Chingford Reservoir was an increase in the demand for the site laboratory in solving other people's problems. The work on the reservoir had plodded along slowly but steadily all through the war, so the laboratory had been kept in action. During the war we became very friendly with the building researchers, in many an evening session they produced their theories and we quoted appropriate chapter and verse from practical experience, for our mutual education.

It became obvious that the theory of soil mechanics, the practice of site investigation and the execution of geotechnical processes were all part of the same science. During an air raid in 1941, I stayed the night in Glossop's caravan at Chingford. We drew up a prospectus for a company to exploit all these features, with estimates and anticipated turnover to make it viable. We infiltrated the idea into the directorate but nothing more was heard of it. Glossop went off to build various airfields, and Golder left the BRS to join him for practical experience. Then after D-Day in 1944 we found that a company had been formed called Soil Mechanics Ltd, but the directors had forgotten to tell us, so we just took charge of it with Hugh Golder as our leading expert. Sir George Burt had rented the mezzanine floor of his block of flats at 123 Victoria as his air raid shelter and we took it over as our laboratory. Later, as the firm expanded, we inherited his spacious fourth floor flat as our offices.

The firm had always employed male secretaries. Just before the war, I paid an infrequent visit to head office, where I retired to the cloakroom reserved for senior staff. While I was sitting 'brooding on life' in decent obscurity, I heard the outer door open and the voices of the secretaries of Sir George and Sir Henry Japp.

'We don't see much of young Harding nowadays.'
'No. Pity. Artful sort of cove, I should think'.

I like to think that this was not so much rudeness but a tribute to my 'tactical handling of directors', who were their masters. Of necessity one had to become artful, this was proved as I had artfully kept my nucleus organisation quietly in full operation at Marshgate Lane throughout the war. We at once transformed the site for Soil Mechanics Ltd as our depot and this provided us with the timekeepers, clerks, leading carpenter, blacksmith, fitters and other intelligent types. It is easy to forget that for several years after the war paper and steel were strictly rationed. If steel was needed for plant or for making equipment an M form had to be obtained and processed through a Ministry. This again called for artfulness.

When the Bow–Leyton tube had started in 1936, the old-time tunnel chainmen attempted a form of blackmail. I went straight off to the Royal Engineers Old Comrades Association and came back with highly intelligent young sappers, who rapidly mastered the mysteries of support in tunnel surveying so the old-time chainmen went elsewhere. After the war these reservists returned to us, as did Sub Lt Birkett, who quickly trained them in the specialist work as foremen-operators, together with a few other faithful followers. Civil Engineering work was slack for several years, so I was able to get back two of my former assistant engineers, by now equipped with varied experience. In this way we were able to develop our site investigation reports under the control of men who knew what information was needed for such future work.

The XXI Club letters which cover the period 1945 to 1950 bring back many reminders of this transitory period. In October 1945, members were informed that the abandoned Putney house had been re-occupied and was open to receive callers.

> After six years of lying empty much attention was needed especially in the garden and among the many fruit trees. Luckily only a few windows had blown out with near-misses from bombs and incendiaries.

The moment the war ended, the Government despatched many missions of experts to visit all parts of Germany to find out all that they could about German methods and actions and Glossop and Golder were both co-opted. Members of the missions were given uniforms and transitory ranks as Colonels or Majors, so that they could move more easily among the armies of occupation while studying all that they could. Glossop's groups went to inspect underground factories, where rockets and much else had been made by slave labour. He reported that we had only just won the war in time. Golder investigated soil mechanics and any appropriate matter in the same category. The Siemens-Bauunion office in Frankfurt was important, as the firm was the leading civil engineering firm of designers and constructors in Germany. They found a vast drawing office quite intact, with drawing boards littered with drawing instruments lying loose on them, exactly as if the staff had just gone out to lunch. There must have been a panic evacuation and it left a curious impression. They both commented on the fact that the Germans whom they interviewed did not seem to consider that they had behaved in any unusual ways and they were most talkative in giving information. My comment, in the letter, was not polite.

> The Germans thought that they were the salt-of-the-earth and so, up to a point, they were – but unfortunately they were fruit-salts so did not act so smoothly as they hoped.[30]
> My daughter is now in full swing at the AA. This does not mean that she wears a badge in her bonnet and is saluted by smart men in gaiters, but that she is a student at the Architectural Association School.[31]
> We all went down to Blandford last term to Bryanston half term and saw a most remarkable performance of St Joan. We also hired a Morris 8 so six of us crowded in and went to Swanage, where we saluted John Mowlem's (cannon)balls and as it poured with rain attended a marionette show

---

[30] It seems odd to refer to the Germans as stomach settlers, so presumably there was a further use as laxative, given HJBH interest in those matters.

[31] At the age of 17.

in the Mowlem Memorial Hall, complete with bust of John Mowlem in terrific whiskers. The family were taken behind the scenes to study the objects and in the summer holidays a great bout of puppet making took place.

My larger son passed his school certificate at 14, so padded off to see John Walker, the Registrar, to see what subjects he should take and so on for entering City and Guilds College in 1948. When they found his age they threw him out as they were too busy with this year's problems to cope with him.

In November 1945 I was sent on a different mission, to help the French in repairing devastation. The firm of Paulings were to assist Batignolles in rebuilding the docks at Dunkirk, the third largest port in France. The big entrance lock to the impounded area had been rebuilt in the 1930s by Siemens-Bauunion as part of German reparations after the first war. As the whole area is built in fine sand full of ground water, Siemens had used the deep well system for the work and I had been included to advise on the use of the method in the rebuilding. Colonel Thompson of Paulings and I flew to Paris by the newly revived commercial flights which were run by the RAF with wartime Dakotas. After we had sat for several hours in a building in St James's SW1, we were told to em-bus. Unexpectedly, a WRAF girl called for me to return to the desk where she asked in dulcet tones, 'Excuse me Sir, your religion?'. This was my first flight but any nervousness evaporated at this kind thought, that if I was not incinerated I should be buried in the right quarter with the correct rites. In due course we bumped over the grass of Croydon Airport and took off, sitting facing each other on metal tip-up seats clasping a cardboard box containing sandwiches for lunch. We arrived at Le Bourget at 5pm, having arrived at St James's at 8am that morning.

Not surprisingly Paris looked very dilapidated and hardly visible behind the masses of Communist Party posters. We drove at once to Batignolles office, where we found that no-one spoke English. As my French had lain dormant since school, it was pretty hard going, but Colonel Thompson spoke a bold simple French in a bold English accent so we got on. Luckily I had brought with me a photostat article from the magazine *Travaux* 1937 on the 'new work' at Dunkirk and, as this had a more-up-to-date plan of the docks than the French had been able to find, a good impression was made.

Next day, we set out for Dunkirk in a 10 horse-power car with broken springs. Whenever the chauffeur saw another car approach on the horizon, he immediately began to sound his horn in blasts until we were 100 yards past. We went through Arras, then Amiens and Hazebrouck, which took us from south to north and so cut across the line of the Allied advance. Every now and then, wrecked tanks and guns still lay beside the road, but the cobbled roads were in remarkable shape, except where they were not. Hardly an original bridge remained and in places the towns had two periods of destruction as could be seen in London. Siepmanns Primary French Course of my youth used to refer to 'les plaines monotones de Picardie' and how true that was, almost entirely planted with sugar beet. All the peasants were hard at it with bullock carts or ploughing, sometimes with a horse and bullock pulling a plough together. At intervals German prisoners were repairing the roads.

We got to Dunkirk by 12:30 and had a very French lunch with vino which, together with the driver's continuous hooting and the bumping left a considerable headache to be coped with. The area around the docks had been very much knocked about and was a depressing sight. We then inspected the Freyssinet docks which had been built after the beginning of the

Third Republic in France in 1870. These were long lengths of water with walls and sheds on each side, leading from an area of water which led to two entrance locks. The whole system is impounded and in peacetime is permanently full of water. When the Germans had to retreat, out of spite, they scuttled many craft in the entrance locks and hung barrels of explosives at intervals all along each dock wall, except one. They then blew the lock gates and let out all of the water. As a result the only unbreached wall, not surprisingly, collapsed into the dock from the unmatched water pressure in the ground behind. The other walls had been breached, so the ground water had drained away. The walls survived, which was the opposite of the German intention, so they were not always as bright as people think. The Germans had also tipped every dockside crane and whatever else was useful into the docks and blown up buildings. It was a sorry sight and made no more cheerful for the appearance of many gangs of mournful German prisoners being made to clear up the mess of their own making.

The upshot of the visit was positive for Batignolles and negative for us, as after much thought I advised Batignolles that as the Germans had been so considerate as to breach the walls before letting out the water so that the ground water had drained away through the openings, there would be no need for deep-well ground-water lowering. As long as the walls were rebuilt before the dock was impounded again, the need would not arise. This saved the gallant French much money. Ten years later a party from the Institution of Civil Engineers were invited to inspect the docks after their renewal. The previous devastation had been so great that I found the place unrecognisable and it was a heartening sight to see how life can survive misfortune.

On the way back to Paris we stayed the night in an 'estaminet' in Armentières and were invited to dine with the relations of the French engineer who was conducting us around. They had been twice fought over, but the tall terrace house was intact with a lively family and servants. After three courses I felt better fed than for years, but then in came six more courses of one sort or another! It is a pity that we are not such a self-supporting country. In Paris we were not so lucky, we lunched at Claridges on half a mackerel for the equivalent of three pounds and the British Embassy with difficulty found us beds in a hotel commandeered for the British. Next day we arose at 4:30am and took the train to Dieppe over scenes of devastation and shaky bridges. The boat was packed and a gale blowing and passengers had to wear lifejackets, as mines were about. After taking two hours to shuffle off the boat and through the Customs at Newhaven, at least there was a British cup of tea once aboard the train.

During 1946 much varied activity went on all over the country and in the Docks. Mowlem did a great deal of work for Tate & Lyle in the Silvertown and Plaistow Wharf works and the October letter gave a detailed account of the processes of sugar-making on one visit and records:

> We then went into the packaging room, where lots of girls guide the machines which fill the well-known cartons. I noticed that everyone looked at the girls first and then the machines.
> We were told that the local girl workers were very tough. One brash foreman on the top floor got too fresh or otherwise offended them. So they took off his clothes, bound him hand and foot, tied a blue ribbon to his person and placed him on the belt conveyor. In due course he was delivered at the feet of the dockers on the loading wharf.

The winter of 1946-7 was very bitter indeed and as only half our meagre ration of coke and coal had been delivered, we were reduced to queuing at the local gas works in the hope of being able to shovel a sackful or so of coke. I was lucky to find long distance trains were heated, but some others were at risk of dying of exposure. On returning from a trip to Manchester, where I was presenting a paper, my train left at 9am and got to Euston by 3am. Luckily it was heated, but as there was no food or drink, London, Midland and Scottish Railway kindly stopped it at midnight at Bletchley Junction, where the tea room staff had stayed on so that they could give us a sandwich and a hot drink. The drifts of snow were up to 8ft deep and German prisoners clearing the line had cut out refuges in the drifts, in which they stood to let trains pass. I remember another day during this bitter weather where I had to go to Middlesbrough to visit the site for ICI at Wilton Castle. After the taxi had gone a few miles the road was so deep in snow that the only solution was to go back to London.

For the first twenty years of my working life the nature of the work had pinned me down day after day for long hours on fixed sites, with only two weeks holiday. The exception was the specialist period, but this was a time of anxiety as to the future. Now, we were free from the traditional bonds, free to move about the world and in the words of Stanley Baldwin, 'to bring to bear such intelligence of which we are capable'.

Now that the war was over, there was a great expansion of interest in various sciences. International conferences became popular, as air travel expanded. More conferences in several countries were also held and we made many friends among fellow workers in these different countries. What we were doing was new, we did not have to follow our elders but had to find our own way with no traditional barriers. It was exciting to build up a firm from nothing, in but a few years, to an international reputation, while we expanded our staff, our equipment and our experience.

Golder combined a fine theoretical brain with sound common sense. He was soon awarded the degree of Doctor of Engineering by Liverpool University for publishing a number of papers on original work. Glossop as a mining engineer 'knew about rocks' from bitter experience, while I had had a basinful of 'soft ground'. So we found that our total was greater than the sum of the parts. In addition to helping to run Soil Mechanics Ltd, Glossop took over the Basin South sphere of influence. This organisation not only did much work for the Port of London Authority (of which there was plenty because of the war) but also much other work on or around the Thames. I soon also took over the tunnel work and a number of other works, some of which appear in the following chapters.

The rest of the parent firm was in a curious state at this time. All good plants have to struggle to overcome the stranglehold of bindweed. This took the appropriate form of the plant department, which might form a tragi-comic thesis on its own. We were free from it with our own organisation, both in Soil Mechanics Ltd and at Basin South. Our hereditary chief, Sir George Burt, who had been knighted for his many services to the country, had been a prime mover in the setting up of the Building Research Station and the Roads Research Laboratory. At this time he was Chairman of the Building Research Board which advised and encouraged the BRS on behalf of the Department of Science and Industrial Research. He was thus our enthusiastic supporter. It is an odd fact that the City and Guilds College in my day was one of the few colleges who gave a course on geology to civil engineers, perhaps because this was available from the School of Mines which was part of the Imperial College. Consulting engineers seemed rarely to consult geologists but we were almost the first to engage engineering geologists on our staff. This reluctance to use such useful men may have been 'fundamental'. A friend told me that he took an antler to his boss, an engineer so distinguished that he should be nameless, and said that he would be interested as it had been found in their tunnel and a Professor had dated it to several million years old. His reception was not what he expected: 'How dare you talk like that. Do you not know that God made the Earth in 4004BC?'. Indeed Mowlem had an elderly director who was an Irvingite – a fundamentalist faith, so when we set a precedent by actually engaging geologists, their names were discreetly buried in the staff list among the civil engineers.

Sweden

The next XXI Club letter brings us to Sweden, which was one of the countries which pioneered soil mechanics. The bearded Professor Fellenius defined the Swedish classical circular slip because such slips were of constant occurrence in their soft glacial clays. When I told my colleagues that we were going to Sweden for our holiday, they decided to fly there first to contact fellow pundits. Among others they met Sven Platzer and came back saying I should find him a kindred spirit as he had a silly sense of humour. We were to become close friends.

> September 30th, 1947
> The next item is to announce the first holiday for years that has not been pursued by gremlins and the first in which I resolutely took three weeks, even though I spent 3 or 4 days visiting works.
> In order to assure a combined holiday prior to call-up[32] and other eventualities I decided to invest in a holiday in Sweden for all five of us and the car. So we embarked at the end of August – as no prior accommodation was available – on a whited sepulchre of a concrete vibrator called MS Saga, from Tilbury.
> It was beautifully appointed and the only one of six Swedish Lloyds with a second class, but her diesel engines (as the steam turbines had failed to be delivered) and their vibration has to be felt to be believed. However, I at least emerged undismayed and intact, but many a passenger was laid temporarily low for no other reason than the frequency cycle.
> After two nights and a day we landed at Goteborg or Gothenburg which is a magnificent harbour with a fine town all round it. The car was carried free as we were more than four persons, so having learnt the signs including Ej Genomfart, which means No Entry, we set out over the heavily glaciated slopes towards Stockholm.
> The roads were mostly gravel being maintained by blade grinders but with proper banking at bends and it was like being back in 1913 to be smothered in your own dust. However, we sailed on to a little town called Granna where we slept in the bedroom previously occupied, not by Queen Elizabeth, but by 'Queen Christina' in the ample shape of Greta Garbo.
> Next day we pushed on 230 miles to Stockholm over somewhat monotonous variations of pines, beeches, granite, roche-moutonnées and glacial eskers and arrived at the Hotel Reisen in the Old Town, by nightfall. We were fortunate to find by accident the best restaurant in the

---

32 Edmund was by now seventeen.

10. WIDENING VISTAS - THE RISE OF SOIL MECHANICS LTD

town, 'Bacchi Wapen' where we had a superb dinner with schnapps and ice cream with wild strawberries.

Stockholm is beautifully situated among an extraordinary mix-up of lakes and arms of the sea and is a very pleasant place. I was let in for visiting an airfield to see a new method of pushing down cardboard drains for helping to consolidate soft clay and also explored the new Underground being made in rock tunnels under the City.

I had met a few of the engineers before and had a very pleasant time, taking my senior son with me. We also had supper in a Swedish family with songs before each drink (Sven Platzer). After a week our hotel could hold us no longer, so we were advised to go to Saltsjobaden Grand Hotel which is where all the English are sent by tourist agencies. We found seven cars all with GB so nearly left but it was not as bad as all that.

We had three very cold bathes, one at a place called Flaten but the season was really over. However we relaxed and did nothing, which was all there was to do except watch other peoples yachts and a film company on location.

We returned in very good form to England having spent an appalling amount of money but no duty asked for or given. Work still goes on with variety and we still do many queer things. Mowlems have been given an acreage of ground-nuts in Tanganyika and are now wondering what to do with them. But at least they have engaged an Old Centralian to assist them in their difficult operation so we hope it will be successful.

There was certainly much variety at this period. Wynne-Edwards soon founded a similar specialist subsidiary in Costains, called Foundation Engineering. After we had trained his expert, Sir Godfrey Mitchell, the wise old Chief of Wimpeys then did the same with their Central Laboratory, after seducing Doctor Murdoch from the Building Research Station. Other firms followed suit, but we continued to set the pace. There was great enthusiasm among our increasing number of engineers, laboratory staff and foremen. One or two of the more theoretical were inclined to be 'sorcerer's apprentices', able to start the magic but not knowing how to stop. In those cases the outside experience of an elder could be a useful damper.

During the war the Institution had formed a Works Construction Division. We members of the Divisional Board decided on subjects and invited authors, often among our assistants and younger men, who otherwise might have been shy to put themselves forward. The papers dealt with methods of doing work and not descriptions of single projects and the meetings were livelier than the ordinary run of discussions and much valuable information was exchanged. All of our 'foursome' wrote a number of papers, mostly by invitation, not only to this and other divisions, but also to the Geotechnical Society. We had persuaded the Institution to support this new society as well as all the other activities.

Soon after the war the vast Scottish Hydro Electric Scheme began to burgeon and this called for a lot of exploration in those rocky fastnesses. We were first in the field, introducing diamond drilling to prove what was rock and what was only a large boulder. This is vital for examining sites for dams. Site investigation boring tools are by necessity and their nature simple and easy to handle in units. One dam was so inaccessible before any roads were made to it that the equipment was carried 'oer the mountain' to the next valley by ponies which usually carry food (and drink) for grouse shooting parties. As the pack train started off, the landlord of a small inn said to the leader of the expedition 'Whist mon, will ye nae tak yon wee still with ye? It will not be seen over the hill'. Shades of 'Annie get Your Gun'.

The next letter, No 60, briefly touches on two affairs of considerable import to the writer, who did not foresee their after-effects which will be enlarged upon in due course.

> Soil Mechanics Ltd continues to expand in its peculiar machinations in spite of cuts, scarcities and scares. At the same time, Glossop and I have inherited other spheres of interest to look after at the same time. I now have to make periodic excursions to Renfrew to watch the building of a power station. I am also due to go to Sweden in June to the Conference on Large Dams.
>
>   Many of the Codes of Practice are coming to a head and on these the Twenty-One Club has been well represented. It is hard to say if these will be a good or a bad thing but, if they are to be done, it is as well to have some of us in on them to try to control the Civil Service boys and others. Always remember however that the voting is by majority so do not blame us if some follies get past our protests.

The first affair was Braehead Power Station on the not-so-bonny banks of the Clyde, west of Glasgow. Mowlem's man running the contract had got into dispute and incurred the wrath of the consulting engineers, Sir Alexander Gibb and Partners. They demanded that the works director and the agent should be removed as allowed for in the contract. So, twenty years after Fords, I found myself once more sent to make the peace and clear up the troubles, this time as visiting works director. Two years before, Soil Mechanics Ltd had carried out the site investigations for Gibb and found that there was a deep bed of glacial gravel overlying very soft glacial silt. I had written and signed the report in which we insisted that it must be specified that concrete piles must be driven right through the gravel and through the silt beneath to rock, or else there would be dangerous subsidence of the power station. This was the chief source of the trouble which had not been handled intelligently, but it was eventually done after much effort. Much else happened including our scheme for making it possible to complete the deep screen chamber, which was within sheet steel piling, in this soft silt. This followed our lesson at Dagenham and was completed in compressed air under a steel deck welded to the piles.

For several years, I travelled by sleeper every two weeks to Glasgow, going up on Tuesday night and back on Thursday night. At that time, only Dakotas were flying to Renfrew and half a day would be lost if they were used. The boiler in my house was giving trouble and one night I was suddenly woken by the noise of rushing steam. I leaped out of bed and dashed for the kitchen. After covering two feet, I fell back stunned – on to my bunk. The train was stopped at Carlisle, a locomotive was letting off steam and I had collided with the partition, head on. On another occasion, I had a spirited encounter with a man with a large moustache who tried to take over my sleeper, without success. *De mortuis* etc but the attendant told me 'Yon man is Gilbert Harding. Yon is a difficult man when he has a drink taken'. I must say that constant travelling to Glasgow is an understandable excuse for this.

> October 5th, 1948
>
> For once I have some fresh news to put in my letter as it covers my trip to Sweden as a delegate to the Third

10. WIDENING VISTAS - THE RISE OF SOIL MECHANICS LTD

Congress on Large Dams in Stockholm in June, or Les Grandes Barrages as it was sonorously called. I found myself being flown to Sweden via Copenhagen and fed by Nordic Air Hostesses one summer afternoon. We spent three days in Stockholm attending dull meetings in the Concert House and then left on a 6 day tour up to the Arctic Circle, living in wagons-lits and travelling by night and visiting power stations and tunnels until we were dizzy.

I stayed near Saltsjobaden with the family of a Swedish engineer, Sven Platzer, whom I met last year, instead of a hotel with the other English members and so went native. I spent most of the first three days being run round in his car looking at jobs and getting to bed about 2:30am being mixed up with their family parties.

The meetings themselves were not up to much. On the third night all the delegates and such wives as were present were entertained to a Banquet and Ball in the famous Stockholm Town Hall by the Municipality. It was a magnificent function in a perfect setting. Before the event my hostess said 'Can I help you with your smoking?' to which I replied, 'Please do not bother, I do not smoke'. It turned out that she wanted to iron my dinner jacket. We went in a party as the Platzers and relations and a Finnish engineer friend and his wife were all invited.

We assembled in the famous Blue Hall, which has a blue ceiling, red brick walls and bluish-white stone cloisters. After a fanfare of trumpeters in Cromwellian dress we proceeded up the stairs led by students in their Baccalaureate Yachting caps to the Banquet Hall. This is all in gold mosaic with inlaid mosaic pictures of a most amusing kind. There were over 250 delegates from 27 nations and with the hosts made up to 400. All guests were very cleverly mixed up in the seating. Trumpeters okayed a fanfare before each course. The best idea was that the speeches were made (in English or French) between each course, after which the doors opened and in rushed the waiters with the next course.

At the end the lights were turned out and the waiters entered in procession with dishes containing blocks of ice lit up by electric torches below and covered with ice cream in tall cones with strawberries. It was an impressive and mouth-watering sight.

After the dinner at which there were few and mature ladies, there infiltrated into our midst a lot of girl students and younger wives of Swedish engineers in order to restore the balance for the dance. This was in the Blue Hall and the whole setting was more beautiful than one could imagine. The Town Hall is a masterly building and this hall had the effect of enhancing by light and colour the beauty of dresses and figures and made the most unlikely people look well.

On the next night we set off in two trains. I was Berth No 1 Compt No 1 Coach No 1 which was an unusual position. I shared it with another Swedish friend, Hagrup, who is a high-up in the Board of Waterfalls who organised the tour.

It was a triumph of organisation and nothing went wrong. It was hard going but the relief of being looked after, told when and where to go and not having to think at all was most refreshing.

There were a very mixed lot. I was the only Englishman in my coach and bus, mostly Finns, Swedes, Indians, Czechs and Yugoslavs. The most interesting places were the tunnels at Hjaalte about 45ft square and 3 miles long and the rock-filled dam and tunnel at Harspranget, beyond the Arctic Circle. Here it was warm and at one time we had to take our coats off. At each place we lunched and dined and were given schnapps for each meal as well as drinks. At this place which has nothing more than a works camp they dined nearly 400 people, each train carried two hundred.

The Americans were the most numerous and noticeable but very pleasant. The Swedish engineers played guitar and accordion and community singing in all languages took place each evening. At Harspranget we were taken up a local mountain to see the midnight sun. This was behind a cloud, but we took photos with 1 sec exposures.

At a little town called Solfetace, we had dinner in the Stadspolleet in the evening and before the singing were a few amusing speeches. M. Coyne, the Chairman spoke in French and his speech was repeated with witty variations in English by Doctor Lea, Director of the Building Research Station. It was to propose the health of the lady organiser of the trains. At the end all the Continentals chanted 'Embrassé – embrassé'. So Doctor Lea, after being chased round the room by Dr Helmstrom, rushed up and gave the lady organiser a terrific embrace among universal cheers.

The usual singing then took place with most groups, except the British, singing national songs and joining in Aluette and John Peel and so on. I tried to interest the other British into doing something, but no one would move. The last night we had a lovely dinner on the King's Birthday at Karlstad, as guests of a big Celanese firm who provided a solemn baritone. Dr Westerberg...then called in his singing engineers and started up our community singing as by this time we were getting very matey.

After Czechs had sung and Indians and Americans and French, under the influence of the schnapps, I bobbed up and found myself standing on one of the tables with the musicians. So I made a little speech, to the effect that 'I was the junior member of the British Delegation and, following our national custom, we found ourselves unprepared and disorganised'. Loud cheers and appreciative noises. 'But I have pulled myself together and would sing to them. First I would sing the first verse of a Devonshire song about seven men who wanted to go to a fair, fête or fiesta on a female horse or jument and the second verse I had composed in honour of the Congress'.

So after giving them Widdicombe Fair I gave them a verse bringing in all the leaders of our train group and Herr Ilrog who organised them so well.

It went like this:

> 'Herr Ilrog, Herr Ilrog lend us Wagons Lit
> All along, down along out along lea
> For we have alot of power stations to see
> With Monsieur Coyne, McHerrur, Westerberg, Mister Koschla
> Doctor Helmstrom, Professor Fellanius, Mr Justin
>
> And Uncle Terzaghi and all
> Old Uncle Terzaghi and all'

The effect was most surprising. Dr Terzaghi who is a terrific character in the civil engineering world and the 'Father of Soil Mechanics' rocked in his chair and yelled and called for a repeat. I was most shaken at the effect of such a simple recipe on the assembled nations. At any rate the British prestige was considerably restored even if it was not a scientific contribution to such a Congress.

I asked Terzaghi's pardon after the dinner and he said 'My dear boy I was delighted – you and Doctor Lea are the only British here who are not "stuffed shirts"'. He is now Professor of Civil Engineering at Harvard University.

After the dinner we had a very enjoyable dance in spite of 'too many men chasing too few goods'. After the ten days I flew back from Gottenburg. Never had so many slept so little for so long but it is a fact that in Sweden while the sun sets at 2am and rises at 3am you feel extraordinarily vigorous. The hospitality was superb. As train staffs were short, ten Swedish engineers' wives took a course with the State railways and put on uniforms and waited on us in the Dining Cars. Dr Justin (USA) Vice president of the Congress was brought a cup of tea. He tipped the waitress. Mr Westerberg said 'Do you know, that is my daughter?'

My senior son Edmund was two weeks too young for the City and Guilds so he is now a pupil of Charles Brand and Sons at Kingston Power Station. It turns out that the chainman there, Curly Bennett was chainman on my first job at Camden Town.

Ground Nuts

The reference in the letter on page x to ground nuts conjures up memories of a series of contacts. When I had joined Mowlem in 1922, I had made a lasting but intermittent friendship with Raymond Head. He was a year or so older and after a spell in the RAF, he had not spent time on a University education but had joined the firm as a trainee, so was four years ahead in experience. He had been made personal assistant to W. Rowell, our chief engineer. Head advanced rapidly, became second man to Sir Henry Japp and later reached a peculiar sphere of influence at the top of the firm, all of his own making. As a young man his motto was 'go out and get the work', this he did with success and charm of manner. He had devised, whether unconsciously or not, a gambit for disarming the habitual criticism of resident engineers and their like. When they would say 'as RE's do from day-to-day' that his men had done something bad or the firm had done something or other, Head would only say 'Is that so?', and then go on to the next item.

During the war, Head had brought in several airfield contracts, and after it finished he used his own initiative to penetrate and gain the confidence of the Shell company, who confided the Shellhaven Refinery to the firm. Before the war, he had gone to Trinidad to start the Nuttall-

Mowlem contract to build harbour works and he used the time he spent there to study the potential of the West Indies and after the war tried to interest the directors. This clashed with the suggestions of a post-war member of the firm who inevitably comes into the picture.

Before the war Ruston-Bucyrus, who built excavating plant, had a successful chief salesman, Major Westrop, who had won a DSO in the first war. There is a story of his working hard on Sir Henry Japp by saying that his machine could do so much an hour, which was so much a week and so many thousand cubic yards in a month. Sir Henry put on his best Scottish accent and said 'Westrop, does your driver never get off to pee?'. On recall to the army, Westrop had helped to extend the Maginot Line along the Belgian frontier by trenches and pill-boxes. As things advanced into Belgium (and hastily retreated) this was never tested, but with this expertise we first met him when he started us off on the defence lines, which I have described in Chapter Nine. Westrop had then departed for India, where he soon became a brigadier. While there he had explored the potential of India. He joined the firm on his return. Thus the Board were faced with the choice of the West Indies versus the East Indies. For the moment, India won, but the effort collapsed, partly due to the impact of the chaos taking place over partition and Westrop left to set up as a consultant on plant. This meant that Head was at last allowed to exploit the Caribbean with much success in a number of works.

Head had been very helpful in getting Soil Mechanics Ltd 'sold' to the Board, even though he had a most peculiar position within the company. In any less feudal firm he would have been welcomed as a director, but there was prejudice against him from E. B. Beck as he had been unlucky with two marriages, through no fault of his. Over-work is a nuptial risk in our profession. Around this time the firm bought a half interest in a firm working from Nairobi, run by the United Africa Company, which later became the wholly owned subsidiary of the firm under the name of the Mowlem Constructing Company. As an off-shoot of this, the firm sent an expedition to the notorious 'ground nuts' affair, which was an unfortunate and expensive British Government scheme to plant peanuts, which need water in large quantities, in an arid area of Tanganyika (now Tanzania). The best thing that Westrop did was to introduce Brigadier Horsfield, who took this over. He had been Chief Engineer of the Fourteenth Army in Burma, after years in the Bombay Sappers and Miners. Under him, the Mowlem contingent was the only one which emerged with credit from the ill-managed pantomime produced by Strachey and the then Labour Government.

Tactical Handling

Glossop was anxious to get hold of Sir George Burt for a quiet talk before a Board Meeting, to persuade him to some new course of action. He told this to Head, who occupied a large room on the ground floor beside the front door of the Head Office at 91 Ebury Bridge Road and here took place an admirable example of the 'tactical handling of directors'. In 1949, Sir George Burt was becoming older and increasingly deaf. He also had an old and disagreeable Sealyham, Joey, who dominated his private life and also came to the office with him. Head said to Glossop to come to the office at 9:15 the next morning, because at 9:20 Sir George would be driven up in his ancient Rolls, Joey would get out and Sir George would follow up the steps in a sort of trance. Head would then open his office door, Joey would see the electric fire was on, come in and lie down in front of it and George would follow automatically. Head would then close the door and George would be at Glossop's mercy. In this case it all took place, exactly as he predicted. This counts as a facet of management.

Houses of Parliament

The excavation for the Houses of Parliament Boiler House interested more than the Lord Chancellor, Lord Jowett. We had to make a bridge over one end, so that fire engines could gallop bravely into the inner courtyards in case of fire. This bridge was generally packed with Members of Parliament gazing into the hole. At a meeting with the Ministry of Works, their surveyor said that the Members had complained to him that whenever they looked down, the men in the bottom were always standing around doing nothing, whereby my building manager retorted 'We have a serious complaint from our men, they say that whenever they look up from their work, there are always dozens of Members of Parliament who seem to have nothing to do but stand watching them, and the men complain that they are helping to pay the MP's salaries' and that was the end of the matter. Throughout this period, 1945-1950, the Labour Government governed with four Ministers of Works in four years. Rationing continued, not only in food and clothing but also of steel, paper, timber, petrol and other vital needs. At one time petrol ceased entirely, except for special exemptions.

In 1948, the ICE nominated me as their Governor for Westminster Technical College, run by the LCC. It also included a famous Hotel School and some of the Governors were Directors of the Savoy, Gordon and other hotel groups. This added spice to life, especially as the students ran a pleasant restaurant in which to test out their efforts and practice the arts of waiting and taking the money. In 1949, I was elected to the Council of the Institution of Civil Engineers. This was a surprise, as at that time the only other contractor's man was Colonel Norris, who wrote *Bridging the Years*, a brief history of civil engineering. To my great pleasure Wynne-Edwards was elected in 1950 and we both served for 20 years, during which we tried to liven things up, but in a constructive manner.

1950 marked the culmination of a period of much change, both professionally and industrially. During the war the ICE formed a panel of senior Members willing to lecture to colleges and Universities on the lessons gained by experience. I was co-opted and for a number of years lectured more than once to every college and polytechnic around London and also to Cambridge, Glasgow, Birmingham and Southampton Universities. My three papers presented to the ICE at this time once again gave away practical experience for the benefit of members. These were repeated by request at Stockton, Birmingham, Manchester, Southampton, Glasgow and other local associations of the Institution. Writing papers and lectures is harder work than realised by those who have not tried it, most of mine were by request and not due to any urge to eloquence. As a result of this conscripted work I was awarded the Fellowship of the City and Guilds Institute, which is awarded 'to those who have been trained at the City and Guilds College and have in the opinion of the Institute contributed to the advancement of the industry in which they have been engaged in the form of professional practice or in the form of original research'.

In private life, we were particularly fortunate in escaping the 'generation conflict'. So much is written and talked about this that rising generations are made to feel that conflict is an obligatory stance that must be adopted. At this epoch, daughter and elder son were both students, of Architecture and Civil Engineering respectively and we (or they) kept open house for their many friends, especially for those 'in digs'. We had parquet floors in both 'reception' rooms and many a dance was held, in the days when dancers clutched each other. There was Scottish dancing on winter Sundays as well. Sometimes breakfast was crowded by many overnight sleepers and their friends were accepted as our friends without criticism and remained so. My daughter was asked 'Do you get rid of your parents when you have a party?', she answered 'Good heavens no! They do all the work!'.

While at college Edmund spent three years in my garage building a 14ft day-boat with beautiful craftsmanship, while my car stood out in the snow, but it was worth it. The radiogram was wired to loud speakers, in the kitchen and other rooms, and records were filed under dance, classical and washing-up music.

11

DOWN TO THE SEA IN SHIPS

One of the pleasures of contracting work is the call for versatility, as the work can vary so much. Until 1930, I was concerned with the 'earth beneath and the waters under the earth'. At Battersea came involvement with the Port of London Authority, which lasted off and on for many years. At Dagenham there was much activity afloat with piling barges, floating cranes, tugs and diving boats, while sailing barges and dumb barges brought in almost all our supplies, so the force and habits of tides and currents had to be studied. When Soil Mechanics Ltd got going, much drilling had to be carried out in rivers and estuaries. This produced mooring problems and the hiring of pontoons and tugs from local pirates. When we were boring in the Usk at Newport, the tidal range was so great that the gang had to keep taking off lining tubes as the tide fell, as the water was going down much faster than the borehole!

Then, one day in 1948, the Iraq Petroleum Company called for an investigation for a new oil terminal off the coast of Syria at Baniyas. This was pompously described in my paper on site investigations:

> A site exploration for a new harbour was required off the Syrian Coast for the Iraq Petroleum Company, Ltd. The conditions called for drilling from floating craft to a depth of 70ft below sea-level in water of up to 40ft depth. The coast is exposed and if wind springs up from the south-west it is necessary to run 25 miles to shelter. The proposition of pontoons with tugs in attendance was considered. It had been found that a pontoon of 75ft by 35ft, as at Southampton, would have proved too small and, in order to avoid the hazards of towage, it seemed more logical to use a self-propelling platform. Two tank landing craft, ML3 and ML4 were purchased, each 180ft by 38ft by 10ft deep, with 4ft draught.... The boring rigs were welded to the ramp so that, when lowered, they were immediately in position.
>
> The crews and foremen all lived aboard and the Lebanese crew carried out the boring work. Three-inch wash borings were used to prove the depth of soft material and then 6-inch shell-auger borings were carried out to confirm results in selected places.
>
> The sea will never allow 'temporary' work and by the provision of these craft it was possible to work safely and rapidly. Within 18 weeks 127 wash-bores and forty-two 6-inch borings were carried out.

The expedition was also reported to the XXI Club in a cryptic but lighter vein:

> It would take too long to describe the greatest naval affair since Matapan,[33] whereby Soil Mechanics Ltd bought two tank landing craft from an Egyptian in Suez for a vast sum provided by the IPC, for carrying out borings in the open sea off Syria. Appointment of 'ships husbands' and the obtaining of 'carving notes' are mere trifles in a two-month battle by cable with Banks and Ministries.

---

[33] 27-29th March 1941 - when the Allies and Italians engaged in a naval battle off the Cape of Matapan, Greece and the Allies carried the day.

> We now have two White Russian skippers and only fear capture by the Irgun Navy or commandeering by the Syrian Government for the invasion of Tel-a-Viv. Fancy living to see the day of serious discussion that British Registration [of ships] may be less protection than Lebanese.

Ex Sub-Lieutenant Birkett took charge and bargained for the ships and engaged the crews. He got to sea the day before the 1948 war broke out between Israel and Egypt, which as usual closed the Suez Canal, and on the way to Baniyas evacuated the Iraq Petroleum Co. staff from Haifa. Irgun was the (now forgotten) Israel terrorist organisation for terrorising the British Authorities and I became 'ship's husband', who, I am told, goes to prison if his spouses misbehave. At the same time a rival firm were attempting to bore higher up the Syrian coast by pontoon but were cast ashore in a gale. When I congratulated Birkett on his remarkable success, he revealed an unsuspected talent. He had started working as a Cadet-Officer with the Union Castle Line at sea, but when, after two years, he fell into a hold and was badly damaged, the Line paid him off. I was able to use his talents again on sea-going work some years later, based on the Baniyas experience.

In 1950 Uncle Joe Stalin in Russia had become so threatening that re-armament became vital and new defences were put in hand. Just before I became a director, Mowlem got the contract to make a new anti-submarine boom from Southsea across Spithead to Seaview, Isle of Wight. This was highly confidential, although extremely visible. Steel box piles had to be driven in line at precisely 12ft apart, so that nets could be hung from them. The works director who had originally tendered for this contract had omitted to ascertain that all the floating plant, run from Basin South branch, was fully committed in the Thames and so two timber barges were found in a hurry which were almost useless. The intention had been to use one tug to tow out the pile barge and then go back for a second barge full of steel piles but it took two tugs and alot of time to overcome the tide-race at the mouth of Portsmouth Harbour and trouble arose over mooring. The men who were huddled on the deck were miserable. The first ten piles cost £1,000 each and the Admiralty were about to cancel the contract.

My first job as director was to take over this forlorn hope. Having taken one look, the experience at Baniyas came to mind and self-propelled craft were the obvious answer. With the help of the faithful Birkett, we found a German seaplane rescue ship near Borstal in full working order. This had a thirty-ton crane which moved back and forth along a flat deck, and it also had separate cabins for twelve crew. We named this craft *Ebury*. Next we traced a tank landing craft Mark 4, almost new, in the yard of a well-known individualist, Vernon, in Chichester Harbour. After inspection both craft were bought at once for £10,000 each and the second one named *Grosvenor*. My wife reminded me about a Captain Clayton who had taught our son Robert to sail near Teignmouth the year before. He came on a three month expectation, twenty years later he was still in command of Mowlem's fleet. Clayton brought along with him another with a Master's ticket, so we paid off all the disgruntled civil engineering labour and they produced young Devon seamen in their stead.

The ships were sailed round to the Camber Dock at Portsmouth, where they were brought up to Board of Trade requirements and twin piling frames were fitted to the sterns at exactly 12ft apart. The crews were able to sleep aboard and have hot meals in comfort and so were available to suit the tides. Each ship carried its quota of piles aboard, the first time out each drove fifteen piles in one shift and later got up to thirty piles a shift. Each ship had its own mechanical mooring winches, as mooring is the secret in such work. Thus the expected loss of £100,000 was turned into a profit by spending money in order to save it.

The affair was so 'secret' that we were forbidden to show photographs of the ships if the row of piles was visible. However in spite of the crisis, we were told to leave a gap

37. M.V. Grosvenor *driving box piles for an anti-submarine boom, Spithead, 1951*

of three piles for the Southsea Regatta. Three years later, at the Coronation Naval Review, the sister ship of the Russian cruiser with which the unhappy Commander Crabb[34] was involved, could be seen moored alongside this secret affair.

These ships bred much work, such as similar booms in the Thames from the Isle of Sheppey, piling for the new Esso jetty at Fawley and alot more beside. On another occasion Shell wanted to install heavy concrete fenders in front of their installation in Southampton Water. These were cast in Basin South Yard and *Ebury* loaded them onto her own deck, sailed round from the Thames and in 48 hours placed them with her own 30 ton crane. Later on, the Admiralty wanted to place radar beacons forty miles out to sea on the sand banks beyond the Thames Estuary. After discussions they designed them around the *Grosvenor,* which we equipped with shear-legs made up from parts of a 15-ton derrick to do the lifting. The beacons were built up from welded steel tubing with cross bracing and with tube piles in guides sliding along each leg of the tripod. *Grosvenor* carried three steam boilers in her hold as well as the final top section. A steam piling hammer was placed on each of the tube piles and connected to the boilers. The ship then put to sea carrying the tripod suspended from the shear-legs, until the correct spot was reached - indicated by an Admiralty launch equipped with Decca[35] positioning instruments. The steam valves were opened and all three piles driven into the sand to anchor the tripod. The shear-legs then placed the top section and so back to Sheerness all in one day. Eleven of these beacons were placed in the Thames and then *Grosvenor* sailed round to Liverpool, where she placed some more beacons in the Mersey Estuary.

38. M.V. Ebury *off Milford Haven*

In 1958, by which time I was in private practice, while we were organising the first investigation in the Channel (see Chapter Fifteen), I took M. Malcor to see Wimpey boring at sea off Dungeness from a tank landing craft, which we used later for our Channel borings. In 1963 Wimpey bought the *Grosvenor*, which put up a fine performance in the Channel for the large scale investigation. She was a welcome sight. In 1959, the insurance assessors, Toplis and Harding (no relation) asked me to inspect fire damage to the new Esso jetty at Milford Haven. Here, when the very first tanker

---

34 A naval diver who disappeared in 1956 in unexplained circumstances: http://news.bbc.co.uk/onthisday/hi/dates/stories/may/9/newsid_4741000/4741060.stm
35 The low-frequency radio navigation system in use in the English Channel from the war years to 2000.

had come alongside, a hose had broken and burning oil floated around the ship and under the jetty. Luckily the tide was in, so only the underside and the top of the piles had been damaged. Precast concrete slabs had been used instead of the customary timber shutters and the reinforced concrete deck laid upon them. This meant that these slabs accepted the rage of the fire and the deck was undamaged. Mowlem had designed and built the jetty and had used 'my' ships to drive 120ft long concrete pre-stressed tube piles. Through this, I met up once more with Captain Clayton and had a most enjoyable cruise in *Ebury* around Milford Haven.

12

COMPANY DIRECTOR

Back to 1949, when Glossop, Golder and I were legally appointed Directors of Soil Mechanics Ltd, a most unusual step in our feudal system, as it had been generally accepted that director's names had to begin with 'B'. In the summer of 1949 Eric Burt, who was only in his fifties, then died from a stroke. He was a charming man, even if one disagreed with his ideas, especially on mechanical plant. Subsequently, in the spring of 1950, to our astonishment, Glossop and I were also invited to join the Board of John Mowlem and Co. This produced a major change in life.

The boards of many civil engineering contractors are usually composed of experienced members of the firm who are really full-time managers of various sections of the work of a firm. This is quite different from a board made up chiefly of directors who appear once a month to draw a fee and add a name to the prospectus. Becoming a director brought many fresh responsibilities, as well as the continuance of existing ones. One was to take over the interviewing and recruiting of engineers and posting staff to fresh contracts as and when they were obtained. One Scottish Secretary had a chartered accountant's theory that, if there was one man too few, then the others worked harder. This is a fallacy. If there is one man to spare, it is then possible to begin to move the 'pieces' about the board as crises arise and postings have to be changed. If a company director has any conscience or imagination, he will remember when he makes decisions that the careers, livelihood and domestic happiness of many families depend upon the continued success of a firm. Civil engineers have to go where the work is and in days of housing shortage and domestic ties a little extra thought can sometimes ease employee problems when deciding who should go where.

There was much work to be done in many forms. Dr Golder managed Soil Mechanics Ltd., but Glossop and I were still involved in its specialities and problems. My parish still continued to cover all tunnelling work:

> I now find myself a Director of John Mowlem and Co Ltd which is an event in its history as it is the first occasion as far as I know in 120 years when one of its staff has been elevated.
>
> This occurred last May and shortly after we had put in tenders for the Camberwell Tube Extension from Elephant and Castle to Camberwell Green. Four contracts were sent out and we were apparently lowest for two of them, amounting to £2,000,000 and we were interviewed and all set to go. It transpired that the engineers' estimate was £3,500,000 for the lot and the lowest tenders in very keen competition totalled £4,500,000. It was then sent to the

> Minister of Transport, Mr Barnes and after sitting in his basket with a cup of tea on it for five months he decided (coupled with Korea etc) to postpone it and our tenders were declined.

It is a serious thought that if the extension mentioned in the copy of the letter above were to have been carried out in the 1980's, it would cost well over ten times as much. In those days, the Federation of Civil Engineering Contractors arranged for all tenderers to object to clauses in the contract which were considered unfair and which should be discussed. This arose in the interview with the new works engineers of London Transport, accompanied by the consulting engineers for the schemes. This dialogue took place:

> 'Mr Harding, your firm has for many years accepted this clause'.
> 'Sir, in those days your Lord was Lord Ashfield. Now that your Lord is Lord Latham that is quite a different state of affairs'.

Near Oban, a large rock tunnel had to be driven to form a chamber to receive the end of the first Trans-Atlantic Telephone cable. Test borings had been made, but as so often happens the tunnel was moved some distance away from the boreholes and no fresh boreholes were made. And again, as so often happens, the rock turned out to be in quite a different condition so the authority was needed for the payment of very heavy steel arch ribs from the consulting engineer. The resident engineer would not take the responsibility, so I rang up the senior partner of Sir William Halcrow and Partners and told him that the ribs were essential. He replied, 'If you say so, that is good enough for me'.

I became much involved in the running of Mowlem (Scotland) Ltd, which called for many trips to that beautiful country. One interest was the driving of a series of approach tunnels for the Scottish Coal Board in a number of coal mines in the Midlothian and other coalfields. At the end of the war, the firm of Marchon Chemicals had been founded at Whitehaven by Sir (later Lord) Frank Schon and Mr Marzillier, who had both come over as refugees from Hitler in about 1938. They asked the Labour miner-Peer Lord Adams where they could find a supply of Anhydrite. He replied, 'You have built your factory on top of the biggest deposit in this part of the country'. So the Marchon firm started to drive twin sloping adits down to reach the deposits, but their rock tunnelers got into deep trouble when they met soft ground in which they were not experienced. Mowlem (Scotland) took over the tunnelling and used rings of bolted concrete segments in the loose ground and then reverted to drilling and blasting with steel ribs as supports when the ground changed back.

I became involved in Rattee and Kett, an old firm that had been long established in Cambridge and carried out expert building work in medieval buildings, colleges, ecclesiastical buildings and cathedrals. Both Rattee and Kett were said to have been Territorials and when 1914 had come and they were mobilised, they shut down the firm except for one young medically exempt carver. He was told to spend the war travelling around and studying all that could be learned about medieval and later methods of construction and use of materials. John Mowlem and Co duly bought up the firm and this man became the acknowledged expert on such matters. Rattee and Kett repaired such monumental work as King's College Chapel and carved the new Altar piece for St Paul's Cathedral, as well as other famous works. This brought about many visits to their work and contacts with Mr Dykes-Bower, the architect famous for this class of work and the Guardian of Westminster Abbey. Management courses and systems are now the fashion, but we had to steer our course by the light of nature and the experience of what a board looks like from below-stairs.

One function of directors was the 'taking' of tenders. At that time, Mowlem had no estimating department and handed out tenders to one or two elder men who did nothing much else. Often, tenders were sent to agents in charge of existing contracts to do as best they could, while still trying to control works, often with day and night shifts. Just before I left the firm, an embryo tender department was being developed.

The 'taking' of a tender took the form of discussions by directors with the estimator at almost the last moment, as often only three weeks or even less was allowed by the client for submission. Decisions would be made over the choice of plant and methods of doing the work and whether some prices should be raised or lowered, thus a director takes the responsibility from the shoulders of the estimator. One hundred tenders of all sorts and sizes were dealt with in a year and usually only ten would be successful. Clients would sometimes spend weeks or months before deciding and there was no way of knowing what future work might be. Even if tenders are unsuccessful they form an 'Exercise without Troops' and provide experience and an idea of what is going on.

As well as the other companies mentioned, Soil Mechanics Ltd still had to be directed. The very large building department for carrying out such major works as the New Lloyds Building[36] or rebuilding the House of Commons also needed direction, especially when deep foundations were involved. Then there was the organisation which worked from Basin South, Royal Albert Docks and later from Byng Street on the Isle of Dogs. This ran a widespread maintenance contract for the Port of London Authority, a great deal of works in the docks and up and down the River and also had a considerable fleet of floating cranes and piling craft. This fell into Glossop's parish, but when he spent many months in Persia I looked after this area. Over the years, one has worked in every dock from St Katherine's to Tilbury on many occasions. To cover this I did not aspire to the customary Bentley of other firms but drove myself for many miles all over the country in a Humber Hawk, although a car with a driver was available when needed.

Voluntary Work

If people are unwise enough to prove capable of making helpful suggestions, they become fair game and fodder for committees. Sir George Burt had carried out much voluntary work such as Chairman of several Government Committees, president of the Federation of Civil Engineering Contractors and Chairman of the Building Research Board, of which he had been one of the prime movers. He encouraged me to accept requests for help, although Mr E. B. Beck deplored such extra-mural activities.

So, from the start of the war and afterwards, I was involved in much voluntary work, much of it on behalf of the Institution. In 1950, I was asked to become a Governor of the Northampton Engineering College, which is in Northampton Square, Islington and was later lifted to become the City University. I also gave a number of lectures there on practical experiences and found the staff and students most receptive, so it was a very worth-while affair. The Governors met in various City Halls, which was better than the glazed brick basements of the Westminster Tech. At one meeting, the Principal asked for our agreement for the engagement of another lecturer. From the depths of personal experience, I asked, 'What steps will the Principal take to ensure that the candidate can deliver a simple, comprehensible lecture?'. He replied indignantly, 'I have never been asked such a question

---

36 An earlier building than the Richard Rogers one, it opened in 1958.

before!'. All the other Governors looked at me in shocked silence. When I told this to a Professor-friend he said, 'We always make our lectures obscure in order to make the student think'. But if the student cannot understand what is being pontifically uttered, what is he to think about, especially in a technical subject? Representing the ICE on the Regional Advisory Committee on Technical Education brought me useful contacts with heads of lesser Colleges and Polytechnics, who proved to be sensible and devoted to their function of teaching and not side-tracked by ambition to research. I do not think that our advice was ever taken. but that is the fate of committees who only advise. In 1953, I was asked to join the delegacy which acted for the City and Guilds College and in 1955 became a Governor of the Imperial College of Science and Technology, which includes the City and Guilds, as well as the Royal School of Mines and the Royal College of Science. This is not a position I should ever have visualised when I was there as a struggling student. I resigned from the Northampton, as the Imperial College was starting on a vast re-building programme.

My workload kept increasing. I had given lectures to all the London colleges more than once and also to Cambridge, Southampton and Birmingham on behalf of the Institution, as well as writing papers, so the XXI Club letter of May 1951 shows a frame of mind:

> Having determined to give up vocal performances and written ones as well, I have been forced into writing a joint paper with Glossop for the Building Research International Conference and another for the Conference of the three Institutions on the practical training of civil engineers.
> I have also succeeded Holbein[37] on the Building Research Board and spent a bewildered morning listening to John Laing and Professor Bernal, the well-know 'commy-boy' carrying on the business of the thirty or so people present almost single-handed.

This letter also said that I had spent an hour at Birmingham University delivering a joint paper with Dr Golder to the British Association. In the discussion, I got carried away and found myself recorded as saying that 'Nature abhors a differential equation'. This lines up with the saying of Doctor Johnson, 'Human experience, which is constantly contradicting theory, is the great test of truth'. I quoted this in my James Forrest Lecture, and it annoyed Professor Pippard quite alot. In 1952, at a Council meeting, the minutes of the Publications Committee had recommended that Dr Karl Terzaghi should be invited to deliver the 1953 James Forrest Lecture on the 'Development of the Science of Soil Mechanics in the Last Decade' – or if he could not do so, to invite Mr Harding. On being confronted with this I exclaimed that, 'This is the sublime to the ridiculous,' but in the end I had to do it. What this meant can be shown by Lord Penney, who was later asked to speak at the ICE Dinner. When told to speak for ten minutes, he replied, 'That means ten hours work'. This is not a bad estimate of the labour required to compose something acceptable.

About this time, that vigorous body of City and Guilds ex-students, the Old Centralians, had their own Symposium, with Sir Arthur Fleming talking about electrical and Sir Frederick Handley-Page talking about mechanical engineers. Sir Frederick was a terrific and cheerful character and he spent much of his racy talk in trying to take the micky out of civil engineers. When my turn came, I said that 'We are one of the two oldest professions, without which Sir Frederick could not rise from the ground'. So honours were easy.

---

37 Arthur Montagu Holbein - nickname 'Bean'.

12. COMPANY DIRECTOR

I also sat on the Contracts Committee of the Federation of Civil Engineering Contractors and so met the leaders of the top firms, who showed a high standard of integrity in all the discussions. The ICE Council used to meet every month and members graduated through the many committees, ending with the General Purposes Committee. This work and also many stints on the Benevolent Fund Committee, took up time which had to be made up after hours. However, this work brought me many outside contacts and also helped the reputation of the firm. In 1955 I was elected a member of that rarefied body, the Smeatonian Society of Civil Engineers, founded in 1772 and which was restricted to 48 Members and 12 Gentlemen; the latter included Lord Mountbatten and the Duke of Edinburgh. There are six dinners a year with toasts and no speeches.[38]

Now we come to the Engineer and Railway Staff Corps,[39] Royal Engineers, Territorial Army, which can be found in the Army List. This unique body was formed in 1865, when the volunteers were being enrolled to repel a possible French invasion. The corps is 'all rank and no file' as it consists of 10 full Colonels, 20 Lieut-Colonels and 30 Majors. These are chosen (in the days when enterprise was free) from general managers and chief engineers from railway companies and dock companies, partners or consulting engineers and directors of contractors. These are given honorary ranks and in the return, from its collective wisdom and experience, the corps gives free advice to the War Office whenever asked. The corps drew up the mobilisation plan for the 1914 Expeditionary Force well before it was put into effect. Later, I headed a sub-committee to advise what mechanical and other plant should be bought and stock-piled for handling bulk cement for stabilising airfields rapidly while advancing. We advised that such plant would be cumbersome and subject to rapid decay and obsolescence and the most useful way to spend the money would be to stock waterproof bags as these would be of use to the troops after they had been used. We received the thanks of the Engineer-in-Chief for saving the country many thousands of pounds. I held the dizzy rank of full Colonel (strictly Hon).

There is no factory floor in civil engineering construction but there is a military analogy, as the 'troops' move from job to job, directors become generals and the engineers are the officers, while the foremen and gangers are the NCO's. Shop stewards are not so numerous and they might be likened to the political commissars so popular in the Russian Army. Field Marshal Lyautey once said that the essential quality in an officer is – gaiety. This is sound advice, as there are two ways of directing or being an agent. One, which I prefer, is to try to inspire a feeling of cheerfulness by encouragement and only scolding when called for. The other, which I have seen used and dislike, is to put the engineers on the job at a disadvantage immediately on entering the site by saying 'You have too many men' without knowing how many, in order to exert superiority and cover up ignorance.

Problems of Plant

In May 1951 one of my letters says that 'one is spending much time in brooding on the question of plant, on which no two experts think alike and on which mechanical engineers mislead themselves and others quite as much as civil engineers'. This cynical view, based on many years of observation, could trigger off a thesis on 'Contractors Plant and How Not to Handle It'. A mechanical engineer can be so deeply interested in what makes the wheels go round, that he has less interest and experience in deployments, whereas the civil engineer

---

38 HJBH was President in 1973.
39 Now called the Engineer and Logistics Staff Corps (from 1993).

is dependent upon their proper use and exploitation.

Over the last century, there has been an accelerating revolution in mechanical plant. Before 1914, steam was the great and almost the only prime mover, if we dare to forget the faithful horse and the use of massed man-power. Steam was beautifully simple. It was used to drive pumps, locomotives, cranes and steam shovels. All over the world, vast works like the Nile and Indian dams and the Panama Canal were built with no other plant. This meant that 'steam bosses' and their 'black gangs' only needed to know how to fit pipes, scrape bearings, re-tube boilers and stoke. Electric power was seldom used, except in cities. After 1918 changes started, hastened by the war. The petrol engine and, later on, the diesel engine, made new types of machines and methods possible. They drove portable generators so that electric drives could be used far from permanent supplies and the spread of oxy-acetylene burning and electric welding changed designs and demolitions. Hydraulic power, before the spread of electricity, had been widely used in cities and docks from high pressure mains, which were laid in city streets. This form of power subsequently revived with the change from water to oil in the systems and once more is used to actuate many machines. The development of the diesel engine wrought even greater changes, with a continuing evolution in earth moving plant from the caterpillar tractor to the multitude of modern equipment and vast Euclid-type trucks. Before the first war, Mowlem had a steam boss who ruled over a small and simple kingdom. His son succeeded him, and after the second war, so did his grandson, but to a rapidly growing empire. This meant that the third in the dynasty had a difficult task in trying to keep up with such fast moving changes. None of the three had any workshop experience or mechanical education, but had had to learn the hard way. Dynastic successions are not very desirable and can be no kindness to the later successors.

During the second war an American colonel said at a meeting at the ICE, 'When your machine begins to go clonkety-clonk, sell it to a rival'. Eric Burt, the director in charge of plant, was often the rival who bought it, out of misplaced economy. Then, Eric Burt died and the directors decided to have a Plant Committee made up of Kenneth and Stewart Burt, with myself as chairman. We decided that two men, not one, were needed to manage the Plant Department. One was to be a civil engineer, with experience in running contracts and who should be at the centre, administering, and to whom contract managers could confide their troubles. This was John Brass, who in the war had got away from Calais with the remnants of Tommy Burt's company of sappers. He had then become a Lt. Colonel in the Western Desert, maintaining Royal Engineers plant. It also turned out that, after taking his degree at the City and Guilds College, he had served a term in the works of Reavell at Ipswich, so was experienced in shop floor problems. The second was McGibbon, a highly qualified mechanical and aeronautical engineer, who had also designed for Frazer-Nash. He was free to visit sites, to see for himself that the right plant was used and to make sure that the wheels went round. We then set about quietly revolutionising the whole set-up. That wise man, Sir Godfrey Mitchell, who built up the firm of George Wimpey from nothing, once said that the unkindest thing to do to a man was to promote him above his ceiling. So the 'grandson' was given a handsome rise in salary and the title of head of the newly organised corps of plant inspectors and face was saved. He was better able to use his considerable store of energy and was relieved of an impossible task. Within a few years the firm had seven fully qualified mechanical engineers and an equally valuable electrical one, where no qualified man had worked before.

We drew up a five-year plan for purchasing a balance of new machines and, as this endeared us to the manufacturers, we were able to rise from the bottom of the list of customers. The increase in morale on contracts when they received brand-new machines and also good support from the home base was a pleasure to see. Then the future of plant depots became urgent, a new properly designed depot was vital. My basic decision was, 'If you must build, then build a property, not a native-village of tumble-down valueless huts'. So a study was made of the London Transport bus and train repair shops, contractor's depots of Laing and McAlpine and the shops of the United Steel Company, among others.

In order to keep discipline and good house-keeping in a depot, it is essential to be generous in space. Not only machines, but many types of inanimate materials have to come and go and space must be left for their return. We found an ideal site of twenty-five acres freehold at Leighton Buzzard, where the ground was a 'natural' for soil stabilisation to provide hard standings which are vital. Unfortunately the board insisted on transforming their existing leasehold area at Welham Green, which made matters much more difficult and expensive.

The design and building of Welham Green Depot is a saga of its own, but for this story the bones of the situation will be enough. We appointed Hammett and Norton as our architects and they dealt with the Hatfield New Town architect over permissions. We built a large main repair shop with sliding doors along a length of each side, so that massive machines could move into each bay without blocking up the entry, with powerful overhead cranes to lift the bodies of such machines, a lesson learned at Dagenham. There was also a platers shop for fabricating steel work, storage sheds to keep machines out of the rain and a jenolite type cleaning shop, where concrete mixers and the like could be dipped in baths of chemical which cleared off concrete and left a priming coat. Ten twelve-ton Scottish derricks, with 120ft jibs, travelled along a raised track and these unloaded and stacked steel piles, timbers, airlocks and all the innumerable articles which have to be handled. The ground was made of glacial clay so the area to be stabilised had to be dealt with by importing many yards of sand and gravel (unlike the forementioned area at Leighton Buzzard).

As the adjoining hill drained off into our sloping area and in order to protect ourselves, we appointed John Balfour and Partners as consulting engineers for the big drainage problem. To the anguish of our accountant, they specified 2ft diameter pipes, but in the first storm the pipes were running full bore, which once more shows the need to avoid reckless economy. One of the side-lines or spin-offs was the fact that there was space in the main building for a training workshop for apprentice fitters, who could also attend Hatfield Technical College. The hostility which the creation of this depot raised in the board room transformed into pride in the finest contractor's depot in the country, as described in a copy of Project South from 1973. Its value had trebled and I take pride in having been responsible for its construction, with the help of collaborators who brought intelligence and logical study to the problems to be solved and the wisdom to realise that the way to save is to spend wisely.

Coronation

In 1953 the Duke of Edinburgh agreed to chair the annual meeting of the City and Guilds Institute and to hand Diplomas to those of us who had been elected Fellows (FCGI). He came up the stairs of Goldsmith Hall with Sir Frederick Handley Page, the Presiding Chairman, and Holbein sat at the top in a wheelchair because he had broken a leg.

> Sir Frederick Handley Page - 'Sir, this is Mr Holbein, the Chairman of our Education Committee'.
> HRH - 'In a very suitable position if I may say so!'

Miss Letitia Chitty, a famous researcher at the CGI went back to her colleagues to say 'I have been watching a student hand diplomas to professors'.

XXI Club letter, April 17th 1953

It was most impressive to observe and listen to the Duke of Edinburgh at the City and Guilds Institute meeting, brisk and to the point and full of wit and humanity in off-moments.

Following a slight connection with the Institute, I was invited with my wife to a sherry party at the City and Guilds Art School in Kensington which was pleasant and interesting. I left my coat and hat in a deserted studio. On leaving later in the evening I stepped briskly into the studio to find it brilliantly lit and rows of students working away on a (for an Art School) remarkably beautiful model bathed in light and nothing else. I retreated far too hastily and sent my wife, who as an Old Slade-ian was of sound mind and able to recover my coat and hat.

We have been sold two seats for the Coronation on the centre of Parliament Square as a member of Council of the Institution of Civils and hope to meet Bean among the Elsans in the same stand. These stands are being erected by John Mowlem and Co. Ltd who have also done the work inside the Abbey and the Annexe for the fourth Coronation running. The sanitary arrangements are formidable. Last time the Queen of Somewhere refused to share her throne with anyone else which caused confusion.

I am told that a High Commissioner went in the wrong uniform and was mistaken by a Countess for an attendant. She rushed up to him and said 'I want your help', and dragged him into the 'Noble Ladies' compartment and said 'Look there!' and behold her coronet had fallen into the pan. With great dexterity he drew his sword and with a combined lunge and parry retrieved it and handed it back on the sword's point.

As some of you know, my junior son Robert went on a visit to Longleat in the Christmas term and was shown the Coach of the Marquess of Bath and the footmen's uniforms. Last term he and a friend looked up in Debrett's how to address a Marquess and wrote,

'My dear Lord Marquess. My friend and I are 16 years old, 5ft 10 ½ inches high and used to horses. We would like to offer our services to you as Footmen for the Coronation, paying all our own expenses'. The reply was an invitation to go over for a rehearsal. The dresses fitted them perfectly and they did a two-mile trip on the back of the coach (pulled by brewers horses from Birmingham) and got the job against many other offers. They went over again and had another try-out to complete the equipment and were very well received. The Marquess then wrote, 'Dear Robert, Could you and your friend possibly grow side whiskers and your hair long at the back for powdering? It is, I fear, alot to ask of you as it may not go well with your fellows (or Headmaster)', or something to that effect. This dates it.

12. COMPANY DIRECTOR

October 10th 1953

In my last batch I expressed the hope of meeting Bean among the Elsans at the Coronation. We did better than that. It turns out that he and Martita Hunt had been allotted seats next to us. You can imagine our delight on seeing him loom over the horizon (Martita doesn't loom). Some members may have seen our photo in Picture Post and Life, with the Peers queuing up below and me diving into my wife's handbag for the rum, badly needed.

39. *Robert Harding and friend as footmen to the Marquis of Bath at the Coronation*

Well, the Footmen to the Marquess of Bath managed their duties very successfully. In our last number readers will recollect that my junior son Robert (16) and a friend volunteered to act as his Footmen. We brought them home, with their hair looking frightful but the Union Club Barbers trimmed it and the sidewhiskers to rights. On the Monday Lord Bath arrived at Putney bringing with him large tin cases full of uniforms and also liquid hair powder and was very charming. At 4:45 am on Coronation Day his Bentley arrived and took away the Footmen looking marvellous in yellow cut away coats, Robespierre hats, black breeches, white stockings and silver buttons, ropes, braids and fandangles. They also picked up the Marquess' butler and cook and dropped them in the West End where he had got them seats in a London Club stand! And so to Paddington GWR mews where a large and disgruntled brewery Coachman was helped to harness two very lovely steeds onto the coach. They then drove to Claridges and picked up Lord Bath and standing smartly to attention on the back of the coach, spanked at a smart trot down St James Street, Pall Mall and Whitehall among the cheers of the mob. Going down Whitehall they passed the Duke of Devonshire's coach, the crowd roared with delight and Robert said 'his flunkeys did look sick'. At this juncture Lord Shrewsbury in the third of the private coaches had reached a standstill. His Footmen (a brother-in-law and a horsey Major) had to get out and push but as the horses refused, his Lordship got in his car which was following and did the last 300 yards by car. Not so our side. They spanked round Parliament Square and the Marquess defied the Police and insisted upon being dropped by St Margaret's entrance. We were just able to see the coach drive off back round the Square with the Footmen really looking very smart on the back, and the coach with a magnificent yellow Hammer-cloth over the driver's seat. We were then able to relax, knowing they had got there. After the service we walked with Bean and Martita to the Marsham Street car park where the coaches were parked.

The coaches were directed by the Police to go back via Lambeth Bridge, almost to the Elephant and Castle, back up Waterloo Bridge, Kingsway, Oxford Street, Poland Street, Gt Marlborough Street, so we led them in our car which was labelled 'exempted services'. Lord Bath had footed it from the House of Lords to the car park, so having taken photos and had a crack or two we set off. It was really rather an amusing sight, the streets were deserted until we got to Kingsway.

We stuck in Gt Marlborough Street at a jam in one of the Mowlems barriers and the coach got a terrific cheer. One woman put her head in the window and said 'Oo are you?', Bath replied 'I am the Marquess of Bath'. 'Enjoyed yourself?' 'Yes, thank you – I hope you have too. Isn't this fun?'. Holbein and Martita got loud cheers as well, looking so distinguished!

Next day the Footmen were again picked up to try and get a colour photo but the weather was too bad. So Lord Bath dropped them at the Union Club for a much needed hair cut and shampoo and they returned to school. He sent them each a pair of gold cufflinks, engraved 'Coronation Queen Elizabeth 1953' and autographed the other link, or rather his signature was facsimiled.

We had some telephone brushes with the Press but managed to shake them off as publicity was not the intention of the Footmen. I forgot to say that Mowlems had an emergency gang sitting by in our office in the Abbey Cloisters and we had given them a television set and I got the Footmen into the shed where they saw the whole service and procession. A gold stick in waiting was so impressed with their appearance when arriving that they had great difficulty in not being thrust straight into the Abbey.

My senior son Edmund has now got a scrape pass BSc (what is good enough for father is good enough for him) and joins the Royal Engineers on Thursday. His 14ft Yachting World Dayboat on which he has worked for three years in the vacs is now finished and a really beautiful job. We sailed her yesterday at the Tamesis Club and then towed her back and hoisted her aloft to swing from the garage roof above the car.

I attended the Soil Mechanics Foundation Engineering Third International Conference at Zurich in August in lovely weather. Taking wife, daughter and junior son we flew by Viscount (a lovely thing and up to all that is said of it) and spent a week before the conference at Zurich and travelling about. We were there for 18 days in all. There was a superb swimming bath on the hill on which our hotel stood, overlooking the lake, with a wave machine.

> An enormous attendant disappeared down below into a chamber every half hour and then the waves started for five minutes. We decided that he jumped up and down so displacing the water. It added to the amusement. The conference was hard work, I cut one meeting and arrived later for the afternoon one and found that it had started with the announcement, 'Will Mr Harding of Great Britain please take the Chair of this session'. Mr Harding was in the swimming bath.

Back to Africa

Soon after the war ended, John Mowlem and Co. took a half-share in a contracting firm belonging to the United Africa Company in Nairobi and in due course they took over the whole of it under the name of the Mowlem Construction Company (MCC). One of its early works, as already mentioned (see p. 100), was to take part in John Strachey and the Labour Government's Ground Nuts fiasco in Tanganyika. MCC were the only firm to emerge from this fiasco with credit as their part was under Brigadier Eric Horsfield, who had been Chief Engineer to the Fourteenth Army in Burma, which speaks for itself.

The old directors were too old for travelling out there and E. C. Beck was grounded from flying by asthma so, soon after we became directors in 1952, Glossop made the trip. This was when all was peaceful and a trip to Tree Tops Hotel was *de rigueur*. Then the Mau Mau trouble started, so in January 1954 I set out on a trip to see how the firm was getting on. I had flown on short trips in Europe, but this was to be the first long distance one. I was booked out on the very first Comet service which had just started. However, the Comet fell into the sea two weeks before, so I had to take an Argonaut, a four engine propeller craft of leisurely performance. We stopped in those days at Rome, Cairo and Khartoum before reaching Entebbe. At Cairo, we were joined for the night flight by a group of large, loud-voiced men, who made felt their considerable weight and importance. The Air Hostess told me that they were a mission of Egyptian and Sudanese officials off to argue in Uganda on the water rights for the River Nile and the Owen Falls Dam. The British were leaving the Canal Zone and Nasser was taking over from Naguib, who had ousted King Farouk in 1952. Entebbe Airport was still in embryo form and the plane turned out over the vast space of Lake Victoria in order to come in from the south. This gave the illusion that we were to come down in the Lake, until the edge of the runway appeared below the landing wheels, on the edge of the Lake. The British Colonial atmosphere was in evidence.

Brigadier Horsfield, who was acting as father confessor to the young local directors, flew up to meet me and we duly paid a call on Government House to sign the book after a smart Askari of the Kings African Rifles had presented arms. There was no sign of Sergeant Amin and the Governor had left for England in order to explain why he had ejected the Kabaka of Buganda from the country. All seemed peaceful in my visit of a week, which is longer than some journalists take to pontificate about a country.

I had read up the history of Uganda in my *Encyclopedia Brittanica* (1911) and a rum story it told. It was only in 1862 that Speke was the first European to visit Mutesa the King of Buganda and very odd things followed, among them a civil war between Protestant and Catholic converts. There were frays with a French 'party' and much else before Lugard got it under control. Mutesa was 'regardless of human life and suffering and was consumed by vanity' and this was a prelude to General Idi Amin, a similar case. This may be a good moment to say that I was taken to see a bridge being built over a small river, where I met the Ugandan material clerk. Years later, I learned that not many years after my visit he had become President Obote.

I was shown work being done around Kampala, which was still being built up into a large town. Entebbe seemed to consist of comfortable spread-out bungalows for Colonial Office officials, but in Kampala most of the business and the large cars belonged to Asians, although whether Indian or Pakistani did not arise. I was taken on a special visit to meet an ex Italian-prisoner-of-war who had settled in Kampala and built a prosperous factory to make clay floor and roofing tiles. He was particularly proud of his hollow ones and showed us the special steel mould saying 'You squeeza da clay through da mould – just lika you make da macaroni'.

We were joined by Westacott, the managing director of our subsidiary, who had driven up from Nairobi, and then we proceeded to follow the trail that would later be taken by Idi Amin as he retreated from Kampala. The first stop was the town of Jinja and a visit to the Owen Falls Dam. We were entertained by Dennis Bertlin, who was running the consortium of contractors building the dam. His father had briefly been a director of the short alliance of Nuttall-Mowlem. This proved an exciting visit to a fascinating scheme vital for the progress, not only of Uganda and Kenya but also for the waters of the Nile. As Winston Churchill wrote in 1907 when he was Under-Secretary for the Colonies, 'What fun to make the immemorial Nile begin its journey by diving through a turbine'. The waters had started to dive through the first of ten turbines and as we arrived on the scene we found men sweeping up a lot of plate glass inside a switch room. A hippo had got in during the night and had put his rump through the plate glass door.

The dam had been designed by Sir Alexander Gibb and Partners and was being built by a consortium of eight British, Dutch and Danish firms, in which my close college friend, Jack Pain (of Dorman, Long and Co. and Sydney Bridge) had a leading part. In fact many friends of mine were involved, but I made a new one in Dr Olivier, the resident engineer who showed us around. As in other cases we at once got on together, with pleasurable reunions at later intervals, which led to my next African adventure nearly twenty years later (see Chapter Nineteen). The dam itself has a graceful wandering shape as it follows, and sits upon, the rock of the Owen Falls and there are good descriptions in the papers to the ICE, November 1954. A succession of large timber cribs, filled with rock, were successively laid from a travelling gantry, after their bases had been tailored to the expected shape of the rock. These supported coffer dams of sheet piling, sitting on the rock with murrum filling and rock rubble on either side. Once this was successful, the concrete work followed inside the coffer dams much as usual.

When wondering how to describe the local impressions, I stumbled upon my part in the discussion of the ICE papers which were always reported in *oratio obliqua* and sometimes turned by the Publications Department into Institutionese:

> Mr HJB Harding said that in January 1954 he had paid a flying visit to Uganda and Kenya to see works which his firm were doing there and had been able to visit the works at Owen Falls. They gave at first a sense of incongruity. That pioneering work in the heart of Africa was in the suburbs of a vigorous but rather unattractive town, with modern 'Gentlemen's Residences' on the road to the dam, which was fed by a main line railway and on a main motor road. The first impression was that the work was much closer to civilisation than any valley in the Scottish Hydro-electric Scheme.
>
> However, on second thoughts one remembered the diseases, the flies and the crocodile infested waters and while it was nice to be on a main line railway, much depended on

12. COMPANY DIRECTOR

where the other end of the line was situated….
(Then followed a hang-over from post-war feelings)
To consider these advantages it should be remembered by future readers of these Proceedings that the builders of the Uganda Railway in the early 1900s had strung out a telegraph line as they went along, so that, when 750 miles inland they required additional plant, all they had to do was to send a telegram to Great Britain and it would arrive at the railhead possibly 6 months earlier than it would have done in the conditions of 1950 and after. The fact that English newspapers and even Directors could now be flown out to the site in 2 days from England was a poor consolation for the long periods of waiting for materials to accumulate.

The week was spent in travelling around in what appeared to be a fertile and contented land. It is the custom of left-wingers to deride our colonial effort and talk of failure to improve conditions, but I saw the work of the firm busily engaged in making covered reservoirs for Kampala, buildings of all sorts, bridges, sewerage works, drainage, roads and much else, all the way from Entebbe to Tororo. These later figured momentarily in the exit of General Amin. The staff were pleased to have a visit from a London director, especially one who had 'risen from the ranks' and understood their problems, although he had not worked in outlandish places. The wives felt that their husbands' efforts were at least being reported upon, which is important for relations and for morale. On one interesting occasion I met a young Kikuyu trainee civil engineer, whose father was a Chief and who had sent his son to Uganda to protect him from the Mau Mau.

From Toror, Westacott drove me for 350 miles to his house outside Nairobi. On the way we passed Lake Nakuru which looked very pink in the distance due to all the flamingos, which became a popular feature of television travellers. We then stopped at Nakuru to pay a short call on my old friend from the Camden Tube days, Robert Barnes. I had kept in touch by frequent letters and in 1936 he had become my son Robert's godfather. On the way it seemed that the Kenya landscape was less overgrown and more open than that I had seen in Uganda, with the occasional giraffe wandering about. Then we ran out of petrol in the last few hundred yards of Westacott's farm road so all the Toto's and women folk came out to push us home. I was rather glad that this was not after dark 30 miles back, but neither Barnes nor Westacott seemed to bother much about the Mau Mau.

Next day I took up residence in the Stanley Hotel and spent a week, partly in the office getting down to details, partly looking at works and partly in paying calls on top people on behalf of the firm. I was much impressed with the standard of work in the modern large office building as the MCC had a well-qualified staff of foremen as well as engineers. None of the staff seemed to carry guns and I was told that most of the raiding was in search of food, arms and money. It appeared that it was those on the farms who were the ones at risk.

What really frightened me were the clerical and business types who came to the hotel for lunch and cluttered up their tables with all sorts of weapons, which I was sure they were incompetent to use. I found it a strange coincidence that in Uganda I had been ejected, under protest, from my room as I was told that it was required for the egregious Ernest Hemingway who had just crashed his private plane. I then found him next door to me at the Stanley Hotel. He was making much noise in the local press, but when I encountered him in the corridor outside my room, I took an instant dislike to him. Nairobi seemed calm and while I was there no-one troubled to remove the false nose that had been applied to the statue of Lord Delamere.

I soon found that when staff went around to tour outside jobs they all said that they were going on Safari. I was taken on a particularly interesting trip to see the Sasamua Dam, which was under construction. The French contractor who started it had abandoned the work, so as specified under the contract, all his plant was confiscated. This led to a long arbitration in 1958 in London, which I was to listen to twice as instruction for my emergence as an arbitrator. The Sasamua Dam was on the edge of the Bamboo Forest, which featured so often in the news of the Mau Mau and certainly it gave a very thick cover. The resident engineer was a romantic type, girdled with a large six-shooter on his hip like John Wayne. There were so many people about that it greatly amused our men, who carried no arms. On the way back, as dark was falling, we met a broken down lorry belonging to Africans who tried to waive us down, but I was told that this was a favourite trick, so we passed by on the other side.

The drive down and up the famous escarpment was interesting. As a boy, I had read all about the building of the Uganda Railway in my *Engineering Wonders* and had made a model of the carriage which was first used to carry the trucks up a thirty degree slope. I made it up the stairs using Meccano for my gauge 0 model railway. This Rift Valley is larger but not so spectacular as the Blade Gorge in the North Drakensberg.

I wrote some of my impressions at the time in a XXI Club letter:

> The Kikuyu women are a depressing sight, trudging under frightful burdens as the men make them do all the work. In Uganda, on the other hand, they are all fat, cheerful and riding on the carriers of their men-folk's super-Humber bicycles.
>
> The local English were pretty hot on the subject of the Colonial Office and Mr D. N. Pritt, QC. The natives assume that a Queen's Counsel to defend Kenyatta must have been provided by the Queen and are mystified.

I was very pleased with what I saw of this subsidiary firm's work and wrote back from there with my views. This letter has a little of the local colour of the time:

> We have a lot of work in the Kinangop. Here on the edge of the Aberdares you get a combination of the discomforts of Uganda with the threats hidden in the bamboo forest. We have agreed that if the wives of staff required to work there are not willing to go, they will be housed in Nairobi at the company's charge and protected. In any case we have arranged with the City Council that our staff and any wives who elect to go with them, will live in a camp which is part of the Council's camp. There will be at least 30 Europeans and consequently the likelihood of attack is very remote. The distances are so immense and the cover so good that this is a perpetual threat even if it does not materialise. As far as Nairobi is concerned, I was impressed by the cheerfulness and spirit of the ladies that I met, but they are undoubtedly under strain as they have to guard their children while husbands do their turn of Home Guard or Police Reserve work. The Home Guard, like fire watching, is on rota. The Police Reserve are on call and may be out for several nights running. All this is on top of their ordinary work in Head Office from 8am to 4:30pm and on Saturday mornings.

12. COMPANY DIRECTOR

I entertained the Nairobi staff and wives to dinner and noticed how pleased they were to be visited and to be able to talk about their work. The relations with the labour, both African and Asian were good and to boost morale I found that they had an all-in annual Sports Day at their depot.

On my return I was able to improve the organisation for sending out plant and materials when wanted. Some years later, at a meeting in London of the Geotechnical Society, Dr Golder revived something which I had quite forgotten. He told the meeting that, when our Kenya firm asked for a steam piling hammer for a jetty in Mombasa, I had sent out two. The local director complained that his job was being unnecessarily charged for more plant than he needed. Golder said 'Harding replied that "if you have one hammer it is sure to fall in the 'drink'. With two hammers neither will fall in. That is a well-known scientific fact".' Golder added 'We called this Harding's Law'. The second half of my trip was an individual effort, inspired by a public statement from the newly formed National Coal Board. They announced that their new deep shafts would be carried out by German firms, as no British firm was capable of doing so (except perhaps Cementation Co., because of their South African connection). So I determined to take the chance to fly on to Johannesburg and the Orange Free State's new goldfields, to see for myself how to do the work.

Much has changed in the years since then, so I will recall four trips which I took on Central African Airways in their more primitive aircraft, before dealing with the main objective. First I flew from Nairobi to Lusaka (then Northern Rhodesia, now Zambia) in an old Dakota with a good view of Mt Kilimanjaro, not forgetting that my friend Guthlac Wilson[40] and his wife had perished on the same flight a year before, when their plane hit the mountain. Our plane seemed to have been intended for a different route, as I was amused by a sequence of four signs to be lit up in an increasing air of panic: No Smoking; Fasten Seatbelts; Fasten Lifebelts; Toilet Engaged.

Lusaka was still very much in a Crown colony, as the 'wind of change' had not yet got up speed. Costains had just built a very modern hotel, but the rest of the town was small stuff. I think it was Laings who were bull-dozing out what looked like an attempt to build a wide imitation Champs-Élysée or Mall and modern buildings were under construction. The local papers were full of abuse of Southern Rhodesia. I had stopped off in Lusaka as I had brought Sir William Halcrow's report on a proposed Kafue River hydro-electric scheme (long before Kariba) and I wanted to find out more about it as it was 'up our street'. The Colonial Office engineers were helpful with details and models but advised that a quick visit to the gorge was 'out' as the wet season would prevent it. They were all working in wooden buildings while outside, a few Africans were cutting down the long grass around them to keep down the mosquitoes. They were using strips of hoop-iron and I gathered that they were not allowed to use Pangas, as these customary sword-like objects were much admired by Mau Mau for dealing with their victims. Otherwise all was quiet.

Next I travelled by Vicker's Viking to Johannesburg, with the pilot making heroic attempts to get down before a superb thunderstorm on our port bow caught up with us. I was met by my cousin and spent my visit staying on his small farm outside the city. During this stay I took two nights off to fly to Salisbury to spy out the land. I called on the Southern Rhodesian water engineer, who was their local representative in the Institution of Civil Engineers Council. We discussed such matters as were his pleasure and he showed me matters of

---

40 Co-founder of the engineering company Scott Wilson.

professional interest. I crossed the main street by leaving the pavement, walking through ten feet of unpaved earth (after rain) and then across the central tarmacadam section and then back through more mud to the pavement. This was a form of 'soft standing' possibly to save money. I then found myself outside a building which turned out to be the House of Parliament, as Southern Rhodesia was self-governing, up to a point. I had a sudden idea and asked an attendant if Sir Roy Wilenski (the Premier) was in the House. He went in and brought out Sir Roy, who sat down with me on a sofa for nearly an hour's talk. He thanked the firm for taking an interest in the country. I formed a very good opinion of him, far better than 'Smithy' his eventual UDI successor.

The impression that I got was that Salisbury was a cross between Cheltenham and Ealing. I had an introduction to a United African executive, who took me to his country house some distance outside Salisbury and then back for dinner at the top of the highest (sixth floor) building, with Scotch salmon, flown out from Scotland that day. He explained that the city services of water, electric power and even gas were coming on so well in the city that many were abandoning their widely spread outer houses and coming into the city, as it cost too much to take the services out to them. I did not see any of the much-televised Whites houses with their swimming baths, as I believe that they were a later development. The newspapers were full of abuse of Northern Rhodesia and the local politics were working up to much future trouble. Zimbabwe is the name of the ruins of a former civilisation. Let us hope that this is not an unlucky choice of name.

On my return to Jo'burg I was asked if I had travelled with the famous air hostess. It turned out that an airways official had given a hostess 'six of the best' for not strapping herself in. When this became known and of great public interest and amusement, he had to change to another airline. His name was McCoy. A Jo'burg paper quickly published an advertisement for a furniture store: 'The Real McCoy. Cane Bottom Chairs only 40/-'. Johannesburg is no thing of beauty but I had useful introductions, including those from Holmans of Camborne, to the six leading mining firms. Holmans' man asked me, in the famous long bar of the Rand Club, what I thought of the city. I told him that it had changed greatly since I was there fifty years before and left him baffled as I did not tell him that I had been six at the time. I told him of my remembering that the entrance to the Robinson Mine was a sloping Drift (we had been shown it) and of seeing the turn tables from which sections of ore were picked out, which proved my veracity.

It was particularly pleasant to stay with my cousin and his wife on his farm outside the city, where his farm boasted a home-made dam and where I was happy to find that playing tennis at an altitude of 6,000ft proved no trouble. I went to the local village post office and was told to go out and start again as I had entered by 'Blacks' instead of the one marked 'Blanks'. We also sailed in my cousin's Sharpie on the Hartebeeste Port Dam. His mother had flown out to see them a few years before by the leisurely and comfortable flight of Empire Flying Boats. The flying boats came down each night on convenient water for the passengers to spend the night in a hotel and they would stop off at Italy, in the Nile, on Lake Victoria and finally on this dam near Johannesburg. This is now long forgotten.

My cousin then drove me for a weekend down to Jammersburg Drift on the Celdeon River on the south border of Basutoland (as it still was before becoming Lesotho). This was where I had spent nearly two years (1904-06) with my uncle, Jack Robertson. My cousin was born there 10 years later, when his mother had married another Scot, who ran the mill and also opened up Basutoland. This was a fascinating experience to come back fifty years

12. COMPANY DIRECTOR

40. Bridge at Jammersburg Drift

later to a spot so clearly remembered. We climbed the Kopje, which had been the centre of the Siege of Wepener, as it was called. This had taken place around my uncle's farm and I found four more expended Lee-Metford cartridge cases where, as children, my brother and I had scooped them up in handfuls to use as toy soldiers. On one large boulder, which was part of a breastwork, was scrawled 'Pity the starving – (indecipherable)'. Everything looked the same except half size smaller than memory, but with many more trees.

On the Monday we had a rough drive in pouring rain across the Free State to the new goldfields. Unfortunately all the shafts had been completed, but we were taken at fast speed a mile down the Welkom mine shaft to see the opening up of the work. I took photos in the quarters of the African miners, who had beds, four to a room and seemed well looked after. The kitchens for the dining hall were worthy or better than the Savoy and our host pointed out the high standard of the vegetables and wished that his wife could buy as good. I took a photo of the diners, who gave me a big cheer and looked very happy. The Anglo-American Co. had built a fine modern hospital, which we visited. Only African patients are treated and the nurses were all white nuns. On the way to the Drift we entered Basutoland at Maseru, which revived memories of driving about in a Cape cart on the way to pay a visit to the then Paramount Chief Leratholi on the top of his flat topped mountain of Thaba Bosiu. He had been wearing corduroy trousers with a blanket over his shoulders, while his 100 wives tilled his fields. Today, judging by the pictures, the Ruler of Lesotho wears a Field Marshall's uniform.

I flew back to London by South Africa Airways, with a fine view of the Victoria Falls. The upshot of the visit was our success in winning two contracts to sink deep shafts in the heart of Midlothian. However, planning permission was withdrawn suddenly.

13

THE PERSIAN EXPEDITION 1954-56

In the spring of 1954, soon after my return from South Africa, there was a meeting between Mowlem and the Iranian Ambassador. This was only a few years after the young Shah had ousted Mosaddeq, who had nationalised Anglo-Persian oilfields. It transpired that, through the Foreign Office, we were being asked to carry out the modernisation of their major roads for their Seven Year Plan, run by Mr Ebtehaj. In July 1954, Glossop and W. Gove, the firm's secretary, flew to Tehran. They spent many months investigating the problem and carrying out slow-motion negotiations, until their return home in November. It then happened that the Persian contractors objected to working as sub-contractors to Mowlem, who were the main organiser, so we were asked to take on the work as consulting engineers. I strongly disliked this, but the British Embassy was pressing us and as the rest of the board agreed, I fell in with them.

So, once again, Glossop and Gove spent many weeks in agreeing a contract on a form of fee basis and it was signed early in 1955. It was then necessary to engage a large staff with different qualifications and experience. Luckily the road programme in Britain was at a standstill. It was long before the days of motorways, so a number of county engineers and deputies enlisted. Brigadier Horsfield took over and went off to Persia to establish our base with one of my long-time assistants George Wild as his deputy. In London we set up a Persia Committee run by Major Morgan DSO, who had worked for the firm before the first war. He had won his award in a tunnelling company, becoming county surveyor of Essex and then Middlesex and a Vice President of the ICE. On retirement he became an experienced arbitrator. Another member was my assistant, Roger Newport, who had distinguished himself in Burma in the Fourteenth Army, on the ground nuts scheme under Horsfield and on a rock tunnel at Machkund in India. We met once a week to manage the London end and service the expedition.

A well-balanced staff was taken on of road designers and constructors, surveyors, draught clerics and accountants, also motor mechanics and drivers. We were lucky to obtain the services of Peter Avery, a Persian expert who agreed to act as our interpreter. Later he became a lecturer at Cambridge University. Horsfield insisted that no wives, except in very special cases, should be allowed out for a year. To the annoyance of 'the Plan' the build-up and general organisation took longer than Glossop had promised, so he flew out there during the winter to negotiate and see how things were shaping.

A Spanish Interlude

Raymond Head, the works director, had made close contacts with Raymond International in the USA. In order to get another slant on organisation, he arranged for us to study

13. THE PERSIAN EXPEDITION 1954-56

the methods used by the USA in having their Spanish military bases constructed (to the annoyance of the Soviet). Head and I, with our wives and Major Morgan, flew to Madrid on Tuesday, December 6th 1955 and took up station in the hotel, where there was not a tourist in sight but much activity over Christmas decorations. The city is in the centre of Spain and we were told that, because of its position on a plain between mountains, the temperature could fall by thirty degrees in a few hours, which it did. The buildings along the main street were twice as high as the usual London ones, just before the property tycoons got busy, but the street was twice as wide so the difference became hardly noticeable. That day and the next were spent in meetings with the Americans, finding out what it was all about and how. Then my wife and I took the chance to spend several hours in the Prado, with the stunning display of paintings by Velazquez, Goya and the unforgettable Hieronymus Bosch.

The USA organisation was unusual and in a pyramid with, as far as I remember, the US Navy at the top. Next came a group of consulting engineers to design runways, petrol tanks and works for the naval base, followed by a group of contractors who took over their designs and drew up the specification and contracts to be tendered for by the Spanish contractors and who supervised the work. This was an unusual structure, using poachers as gamekeepers. I may be wrong, but it seemed that, put rather simply, the USA would ship over a load of, say, refrigerators and the Spanish Government would unload and sell them to their citizens. This raised the money to pay the pesetas to the Spanish contractors.

On December 8th my wife and I flew to Seville, while the Heads travelled north to see another series of bases. Previously, we had been heavily insured when flying by BEA, but this time the Spanish plane was of some antiquity. It had a sort of hut on the top, in which sat the pilot, with no visible connection to us below. There was no stewardess and no heating, so we were handed blankets as we boarded. We bumped safely to Seville, where it was mild and sunny with oranges ripening along the roads inside the airport. We cabined up in the superior Hotel Alexandre XIII and made contact with the US Navy, who were to show us around. On the 9th they drove us in a large American car with 'US Navy' obliterating the paint on both sides. We went first to Moran to see an airfield being built. The plant which they had brought over for the Spanish contractors to use was certainly not their latest, but was about the same as we had imported for Chingford Reservoir in 1935 and for airfields in the war. The ground consisted of fairly heavy clay and as it had rained heavily there the day before, it was too soft to be made far worse by the caterpillar machines. Therefore, the only machine in use was a Swedish small vibrating roller which they were testing out. After some discussion we took the road for Cadiz.

On the way we passed a section being widened by local labour, who were mostly women. They were digging away three feet depth of soil and loading it into panniers on each side of a number of patient donkeys and the rain had not stopped this well tried method. They left small columns of earth standing, so that the overseer could measure the depth dug before paying. The country which we passed through seemed very poor and some sour looks were turned upon our opulent progress. The evening turned wet and cold as we approached Cadiz along an extensive causeway bounded on either side by wide ponds for evaporating sea water, this being the local salt industry. We found a small hotel since this was definitely the close season for tourists and next day had a quick look round. On Saturday we drove off to Rota on the coast above Cadiz, where the main work in hand for a naval base was the building of fuel tanks let well into the ground. We were also told how the rest of the work was to be organised. After this, we went to Jerez for lunch in a small local restaurant. The

41. Road building in Spain

menu had helpful translations at the side of the Spanish. We were baffled by 'Soup as at Godfathers', but 'Muffled Hake' became clear when the fish arrived in batter. Jerez is the centre of the sherry industry and we were shown a little of this, after which we headed back to Seville, calling on other fuel tanks being built on the way.

We managed to fit in moments to explore Seville and on Sunday we were able to see a great deal, even though the weather was cool and showery. In the morning we came across an imposing and unexpected group of buildings left over from some international exhibition many years before. The main feature had a massive semi-circular cloistered arcade about 12ft from the ground. The terracotta and stone face below the footway carried coloured panels showing incidents in Spanish history, while the back of the arcade was decorated in much the same striking manner. Some years later, we went to see the film *Lawrence of Arabia*. In it Peter O'Toole entered the imposing headquarters of the British in Cairo…to our surprise the location was actually this arcade in Seville.

The Seville Cathedral has five aisles and is converted from an original mosque. We happened by chance to visit it at the moment when choirboys, dressed in Velazquez style boys dresses, were carrying out stately dances before the altar, accompanied by a priest playing a harmonium. Apparently this is a twice yearly event. That afternoon we encountered crowds making for motor coaches labelled 'Futba', which seemed highly popular, and then much smaller crowds who were making for the bull ring. My wife said that although we were both strongly anti-bullfights, it is wrong to condemn such affairs without at least seeing what it is really like. So, we found ourselves among a motley crowd in the inner circle and watched two fights, which was quite enough and then pushed our way out. As it was out of season most of the performers were in ordinary clothes and cloth caps, the piccadores and the torreador wore partial finery and the team of horses who dragged out the victims had a few bits of colour, so it was not much of a romantic spectacle. Our only regret was that each time, the bull failed to finish off the toreador, which was a really disgusting affair. No wonder most of the population prefer to go to the football. On the Monday we travelled back by rattletrap plane for two more days in Madrid. There were more discussions with our hosts and a look around at the costly Christmas decorations, before flying back to London.

The Persian Visit

Since the beginning of the expedition to Persia, Glossop had paid the occasional visit, but Ebtehaj now asked to see other directors. The two senior directors were far too old for this and, although he was nominally in charge of this Persian affair, E. C. Beck did not like to fly due to asthma, so on April 28th 1956 I flew out to Abadan for a month's stay that changed my career. I was met by a small deputation. Brigadier Horsfield had driven down the road with Wild to seize the chance to see the problems for himself, as until then organising and having to attend constant meetings with 'the Plan' took up all his time. Peter Avery, our interpreter

13. THE PERSIAN EXPEDITION 1954-56

was also there. We crossed by ferry to Khormashar and stayed for a day or two in the house which we had taken over at that end of the programme.

The main road to Tehran was still only the two lane rough affair made during the war by the Royal Engineers to carry supplies across the Caspian Sea to Russia. To modernise this road was one of our priorities. This was at the time when the British and others removed Rezeh Shah from the throne that he had seized from the old dynasty. After briefing me on the position and seeing a little of the work being started at Khormashar, Horsfield and Avery flew back to Tehran, while Wild and I set out to be driven up the road for five days, to see it for ourselves and to visit our survey parties strung along its length. Apparently, if an accident occurs, both drivers are automatically arrested, so we wisely only employed Iranian drivers. I was told that if anyone got killed, the police left him at the side of the road for the relatives to collect, but I do not know how true that might be in that peculiar country.

We left Abadan to drive over fifty or so miles of dead flat mud-flats, which seemed to sustain the traffic up to a point, but was causing much headsearching in the design for a stronger version. We then drove through Canyon country, made of hills of heavily eroded sand and gravel. At Ahwaz we stayed the night, in the house with small garden which was rented by the newly recruited engineer (who was to be responsible for this area) and his wife. This is beside the Karun River crossed by a many-spanned bridge, which featured in the pictures of the 1981 Iraq-Iran war. Next day, we left the meandering Karun River to reach the River Kharkeh, where a sharp bend was continually eating away the sandy ground and making it necessary to keep moving the road away from it. This problem had already been passed on to Sir Claude Inglis and his Hydraulic Research Laboratory to suggest a cure.

42. Canyon country in Iran/Persia

The road then eventually entered the mountainous country, with hills over 5,000ft and deep ravines and general distorted strata to add to the problems. Traffic came along in spasms, with stretches of nothing in between. The convoys were made up of twelve ton trucks carrying thirty ton loads or more. There was supposed to be a weight restriction, but no-one took any notice, as the idea was to carry far too much and to make a quick profit and throw away the exhausted truck. Most of the trucks carried loads of steel joists, about 6 x 4 inches and long enough to extend 6ft or more ahead and behind the vehicle. Between these, which were on the outsides of the trucks, lay a mixture of every conceivable kind of good likely to be wanted in the cities. As Iran is very destitute of timber, thanks to all those goats as well as men, floors in nearly every building were made by laying the steel joists 2-3ft apart and filling in between with flat masonry arches.

Before entering the series of mountain ranges which run parallel from north west to south east, we stopped for our usual picnic lunch at Shush. This is the site of the former capital of the Persian Empire and which features in the Bible in the Books of Esther and Daniel. It is there that King Ahasuerus kept his court, where he reigned from India even unto Ethiopia. In subsequent years, Daniel entered his Den of Lions at Shushan after defying Darius, who had taken over the country from the Babylonians. All that was visible was a series of mounds, which are said to have been excavated at times before the first war. There were a few marble or stone columns lying about, but little else. Not a place for a special visit but interesting to encounter on the way.

Either on that day or the next, we came across my friend Wynne-Edwards' firm, Constructors John Brown, laying an oil pipeline from Abadan to the Caspian and we could see some US equipment and operators were visible. Although the country seemed uninhabited, they had laid one mile of pipe on the previous day. Excavators dug a trench ahead, others laid out steel pipes heavily wrapped along the edge and gangs of welders followed joining the pipes. Then came a number of caterpillar tractors fitted with short crane jibs. They picked up the welded pipe many hundreds of feet in length and laid it to rest in the trench, while behind came others to backfill the trench. This was in sharp contrast to Mowlem's efforts to lay a pipeline in Hampshire through densely populated areas, with all the permits and local obstructions to be overcome.

As there was much to study, we only drove about a hundred miles for the first four days. On the second night we slept in one of the camps of our survey parties, which was most enjoyable. The tents had been ordered by Horsfield through his Indian experience and were new to my eyes, they were capacious and inside were divided into rooms by woven curtains. Next day we went off into the mountains, where the road zig-zagged round sharp promontories as it followed the line of a river which, from time to time, breached the line of mountains into the next valley. This presented problems for the redesign of the road from an improved mule track to a modern highway, especially in avoiding the sharp turns to cross rivers by narrow bridges. In one valley, which was wider than others, the whole side of a mountain (to be seen across the valley) had careered across, leaving boulders the size of a two story cottage among the debris.

That evening we reached the town of Khorramabad, a mixture of the old with the occasional modern avenue, and this was where we stayed with our unit who were working from there. On the way we had met and passed the occasional firm's car, which ran up and down the road to service and keep in touch with our scattered units. In the centre of the town was a large citadel on a hill, which was said to be full of political prisoners. Next day we went off to Hamedan, along the road where it left the rivers and cut across more rolling ground and past a snow capped mountain, which according to the map was 11,719ft. The altitude at the road must have been high. We held back discretely while a whole tribe of gypsy-like people, men, women, children, goats, mules and horses trekked across. There must have been two hundred or more. The men looked dark, swarthy and sinister and best left alone.

Hamadan looked more modern than Khorramabad, as it is on the main road through Iraq to Tehran. We found a modern hotel and gratefully filled up hot baths, but unfortunately we found that, being Persia where nothing works very well, no cold water came out of the pipe, so it took a very long time to cool off enough to use. Next day we had a longer pull of about 200 miles to Tehran, with the last few hours in the dark. Most of the trucks carried coloured fairy lights strung across them, which made it hard to see which way they were going. When we arrived, we went to a fairly modern hotel where a room had been booked for me. This room with bathroom and precariously working plumbing faced the brown level topped Elburz Mountains with a carpet of snow, a view which soon palled.

Two weeks in Tehran

My stay in Tehran should have been longer but was shortened for reasons that will become apparent. We reached Tehran on Saturday evening, May 5th, and spent Sunday in the office as the Moslems celebrate their day on Fridays. I went into all the details and toured round the various departments to meet and also become known to many new faces, British and Persian. The two nationalities were 100% integrated according to their rank. The Indian custom of having a messenger stationed outside the boss's door was followed and was useful.

I found a surprisingly good organisation, with a soil mechanics laboratory in full swing. In Iran, many young men go abroad and obtain degrees. They then feel so important that they must have a desk as a top manager, without going through our usual training period as juniors. In this case, they felt that they were scientists in the laboratory and got down to work. Their mothers thanked Mrs Horsfield for the fact that they were actually working. Engineers were designing, draughtsmen were draughting, accountants were accounting, transport was being organised, quantity surveyors were taking off quantities and others were drafting contracts and so on.

On Monday I was taken to a meeting with Mr Ebtehaj's assistant, Mr Daftarian, with whom we were to have many discussions. Most of these were carried out in French and he started every sentence with the word 'Donc'. Horsfield, Wild and I were invited to lunch with the British Ambassador, which went off very pleasantly. Then after lunch we were asked to meet, in the Embassy Garden, two of the Seven Year Plan foreign advisors. 'The Plan' seemed to be constantly calling in such advisors. One was Monsieur Prudhomme and the other an American whose name escapes me, but whom Wild described as a fisher in troubled waters. The meeting started off with an attack. First, they challenged Glossop's promise to do 1,000km of road a year (I had objected at the time, but Glossop had said that it did not matter as 'the Plan' had no money to do so much anyway). They next attacked his promise to have staff in Iran within two weeks, which was manifestly absurd. Their next complaint was that, on his previous visits, he could make no decisions without telephoning (difficult operation) to London. They said that now I was there as a director, Mr Ebtehaj insisted that, unless I answered 'Yes' or 'No' to the next question, without referring to London, he would cancel our contract. This was all in the presence of the Ambassador and we were being urged into the contract by the Embassy. I asked what the question was and they replied that as progress was so slow, Mr Ebtehaj intended to give half our contract to Kamsax, a Scandinavian organisation who were doing other work for 'the Plan'. In fact we had discussed this as a possibility in London. I replied that we could not prevent him giving half the notional mileage of roads to Kamsax and that as they would suffer the same constraints with which we had met, it would show how well we had done.

The meeting then broke up and later Mr Ebtehaj thanked the Ambassador for helping to bring about such a happy solution, which would help him politically. Next day I met Mr Ebtehaj and we had a most constructive meeting. It was obvious that this would not materially affect our contract, as there was no idea at the time how much work could be financed. I then wrote a long letter to Sir George Burt as the head of the firm, describing what had occurred and got down to proper work.

One of the fallacies was that whatever we put forward, agreed by 'the Plan', had to be agreed by the Minister of Roads. He hated Ebtehaj's guts, to put it crudely, so if a culvert was designed as a square, he wanted it circular or vice versa. So, six contracts were lying ready to send out for Iranian contractors to tender, but were frozen by the failure of the Minister to agree. I found that among our staff was a Persian graduate, whose father was a man of power, so I took him and Peter Avery with me and bearded the Minister in his office. I set about him firmly, but forcefully, backed by my support and came away with his agreement to pass the contracts. I therefore felt that I had achieved something in a few days. A house had been rented with a pool, as a club for the staff, all nations and ranks welcome, so I visited it on a Friday off day and found it very well run.

Another expedition was to visit the American firm of Morrison-Knudsen (with whom I was to have unforeseen future connections). They were attempting to start to build the Karaj Dam in the Elburz Mountains, but were waiting for the Government to make up their mind. This dam was designed by Sir Alexander Gibb and Partners, who were engaged in installing the first ever piped supply of water to this sprawling city. The contractor had built the diversion of roads above the future water level, and had tunnelled through the spurs of rock which marred the way, instead of wandering all round them. In Tehran, the water supply was still provided from the Kanuts or long tunnels driven, often by children, through the foothills to collect groundwater. Each main street had an appointed day when water would be diverted to flow down channels on each side of the street and water bailiffs controlled the moment when each succeeding house could divert the water into the cellar below for future use. The piped supply was finally provided along the streets long before any main drainage was installed. The pictures shown on television by 1982 bore no resemblance to the city as it was in 1956.

Horsfield and I were driven along the road to the so-called Holy City of Qum, with the engineer who was to be in charge. We had strictly been warned not to enter the city as, even in those days of the young Shah, it was a hotbed of Ayatollahs. We stopped for our sandwich lunch in view of the main gate. Soon a procession of Mullahs on muleback emerged, surrounded by their excited followers. They gave one a very creepy feeling, so unlike our dear C of E. I was pleased to feel that I had accomplished something and on the next rest day Horsfield and I were due to fly to Ispahan for a visit. This was not to be. I have a draft of my own conclusions on the Persian affair scribbled in my bedroom at the time:

> The contract as a document is a model of perfection on paper, with intense thought given to many details, weakened at the last moment by Persian demands. But – it is based on several fallacies:
>
> 1. The Plan **shall** obtain the Minister of Roads signature. There are no powers to ensure that.
>
> 2. The Company shall satisfy the Plan **and** the Minister. This is a duality which will lead to endless trouble.
>
> Also the provisions are such that Mowlems have to go cap in hand to the Plan for every necessity.
>
> **Execution of the Contract**
>
> The organisation built up, started like all British military operations very badly but now has a very good general skeleton and on the whole is very efficient, more so in many ways than the JM home organisation. The spirit is excellent in spite of all set-backs. The amount of work carried out is quite remarkable but invisible to the outsider. Horsfield has done a very fine job. The less he is bombarded from London, the better the job will go. Wild and Robinson, who runs the lab, have worked themselves exhausted. To these three must go the credit for our survival.
>
> Considering that it is Mowlem's first adventure of this kind we have much to learn. The extent and difficulties of the contract are sometimes overlooked at home. The

Persians are almost impossible to deal with but the only hope is to keep pressing on and not panic over contract and agreement details. They will certainly not throw us out now and at the end of the second year will be the crisis. We can produce 1,000km of roads 'on paper' by then. Whether they are built or not is beyond our powers. The figure of 1,000km is just an algebraic expression. The Plan do not know if they have the money nor can anyone tell if the contractors will have the plant for such a high rate of progress. It will be a surprise if the actual work ever gets voted or carried out at such a rate.

The second struggle will be to see how the contractors perform and what pinpricks will be inflicted by the Ministry of Roads during performance. The third struggle will be over the final accounting of costs. The whole affair is riddled with dilemmas which can only be solved on the spot by the judgement of the man in charge.

If we take on such extraordinary undertakings at such a distance we must obtain the best possible man and leave him to act as he thinks best. If we must run such an affair from London then we have no business to embark upon it.

Brigadier Horsfield was to return to London in a few weeks as he was in great pain from an arthritic hip. He had obtained as his successor one of his former officers, Brigadier Cavendish, who incidentally, had been a contemporary of mine at Christ's Hospital, before going to the Royal Military Academy at Woolwich for the last year of the first war. Cavendish arrived towards the end of my stay and I spent some time briefing him on the contract and upon the habits and customs of John Mowlem and Co.

Just before we were due to fly to Ispahan [Isfahan], the British Embassy sent me a telegram put through them from E. C. Beck. It was abusive and demanded my immediate return. I sent a dusty reply, but after further exchanges I flew back by Air France via Beirut, Istanbul and Frankfurt. I was not in a happy frame of mind, which did not improve at the airport. Avery had begged off, so Horsfield and Cavendish alone came to see me off. Just before entering the car, the secretary of our office handed me an envelope to take to Mr Gove, the company secretary. Once parted from my friends and in the Customs, an official opened my briefcase and picked out the envelope. 'You are defrauding His Majesty's Customs smuggling letters and robbing the Post Office'. He opened it and out came, not the letter I had expected but a number of travellers cheques, which in Persia were treated as currency. 'You are smuggling currency' they accused and things looked black. However, I managed to attract Horsfield through a glass partition and after half an hour he had straightened it out and paid a large fine. They wrote a page of very rude words in their Arabic script in my passport. Later it turned out the cheques were out of date and only payable at Beirut. They had been issued to the drivers who were bringing out motor transport for us by driving it from Beirut through Syria and Iraq. Luckily when I had to go to Bombay a few months later, the Passport office kindly gave me a new passport.

On my return from Iran, I was called to head office and found myself in the Board Room with Sir George Burt and the Becks, father and son. Sir George was very deaf and did not seem to gather all that was said and I felt that he had not passed on my letter describing the meetings with Ebtehaj to the others. In brief, E. C. Beck (the son), lost his temper and spoke abusively, saying that I had made a fool of myself, given away half the contract and let down the whole British position in the Middle East and that Denis Wright of the Foreign Office said the same. He then flounced out. I returned to the office which I shared with Glossop, who seemed quite unmoved by the whole affair, as we had expected it before I left for Persia. He agreed that there was no reason why I should not see Denis Wright, who having served there was the senior man on Persia. Later, as Sir Denis, he became British Ambassador in Tehran and rose even higher. So I drove round to the Foreign Office to see him and had a friendly reception. I told him what had been said and that I had been told that he had agreed and he pulled out a memorandum and gave it to me to read. To summarise, it went like this:

> I met Beck at the theatre, he said to me that Harding had made a complete fool of himself, had given away half the contract and let down the whole British position in the Middle East. I said that I did not think he had done anything of the sort, that Harding was entirely justified and on no account should you have a head-on collision with the Seven Year Plan.

This was very comforting and we had a long talk about the matter. Next day I was met by so much abuse that Sir George instructed me to go away for a month's holiday, as I knew how difficult the Becks could be. I learned later that Denis Wright had rung up E. C. Beck and had spoken to the point on the matter, so much so that 'father' Beck could not forgive it. So, my wife and I set out to explore mid-Wales. We found a lane in the crest of the Cotswolds above Broadway leading to a country house hotel for lunch and we stayed there for nearly a month. There was a fine view over the Vale of Evesham and the house had superb gardens with a nine hole golf course nearby. Unfortunately, our state of mind was not very pleasant, which rather spoiled the holiday. I then got a letter from Sir George, instructing me to take another month and we spent this extra time at home. I attended ICE Council meetings and Imperial Governor's meetings, taking care not to show that there was a crisis. Then after two weeks I received a formal letter from the firm's secretary, asking for my attendance at a meeting to consider the future.

The meeting was only attended by the two Becks and Sir George. I asked to have the rest of the board, with the other three younger directors present, but this was bluntly refused. Briefly E. C. Beck demanded my resignation. I told them flatly that I refused and that I was putting the whole matter in the hands of my solicitors. Immediately I had returned from Persia, I had briefed an old friend, John Buckley, who was a solicitor in Macfarlanes, a big city firm. He and E. C. Beck, with their wives, had been our guests at our house in Putney for dinner only a few months before, so he knew the contestants. John Buckley had warned me never to resign and to say nothing which they could try to use against me.

For some days John Buckley wrestled with Gove the secretary, to no avail. Then he contacted me and said 'Next Monday your leave is up, you will go back and take your seat in your office'. I was met with acclaim by top works directors and by others of the staff, who had been told that I had left, but the chiefs were not in. The next day I again took my seat and, as there was to be a Tuesday Board meeting at 11am, I called in on R. I. Beck and Kenneth Burt who were close friends of mine, which astonished and embarrassed them. I then told the secretary that I should be at the Board meeting which threw him into confusion. I had been advised that I should stay at work until definitely told or asked to go. Sir George came rather shamefacedly into my office and said this, but I insisted that he must say so unequivocally. I then said that I had served him faithfully for thirty-four years and that this was not the end of the matter, as I had been treated disgracefully. I then packed up my papers and went off to see John Buckley before lunch. He went straight off to the Middle Temple to consult a QC, who drafted a letter which went off that evening to Linklater and Paine who represented the firm. It was a strong letter, setting out my position in the profession and went on to this matter 'without a scintilla of reason. It is a pretty story'. He then gave them forty-eight hours to make a proper offer, or else proceedings would be taken. This was on August 14th. The letter reached the solicitors next morning and John

Buckley rang to say that they had telephoned an offer on receipt. He asked me if I wanted to argue for more, but I said accept at once, I want no more to do with them. So I became a free man and my own master. What I most resented was the anxiety caused to my wife and the ingratitude. I had so many letters from all ranks in the firm, as well as from outside, that I had to open a file marked 'Fan Mail'. One particular one from George Wild pleased me greatly, 'the firm will miss most – sanity in times of excitement'.

PART 3

14

CHANGE OF LIFE

On the 16th August 1956 came the start of an entirely new life, with the end of what a friend, echoing the Dreyfus Case, christened *L'Affaire*. Like the gentleman in *Pilgrim's Progress*, I had emerged from the slough of despond and entered a promised land, while a great burden of many responsibilities slipped from my back. I was no longer bound or obliged to anyone and had enough capital to provide ample time to contemplate the future. Many natural emotions were summed up in a letter from my son in Glasgow – 'You can now point the two fingers of scorn.'

A pointer to the way ahead arose in an incident when the contest was reaching its climax, where I was asked by the 'Civils' to represent them at a luncheon being given by the Institute of Mechanicals to a party of Soviet Professors. As we stood around drinking (not vodka), who should suddenly appear but the by now famous Doctor Terzaghi. He came straight across to me and said 'Please, may I sit next to you?'. I was naturally delighted, as he had closely followed the progress of Soil Mechanics Ltd, ever since he had saved Mowlem's reputation over the slip at Chingford Reservoir and I had been adopted as one of his disciples. I told him what was going on and that I should be leaving the firm after 34 years. He said 'That does not surprise me. I know them. You must go it alone. I have read all your papers and know of your wide experience. You have many friends. You must go it alone'. So that is what I did. A year later he sent me a Christmas card of a galleon with the message, 'All good wishes now that you are Captain of your own ship!'. This was most heartening from a busy world famous man, who had 'gone it alone' with outstanding success.

Things at once began to happen. My old friend Robert Wynne-Edwards who alone knew what was going on said, 'As soon as you are free, will you help a consortium of your friends who are bidding for the Rocket Range at Spadeam? There are six firms involved, you know all the directors. I want you to take the chair and knock all our heads together.' So I set out to his office armed only with a fountain pen. To my relief I found that I could play a useful part, although the bid was not successful.

Field Marshall Lord Alexander once told a friend of mine that the success of a man in any profession varied as to the square of the number of people whom he knew. So, when starting to review my assets, I was surprised to realise how many friends I had accumulated over the years. This was first brought home by the many letters which I received from friends, officials, consultants and contractors, as well as over twenty from colleagues and disciples in the firm. There was one man to whom I wanted to report as soon as possible, Vernon Robertson. He had been President of the ICE for my first year on the Council and I had worked for him when he was District Engineer on the North Eastern Railway,

14. CHANGE OF LIFE

Chief Engineer of London Transport then Chief Engineer of the Southern Region and, after nationalisation, became a partner of the consulting firm Sir William Halcrow and Partners. He at once said, 'Can you possibly visit Bombay on our behalf in October, as we are very stretched? We have been asked among other firms for our terms for designing an Underground for Bombay and you are just the man to find out what they want.' So I had this assignment lined up before taking my wife for a revival holiday in Italy to Levanto, Florence and Rome.

Before leaving for Bombay, I naturally headed for the Geological Museum to find out what was below, vital for an underground railway. I found that the city was on a peninsular made of several islands connected by filling, with a causeway to the mainland. The original land was mostly of Deccan trap rock, a form of basalt, so there would be quick changes of ground from soft to hard in any work below ground. I also learned the layout, a flat plain between two lines of low hills. These embraced Back Bay, nearly two miles wide, where on its wide sands Ghandi had started his campaign by 'making salt' among his throngs of disciples. The east side forms a narrow peninsular covered with Government and prosperous buildings and round the bottom corner can be found a large bay five to seven miles wide with the hilly shore opposite.

I flew out to Bombay in the company of Lord and Lady Attlee. He was very quiet, but she made much fuss at touch-down airports. Then we sat for a long time in the plane at Bombay, while the Attlees were received and covered with garlands, in token of his work in granting independence. I spent ten days mostly with the engineer, Mr Rao, of BEST (Bombay Electric Supply and Transport Undertaking) who was masterminding the general plan, which needed a consulting firm to bring it into proper modern design and contract conditions. Mr Rao showed me the city, sites of stations and so on. He had just returned from Japan, where he had seen their new cut-and-cover Metro and thought that this was the very newest method. I had to disillusion him by quoting the London Inner Circle (1884), although it might well be their solution.

Among introductions was one from a firm (their name escapes me), but as their top man had to be away, he sent his Indian deputy in a beautiful turban. I was lodged at the Taj Mahal Hotel, so he took charge, filled in the forms and got me two half bottles of whisky, as I understand it, by proving that I was an alcoholic (Bombay being in the throes of prohibition). This was the legal way of helping us foreigners. There was air conditioning in the bedroom but not in the bathroom, so one had to strip to the buff to enter the latter and then use a towel, thus I caught a roaring cold. However, by staying in bed and consuming one half bottle of whisky, I quickly recovered and gave my new friend the other half out of gratitude. He took me to lunch in the Bombay Cricket pavilion with the Indian equivalent of the MCC and also to his own flat. While I was in Bombay, I had to talk to the chief engineers of the railway, who were impressive men. They showed me the photographs hung around the room of their British predecessors and said that they kept them because they were proud to have worked under such men.

I also called on a man who was Gibb's resident engineer on the harbour works, because he had been their man on the Iranian and Teheran Water Supply a few years before and I wanted to talk about Persia. He had been resident engineer on an airfield which Mowlem had built in the war and he had been described as very fair and competent. Later, this was a bonus when it came to being involved unexpectedly on the dockyard case which, unknown to me, was just coming to a standstill during my first visit.

To study the problems I made several tours round the route by myself. As Mr Rao belonged to a surface transport undertaking, he knew more about the surface than the underworld beneath, so little could be learned from him. The line was to form a circle around the densely populated middle area, connecting the Great Indian Peninsular Railway terminus and other points. To the south, it ended in the narrow peninsular between Back Bay and the Harbour, which is covered in imposing Victorian buildings in all the pomp of the Raj. Inland the streets grew narrower and were lined with varied buildings of entirely Indian conception, gaudily decorated. I was warned that these houses collapsed at a rate of over one a year. I took photographs of potential station areas and places which might need thought. One of these areas was in the unfinished part of the Back Bay Reclamation, a once famous but now little known case of aborted work. In 1922, we embryo engineers were excited by the much trumpeted scheme backed by Lord Lloyd, who was the Governor of Bombay of whom the then Prince of Wales spoke on return from his Indian Tour… 'I never knew how royalty lived until I stayed with Lord Lloyd in Government House.' Sir Robert Buchanan had formed a firm of consulting engineers when he retired from many years of Indian Government work and his firm was briefed. The scheme was admirable. In order to provide more land in this densely populated city, banks of rock-fill were tipped as needed along the Back Bay to hold the material. This was to be laterite, dredged from the harbour just across the peninsular, and pumped through twenty 4-inch diameter pipes laid in streets. This was also to improve the harbour approach. Laterite is a clay-like material formed from decomposition of rocks which abound in India like the Deccan Trap or basalt under tropical conditions. However, no one noticed the small print on the quotation for a cutter-suction dredger. It was offered as capable of many cubic yards per hour – but in soft sand! The laterite was admirable for its purpose but slow to dredge, so progress was falling far behind programme. The final blow arose from the primitive ways of site investigation over water in those days. The estimate of available laterite as a deposit in the harbour proved quite wrong and the supply petered out. There was no other suitable material for miles and the extra cost could not be faced and as the scheme was only partially finished, many lakhs and rupees were lost.

On returning to London, I borrowed the verbatim report of the Commission of Enquiry which sat for many months to look into the matter and search for scapegoats. The members of the Tribunal were mostly distinguished Indian lawyers and most of the cross-examination of the many witnesses was by Indian QC's, although some British were represented by their own counsel. Few of us in our professional lives become exposed to relentless cross-examination by QC's or even attend the daunting performance of such a body. Little did I realise that reading the report was to prove a valuable guide to my unforeseen future involvement with so many barristers in arbitration and in the Aberfan Tribunal. This was no disaster, no lives were lost, only money, but it was interesting to read how witnesses, some of whom had no real responsibility, had to weave and loop to avoid the zooming attack of a ruthless lawyer. Indian independence was by then being hotly pursued, so few holds were barred where the British were concerned, while Indian witnesses escaped more lightly. In the end Sir Robert, as head of the consultants, had to carry the can as responsible for the actions of those under him. In due course, I passed on all that I found to Halcrows, but nothing further seemed to happen as far as I heard.

15

CHANNEL TUNNEL 1958-1972

On February 1st 1958, after some exploratory talks with Leo d'Erlanger, the Chairman of the British Channel Tunnel Company, I found myself invited to act as consultant to the Channel Tunnel Study Group, as British opposite number to René Malcor. Malcor was Ingenieur en Chef des Ponts et Chaussées, who had been lent by the French Government to act as Chef de Delegation. This means that he was to direct the studies and handle such cash as was available and as part of this he had asked for a British opposite number. I was to be paid on a basic hourly rate. but with a limit of hours-per-week to save the finances of the Group, which meant that I was able to carry on the rest of my practice at the same time. I was impressed with René Malcor at our first meeting and formed an instant liking for him. I told him that only one man can stroke the boat, so I was happy to row bow, which led to a close and lasting friendship. We had many outside interests in common and as his good sense of humour included an accurate appraisal of the national idiosyncrasies of the British, Americans and French we thus shared no illusions. René Malcor had behind him a long and varied experience, this included being wounded and taken prisoner in 1940, escaping from hospital, swimming a river into Vichy France and making his way to Algeria.

One of my duties was to attend meetings of the Study Group, held alternately in France and England. There were four component organisations in the Group:
- the British Channel Tunnel Company including Sir Alec Valentine and others from the British Transport Commission;
- the Societé Concessionaire du Chemin de Fer Sous-marin entre la France et l'Angleterre who also brought in the French railways, the SNCF;
- then importantly came the Suez Canal Company who had lost their empire in the Suez debacle of 1956 (but had kept their cash). They were led by Jacques Georges-Picot and included the services of the formidable Louis Armand, who had revolutionised the French railways after their destruction during the war (how different this was from the purge prescribed by Dr Beeching!);
- the fourth member was Technical Studies Inc., run by two active Americans who had married Schlumberger heiresses and who had become interested in the idea of a tunnel when their wives suffered on the sea crossing. This experience was also what had given Queen Victoria the same idea. Frank and Al Davidson were Harvard lawyers, whose father had been a friend of F. D. Roosevelt and they co-opted American bankers, diplomats and others.

The alternate chairmen were M. Massigli, the charming former French Ambassador to Great Britain and Sir Ivone Kirkpatrick, former pillar of the Foreign Office. When Sir Ivonne died in 1964, Lord Harcourt, who was also a director of the Suez Canal Company, took his place. The most enjoyable meetings were those held in the resplendent 'second empire' cubical chamber in the re-named Compagnie Financière de Suez, which included lunch with the highly talented and friendly members of the Group. Malcor had been given an office in the building, but was not under Georges-Picot for 'discipline, pay and rations', a fact which the latter found hard to swallow. In one corner of the chamber was a collection of vast volumes of steel engravings, showing every detail of archaeological, architectural and naturalist interest collected by the bodies of savants whom Napoleon had carried with him to conquer Egypt and who had been marooned there by the Battle of the Nile. Thus began my thirteen years of spasmodic involvement in the enterprise. I was to receive several invitations to write a book on the story, but turned them down, as there were already too many bad ones. There were three distinct periods to this project: the first was the 1958 – 1960 study and report which could have led to the actual start, but Government ministries became intermittently involved. The second large scale site investigation in 1964 – 1965 was followed by many ups and downs, until work actually started in 1973, by which time a fresh group had been formed. Malcor and I had fulfilled our function and faded out at this stage.

The First Study 1958 – 1960

Members of the Study Group included bankers, lawyers, diplomats, government officials, and managers, so they were able to set about studying the international, legal and financial problems themselves. The feasibility study was entrusted to Malcor, with a sum of money put up by the Group members. Half the money was spent under Malcor's direction on a traffic study of considerable depth by three expert firms, French, British and American. The remainder was spent from 1958 to 1960 investigating the conditions above and below ground and sea for the feasibility of a rail tunnel, a road tunnel, an immersed tunnel and a bridge. Malcor's Paris deputy, Charles Ribeyre had been a top engineer on the Suez Canal, until he had had to abandon all his possessions during the invasion of Egypt in 1956. We also drew in J. M. Bruckshaw, Professor of Applied Geophysics at Imperial College and Jean Goguel, Head of the French Geological Survey. They had both been Presidents of the International Geophysical Society and were professional friends. The exploration was directed by this tight coterie of kindred spirits, which led to the definitive paper to the Institution of Civil Engineers in 1961, 'The Work of the Channel Tunnel Study Group, 1958 – 1960' by Bruckshaw, Goguel, Harding and Malcor. My diary records more than fifty trips to Paris over the years, mostly by night ferry. This wasted no time, with supper on the train, a night's sleep, breakfast and arrival at the Gare du Nord promptly at 9 am (but not so promptly at Victoria coming the other way).

Malcor had assembled all available data from previous work, as the French had ceaselessly campaigned for a tunnel and De Gammond had kept the Napoleonic idea alive from 1832 to 1869 with an enthusiasm which led him personally to dive to the sea bed. The major basis for the future exploration consisted of two large steel-engraved charts, which showed the numbered position of each of many thousands of samples from the sea-bed, picked up during 1875 and 1876 by three French engineers. These samples had been discovered in a disused French suburban railway waiting room. The samples had been obtained by dropping a heavy iron or lead weight from a crane, a short 1 inch diameter steel tube was embedded in the bottom and, with luck brought up samples. The position of each sample had been fixed by sextant. The log of a borehole survey carried out earlier in 1866 by Sir John

Hawkshaw was also obtained. An article in *Marine Geology* in 1966 subsequently disclosed the extent of a prototype testing of the sea bed at Dover by Hawkshaw's young assistant, Henry Brunel. Henry was the son of Isambard (Great Western) Brunel and the grandson of Sir Marc Brunel, who drove the famous Thames Tunnel. Henry had invented the drop sampler, which was later used by both the British and the French in 1959. He too fixed the position of his 207 drops by sextant, sometimes working at night using a light on a buoy. The beds of the Upper, Middle and Lower Chalk, resting on the Gault Clay can be seen in the cliffs on both sides of the Channel, dipping to the east. Examination of the samples made it possible to deduce a rough outline of the outcrops of each bed in the Channel. Between 1882 and 1883 a pilot tunnel of 7ft diameter had been driven from both coasts for a mile out to sea in the Lower Chalk, using one of the first ever tunnelling machines developed by Colonel Beaumont. The French were disappointed to find that he was a British sapper, in spite of his name. Lord Wolesley then prevailed upon the British Government to abandon the work – for fear of future French invasions. Toujours la Politesse!

The science of geophysics had advanced on two fronts – oil exploration, mostly by seismic methods, and anti-submarine warfare. A few years earlier, the US Navy had removed employment restrictions, so that their liberated researchers rushed to the former rival firms to exploit such methods as Sonoprobe, Sparker, Boomer and Thumper. Each method involved towing an electric device that made intermittent rude noises under water, which were recorded on rolls of perforated paper. Interpretation of these multiple reflections is strictly for the experts, but even then they need calibrating by occasional borings. The first step was to carry out boreholes on both land approaches. The cliffs on either coast are so high that long approach tunnels under land were needed to reach the Lower Chalk below the sea. These boreholes were to test out the effect of seismic methods on the strata by explosives, before going out to sea. The army allocated us a site on a bombing range so that no one could complain of the noise. Rotary boring was by diamond drills, with tungsten carbide bits in place of industrial diamonds, producing cores of 4 inches diameter. The 'experts' found that seismic explosive methods were not suitable, as the velocity contrasts in the layers of strata were the wrong way up for interpretation.

The Straits of Dover have been covered by the Decca Navigation System,[41] which emits continuous radio signals on different lanes. It consists of a master station and three slave stations: red, green and purple, which emit signals of definite and neighbouring frequencies. As the slave stations are on a triangle around the master station, only two colour lines are recorded in any particular area. The Dover area Admiralty Chart NL (D5) 1875 showed red and green Decca lanes which, for fixing positions, could act as a substitute for latitude and longitude. A young American expert installed his gear in an old life boat named *King John II* and set out with the skipper to quarter the area. The currents were so strong that when weather allowed work, the *King John* had to travel where she was driven by the conditions, rather than follow a predetermined course. A thousand kilometres were covered with the appropriate length of paper record. Hired equipment, installed in a vessel, recorded the number and decimal fraction of each lane crossed both visually and on paper, so that it was possible to plot the course followed and the time of each position. The Decca lanes were traced and blown up to three times the chart scale and then my son and I plotted the track of *King John II* on a bedroom floor, while the original records were flown to Telephonic's office

---

[41] Retired in 2000.

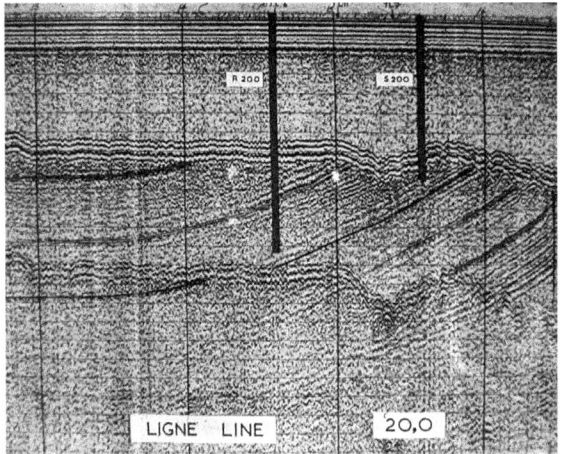

43. *Investigation work for the Channel Tunnel. Sparker plot*

in the USA. The final plotted results did not arrive back from them until June 1959! We authorised two surveys, Sonar in 1958 covering 1,000 kms and a better Sparker one over 500kms in 1959.

There were several highlights during the first year. In June 1958, after leaving a cocktail party at the French Embassy which had been arranged to meet British MPs, we slept at Folkestone and then set out to walk along the railway at the base of the cliffs. This included a visit to the much photographed 1882 gallery and Shakespeare Cliff, where the original shaft had been flooded by a subsequent, now abandoned, mine shaft. We saw the tail of the crude Whitaker tunnelling machine protruding from the cliff where test runs had been made in 1922. An early boss had attended and I still had my notes on the results. Next day at low tide we walked along the base of the French cliffs to Wissant, getting a good idea of the beds dipping east. After a French *fruits de mer* lunch we examined the vast area of Boulogne Harbour by launch, as a potential assembling base for units of a possible immersed tunnel. For good measure, we boarded the concrete, compressed air caissons which, forty years after the first war damage to the harbour, were being sunk to close the east breakwater. Almost in the English tradition.

We then drove to Bethune for the night and next day went down a coal mine to see a Marietta tunnelling machine. At the shaft bottom we saw two shifts of French miners meet, where they shook hands all around as is the French habit. After a lunch with local dignitaries in Arras, we took the train to Paris, and on to Rennes, arriving in sheets of rain at 23:30 hours. After driving next day via St Malo to Cancale, we went five miles out to sea to inspect a steel tower where borings were being made. These were for a vast breakwater which would enclose about ten miles square in order to produce tidal power from the high tidal range. We were shown samples of rock taken from the sea bed by engineer frogmen. These were being examined by their geologist, who interestingly was also a Jesuit priest. This visit led us to try out frogmen divers in the Straits, but the swirling currents proved too strong for useful work. In the end the tidal power scheme was reduced to the mouth of the River Rance, which I was taken to see during construction a little while later.

A highly interesting exercise was the re-opening of the French pilot tunnel at Sangatte. When the contractors removed the concrete capping to the shaft, they found that the Germans had filled it with scrapped machines and other rubbish. The 90m deep shaft had been lined through the waterbearing middle chalk with wooden blocks caulked with oakum. This had perished, so that there was 70m of water in the shaft, which needed to be pumped out once the divers had helped to sling the obstacles to a crane. We found the unlined tunnel, which had been under water for 77 years, to be quite 'dry' (watertight) once the shaft had been emptied. All the many tests and visits by important people had to be hurried so that pumping could stop and the shaft be abandoned once more. By autumn, we had collected enough data and Professor Bruckshaw had interpreted the sonar readings to

15. CHANNEL TUNNEL 1958-1972

44. *M. Malcor* et al *down the 1880 Sangatte tunnel*

a sufficient extent that we could advise that feasibility was in sight. The Group then mobilised their consulting firms, who were to co-operate under the direction of Malcor and myself.

Malcor appointed the American firm of Parsons Brinkerhoff to report on a bridge and also on an immersed tunnel, or a combination of both. On the British side there had been three firms over the years who had reported from time to time to the Channel Tunnel Company, including the report of 1929 which had been defeated in Parliament by only four votes. However, the Study Group did not want to start paying fees to the firms until we had made our first report. At the same time the French appointed the firm of Société Generale Exploitations Industrielles (SOGEI) founded by M. Grange who had been Chief Engineer to the Suez Canal Company. With French logic, they also involved M. Ischy, the head of Soletanche, the soil mechanics firm and M. Bouvier of the contracting firm Fougerolle, who had performed at Sangatte in 1882 and again in 1958. In addition, Malcor appointed three firms, one British, one French and one American to carry out a traffic and revenue study. However, the American bankers who were involved with Technical Studies Inc. said that they preferred reports from advisors whom they knew and they chose three further contractors to work together: Morrison-Knudsen, Brown and Root and Bechtel. This embarrassed the others so it was arranged that these three should be responsible to Technical Studies Inc. and not the Group. Malcor and I had to brief all these bodies and keep feeding them with all the information coming out of our site investigations. This process went on at intervals until our final report in 1960. The British consulting firms, Rendel Palmer and Tritton, Livesey and Henderson and Sir William Halcrow and Partners each provided two engineers and their French counterparts did the same. By a series of meetings under Malcor's chairmanship (held alternately in Paris and London) the whole design for twin rail tunnels with a service tunnel and, in addition, a road tunnel was agreed and an estimate made towards the end of 1959. The Americans caused a little embarrassment by making a bid and not a report. The costs were much the same, but the British time was longer while the American seemed optimistically short. On my visit to the USA, I was able to make both sides compromise on an agreed time.

The winter was spent in preparing for the 1959 spring offensive on the Channel, in between dealing with the press and attending meetings with many bodies. We examined offers for boring at sea and more geophysical work. Jack-up platforms (see below) were few and prohibitively costly and I took Malcor to see Wimpey carrying out diamond drilling from a tank landing craft (following my example at Baniyas) for the design of the cooling water tunnels at Dungeness. In the spring, they were able to transfer their craft to Dover for us. The really novel addition to our study was the work of Dr Carter of the Imperial College, who had made a study of the micro-fossils in the chalk. He was able to divide the study area into zones of measurable depth, defined by the type of fossils in each zone. He applied this not only to the borehole cores, but to the examples from the sea bed. We had steel engravings giving the exact positions of several thousand samples from the French expedition of 1875-76. He was able to carry out tests on some of these and some of the bores that we took. From these, he was able to define the dip of strata and their relation to zones in nearby boreholes.

In May 1959, we engaged Alpine Geophysical Services to carry out another 500km survey with the more powerful sparker method. By using an ex-RAF tender launch, it was possible to breast the tides and follow a prescribed path. Unfortunately, from April 9th to May 9th eleven and a half days were lost through bad weather. This survey loosely agreed with the sonar, but with deeper penetration. In both cases, where known faults existed, these were recorded clearly, so the assumption was made that where neither method showed serious faults, they would not exist. They also proved that no buried channel existed and that the Lower Chalk was continuous across the Channel.

Between April and September, Wimpey carried out eight successful and two abortive borings at sea. The Channel is no health resort even in summer (which promoters of other methods of crossing are apt to ignore) and for seventy-seven days, out of one hundred and forty, the summer weather was too rough to work. Twenty-four days were spent setting the large Trinity House lightbuoys and setting moorings, leaving twenty-seven days spent in boring 310 metres. Other tests were made by instruments down the boreholes, when time and tide allowed. Drop samples were also obtained and the position of each borehole surveyed by a Major of the Ordnance Survey, using the permanent OS stations on the cliffs. We had very little money for this work, so boreholes were chosen from the Sparker Survey to calibrate and give greater detail where it was most needed. However, it was a benefit that Dr Murdoch of Wimpey and his staff were all fellow members with us of the Institution Geotechnical Society, so we were all working to the same purpose and happily avoided money conflicts. My relationship with Malcor had grown to such an extent, and he and I were in harmony, so that we were able to make instant decisions by telephone if he was in Paris and not on site. For the first year our only 'office' for meetings in Dover was the bar of the White Cliffs Hotel. This was until the District engineer lent us an old railway coach, which was fitted out as a drawing office and stabled in a railway siding in the Harbour. Looking back on this 1958-60 survey in comparison with our later one under the Franco-British Government Commission of Surveillance, we were able to boast that 'Never in the history of site investigation has so much been found for so little'.

During these two years my diary shows that I had many other irons in the fire, or, in another metaphor, keeping 16 other balls in the air at the same time. One case that was instructive, was a visit to Malta in February 1959 on behalf of Binney and Partners, to do with water supply. Part of Malta is covered by a permeable limestone, which can collect water, but as it outcrops in cliffs on both the north and south coast, the water is apt to escape. In an attempt to collect the water, a series of small tunnels were being driven in this self-supporting ground and I was asked to inspect them and to advise on other problems. I found that the limestone was heavily fissured and that many solution cavities had been revealed, like spherical chambers. However the fissures had become bunged up by soft clay-like gunge and over the years the spherical chambers had completely filled. This was disappointing from the angle of water supply, but helpful to me because, although the limestone was not precisely the same as the Lower Chalk, it was of the same age, so that the chance of any fissures in the Lower Chalk also being choked seemed very likely, although few such had so far been revealed.

15. CHANNEL TUNNEL 1958-1972

Visit to the USA (11th to 22nd October, 1959)

This trip was my first visit to the USA, to cover several different aspects of the study. Colonel DeLong had invented several types of jack for use in oil-well drilling platforms, a new art form that was just developing in the sea off Louisiana. His firm, with Raymond Piles and Tibbetts, had teamed up with the British firm of Costain in the hope of a contract for an immersed tunnel across the Channel, a task which they sought relentlessly for a dozen years. They were engaged in laying the Hyperion Outfall Sewer for Los Angeles by dropping 14ft diameter concrete tubes in 200ft lengths for 6 miles out to sea. These lay on a flat sea bed after being dropped from a vast platform on jack-up legs and then surrounded by broken stone. This was hardly the same as a Channel tunnel but the method of working was worth seeing. DeLong's second-in-command, Suderow, flew with me to Los Angeles, leaving on the evening of October 12th by what TWA called a trans-polar jet flight. It went nowhere near the Pole and involved trying to sleep for fourteen hours in the tail end of a Constellation, until touching down in snow at Winnipeg and then on to Customs at San Francisco Harbour. A curious dialogue took place.

> Harding: 'This is interesting. I have read so much about it in Jack London.'
> Suderow: 'Who is Jack London?'
> Harding: 'Why, one of your best-known authors.'
> Suderow: 'Mr Harding, you read a great deal, don't you?'
> Harding: 'Yes.'
> Suderow: 'D'you know, I've never read a book.'
> Harding: 'Then what do you do when you are stranded in an airport?'
> Suderow: 'I think about me woik.'
> Harding: 'I read to stop thinking about my work.'

Suderow had degrees not only in civil engineering, but also in naval architecture. Perhaps his reaction showed the influence of the Reader's Digest. At the end of twenty-four hours we found ourselves in a hotel outside Los Angeles and sat up until after midnight (their time) talking to Suderow's colleagues. He then said, 'I forgot to tell you. We get up at 4am to go to sea.' A few hours later we had breakfast in an all-night drug store and drove to the beach, where we embarked in the dark on a large motor-barge, with about two dozen hefty Yank artisans more of the calibre of NCO's than of plain workmen. We completed our six mile trip out to the platform, just as the sun was rising. A XXI Club letter written on my return described the scene:

```
The main platform is a steel box 200ft long by 90ft wide
and 18ft deep. Four square lattice towers pass through the
corners of this box. They are 14ft x 14ft and are 270ft
long. The plated corners carry square holes every 2ft.
Sixteen hydraulic jacks each lifting 600 tons, all worked by
one man with press-buttons, lift or lower the platform at
the rate of 3 minutes per foot. Horizontal square pins above
and below the jacks are pressed alternately into the square
holes to grip the legs while jacking proceeds.
  Between decks are engine rooms, generator rooms,
hydraulic pump rooms, workshops, bunkhouse, mess rooms
with cooker, with pornographic ash trays and calendars.
On deck a 100 ton crawler crane wanders about. Along the
centre is a trough full of broken stone. At each end there
are 30 ton winches for grasping the pontoon carrying the
```

pipes and lowering them. Another ten winches of 30 tons are spread about, for moorings and pulling the platform ahead, all worked by one man.

To get aboard I had to climb 50ft up a Commando scaling net encumbered by my 'Mae West' aluminium helmet, camera and my own vital statistics. The legs of the platform were standing in 200ft of water and the underside of the platform was 30ft clear to avoid waves. Large buoys marked the moorings, each with 4 or 5 seals lounging on them. The pontoon, which had the pipes slung below is lowered to the sea bed and the pipes joined by pulling socket over spigot. They are then inspected by underwater television and by diver. The whole operation took three hours and I was then allowed to jack down to the platform. It was then pulled ahead and we re-embarked for the next episode.

The next step was to have lunch in a bar so dark as to be almost invisible and then to drive down a freeway to Long Beach, which is a vast area enclosed by sea walls far larger than Dover Harbour. There we saw a simple form of DeLong platform, with pneumatic jacks gripping tubular legs. The concrete pipes were cast vertically in 25ft lengths and then joined, lying on the platform to make 200ft lengths. After this, the platform was jacked down below water level and the pontoon floated on top, following which it was then jacked up and the pipes secured to it for towing out. Our drive there and back to the hotel was along the coast and not through the centre of Los Angeles, so I never saw it. In late afternoon I was driven to Los Angeles Airport for the flight to San Francisco. I had dinner on the plane and was then met by one of my hosts who installed me in the Sheraton Palace Hotel for a hard earned night's rest. Next day was spent consorting with the top brass of Morrison-Knudsen, Brown and Root and Bechtel and discussing their report and ideas.

Charlie Dunn, of Morrison-Knudsen, had previously been sent over to interview us and I became very friendly with him. Like many great men, he was small in stature but impressive as well as amusing. His fame rested on his method of sinking the vast concrete monoliths for the piers of the Oakland Bridge through unprecedented depths of soft mud to rock. The surface of the bedrock had been carefully examined by probing and the bottom edges of the caissons tailored to suit. Each caisson was made up of many compartments and each of these could be covered by a steel dome. Half of them could be pumped full of compressed air, so that the caisson could float, while the walls of the other compartments could be raised by another lift of the concrete then, by changing over the domes, the other half of the lift could be completed. Grabbing could be carried out in the open compartments. At my request a car was laid on and one of my hosts drove me round to cross the Golden Gate Bridge and the more remarkable but less photogenic Oakland Bridge. This in fact consists of two suspension bridges in series, leading into a short rock tunnel from which we emerged onto another long length of normal steel girder bridges. We also covered Nob Hill and other interesting points and gazed at Alcatraz Prison Island.

Next day I was up at 6am and with Charlie Dunn and two others we boarded Morrison-Knudsen's private Dakota and flew off to Pierre, the state capital of South Dakota. We crossed the Rockies, then miles of red desert and, after touching down for fuel, crossed the Great Salt Lake in Utah. This excited Charlie Dunn, as one of his major works had been to build a vast embankment across the lake to carry the railroad. This was in place of the trestle

15. CHANNEL TUNNEL 1958-1972

bridge built for the first trans-continental railroad in 1869. I surprised him by showing interest and told him that I remembered reading about it as a boy in my *Engineering Wonders of the World* and described from memory photographs of it. Both this and Charlie Dunn's embankment are astonishing feats as, on referring back to my book, I find that the distance from shore to shore is about 22 miles, with considerable depths of mud in many places. The vast quarry in the mountain on the east shore, which provided most of the filling, was an impressive sight, as well as the length of the bank. The old trestle bridge was still visible. We then flew past Mount Rushmore where the heads of four Presidents have been carved out of the Mountain and which was also the scene of a Hitchcock film. That evening we clocked in at a hotel in St Pierre, where I got slightly detached from the others and the receptionist asked me if I had come for the hunting. I was puzzled by this question and replied that I did not ride. It turned out that St Pierre is a great centre for pheasant shooting, which is called hunting in the USA.

Next day came the object of the exercise, to inspect the Robbins Tunnelling machines which were being used to drive tunnels under the large earth-filled Oahe Dam to carry water for power. In 1959 these Oahe machines were almost the first of their kind. There were two in action one on either side. The first one was being driven by Morrison-Knudsen and the other by a rival contractor, so that they joined up under the centre of the dam. Tunnelling machines today are two a penny (only metaphorically – they are vastly costly) as great strides have occurred in their development. On the Morrison-Knudsen side there were to be seven drives of 1,500ft from seven shafts, 42ft in diameter and 150ft deep. Two gantry cranes straddled the row of shafts and these could handle the 100 ton cutter assembly. The tunnel was supported by steel circular ribs at 4ft centres, with random steel plates between to hold the ground until it was eventually lined with concrete. The circular cutting assembly was bolted to a 'jumbo' or platform of steel frames which ran on rails welded to the steel ribs. This jumbo was about 14ft square divided into two floors and was 40ft long. The lower floor carried all electrical equipment, transformers, air compressors, switch gear, automatic controls and the hydraulic pumps for working the many hydraulic jacks. On the floor above were the conveyor belts for removing the spoil and devices for erecting the four quadrants which built up into the lining rings. The mole was propelled ahead every 4ft by means of jacks pressing on the ribs at the side. It had just come to the end of one drive and we were able to crawl all over it as it was about to be taken down for transfer to the next shaft. The machine had taken ten months to make and a month for its first assembly at a cost of $50,000 to erect. They needed to work fast to beat old-fashioned drilling and blasting. We then went to see the rival machine at work. This had longer drives but from open ends. Unfortunately, as we were nearing the face, the machine stopped as something had gone wrong, so its progress had to be believed and not witnessed.

While my hosts were gossiping with the project manager, I got hold of his assistant and by relentless questioning and persuasion made him pass over copies of drawings, so that, when back home, I was able to compile a report of ten pages and six drawings with full details of methods and progress in broken shale. The outside diameter was 29ft 6 inches while our choice for the Channel was 23ft 10inches, so the comparison was valuable for our estimates and my visit well worthwhile. After lunch, we re-embarked the Dakota and set out for Denver, Colorado. As we were getting nearer, Charlie Dunn felt ill and I also noticed that the pilot was taking oxygen. I could see the ground not particularly far below us, so supposed that this was because the pilot was an elderly man. I later found out that the flat land below us was a mile above sea level, so our altitude was higher than I realised. While the rest of our party flew back to San Francisco, Charlie and I stayed the night in Denver in a fine specimen of a Mid-West hotel. The next day, Charlie remained while I boarded a plane for New York, where I was met by Professor Means of Technical Studies Inc., who installed me in the St Regis Hotel, immortalised as the home of *Auntie Mame*.[42] By then it was Saturday night.

On the Sunday José Machado[43] took me to his apartment high up over posh Gracie Square and then, with his wife and daughter, drove me 90 miles out to Cornwall, Connecticut to meet his sister, Dodie Prentice. We were having a pleasant chat in the garden when we were joined by a man in jeans, with the typical American peaked workman's cap. As he joined in the conversation, I thought that he was the 'hired man' but soon found that he was my host, a highly distinguished architect from Hartford, Connecticut. Cornwall is a weekend hideout of old white timber New England houses, the area had been farmed by descendants of the Pilgrim Fathers who later went to the prairies, so scrubby timber had taken over the previously ploughed fields. The next few days were spent hob-nobbing with bankers as well as other interested parties. On the Monday I found my way to Technical Studies Inc. which was up a skyscraper off Fifth Avenue. Here I consulted with Frank Davidson and brought him up to date with what was happening. We then went off to meet Colonel DeLong at 29 Broadway who took us, with Mr Bickel of Parsons Brinkerhoff, to lunch at the Down Town Athletic Club. I then met Mr Kingman Douglas, partner of the bankers Dillon Read, at the Harvard Club and ended the day with a sole supper at the hotel and an early night.

The next day saw me again at DeLong's with Mr Bickel for a discussion on the problems and pros and cons of an immersed tunnel. DeLong had a scheme for a formidable platform for dredging a trench and another for handling the pre-cast sections. I had to point out the need to continue tunnelling for seven miles or more once ashore, as the Folkestone Warren was too unstable a landing ground and also the need for fourteen changes in grade to suit the sea bed. We then went off to lunch with Lamont, the head of Morgan Guarantee Trust, on the top floor of 23 Wall Street in the directors' dining room, where we met Steve Bechtel who was also a director. Following this, I went to 10 Gracie Square to meet John Young, partner in Morgan Stanley and Co., this conveniently was in the same block as the Machado's, with whom I then had supper, followed by a night drive round New York, before they dropped me at the St Regis.

It was now Wednesday and I spent this with Frank Davidson and Means discussing our efforts. They seemed to think that we had made a good impression on the bankers regarding feasibility and methods of working. At my request, they took me on a more cultural tour, visiting the new Museum of Modern Art (Frank Lloyd Wright's latest work), the Metropolitan Museum and Art Gallery, the Natural History Museum (where the diplodocus collapsed in the 1949 Gene Kelly film, *On the Town*). I found there that I was able to give my companions a short lecture on geology, so that they could understand our efforts with advantage and then inevitably I asked to go to the top of the Empire State Building with its surprising view of New York. We rounded off the day with a short experimental trip on the New York Subway and a drive through the newly completed Battery under-river tunnel to Idlewild Airport[44] for a British Airways flight back to Heathrow. I found myself seated beside an engineer from Raymond Pile Co., who was also hoping for the immersed tunnel.

---

42 This refers to the best-selling book published in 1955, rather than the 1958 film.
43 A friend and also HJBH's son-in-law's uncle.
44 Since renamed Kennedy.

His father and his son were both Mormons, although he was not, but he told me much about them to pass the time. The plane left at 9pm and got to Heathrow at 8:45am GMT but the flight was only 6.5 hours. We were served a long and leisurely dinner, but were woken up again after two hours sleep for breakfast. After completing reports and writing many letters I took a day trip to Paris to provide a quick update on the visit. Following that, activity was fairly intense, until the Group unveiled their ideas at a press conference on April 20th, 1960 (more on this follows below). In the intervening periods, there were interviews at my house with Ed Murrow and, among others, a young man named Reginald Bosanquet, who admitted that it was his father who had introduced googly bowling. He later rose to delivering News at Ten on ITV.

Leo d'Erlanger, the most courteous and kindly of men, would pass over to me the many enquiries from the press and also the freak ideas which poured into him as Chairman of the British Channel Tunnel Company. The members of the Group, great men though they were, had no idea of conducting public relations and I found that this duty fell upon Malcor in France and on me in the UK. They only allowed money for Malcor to make 200 copies of the internal report and only 100 copies of an addendum, which contained a closely written and valuable summary of material that had emerged, which ranged from claustrophobia to tidal effects. If they had had the wisdom to publish this 'Report of the Delegate' it would have sold many copies, possibly even becoming a best seller. The preparations involved many letters and hours of translation, as the agreed version was to be in French and English.

In January, Malcor gave a formal lecture to the Societé des Ingenieurs Civils de France, in which I took part and for which, in due course, I was given the Prix Croisseau of the Societé. Following this, I was invited to lecture to the Royal Geographical Society on February 8th. For this lecture I had prepared many slides with coloured photos of our work and many maps and diagrams. By coincidence, the lecture was on the evening of the day when the whole Group were meeting in London, so they all came along, M. Massigli, Louis Armand, Georges Picot, the Davidsons and the British contingent. For the first time they were able to see for themselves what they had been paying for and were very pleased. I had found a pamphlet dated 1908 by 'A Military Expert' so had made a slide of his plan showing the tunnel on the French side emerging onto a viaduct and then disappearing into the cliff. This also showed the British Navy shelling this as a means of putting the tunnel out of action in case of war. My comments on this and my slide of a published drawing of a lavatory plug hanging from the tunnel roof with a French and British General reading the inscription 'En cas de guerre, tirez!' led Georges Picot to say to me, 'Now I understand what is meant by the British sense of humour'. Sir Ivone Kirkpatrick was all set to make a speech at the end of the lecture, but after the chairman had invited Malcor to speak, the chairman closed the meeting. Sir Ivonne was very cross and turned to the lady sitting next to him and said, 'These engineers talk too much', to this my wife replied 'Yes, don't they!'. He never guessed that she was my wife. Sophie and I then went off to dinner as guests of the Society. This was the first of nearly fifty lectures that I was asked to give on the tunnel over the next ten years.

While Malcor was writing his internal report, the Study Group composed their own report to be presented at the press conference. I was appalled when I saw it. It was a small grey booklet of 32 pages, the size of a paperback novel, with no maps, pictures or diagrams and most unimpressive. Malcor's report was very different and apparently the Group, being busy men, thought that journalists too would only want the barest summary. The press conference took place in Church House, Westminster without a drink in sight!

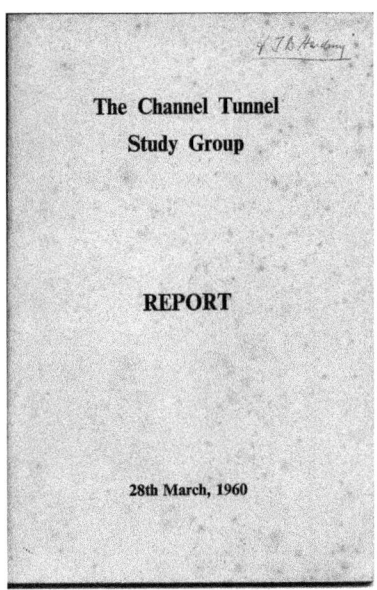

45. *Original stark report produced by the Channel Tunnel Study Group*

Unfortunately, it had clashed with the first announcement of Cockerill's hovercraft, otherwise we could have shown fifteen impressive bound reports from all the various firms and people who had worked with and for us. For some reason, this was not done and the chance was lost of rubbing in the fact that these were all at the top of their professions in Britain, France and America and that the whole study had been thorough and impartial in arriving at its conclusions. These conclusions were that the cheapest form of construction, yielding the quickest return on the invested capital, was a twin bored railway tunnel, using 25,000 volt electric traction, with modern signalling which could pass one train every five minutes. Not only could it deal with all the rail traffic for many years to come but could also carry 1800 vehicles per hour in either direction, compared with 1300 in a much more expensive road tunnel.

For many years there had been a Parliamentary Group of Members from all parties that supported a Channel Tunnel and whose Secretary, Commander Powell, was also the Secretary of the Parliamentary and Scientific Committee on which I represented the Institution of Civil Engineers. Some Members recommended the Study Group to engage the public relations firm of Whittaker, Hunt and Company to represent their case, as they had found them competent. Public relations officers (PROs) had not come my way and, like many others, I confused their activity with advertising. I discovered that a good PRO works to ensure the spread of truth and the correcting of false versions. I quickly came to admire the work of Donald Hunt and the value of his contacts with leading editors and journalists. I found that once we had filled him with the facts, which he quickly grasped, he organised an appropriate press conference at Brown's Hotel, which included the proper and vital supply of fluid to wash down the message. There was also the chance for journalists to wander about during the buffet lunch and talk to those of us who had been involved in the work, so as to fill in the picture. Don thus relieved me of my own unwilling PRO efforts and we continued to enjoy working together for many years.

Next came a four year interval of spasmodic efforts, until our major investigation in 1964-65 and the after-effects of that historic work. In the next section, there is more detail of some of the other incidents which were interwoven with much other work in those four years, including twenty flights abroad, several arbitrations and four years as Vice-President of the Institution of Civil Engineers, becoming President in November 1963.

Other Work

The first incident was the writing and presentation to the Institution of our paper, 'The Work of the Channel Tunnel Study Group 1958-1960' by Bruckshaw, Goguel, Harding and Malcor. This presentation was held in London at a joint meeting with the British Section of the Societé des Ingeniours Civils de France on Tuesday 21st March 1961. Only those who have suffered the pangs of conceiving and giving birth to a major professional paper can sympathise with the period of labour, more so if it is a multiple birth by four authors of two nationalities. The resulting progeny proved so popular that over 450 members turned up for the christening discussion, so this was held in the ICE Great Hall as the normal Lecture Room only held 100 or so people. The other three authors of our paper were awarded the George Stephenson Gold Medal (Vice Presidents are, quite rightly, disqualified, but get a letter of thanks). We emphasised that our studies had been impartial and declared that tunnels, whether for rail or road, bored or immersed, were feasible at greatly differing costs. A bridge, as well as a composite of bridge-immersed tunnel solutions, was also feasible but with formidable problems. As expected, those with road interests could not bear the idea of driving cars onto flat rail-cars, although they are prepared to drive onto ferry-ships or even aircraft. So they pressed for road tunnels, even if 34 miles long, or preferably a bridge as they feel the urge never to stop! The discussion was lengthy, we had described how cars and any truck smaller than a double-decker bus could be loaded and whisked through in 50 minutes, of which only 30 minutes of this was in transit, regardless of the weather. Sir Owen Williams, the prophet of reinforced concrete, was advocating a (concrete) bridge, so he asked how people could be expected to spend 50 minutes with no lavatory accommodation, I could not resist saying in reply, 'We have all been sitting here in the chamber for three hours'.

My old college friend, Jack Pain, by then a director of Dorman Long, had been inveigled into a consortium of British and French firms who were seeking a contract for a bridge. He got a rather unkind answer, 'It ill became Mr Pain to complain that bridge foundations were not mentioned in the investigation of the Channel bed, when he himself was sponsoring the design of a bridge with no investigation whatsoever'. A year later I debated this bridge-tunnel matter with Pain before a meeting of the Parliamentary and Scientific Committee at the House of Commons where this body was made up of MPs sufficiently keen to keep up with the science and two non-MP members from each important but immensely varied scientific body or society. I represented the Civils and for some years was Vice President of the Committee. Their custom was to have periodic meetings in one of the committee rooms at the House in the late afternoon, to listen to a speaker and discuss his subject. Then a dinner would be held in a private room for ten or so MPs, ten other members and the speakers, during which the discussion would continue, but on more robust and amusing lines. The non-MP members were invited in a sort of rota and the programme was chosen by a sub-committee. I had one amusing battle with Mrs René Short (Labour MP) over the tunnel and Commander Powell considered that I won handsomely. Although her politics were the opposite of mine, we both enjoyed it.

At fitful intervals there were meetings of the Study Group, sometimes in different merchant banks in London and at Suez in Paris. Many interesting people attended, such as George Ball (who became a famous USA diplomat), and for these meetings M. Ambassador Massigli lunched us in several spots off the Champs Elysées and at one or two embassies. Two 'experiences' were organised on separate occasions by the French railways, SNCF. In each case, French engineers provided their own cars. In the first case, there was a special train of six coaches with three double-decked car-carrying wagons at the rear, with the lower decks loaded with new Dauphines for delivery. The French drove their own cars up a ramp and along the upper decks, where they were left in gear with engines off and brakes on with chocks applied. We were then all invited to seat ourselves in the cars and away we went at 140km (87 miles) per hour, for 50km to Montereau. It was a lovely smooth ride and a fine view of the countryside. We returned inside the coaches and were then entertained to an excellent lunch by SNCF in the Gâre de Lyon. It is a good idea to work with Frenchmen.

A year later I was able to take a day off from an arbitration for a quick trip to Paris on the night ferry, for a meeting of the Study Group, followed by 'une autre experience'. This consisted of sixty cars being driven, this time along the length of a number of flat trucks, at the Gâre de Lyon. We applied the hand brake, this time with no chocks, and departed at 85mph to Fontainbleau, where the sixty cars were driven off in less than 2 minutes to a temporary platform. We then had drinks and snacks with much French speechifying and drove back onto the flat trucks and were whistled back to Paris in style. This was before the British car-carrying trains had become popular.[45]

The idea of the Study Group was to proceed with the necessary steps to bring about the bored railway tunnel by private enterprise. This was to be by the issue of shares, as the support of finance houses was forthcoming. In addition, it was necessary to get the support of the British and French Governments, as treaties, agreements and so on would be needed. The two governments set up their own working group in 1961, which proceeded to tramp through the problem of tunnels, bridges and combinations all over again and they came to exactly the same conclusions, but upped the costs and reduced the revenue expectation. Because of all this, the meetings of our Study Group became more involved, bringing in other important worldly figures, which meant that some meetings were carried out with the aid of a young lady interpreter-reporter used to working at international conferences. Her technique was to scribble down short sentences on separate pieces of paper and discard each one on to the floor as she interpreted its message. British and French Railway staff joined in our meetings as they were keen on the project and had many of their own interpreters. The title of British Transport Commission, British Rail etc. changed every two years, as did their organisation, and we lost Lord Robertson and gained, temporarily, Dr Beeching. The idea of a model of the terminal and tunnel was mooted and so on the Saturday morning after a Friday meeting in Paris, M. Hutter, the top man of SNCF, took Mr Quicksmith of British Rail and me to the best toy shop in Paris. Here, we

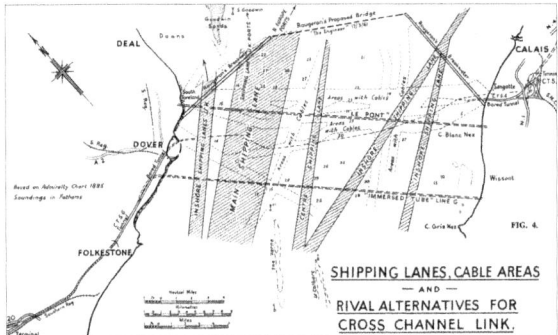
46. Shipping lanes in the English Channel, published by the Dock and Harbour Authority, 1961 - part of HJBH's scheme review

47. Artist's impression of the finished tunnels as proposed in the 1961 version, published by the Dock and Harbour Authority

---

45 Blink and you miss that one, apart from Eurotunnel, they ceased in 1995.

astonished the sales girls by measuring up both British and French rolling stock and the gauge of the rails for use in a working model. Our chance of playing trains was diminished when we found that the scales for British and French stock did not agree and the rail gauge was out of proportion, so in the end the large model which went on show at Marylebone was equipped with rolling stock made in British Rail HQ. It was all great fun at the time. It had always been our stated intention, once a start could be made, to spend a million or so pounds (1960 currency) out of the total for the contract on an intensive detailed survey of the sea bed, with the proviso that, if the Government carried out their usual volte-face after the start, then the Governments should reimburse the group of companies for this cost.

Survey

The next real crunch was the large-scale survey of 1964-5. During the previous four-year interlude, my work on the Channel Tunnel was fitful, interweaved with much other work and travels abroad. It consisted of meetings with civil servants, lecture engagements, meetings with the Group and fending off contractors eager to get a contract for tunnels whether bored, or especially immersed, as well as pressures for a bridge. In order to advance the project by educating our rulers, Leo d'Erlanger would have occasional lunches at his house in Upper Grosvenor Street for five or six Members of Parliament, one party at a time, and he would ask me to stand in to assist him. There was one occasion, after the 1964 election, during which Gordon Walker lost his seat although Secretary of State for Foreign Affairs, where I found myself sitting opposite Mr Enoch Powell MP at lunch. In the general talk I said that I thought that if Gordon Walker had stood for another constituency (whose name I now forget) he would have been elected, to which a partner of Mr d'Erlanger agreed. Enoch fixed me with his glassy stare and sneered, 'We have an engineer venturing an opinion on a political matter.' Unfortunately, as I was acting as a host, I could not metaphorically kick his shins as he deserved, as politicians give opinions on every conceivable matter, upon most of which they are ignorant. Some years later on a BBC satirical programme Enoch Powell was described as 'The creep that makes your flesh crawl'.

The Governments had continued to shilly-shally until in September 1963, Mr Marples, Minister of Transport, presented Command Paper No 2137 to Parliament – *Proposals for a Fixed Channel Link*, stating that a rail tunnel was technically feasible and that the Governments had decided to proceed. They proceeded so slowly that, after many more talks, another Command Paper appeared ten months later (dated June 3rd 1964). This paper, No 2416 Treaty Series No 35, established a joint Franco-British Commission of Surveillance for the survey with detailed terms of reference. The Study Group was then drawn into contracts with the Ministries on July 7th 1964. This signalled the start of the major survey of 1964-5, which unfortunately, by procrastination, had lost the calm summer of 1964. The Commission of Surveillance was formed because the Governments had decided to finance the cost of the survey themselves. It consisted of six (or at times more) civil servants from each Government, admirable and agreeable gentlemen, but so keen on keeping down costs due to their restrictive terms of reference, that they succeeded in helping the costs to escalate. Colonel Dennis McMullen RE, Chief Inspecting Officer of Railways was the British chairman and M. Mathieu, a product of the École Polytechnique in the same year as M. Malcor and a leading light in the Ponts et Chaussées, was the French chairman from their ministry. Some of the others knew little of site investigation and even less of working at sea. In spite of the words in the Command Paper, the Study Group, being an intermittent gathering of individuals, could not organise such work, which became a matter for the consulting engineers. The general scope was described in the paper on 'The organisational framework and principal nature of the work'. Grange and Muir Wood write advisedly:

> The administrative machine for the work was complex on account of the unusual position of the Study Group as client (to the consulting engineers) and suppliant (to the Governments) with the additional feature that on the French side, for reasons of budgetary control, all finance was handled through the railways (SNCF).

The work itself was a major sea-going expedition in one of the world's busiest sea-lanes. Our own intention was to follow the lines of our 1958-60 expedition, as I have described, but multiplied twenty-fold in scope. In addition numerous other tests were to be carried out. No previous tunnel of any sort had been so intensively investigated throughout its length, with results recorded in such intricate detail.[46] Unlike 1958, the consulting engineers were fully mobilised at the start, so they provided the considerable staff required, which meant that Malcor and I were spared our previous role of 'pushing the barrow all by ourselves'. This was just as well for my term as President of the Institution of Civil Engineers started on Guy Fawkes Day 1963 and ended on the day of the Opening of Parliament a year later, so it overlapped with the start of the survey.

The Government provided us with a site office by taking over the abandoned Officers Mess in Dover Castle, with a subsidiary office in Calais and this was where Joint Franco-British staffs were gradually assembled. The most vital member was a bi-lingual secretary-typist and an agency sent us a charming and highly intelligent young girl, who was at first disconcerted by the shambles of a partly furnished office. Then one of our engineers told her that the place was haunted so she decided to stay. Half way through the contract she married one of our younger engineers and towards the end, as our report began to take shape, she herself showed the shape of things to come – the happiest result of our survey.

The first boring at sea started in style on September 14th 1964. Mr Marples, the Minister of Transport, and M. Jacquet his French opposite number, followed by 50 gentlemen of the press, drove to Dover Castle and inspected the 'troops' drawn up on parade outside the office on the edge of the cliff. After the Minister had shaken hands with the members of the Commission, the members of the Study Group and with condescension the rest of us, we all proceeded to lunch at the Maison Dieu. Selected members of the party then embarked, Mr Marples with us, on the Trinity House flagship *Patricia* and M. Jacquet on the equivalent *Les Phares et Balises*. Just as we were embarking, our lovely bi-lingual secretary and her young typist appeared on the wharf in trousers and sweater all ready for the trip. Colonel McMullen sternly forbade them to come aboard, but they looked so crestfallen that, as soon as he had gone below, I called out to them to come aboard quickly, and they were just able to scramble on as the gangways were withdrawn, so cheering us on our way. We put to sea followed by our two boom defence vessels, each craft dressed overall, until we came close to the first boring vessel off the French coast. The two Ministers then made parting signals and M. Jacquet disappeared to Calais and we returned to Dover, with the mainbrace well and truly spliced.

A bookseller friend once told me that when a biography is published he sees people turn first to the index to see if their name is there. It would be invidious to enter all the names

---

46 Anyone who would seek professional details can find them in the paper in Volume 45 of the *Proceedings of the Institution of Civil Engineers* for January 1970 entitled, 'The site investigation for a Channel Tunnel – 1964-65' by A. Grange and A. M. Muir Wood.

involved in the sort of work now described but I must mention my friend and future ICE President, Alan Muir Wood. He had led the team of six who made the 1960 design, report and estimate, but by 1964 he had risen to be a partner in his firm of Sir William Halcrow and Partners. The first man sent to Dover was quite out of his element, so I asked Leo d'Erlanger to persuade the firm to liberate Muir Wood to spend his full time on the work at Dover, with most successful results helped by his fluent public school French which was fully understood.

We found all along that there is something about a Channel Tunnel which seems to mesmerise people out of all sense of proportion and reality. On any normal civil engineering contract, any leading consulting firm is accustomed to simultaneously letting and supervising several contracts worth considerably more than our survey, which eventually cost little more than two million (1965 currency) British pounds. This is a small sum, considering the scope and siting of the work and the cost of the future tunnel. All the work was done in twelve months from September 1964 to September 1965. We had to place over eight major contracts for chartering vessels and platforms, for boring on land and at sea, installing Hi-fx Decca navigation aids and other matters, also ten medium sized contracts for other equipment, not forgetting nineteen other small contracts or negotiations. If we had been on our own much of this could have been done by talk and exchange of letters, but we had the Commission of Surveillance of twelve or more personalities breathing down our necks. They in turn suffered from their own senior Permanent Secretaries, who were being bullied by Ministers, who were frightened of their MPs and the Parliamentary Estimates Committee, while all the time the press lurked in the undergrowth on hopeful look-out for bad news. In their zeal for economy the Commission insisted upon scrutinising every word of these many contracts, dotting every T and splitting every infinitive, with the delay landing us into the winter.

A Hi-fx Decca system which could locate craft within 10 metres and Hydrodist, which could penetrate through poor visibility for survey, were installed with a master station in Dover Castle and slave stations at Dungeness and Bleriot Plage near Calais. One of the two major items was the fresh geophysical survey by the EGG firm from the USA using Sparker and other means, utilising powerful launches. This was completed within six weeks from early September. In all a considerable fleet of vessels was employed, with intricate insurance problems. We used four tank landing craft (Mk IV), two platforms with their attendant tugs and supply vessels, two Royal Navy Boom Defence vessels for moorings and buoys, 6 or 8 launches and a few other non-descript craft for special uses, not forgetting much diving which was only possible for about an hour around slack water. The large regimental mess room was used to store the cores, which 'if laid out end to end' would have covered 6,000 meters. In addition a well-equipped laboratory was constructed in this hall.[47]

So let us get on with the action. We had made enquiries early in the year for possible boring ships but, just when the Government were summoning up their courage to move forward, the rush to the North Sea started. The vessels which we had earmarked were snapped up, not to drill for oil but to test out sites for oil drilling platforms. It is as well to remember that there is a vast difference between boring for valuable oil at great depth with expensive equipment and boring relatively shallow holes at a cost which shows no immediate cash

---

47 This can be seen in a Pathé newsreel of the time http://www.britishpathe.com/video/french-channel-tunnel-survey-under-way-dover

return. We sent out 17 tenders for contracts for boring at sea and only five firms responded. A contract was agreed with Wimpey-Forasol (a consortium of Wimpey and Soletanche) and with Forzky, all of whom I had been previously involved on earlier projects. Forasol provided their own craft *Caulville* and Wimpey surprised and pleased me by buying 'my' old ship *Grosvenor* which had been sold on by Mowlem in 1958. Unfortunately the *Grosvenor* had to have her bottom replated, which kept her out of action until the end of the year, while *Caulville* had to complete a French assignment, so two other chartered vessels were commanded and manned by crews that did not belong to the consortium. One crew was Spanish which did not help communication.

As a result of the predicted late arrival of our two best ships, we had to start the boreholes with these two weaker vessels. The Commission and some others doubted the ability of the ships to bore in the 60 metres of deep water in mid-channel, so made them shake down their crews and teams in shallow water. This meant boring through 120 metres instead of only 80 metres in deep water to reach the tunnel line. As a result, a vessel had to remain moored over each hole for a much longer period. As the winter weather turned wet and windy, it was found that they were unable to bore successfully in a swell over 4ft or a wind force of more than Force 4. This meant that the ships had to return to harbour several times and then seek out the partially drilled hole when the wind dropped. An unexpected hazard arose where twelve inch diameter guide tubes were drilled into the sea bed and, when a vessel returned to port, would be left standing in the water. On several occasions, when the ship returned to its moorings, the tube was found to have broken off and was lying on the sea bed. The cause was found to be the swirling currents that had set up vibrations of a critical period of oscillation, which caused the tubes to snap (this destructive resonance was later reported as affecting temporary tube structures at the Severn Bridge and on tube piles in a jetty at Immingham). It seems an appropriate moment to quote from Malcor and my report to the Study Group on completion of the work:

> A survey of this kind at sea is completely different from the same number of borings on land. There are four main variables – Finance – Availability of craft – Time (whether Summer or Winter) – and above all, Weather. A hard winter may coincide with a calm sea at the risk of fog. A wet and windy Summer or Winter may prevent fog but can completely alter working conditions.
>
> The weather in the Straits of Dover is notorious for its fickle nature and sudden severities, due to the funnelling between two large areas of water and the change from Island to Continental weather conditions. The currents are powerful, up to 5 knots, and these continually change direction. Many operations can only be carried out at times of slack water.

The intended position of each borehole had to be notified to the Admiralty, Ministry of Transport, Trinity House, Lloyds, the Ministry of Agriculture, Fisheries and Food and the French equivalents, so that the precautionary 'Notice to Mariners' could be issued. Only once this had been done could the invaluable boom defence vessels put to sea with a large load. First, they had to place three of the large Trinity House lighted buoys, two forward and one aft and then distribute six Stanforth anchors mostly four to five tons each, with 16 metres of chain cable to each as extra drag weights. These would be set around the position fore, aft and on the quarters. All these had to be buoyed for picking up and recovery. The boring ships had to have six winches at work, correcting the length of mooring cables to hold the boring gear over the holes as the tide and currents changed. Before a ship went out the Meteorological Office had to be rung up for a forecast, their reply, 'what is it like with

15. CHANNEL TUNNEL 1958-72

you?' is not as silly as it sounds, as it would help them to pinpoint our area.

The early set-backs caused alarm and despondency to the Commission, which they gave vent to at frequent meetings. They would summon us to a meeting at Dover in the morning and keep us kicking heels while they wrangled amongst themselves. On one occasion they kept us from 2:45pm until 7:45pm while they scolded us. This made us all pretty mad, as most of our troubles had been due to the Commission's own actions and inactions over important decisions. After all, in terms of 'professional pecking order' our side had more first division players than they had. The brunt of these scoldings fell upon Malcor and me, as leaders of the Group's side, but they failed to break our customary united front. At one meeting Malcor retaliated with a torrent of voluble French, I added, 'I entirely agree with M. Malcor' and then in a suitable quiet moment I asked him what he had said! After one period of bellyaching I decided on a different tactic, I said 'Gentlemen, we come before you like the Burghers of Calais, with halters around our necks', the French members then cried out, 'Ah, zee Burghers of Calais, zee Burghers of Calais' and became wreathed in smiles, this relieved the tensions and after that things got better. I told Colonel McMullen that we were not witnesses in one of his railway accident cases to be interrogated and cross-examined and that the Commission should look back over the way things had happened. After that we had little trouble.

By December only eight holes had been completed by the two inferior ships and now that the two good ones had come alongside, the decision was made to pay off the first two. When M. Louis Armand heard that the ship with the Spanish crew was to go he made a classic remark, 'At Trafalgar half our ships had Spanish crews and that is why the battle was lost!'. The *Grosvenor*, renamed *GW14*, and *Caulville* were each wholly owned by their respective firms, who also 'owned' the crews and boring gangs, so each ship could work as a self-supporting unit and their performance was in strong contrast to the first two. My idea had always been to spend the winter in trying to find two good vessels, or rather helping the contractors to do so, and complete the programme by this proven cheap means, but others lost heart, so a search was made for platforms instead. Our previous search for platforms and large drilling vessels had proved abortive, I quote from our report:

> It must be remembered that in the oil-rush conditions of 1964 such valuable properties were not lying idle for our convenience, to be available for hire at the wave of a hand.

A start was made by asking the consultants, Posford and Pavry, to report on the possibility and the suitability of using the submergible platform which they had designed for the National Coal Board. The answer was 'No' and so on February 11th I met a select party at Copenhagen Airport – M. Mathieu and Colonel McNaughton of the Commission and Malcor, Muir Wood and Grange. Two Danish engineers took us for a chilly drive across Zeeland to inspect a floating platform which was being used to survey the 15 mile ferry crossing of the Storebælt for a possible immersed tube or bridge. Next day we watched a model of this platform being tested in simulated Channel conditions in the tank of the University in Copenhagen. It behaved badly so we flew home empty-handed. On February 23rd I travelled by night ferry to Paris to attend a full day meeting of the Commission and the next day, with Malcor and Ribeyre, caught the 7am train from St Lazarre to Rouen. There we inspected a full-size oil drilling platform being built on launching ways, which was to be named *Neptune* (she later made news in the North Sea) and we were told that she might be available in June, if her vast hire rate could be accepted.

The following day was a complete change. I fulfilled an invitation to lecture to the Societé des Ingenieurs Civils de France at their institution building in Rue Blanche. As we had paid several visits to the extension of the Paris Metro, including inspection of the large Robbins tunnelling machine, I chose as my subject *The Building of the London Metro*. I had typed a script to match my slides and had tried to turn it into French and I opened by saying, in French, that I had tried my script on M. Malcor and he had said, 'C'est claire, mais malheuresment, c'est mauvais francais'. This went down well and Malcor translated my script to match the slides. First I showed tube construction, including part of a pile of the Victoria Line, and then station construction. I showed slides of Piccadilly Circus Station, including the booking hall, being carried out in difficult tunnelling under the road concrete and then the London Transport film of Oxford Circus. Bored piles were shown being done at night and then, over one holiday week-end, prefabricated steel deck units were brought in during the night and placed on the piles and bolted together. This formed a roof over the booking hall so that it could be dug out in the open – so to speak. To my surprise, when the film showed the first bus driving over the platform early that morning, the French audience burst out into long and sustained clapping before the film had finished. The meeting lasted from 11:30 until 1:30pm as, very sensibly, it ended with a buffet lunch.

Following this it was back to the search for floating platforms. We found that the only other possible one would not be available until the last week in June, as it was in the course of being constructed by cutting off one end of a large DeLong platform that had many legs. The longer section (named *Sea Gem*) eventually went off to the North Sea where, in 1966, it collapsed through brittle fracture of jack tie-bars with tragic loss of life that led to a public inquiry. The smaller four-legged platform was named *Gem III*, which was short for the name of its owners who eventually hired it to us, Compagnie Generale d'Equipment pour les Travaux Maritimes. It needs no stroke of genius to realise that it is much 'easier' to bore from a platform, but later we had to remind our critics that in 1964 they did not exist. Also that although the hire of tank landing craft had advanced from £125 per day in 1959 to £362 in 1964, the hire of *Gem III* was £1,740 per day, while *Neptune* was very much higher. Having said that, on March 31st the Commission were given clearance by the Governments to spend money on hiring these two platforms.

While awaiting *Gem III* and *Neptune*, our ships *GW14* and *Caulville* went on drilling with ever increasing success. This seems a good moment to list the tests which were carried out on completed boreholes after each had been cored from top to bottom. Our team included the appropriate boffins to interpret these mysteries, led by Professor Bruckshaw, a charming and invaluable man. Unhappily, he did not live to see the start of the tunnel work but he also did not suffer the frustration of seeing it aborted (to the wrath of the French) by a timid Government. As each borehole was nearing completion, the appropriate expert had to be ferried out to carry out the many tests while the ship was still at its moorings. These tests included verticality, permeability (by pumping in or out through Wimpey of Lynn packers), television mini camera inspection, sonic, caliper, SSL logging, Gamma logging, gamma-gamma (whatever they are) which went to show that we were doing our best and that organisation was needed. Towards the end of our time, five other odd vessels were taken on short term hire for tests by oblique asdic, acoustic surveys, drop sampling, transport of sea bed material (for its effect on immersed tunnelling) water sampling and the use of Kelvin Hughes direct recording current meters. The cores themselves were subject to many tests in the site laboratory such as triaxial strength tests, micro-fossils and calcimetry and these were all recorded in the final reports. No stone was left unturned. In addition, we must not forget the elaborate recording tide guages that we installed in Dover and Calais Harbours, from which better brains than ours deduced the relative datum levels between our Newlyn Datum and the French NGF (Nivelle General de la France). This was done from 160 readings of high and low water on particularly calm days and then 'making allowance for accelerations, surface and bottom stresses, atmospheric pressure and Coriolis effect'. Naturally this intricate performance was organised from Dover by the day-to-day work of Muir Wood, Grange and the consulting firms, with René Malcor acting as a form of visiting circus ring-master and me as his part-time alter ego. At first, the Commission were a critical audience, until we had educated them. Meantime, Don Hunt was occupied in educating the Press, who were trying to announce that we were behind schedule when we had no schedule.

Returning to the matter of the platforms, *Gem III* was towed to the site on June 19th, 1965. Her square hull was 50 x 27 metres and her four legs of 1.8m diameter were 59m long. The underwriters would only allow her to work in 30m at low tide, in order to keep the double-decked hull clear above water by 8.5m, which would protect her from tides, waves and surges. However, many of the boreholes had to be made in 60m of water. A platform needs a tug in attendance to move it from site to site and a special crew to work the jacks and moorings until the platform has been elevated on its legs. If no cores are needed, holes can be drilled rapidly, however when coring, the string of rods has to be drawn up to empty the core barrel. *Gem III* had a flat deck with no equipment except a crane, as the crew had quarters below deck, so we mounted two drilling rigs, one on each side. One cored steadily all the way, while the other drilled rapidly down to the tunnel zone. Tests were then carried out while the main hole was still coring, which saved much time. *Gem III* also was unaffected by the perverse weather of June and July, so began, almost, to pay her way, but the ships still remained the only way of boring in the depths from 30-60m.

One day we crowded a bunch of assorted press men onto the tug and took them out in half a gale to visit *Gem III*. To get aboard, the crane lowered a contraption made up of a large rubbery ring of 9ft in diameter, which was suspended 8ft below a 6ft ring by Commando type landing net. As the crane waved this contraption to and fro above the heaving tug, we leaped nimbly on to the lower ring, clutching tightly to the net while we were hoisted up on board – all except a correspondent from the USA who did not like it. Don Hunt remarked, 'I have not been so frightened since I towed a glider into Arnhem'.[48] Luckily the press were duly impressed and took a kinder line. *Gem III* carried out 22 double bores in 110 days and left for Le Havre on October 10th. On June 5th the large *Neptune* had started on a month's work, at a price reduced from its normal astronomic heights, so that the French drilling crew could get into training before battling with the North Sea. *Neptune* had a triangular hull 10m high to provide space for machinery and three decks for all home comforts for the crew. Although her triangular lattice legs were 76m long, she could only drill in 46m at low tide and it is interesting that she was almost the first to start boring in the North Sea before television had made us all familiar with such craft. The Commission of Surveillance became more sympathetic as they learned for themselves the art of working at sea.

In mid-June we made a combined sea trip with the Commission in one of our attendant vessels. First we boarded *Caulville* so that they could see for themselves what it was all about

---

48 It is shown in the Pathé Newsreel previously mentioned http://www.britishpathe.com/video/french-channel-tunnel-survey-under-way-dover

and then on to spend some time on *Neptune*. This was the first chance to watch the complex and highly sophisticated equipment used for oil boring, although in this case they were not using mud-flush, but it was still a powerful method of doing simple coring. *Neptune* bored five holes for us in her stay of a month, at a cost of £27,000 per hole. By this time the two ships had completed 20 borings and they went on to finish their total of 46, mostly in the deep water, by the end of September. Out of the total score, the first two ships only completed 7 holes, *Caulville* did 15 and, to my natural pleasure, *GW14* (the ex-*Grosvenor*) had completed 24 (her bores cost about £11,000 each). It made me feel that in 1950 I had backed a good sea-horse when I bought her for only £10,000. I had always pinned my faith on boring from landing craft as the cheapest method, if only two more good ones could have been found by Wimpey-Forasol early in 1965! The Commission and others lost faith and went for platforms but, as already described, no platforms were available until the birth of *Gem III* nine months after the start of the survey. At the end, when we were flying back from the last meeting in Paris, Colonel McMullen came across the plane and said, very gallantly, 'You have justified yourself nine times over. You were the only one who insisted that ships could do it'. I replied that I was the only one who had done it before, several times. During the survey, our travail was lightened by a visit from the House of Commons Conservative Transport Committee in March, followed by a visit in July from the all-party Parliamentary Committee on a Channel Tunnel - at the end we staged an International Press Conference at Dover.

Press Conference

The press conference was arranged for members of the press from many different countries and took place on September 29th marshalled by Don Hunt with his usual efficiency. First there was a photographic session on the cliff-top esplanade in front of the offices, with a clear view over most of the Channel. This was a reminder of the distance involved. After inspecting all the electronic equipment in the Officers Mess building, the party adjourned to the big mess-room, where they were shown the vast number of cores in their boxes and also the equipment in the adjacent laboratory. From this they got a good idea of the nature of the continuous Lower Chalk, ideal for tunnelling. On a special platform were laid out drawings, plans and sections and also a transparent model. The model was made of vertical plastic sheets set at the equivalent distance of 50metres, which matched the cross sections deduced from the survey. The results in strata and the line of the tunnels were visible for study. The press had to be reminded that, as the shores were 24 miles apart and the vertical depth of the tunnel area from sea level was about 400ft, the distortion of vertical scale to horizontal had to be about 1 to 100. From this, we were able to show how the dip of the beds on the English coast was gradual, so that there was a wide choice of alignment. However, on the French side the dip was much sharper, so that movement east to west had to be restricted in order to keep in the Lower Chalk. For this reason the boreholes had to be much closer together on the French side. It also helped to show that there was no difficult ground under the sea and that the only troublesome part was inland on the French approaches, which would be met while driving to reach the Lower Chalk near the coast. Other points were driven home, not only on the number of cables on the sea bed but the 123 identified wrecks between Dungeness and Dover. The Commissioners and our team were ranged on one side, with the press a little distance away on the other. They were busy with notebooks and tape recorders, while the noise of film cameras and television cameras was considerable. We took turns to speak and answer questions, Malcor gave a clear address in French and I held forth for quite a bit in English. Some weeks later my daughter's Swiss brother-in-law took his French wife to the cinema to see Norman Wisdom for a whiff of English humour. As they sat down, the newsreel began and they were startled to see a blown-up view of myself talking away – in fluent German! They found this funnier than Norman Wisdom.[49]

The end of the works in the Channel marked the end of the first part of my work for the Channel Tunnel Study Group and what might be called the practical engineering. This was the period from 1958 to 1965. My total involvement, although intermittent work for the Group, covered 13 years until August 1971 or alternatively a quarter of my professional life which is possibly why I have written about it at such length.

The next six years consisted of spasmodic attacks on the problem of the tunnel. The consultants' London office was in continuous session and so was the body made up of survivors of the Commission with others at the Ministry of Transport. First, the 8 leather bound volumes in which every detail of the survey, with some beautiful printing by SETED one of the French consultants, were enshrined. Each volume was numbered and issued with strict injunctions to secrecy, although we were not told if this involved the Official Secrets Act. Malcor and I also composed a shorter more easily digested report in French and English, with a few of the important drawings for the members of the Study Group, who had been left much in the dark, with no powers of their own under their contract with the Governments, but with notional responsibility. Then followed innumerable meetings and much correspondence with governments, bankers, consultants, civil servants, peddlers of freak ideas, rival schemes, lecturing, writing articles asked for by newspapers and so on.

On reviewing my XXI Club letters, several incidents in this period come to mind including a visit to the Rance Tidal Scheme under construction with 'some French top brass', and a longer trip to the right hand bottom corner of France. Also on March 31st 1966, on the first day of the French school holidays, Malcor and I fought our way onto the Mistral train bound for Lyons, where we changed on to an Italian diesel-coach train to Modena at the foot of the Mont Cenis massif. We had been invited by Electricité de France to visit tunnels for hydro power being driven high up near the top of the Alps. Next day we were driven up into the foothills and then wafted by a large cable car amid snow and sun to see a 7ft diameter Robbins tunnelling machine grinding away in sandstone and shale. After a sustaining vinous French lunch we drove on, to be again carried up towards the mountain top in an even larger cable car to the site of another tunnel. It should be noted that these cars had to be large enough to carry the necessary plant for the job and were not for public pleasure. I had to stumble in mud and snow for several hundred yards to reach the office hut, where we were lent gumboots. As I struggled with the climb, I felt that age was at last coming to roost, until I realised that we were many thousand feet up, after coming up the last 2,000 ft in very quick time, so my morale if not my physique recovered. As I knew that the Alps were made of sedimentary rocks and had seen their tormented shapes in Switzerland, I was surprised to find that the strata in this tunnel were dead horizontal. I was told that this was because they had been completely turned over 180 degrees. In order to get back to Paris, we found that we had to take the train through the famous Mont Cenis Tunnel (the first of the great Alpine rail tunnels) to Turin in Italy, in order to pick up a sleeper back to Paris for

---

49 It is likely that the Pathé News clip at http://www.britishpathe.com/video/mr-john-harding-speaks-on-channel-tunnel-aka-john/query/harold+harding is from this press conference.

a meeting of engineers and the Study Group in the Rothschild Bank. This was a memorable meeting, as the famous old building was to be demolished in the following few days and because we were served Mouton Rothschild with our buffet lunch. Engineering is not all hardship. Another quick trip was a flight to Stuttgart for the night and a visit next day to see a 10ft Robbins machine boring its way through limestone with no need to line the tunnel. This was in the Schwabian Mountains – a lovely part of the world.

In November 1968, three years after the survey finished and possibly because the Permanent Secretary of the Ministry was being scolded by the Parliamentary Estimates Committee, the Study Group were asked for a brief report 'rendering an account of the duties of the Group on the management of the survey, including cost control'. The Group's section ends thus:

> This is a highly technical matter and we, accordingly, have requested our Consultants Messrs Malcor and Harding, to compose this part of the Report and it is attached as an Annexure. The Channel Tunnel Study Group has full confidence in their judgement and experience in this matter'.
> Signed by Viscount Harcourt, British Chairman.

Our Annexure ended with some cogent comments on the effects of working under a Committee and the danger of perfectionism, and that in future work the responsibilities of the owner should be concentrated in one person. It went thus:

> The extra expense was not only due to the weather, but because a great deal more information was also obtained than had been foreseen in the original estimate. One can be proud that in spite of all the difficulties the objective of the Survey had not only been obtained, but often surpassed, and that this had been obtained without the loss of a single life in the course of the work, the only casualty occurred in Calais Harbour out of working hours. We were also fortunate in having no accidents while using the platforms in spite of the experience of others in the North Sea. Most of the difficulties encountered during the Survey were inherent in the unusual nature of the Survey and in the equally unusual bipartite and Governmental sponsorship.
> The formation of rules imposed upon the Commission of Surveillance made initial difficulties for them as well as those who had to act on behalf of the Study Group. But due to the comprehension and wise behaviour of Members and in particular of the two Presidents of the Commission, most of the difficulties were ironed out and so led to the successful conclusion of the Survey.

About this time Westminster Abbey celebrated its 900 years of existence with a big programme which included staging an exhibition in the Great Hall of Westminster School to celebrate 'The Abbey Scientists' which included civil engineers like Telford, who are buried or commemorated in the Abbey. I was appointed by the Civils to serve on the Royal Society committee who were organising the exhibition. It covered many subjects, with references to those who were concerned in them, such as astronomy, biology, radio-telephony, nuclear physics, not forgetting civil and mechanical engineering and appropriately geology (shades of Bishop Wilberforce and Darwin). For this I organised a section on the Channel Tunnel and its surveys which the committee felt was helpful. The meetings with the varied proponents on the committee and the Abbey clergy were most interesting, we were also entertained by the Dean and Chapter, all in their red robes, to a dinner in the Jerusalem Chamber where we had held our meetings. We were also presented our own private *Son et Lumiere* by the Dean in the deserted and newly cleaned up Abbey. It was all very impressive.

In September 1968 I was interviewed, impromptu, on BBC Radio 4 in their series *Scientifically Speaking*. David Wilson asked me about the Channel Tunnel and our report and I said that 'it was locked in Barbara's bosom', Mrs Barbara Castle being the Minister of Transport. I had expected this to be cut out but it was not, I went on to say:

> Harding: 'When we'd finished our second big investigation....we then hoped that the Government would say, yes you can now go ahead, but they had the wonderful idea of asking other banking groups to say how they would raise the money and how they would do it'...
> Wilson: 'But would you like to have a hand in building it, for, in a sense, the fun of it?'
> Harding: 'Yes I would. If we could have started after our first investigation in 1960 you would be using it by now. What I have kept saying, and what people do not understand, is that, from each coast it consists of two running tunnels, the size of two Tube stations and one service tunnel, a little bigger than a Tube running tunnel. Some of us have done jobs of that size while doing others simultaneously. It does not need an enormous consortium on it of other people. It needs engineers, but it needs a lot of money, that's the trouble.'

The wonderful idea of the Governments was to mobilise three groups of international bankers, of which the Study Group formed one, to put forward rival schemes prepared by their own rival consultants. Then they brought about a 'shot-gun wedding' of all three which brought about the birth of Rio Tinto Zinc (RTZ) as 'managers' of a tunnel scheme. RTZ consequently got the sort of job for which Bechtel had lobbied. Presumably bankers are more familiar with mining firms than tunnel contracts. I still believe that consultants and contractors together need no 'managers', their life's work is managing such projects. Mott, Hay and Anderson (MHA), my old friends, had put up the solution that the bankers and Government preferred, although the chief difference was the change of working shaft from Dover Harbour to the base of the Shakespeare Cliff where the aborted 1882 shaft had been sunk. They had been among the three hereditary firms of the old Tunnel Company until for some obscure reason British Rail preferred Halcrow. When the work started, MHA collaborated easily with Halcrows. My elder son Edmund was by then working for Motts, so he found himself for a brief space following in his father's footsteps.

As a result of this, in September 1969 I found myself briefing Sir Val Duncan of RTZ in a four hour meeting and then handing over my material to his Mr Frame, while about the same time René Malcor returned to his post as an Ingenieur en Chef des Ponts et Chaussées, who had loaned him to the Study Group in 1957. I ceased to draw fees, but was consulted at odd intervals, until the Study Group and the Channel Tunnel Company wound themselves up and finally parted with me on May 5th 1971, with very kind vocal and written thanks, including a financial appreciation for efforts 'beyond the course of normal duty' so to speak.

I thus became a spectator from 1969 and can only call on memory. The tunnel was actually started with Edmund Nuttalls Ltd as contractor on the British side. Nuttalls had just made a superb job of the twin Kingsway road tunnels under the Mersey in the face of much Marxist Liverpudlian obstruction, by using the same tunnelling machine which had been used on the Manglea Dam. In September 1971, I had led a party of the Tunnelling Society to visit the Mersey tunnels so was familiar with the work.

On the French side the work was through the difficult ground which we had forecast, which gave pleasure to the pessimists of the press. At Dover, a splendid start was made with a beautifully engineered approach and a tunnelling machine designed and made by Priestleys, a subsidiary of Nuttalls. This was making astonishing progress boring the service tunnel under the sea in ground exactly as forecast. Then the Labour Government, in a panic, stopped the work, although they were not paying for it. The finance was by private enterprise, so had no connection with the problems of Concorde or Maplin, but the press insisted on linking them. The scheme had lasted through Tory, Labour, Tory and Labour cabinets, all

48. HJBH (centre) on site at the Channel Tunnel stage 3 at Bruay

of whom dithered. The Japanese were going ahead with an undersea tunnel in appalling geological conditions just to join up two islands. Here, we had a uniquely easy material through which to drive in order to connect British Rail to a vast, efficient continental system, with the added ability to help road vehicles on their way at the same time. But we panic, the French do not.

There is neither space nor inclination to write of all the rival schemes and wrestles with opponents. The pleasant part is to remember how easy it proved to be to work with men of different nationalities and especially my relationship with René Malcor. We became very close friends and in all the years never had a word amiss or ever differed and the Study Group were kind enough to say how helpful this had been. Two memories surface: at one meeting Malcor and another Frenchman had a torrent of verbal argument, I asked what it was about and Malcor said, 'It is nothing but we are French, so we get excited'. On another day I said, 'I am taking you to lunch at the United Service Club. Now at the top of the stairs you will be confronted by a picture of the Battle of Waterloo and opposite, a picture of Trafalgar. So you must think hard about Joan of Arc'. He replied, 'Yes, I will concentrate on Joan of Arc'.

The last of my public appearances, visible or in writing (apart from later letters), was a filmed interview in my Tunbridge Wells garden for a Horizon programme put out by the BBC. Then on March 13th 1975, I took part in the Last Rites by joining a party of the British Tunnelling Society to visit the machine in the face of the tunnel as the Government had brought about its sentence of death. A wonderfully organised work killed off at the moment of proof of success - a very bitter moment.[50]

---

50 This situation did alter in the last few years of HJBH's life and with all the determination that he had always had he was able to continue the lobbying of politicians in the struggle with the strange array of competing schemes and ideas. For a bit more detail of this period, please see Chapter Twenty-two. He would have been glad to know that in the end things went very much according to their plan as the Channel Tunnel started construction in 1988 and opened in 1994, with involvement from Sir Alan Muir Wood and John Bartlett, an HJBH Mowlem protegé and long-term friend, senior partner at Mott, Hay and Anderson, later Mott McDonald. Douglas Parkes was another of the many friends who were involved with the construction.

16

TRAVELLING BROADENS THE MIND

The saga of the semi-abortive Channel Tunnel deserved a chapter to itself but, as I was engaged on a freelance basis, many other projects were intertwined in those years and can stand on their own, out of time.

The World Bank Mission to the Litani Project in the Lebanon – 20-31 March, 1960

The World Bank invited me by cable to be part of a mission to the Lebanon, to examine and report on sizeable troubles in a hydro-electric tunnel for which the bank was paying. Our team of five was led by J. Donovan Jacobs, a consulting engineer from San Francisco, with a large practice in rock tunnelling. There was also Mansell MacLean, a contractor who had just finished the compressed air Lincoln Tunnel under the Hudson River, New York, and Victor L. Stevens from Boyle Brothers Drilling Co, Utah. The other two were British: one was my old friend Jack Kell (by then senior partner of Mott, Hay and Anderson) who had been deputy resident engineer on the Bow-Leyton Tube. He had the Dartford and Tyne tunnels to his credit and had started life as a mining engineer. Together we covered all possible aspects of the problem, not forgetting drilling and injections. At that time Beirut was flourishing and we all foregathered in a new Hilton-type hotel in a rapidly expanding luxury area to the south of the city centre. No doubt the hotel has long been shelled to bits. Engineers from the Litani River Authority were our hosts. Most of them were Maronite Christians who had trained in American universities, which meant that communication was easy. That night they drove us for some miles up the coast to dine at a brand new restaurant and nightclub, where the performance was straight from 'Place Pigalle'[51] and the French nudes were much appreciated by the numerous Arab sheikhs in traditional dress, complete with burnous (hooded cloak). The occasion helped us all to get to know each other a lot better.

The Litani Project had been designed by French consultants, since the Lebanon has for long been a centre of French interest and culture. The Litani River runs southwards down the rift valley between the Lebanon and Anti-Lebanon (Eastern Lebanon Mountain Range) mountains, until, just before entering Israel it turns smartly westwards and flows into the Mediterranean between Tyre and Sidon.[52] In the peaceful times of 1960, a Jugoslav Communist contractor (curious thought) was building a dam across the Litani between the mountains, while a consortium of French and Italian contractors had driven half of a ten-mile long tunnel under the mountain to the west, which, at its end would serve hydro-electric stations and the water would then irrigate the coastal lowlands. Curiously

---

51 A Paris cabaret centre.
52 This short length of the Litani River became notorious in 1978 as the line along which United Nations Forces tried to separate Palestinian, Syrian, Israeli, Maronite and Moslem Lebanese warriors and prevent them from killing each other and any unfortunate local inhabitants.

49. *HJBH making notes on the Litani Project*

enough, the formation, like that for the Channel Tunnel and Malta, was of the Cretaceous period. My *Encyclopedia Britannica 1911* says that the word Lebanon comes from the Semitic 'laban' meaning white and probably does not refer to snow but to the bare white walls of chalk or limestone which are characteristic features of the range.

The tunnel was being driven from each end and from two adits in the valleys, where it ran under the foothills, so the longest drive was under the Lebanon Mountain. There was trouble in all six faces, but the worst case was in the face driving east under the mountain. This had passed successfully through a syncline of chalky limestone, which had proved to be dry, and on through a bed of impervious marl which had a thirty degree upward slope. Then the drills had tapped the joint between this and a thick bed of weakly cemented cretaceous sandstone. At this point, the tunnel was 1,000 metres below the mountain and this joint contained water (a later estimate guessed that it was under a head of 500 metres). The result was that the water rushed out, eroding the sandstone so successfully that 70,000 cubic metres of sand ran in and filled 3,600 metres of tunnel, as well as flowing out and then swamping the working site at the adit. We were told that the flow of water reached 6,300 gallons per minute. Luckily the few men in the tunnel ran faster than the sand. When it came to rest they dug out a considerable length, but the sand came back and drove them out again.

This was the state of affairs that we found on our arrival, with French and Italian rival specialist firms trying out chemical injections in other faces, which were in the sandstone. Meetings with much talking in French, American and English took place and the French consultants wanted to join up with us, however we upheld our independence. We visited the various tunnel faces, with quite long trips by 'train' sitting in empty concrete skips. The flooded tunnel was different, it was possible to walk some way in, until the gradual slope of the sand led us up to the roof of the 12ft arched tunnel, which was by then fully filled up. It was a little eerie, as some water was still running and we might also have had to run for it. Next day we were taken by jeep to the mountain top, but failed to find any trace of a depression or hole. The outcrop of the sandstone proved to be reasonably hard, which emphasises the effect of water, so maybe one day someone will be surprised by a cavity of many thousands of cubic yards under the summit. While we were wreathed in a fine mist on the top of the mountain, we were passed by a small group of nomads with women and children on donkeys and we were told that these were Turkish gypsies. As we drove back, they were on the narrow path ahead of us and the Lebanese engineer-driver drove the jeep straight into them, scattering the donkeys while we yelled blue murder at him. He merely said, 'Were you afraid that I should damage the Jeep?'. This may explain why, in spite of education, they still like shooting each other.

16. TRAVELLING BROADENS THE MIND

On the Sunday our hosts took us on a trip to Baalbek which proved fascinating. It lies in the valley between the mountains, near the source of the Litani. I took the chance to buy three of their superb brocade dressing gowns for my three grown-up descendants and a length of brocade for my wife. On the way, we were fed with frogs legs, which tasted like tiny chicken legs. At our hotel each night we recovered ourselves, not with strong liquor (entirely) but with long glasses of orange juice pressed straight from the trees.

Don Jacobs managed to borrow a secretary from the US Embassy to cope with our report. We each submitted our versions of appropriate chapters to each other to be agreed and amended, covering history, geology, grouting injections, drilling methods and tunnelling methods in soft ground. Our solution, roughly, was to adopt soft ground methods with injections in parts but, for the main drive where the sand had overwhelmed them, we had a different solution. This was to advance carefully up the completed tunnel for a series of lengths of 300 yards, with due precautions, and then build bulkheads on the same lines as airlocks, with steel doors which could be closed in a crisis. Pipes left through the concrete walls could carry away water and also allow the sand to be pumped out. Then, at a safe distance from the break-in, a by-pass tunnel would be driven to circumvent the disturbed area using long borings ahead to forestall the unexpected. Then, by turning back they could join up with the tunnel driven from the river end. We worked in one of the bed-sitting rooms at the end of a long corridor and by concentrated effort we were able to have our report typed and copied for all concerned before we left on Day 11. We then beat off the attempt of the authority to pay us in Lebanese money, collected our agreed fee in US dollars and flew home.

We heard nothing more for years, except an occasional Christmas card from one Maronite engineer, and then one day I received a copy of the French magazine *Travaux*, which described the affair at length and actually mentioned in a short paragraph that we had visited the site – full stop. The story went on to say that, soon after, a mission from Electricité de France studied the case and then described how the work was done to their recommendation. The description was so remarkably identical to our solution that we can only suppose that the French were clairvoyant! Incidentally, Kell and I had had to work hard to prevent Don Jacobs including harsh criticism in our report of the earlier French supervision, being an American he did not realise that this would have caused much ill-feeling and loss of French *amour-propre*, in an area where the French were sensitive to their historic influence.

Vancouver – November 1960

One evening in November 1960, I was in the Hall of the Institution just before a meeting, when Wynne-Edwards bounced in with a perfect stranger. He introduced him, 'This is Noel Lambert, an old friend from Canada. Can you possibly meet him at the Savoy tomorrow morning as he wants to talk to you?'. I spent next day in Lambert's suite at the Savoy listening to his troubles. He was head of the Northern Construction Company of Vancouver, who specialised in rock tunnelling and had recently blown up the Ripple Rock in Vancouver Straits. I said that I had read of this which cheered him up. His firm were in trouble over a soft ground tunnel in Vancouver, which had collapsed. A leading British consultant happened to have been in the area on the Peace River Scheme and had told the local authority that this work could only be done by a contractor with experience in compressed air work. This annoyed Lambert and, finding himself in Montreal and realising that he was nearer to London than Vancouver, he hopped on a plane to try to find his only British contact, Wynne-Edwards, who had previously worked in Vancouver and with him on the Detroit Tunnel in the 1930's. Wynne-Edwards had in turn passed Lambert on to me, so after a long talk we went out to the nearest travel agent, where Lambert brandished the usual transatlantic handful of credit cards and booked a flight for me for two weeks later.

My plane was due to take off at 1:00pm on Saturday December 3rd to arrive in Vancouver at 10:00pm that evening (plus 8 hours time-lag). However, as is the custom, there was a delay of four hours, so the DC8 gave up trying to go any further than Montreal and they turfed us all out. We then flew by Viscount to Toronto, followed by a leisurely night flight by Constellation. In turn we were turned out in thick snow at Winnipeg, Saskatoon and Edmonton. We got to Vancouver by 10:00am on the Sunday, but missed seeing the spectacular part of the Rockies due to cloud. Hugh Golder was there to meet me, as he had work in the area and gave me my first introduction into local customs. At lunch the first thing the waitress asked was 'What is your beverage?'. This had to be tea or coffee, drunk throughout the lunch. I saw little of Vancouver, as the next three days were spent either in the tunnel or in long meetings, but had the impression that in spite of newer tall buildings, in winter the city had a provincial air and life did not seem unduly hectic. At that time there was no snow in the streets and it was fairly mild. I deduced that the locals could merely cross the Fraser River and take a ski-lift up to the top of the mountain opposite, which was several thousand feet high and, after skiing, come home for tea.

The tunnel was to be a new main sewer 3 miles long under a ridge rising 300ft above it. The ground was composed of a medley of mixed drift from several ice ages. We walked for a mile down the tunnel from the open portal at one end, until we reached the spot where the ground and water had run in and brought everything to a standstill. Up until this point the ground had been dry sand and gravel, changing into dry silt. There had been a thick layer of glacial clay above, which had kept out the considerable amount of water in the gravels above the clay. Unfortunately, the clay had unexpectedly petered out and a cavity had reached the surface in the grounds of the University. With difficulty, a short length of tunnel had been driven from the other portal, while a number of deep wells with submersible pumps held down the water. The ground had been supported by steel horseshoe arches, with boards between in the rock tunnel fashion. An emergency timber heading had been driven for some unexplained reason, but it was so competently driven that I could tell my hosts that the firm would be perfectly able to do the work themselves, with a little tuition on steps needed to finish it in compressed air. The consulting engineers for the contract had a reputation in mining and rock work and very sensibly had formed a united front with the contractor, instead of taking up the more usual adversarial position.

On the Tuesday, Lambert had mustered his leading engineers and the consultants came as well and asked me to express my thoughts from what I had learned of the conditions. I launched into a harangue for nearly two hours on the problem, and every aspect and action needed to finish the work in compressed air. I found out afterwards that they had tape-recorded my off-the-cuff dissertation and they gave me a copy. It was somewhat colloquial, but much to my relief I had no reason to be dissatisfied with it. The question of temporary lining was discussed and the costs and merits of concrete or cast iron segments. Lambert decided eventually on using liner plates, which consist of thin-wall steel pressed pans with narrow flanges that are very popular in the USA but have never broken into British practice due to different conditions. If the ground is very heavy, liner plates need extra arched steel rings to strengthen them and the finished tunnel was to be lined with concrete inside the temporary lining. However, liner plates would be very difficult to caulk to keep out water when the air pressure was taken off and would not be strong enough on their own, so I persuaded Lambert to concrete the lining while the air was on. The extra cost would pay for itself against the cost of trying to caulk and the need for supports. In addition, water would damage the unset concrete lining if it poured through the joints while the pressure was off.

After these talks his staff got busy searching for compressors, air locks and all the other equipment which had been recommended to them. Then on Wednesday morning I was taken into a meeting with a number of fresh faces and asked to talk about the latest British regulations for working in air and the precautions to be taken, as well as the code relating to times of locking out and so on. I did not know who these people were but was told later that they were doctors and members of the corporation and that I had converted them into adopting the British regulations and abandoning their own out-of-date and seldom used ones.

That afternoon we were to have flown to Chicago to confer with Dr Peck at the University of Illinois at Urbana but planes were grounded through fog, so we drove to Seattle through the recently completed immersed Deas (George Massey) Tunnel under the fast flowing Fraser River. In London, Christiani and Neilson had given us a very good account of this, which was of much interest to me. By good luck, I still had a visa for the USA, although I had not expected to be crossing the border. The Canadians had no trouble, but my passport as a limey was closely scrutinised. After a night at the Olympic Hotel, we drove through Seattle to the Airport to take a DC8 to Chicago. I found Seattle seemed much brisker than Vancouver. At that time the DC8 was a novelty and a high flier, so we saw nothing but cloud in contrast to my earlier trip in Morrison-Knudsen's Dakota. At Chicago we picked up a hire car to drive the 120 miles to Urbana. I sat by the hired driver, who asked me to keep my eyes open for the constantly changing speed limits (20, 30, 40 and 50 miles per hour) as he said that the speed cops were very fierce and he did not want to risk his licence. The country was flat and monotonous with small townships covered with hoardings and with bedraggled outskirts.

After a night at my first encounter with an American motel, we spent the next morning with Dr Peck at the University. He is one of the USA high priests of soil mechanics and had been advising my clients on the deep-well borings and pumping on the site. We had a long talk, in which I outlined what my opinion was and how I would try to find Lambert some shields. Dr Peck had told us that he had just come from seeing Dr Terzaghi, who had had his leg amputated a week before and found him, at 80, propelling himself about in a wheelchair. Dr Peck then said, 'I do not consider that a shield is necessary and Dr Terzaghi says the same'. This disconcerted my clients but I replied with a smile, 'I am not afraid of you or of Dr Terzaghi and I say that they do need shields'. This did not cause resentment but sometime later, when the shields had proved the necessity, Lambert said 'We are very grateful to you'. I asked why and he said, 'Well, out here everybody thinks that Dr Peck is God'. I do believe that Dr Peck's advice on the deep wells was excellent and he wrought miracles in his professional career, even if he was not divinely infallible.

We drove back to Chicago that afternoon, so that the others could fly home from one airport, while I flew to Toronto from another. As it was rush hour and time was too short to drive in the dark between the airports, they paid for me to take a helicopter, which ran a shuttle service between the airports – an idea which, much later, caught on at Heathrow and Gatwick. As both Chicago airports are on the southern outskirts of the city, I never actually saw this fabled city but only the suburbs from a low level, where selling second-hand cars seemed to be the activity all the way. Hugh and Mollie Golder met me at Toronto and took me to their comfortable home for a delightful evening and a good night's rest. Next morning the sun was shining on the snow as Hugh drove me out to visit Niagara Falls. I felt slightly superior, as I had previously seen Victoria Falls (but only from the air). The Canadian Horseshoe Falls were most impressive, especially as on that side the visitor can stand almost on the edge. However, the surprise was to see how the American Falls had been reduced to a cascade over heaps of rubble where the rock had fallen away. We lunched high up in a restaurant with a fine view over the falls and the whirlpool, then, following a quick look at the streets of Toronto on the way to the airport, I was home by 10:00am on Sunday. A fairly full week's work.

The next step was to organise an instructional tour for two senior engineers of Northern Construction as well as Mason, the consultant, who arrived in London on Sunday January 22nd. On Monday we drove first to Mowlem's Depot at Welham Green to meet my successor as works director, Vincent Collingridge, who showed them two hooded shields which they could buy and other typical plant. Next day we all flew to Belfast, where another of my former disciples, Nigel Boxer, who by then was a director of Kinnear Moody and Co., had laid on a visit to a 14ft diameter cast iron lined tunnel in air which was close to their size. I believe this was under the notorious Falls Road. We flew back to London the same evening. On Wednesday, after more meetings with Mowlem, who promised to lend them a tunnel foreman, the Canadians flew to Glasgow while I followed by night sleeper after a prior engagement at a Smeatonian Society occasion and dinner. I had been able to arrange for Charles Brand and Son Ltd to show them the method of work on the Clyde Tunnel. After a medical examination, we spent the morning down the tunnel, where they experienced the rigour of long spells in the man-lock, decompressing from over 30lbs per square inch pressure. The size of the tunnelling was out of proportion to their case, but the ground was made of glacial deposits which was appropriate. In the afternoon, Brand's engineers went to great lengths to show them all the charts, forms, regulations and medical care, as well as the gauges and recorders which are now the fashion. They also showed them the way to keep the men on the site for the required medical time after 'locking out' by providing television.

That night we had a convivial dinner with the contractors and resident engineer (I forget which of them were the hosts) and next day we studied the compressor shed and other plant. In the afternoon the Canadians flew back to London, well pleased with their visit. As my son Edmund was working in the tunnel for Brand's, I went back to the site to pick him up and to take him and his fiancée Diana out to dinner. On Saturday I went back to London by air. During the next week we composed a joint scheme for the shields and a report for Lambert. Now that we knew the shields' dimensions, I persuaded Mason, the engineer, that it would never be noticed that the flow through the tunnel had to be two inches smaller than he had designed. The result of this visit was that the shields, which had been designed to pass through air lock doors in sections, were put into shape by Priestley of Gravesend, who had them made and shipped off with every part duly marked with erection instructions and followed by the foreman. In the meantime, air locks had been built and compressors erected on the lines advised to them. After a few crises, the tunnel was successfully completed and lined with concrete. My part was a form of correspondence course with Lambert as advice was asked and the result of this expedition led to several contracts and requests for advice. I had been introduced to Jack Bonny, the head of Morrison-Knudsen (MK), who had paid a quick trip to Vancouver, but I had not realised that the Northern Construction Co. were a subsidiary of MK. As a result, in 1964 I was invited out by MK to spend three days in Lisbon to see them sinking the caissons for the Lisbon suspension bridge, to Charley Dunn's design. So with Jack Bonny, Noel Lambert and Charley Dunn and some others a very good time was had and this was not my last connection with MK.

The Swedish Connection - February 1961

I have already described some of the results of my friendship with Sven and Mai Platzer and much else now followed. By 1961, Sven had gathered around him some of the best young rock-tunnel engineers in Sweden and the firm of Widmark and Platzer, known as WP, was well-established. On several trips to England he had sought my advice and on Sunday February 5th 1961 my wife and I flew to snowy Stockholm on his invitation and were lodged in the Grand Hotel, overlooking the Old Town in this delightful city. For a moment we thought that we were in the wrong country when the lift door opened to reveal a Sheikh in full Arabic costume. We heard later that, when the Sheikh had done his business, he changed into European dress for the night-life. The next two days were spent in discussions in the WP office on many matters. They were working on tunnels in the Philippines, South America and Mexico, so Platzer had installed a Spanish professor in the office and his staff each spent an hour every day learning Spanish. I am not aware that at that time any British firm had done the same. I met the head of a firm of Mexican contractors, Señor Saturnino Suarez, with whom WP were collaborating out there. They were discussing the project to bring the drainage of Mexico City up to date and I was able to bring off a nice example of one-up-manship by saying that the original work had been done by my relative, John Body, and I had his paper to the Institution – *Body on the Drainage of the Valley of Mexico 1900-1912*. Suarez said, 'You must come out one day and help us'.

The Platzers and their four sons were a delightfully happy-go-lucky lot, but one evening we saw a slightly different side to Swedish family life when we were invited by some other friends to a formal dinner party for twelve made up of mixed generations. Schnapps seemed to have given way to sherry, but the host followed the custom of rising at intervals to make a little speech. However, when he rose to address his four young, aged from 16 to 21, on the necessity for polite behaviour, good manners, hard work and so on, our reaction was discomfort on their behalf and admiration for their silent reception. The reaction of our own young, if we had ever been so foolish as to expose them to such admonishments on a public occasion, might not have been so polite. At midnight, other children came to drive their parents home, because of the strictness of Swedish police over driving after any drinking.

Next day we drove over multicoloured ice floes on the Baltic coast to Ustermund, from where we were driven by skilful skidding over icy roads to Korsselbranna a few miles short of the Arctic Circle. Although the snow was deep, the roads were clear. Perhaps because traffic was so thin there was little to impede the snow ploughs, so different from our chaos, where roads are crammed with abandoned cars when we are caught napping. At all the villages the inhabitants moved about by pushing wooden kitchen chairs on runners which acted as prams, shopping baskets, toboggans or merely as support when walking. The temperature was $-10°F$ so Sven had lent me a warm fur hat. These were being sold in Stockholm as 'Macmillans' as it was just after Harold Macmillan had worn one on his visit to Moscow. The object of the visit was to show me WP's skill in driving rock tunnels on this hydro-electric scheme, and especially the method of ladder drilling. They had introduced this method two years before, then passed it on to Atlas-Copco and at that moment it was being used on the Inverawe tunnel in Argyllshire. Instead of using heavy 'jumbo' stages with fixed drilling machines and many changes of drill steel, the drills were mounted on flat ladders and propelled by pneumatic pusher-legs, a kind of jack in the form of a narrow tube. These legs thrust on the rungs of the ladder. At intervals they can be jumped forward a few rungs so the drill steels need not be changed, with the benefit that one man can work two drills.

My wife and I were given a room in the warm and beautifully built timber engineers' quarters, complete with sauna, and 10ft drifts of snow outside. There was a small dinner party that night and we were told that next day there was to be a special demonstration. I said that in England, if a demonstration was organised, either the plant would have broken down or else the men would be sitting around drinking tea. This amused them. The next day, when we drove to the face of one tunnel, all was silent. The plant had broken down. We drove out and then up another tunnel. At the face, the men were all sitting down drinking coffee. We later drove up a third tunnel, where the loader had broken down, but at the fourth all hell broke loose, as all the drills were in full swing. My joke had saved their face. The beauty of a rock tunnel with a flat bottom is the ability to drive a car up to the scene of the action, that is, if rubber tyres and not railtrack have been chosen for transport.

We had to scramble up a large spoil heap following one blast, as the tunnel was about 26ft each way. While my wife was being helped down the heap by a young Swedish engineer, she surprised him by asking questions about the rock and whether the black material was graphite. He told this to Platzer, who announced at supper, 'The truth is out. Mr Harding is only the mouthpiece'. Next day a correspondent from the local paper of the nearby town asked for an interview and after the inevitable questions about the Channel Tunnel, he asked me for my impressions of the temporary quarters built for the workmen. I replied that I was most impressed with the standard of comfort provided. Next day, the paper came out with a large photo of my wife and of me in my Macmillan, the title was 'Tunnelbygger ed sa make' – (roughly translated as 'Tunnelbygger with her husband'). Sven, who had a good command of vulgar English, pointed out that in Sweden, Y is pronounced U! We then drove many miles farther north to see some vertical shafts driven upwards by the Alimsk method on a job for the 'Board of Waterfalls'. Back in Stockholm, Platzer had organised a meeting with a group of his own and other engineers (some previous friends of mine) and Señor Suarez. He had asked me to bring some slides, so I gave a talk on soft ground methods, of which there is little in Sweden because of their beautiful unbroken rock, and, of course on the Channel Tunnel. That night Sven gave a hilarious party in which Señor Suarez provided a strong Mexican flavour. I have noticed that the Swedes whom I have met do not live a life of sordid gloom, unlike the films of Ingmar Bergman. The curious effect of walking about Stockholm, compared with many Latin cities, is the impression that you are among Englishmen who insist on speaking an extraordinary language. Luckily, so many do speak English that life and communication is easy. Thus ended nine full days in Sweden.

50. The headline in Sweden caused some international exhanges of pronunciation

In the following July came a second visit to Spain. As Platzer had work out there, he had purchased a deserted finca or farmhouse north of Alicante for family holidays in the sun. He invited both my wife and I, first to help in discussions on the proposed road tunnel through the Guaderrama Mountains and then to stay nearby at San Juan before it became a 'Costa'. On the drive back from the mountains, we took the chance to visit the vast underground cathedral hewn by General Franco out of the remarkably solid rock to immortalise the dead of both sides in the Civil War, and especially himself. This provided matter for thought, professional, religious and touristical. An extra bonus was a second visit to the Prado to absorb once more the Velazquez, Hieronymous Bosch and much else. We then travelled to Alicante, but as this was the time before it boasted an airport we travelled by primitive, Victorian-era, night sleeper. When we arrived there, we stayed at a simple little Spanish hotel on the shore three miles north near the finca. Unfortunately, a partner of Sven's died suddenly so he had to go home, which left his family with no car and this rather took off the pleasure. However, before he left we found that he had organised a little Spanish chap to drive Sophie and me to Granada. This was a memorable all-day drive on a good road with almost non-existent traffic. We travelled through the cathedral town of Murcia and then through thinly populated areas with changing scenery. As all the houses were dug out of solid rock, presumably limestone, this area attracted many visitors. The harvest was over and in village after village we came across scenes of threshing by the traditional method, where a man rode or drove a horse or mule in a circle around the threshing floor, towing a loaded sledge, while women threw fresh material under the sledge as it travelled to do its work. The little elderly driver of our car spoke no English but we were able to build up a good *modus vivendi*. We spent two nights in Granada, with a whole day spent in the unforgettable Alhambra Palace.

On our way home to London, we spent two nights in Madrid and took the chance to be driven to Toledo, a city in a wonderful situation. We naturally visited the studio of El Greco, which is half underground with poor lighting. My wife deduced from this that it was the reason for the curious pale colours which the astigmatic El Greco adopted. I must admit that I never liked his paintings. In the Chapter House of the Cathedral there are two rows of portraits of the Archbishop of Toledo from earliest times. Most of them looked how you perceive an Archbishop to look, except for two in the centre. These had positively evil appearances, more like Mafia Godfathers. On looking closer we saw their dates, which exactly matched those of Philip II and the height of the Spanish Inquisition.

Mexico – 1962

A visit to Mexico at the end of 1962 came at the joint request of Widmark and Platzer and Saturnino Suarez. This call took the form of a visit by two youngish engineers from some of Señor Suarez's many firms, Felipe Pescador of the contractors ECSA and Enrique Tamez, Director General of Solum, a soil mechanics specialist firm. They spent all day with me in my house discussing their problem, with a break for lunch at the Hurlingham Club, which was a novelty for them. We learned that they had brought their wives with them and so took them all to the Talk of the Town for supper and a show. The wives spoke good English and proved very pleasant young companions and the evening was a success. In due course, when we got to Mexico, we found that the Pescadors and the Tamezes had been detailed to look after us and entertain us during our stay, so this first get-together was fruitful and greatly added to the success of our after-work periods.

When I told the Channel Tunnel Study Group that I was going to Mexico via New York, I was asked to stop off and brief Frank Davidson and Technical Studies Inc. on the situation and also to visit the Chesapeake Bay scheme of combined bridge and immersed tunnels. At the last moment another matter arose. Sven Platzer had been asked by the Swedish Government to look into a tunnel problem in the Bahamas, as they had had a call from a Swede in Nassau for advice. As Sven had to dash off to the Philippines, he asked me to stop off at Nassau on the way home to see this man and investigate the problem. I organised a three week itinerary with flights covering New York, Mexico, Norfolk, Virginia and the Bahamas, all of which came off as intended!

On Thursday November 22nd 1962, Sophie and I flew to New York, leaving on time at 11:15am and arriving at 3:15pm (7:15 GMT). Once there, we clocked in at the Gladstone hotel. Although the flight was by BOAC, we had been served a heavy Thanksgiving Day lunch of turkey and pumpkin pie. That evening we dined with José Machado and family in Gracie Square, where we had another Thanksgiving Dinner of turkey and pumpkin pie. We left early at midnight because of our insides. It was 4am GMT. Next morning we walked up and down part of Fifth Avenue to look for Christmas presents, but everything seemed so inferior to British products that we saved our dollars. At every fifty yards along the 'sidewalk' a Father Christmas alternated with a Salvation Army lad or lass ringing bells. We then went on to lunch with Frank Davidson at the Harvard Club (very British), after which we seized the chance to visit the Metropolitan Museum and Art Gallery, which was a very rewarding sight. That evening we went to dinner with the Machados at the Colony Club. I seem to have eaten at more New York clubs than restaurants.

On the Saturday, Sophie and I got up early to brave the complexities of Grand Central Station to take a little rickety Disney-like train to Cornwall, Connecticut. We were met by Dodie Prentice, sister to José Machado and also to the mother of our son-in-law, Philippe Oboussier. To the surprise of us all, we were confronted by two men in complete Robin Hood outfits in all shades of green and yellow, armed with bows and arrows. It appeared that the fashion was to go hunting in the local woods in that dress and equipment. The traditional white Colonial timber houses are used as holiday and weekend places for those who can afford them. Although it was at the end of the fall, there were still enough autumn colours to enable us to appreciate the beauty of the country and the villages of white timber houses and fascinating timber churches, dating back to the days of Rip-van-Winkle. Among much fraternising we were taken as uninvited guests to a local cocktail party. Our host said, 'We have brought the Hardings, who have just come from England', to which the reply was 'Oh! from England? Tell us, are you going to have a Channel Tunnel?'. By accident they had come to the right man, but one who could only prophesy darkly.

Father Prentice, whom I had met on my previous visit, was a leading architect in Hartford Connecticut and we were driven there on the Monday, from where we took a plane to Boston. After a quick so-called lunch in a drugstore, we persuaded a taxi driver to take us some way to the outskirts to have tea with my hero, Dr Terzaghi. I have told how he had a leg taken off two years before, which had put an end to his worldwide travels. He gave us a great welcome and thanked us for making such a long detour just to see him. Obviously, he missed his outside personal contacts. I was able to tell him that I was President-elect of the ICE and asked his permission to quote him in my address. Sadly he died a few months later. We never saw Boston proper, nor the seats of learning such as MIT and Harvard, but took one of the hourly bus-planes back to New York, from where we flew to Mexico City.

16. TRAVELLING BROADENS THE MIND

We were met at Mexico City airport on the Wednesday by Felipe Pescador and Styg Wikstrom of WP, who took us to the old British-built Hotel Genève. That evening, Suarez took us in a party to the Opera House to see the astonishingly beautiful Ballet Folklorico de Mexico, followed as was the custom by a very late but cheerful dinner. I had met Wykstrom in Platzer's office the year before. We both stared at each other and then realised that twelve years before that, he had been the young guitarist who had joined in with me on 'Widdicombe Fair' during the High Dams visit. Both Wykstrom and Pescador spoke excellent English and Spanish.

Next day I was taken to a building site and asked what I thought of a 6 inch steel tube sticking up 15ft above a pavement. I thought that it was a ventilator shaft but was told that it was an old lining to a well around which the ground had settled for 15ft. This phenomenon of Mexico City is famous among soil mechanics enthusiasts. Their most quoted example of settlement is the Opera House, so it was no surprise to have to walk down ten steps to enter the Foyer, knowing that when it was built visitors had to climb up ten steps to enter. Dr Terzaghi had written some years before:

> ..the most spectacular regional subsidence is going on in Mexico City. This is due to the fact that the clay layers located between the ground surface in this city and a depth of about 300ft below the surface are by far the most compressible clay strata which have so far been encountered in any part of the world..... The subsidence due to pumping alone produced differential settlements by amounts up to ten feet. The effect of these movements on public utilities such as the sewer system is disastrous.

51. *The Temple of Quetzalcoatl (the Feathered Serpent) at Teotihuacan*

Before getting to grips with this portentous problem, the Pescadors and the Tamezes drove us out to see the unpronounceable stepped pyramids at Teotihuacan which television has since made so well known. Their size is less surprising than the immense size of the 'sacred area' in which they stand. We were introduced to an archaeologist who had just discovered a lower below-ground chamber of a greater age than the Toltecs and their Aztec successors. The walls of the chamber had been decorated with paintings but the paintings were not easily visible.

We had lunch at a little inn and then down to work at the ECSA office from 6pm to 9pm, followed by dinner at 11pm. This is a Spanish habit, but we had quickly gathered from Tamez, who was himself Spanish, that one does not call a Mexican Spanish any more than one calls a Yank English, whatever his descent. I produced John Body's paper, which saved them much explanation. It showed that in 1896 the area of Mexico City was 4 square miles and it showed the Grand Canal, which Lord Cowdray's firm of S. Pearson and Son had built and was still running, carrying its content for 30 kilometres to the ring of mountains which encircle the valley. It then discharged through a tunnel built by USA engineers and what then happened to it, I did not enquire. The paper also described how the successors to the Conquistadors set about the problem.

The first systematic attempt to drain the waters of the valley was made in the year 1607, when Enrico Martinez drove a tunnel through the low depression of the mountains at Nochistongo, the northern extremity of the valley. This tunnel had a cross section of 13ft 9 inches by 11ft 6 inches and was completed in the almost incredibly short time of 11 months; this however is explained by the fact that many thousands of Indians were employed throughout the whole of the time, working in a great number of shafts and that the soil offered no great resistance. Unfortunately the great urgency of the Spanish Government to get the work finished did not allow time to line the tunnel with masonry, and consequently, very soon after the first torrential rains passed through, the whole work collapsed.

Modern maps showed the built-up area of the city now measured 10 miles by 6 miles and that the Grand Canal had been widened and deepened, but the problem was becoming serious. I spent the next morning at the City Hall with Pescador and Wykstrom, where they introduced me to Señor Ochoa, who seemed to be the 'hydraulique' engineer. There were two alternatives to be assessed. One option was to build a vast pumping station to lift the discharge from the sewers, existing and future, and let it be carried by the Grand Canal and an additional canal. The other idea was to drive a tunnel all the way to and through the mountains, with the minimum of pumping. The first solution would be the cheaper at first cost, but would entail daily running costs for all the years ahead, with the risk of power failure. The tunnel solution would have a much higher capital cost, but less cost of maintenance. Naturally much study would be needed, both in theory and some practice before a decision could be made. In the meantime, they wanted me to advise them on any possible problems which would have to be foreseen.

After this preliminary skirmish, Sophie and I were taken by the others to Señor Suarez's house where we sat down, fourteen in all, to a lunch which went on for three hours in charming surroundings. We were shown how to imbibe tequila, which is an acquired taste, if acquired at all. Suarez then said that as the weekend was starting he had asked Pescador and his wife to drive us to Acapulco for a rest before our labours. So we set off in the dark at 8pm to drive to Taxco for the night. Next morning, from the old Spanish-style mansion which had been turned into a hotel, we looked across to the little town of Taxco on its hill surrounded by low mountains. It had been a centre of the silver mines for many years, although the inhabitants looked far from prosperous. We spent a cloudy wet morning exploring the old town, its street market and the miniature Baroque cathedral and we were enticed to see some performing fleas, who failed to perform. Pescador then introduced us to a little silversmith and we were able to buy some really authentic and beautiful things for our family.

After lunch at the hotel, so very unlike a Hilton, the sun came out and we drove off to Acapulco on the Pacific Coast. The country was striking and constantly changing, being hilly with occasional plains. Parts of it were so like what we had seen on the road to Granada that we felt that the Conquistadors must have felt at home. In 1962 Acapulco had not been immortalised by the late Elvis Presley and television, but since then it and Taxco became added to the packaged holiday trail, as they are nearer Mexico City than most other attractive places.

> Acapulco is a city and port located in a deep semicircular bay almost landlocked, easy of access and with so secure an anchorage that vessels can safely lie alongside the rocks that fringe the shore. The town is built on a narrow strip of low land scarcely half a mile wide between the shoreline and the lofty mountains that encircle the bay. There is great natural beauty in the surroundings".

Thus quotes the *Encyclopedia Britannica 1911* and I could not say it better. It was like a greatly magnified Lulworth Cove.

Our hotel was on a ridge with its back to the town, overlooking the high rocky coastline to the north. We were given separate little chalets, some of which were placed all down the steep slope, but served by a lift running up the slope. Each chalet had its own little individual swimming pool beside it. When we took our seats for supper, we found that the restaurant overlooked the deep cleft in the rocks where the nightly dive took place. The tide surges in and out of the cleft, which is long and narrow, and it is alleged that there are only a few moments when the water is deep enough for diving into. While the tourists crowd hopefully along one edge of the cleft, a young man climbs up the opposite rock face, for what might be 100ft (though it is easy to overestimate). He then, quite understandably, kneels before a shrine and crosses himself. After a pause to work up the tension and while he judges the right moment, he takes off on a spectacular dive into the tidal flow below. As the crowd peered over the edge to see whether he came to the surface, I wondered where and when did they practice and try it out? However, it seems to be a nightly show.

The Pescadors proved charming companions, amusing, interesting and intelligent. We have so often found that it is not nationality but mutual interest which matters, as long as one has some language contact. On the beach in front of the old town, the usual crop of high-rise American hotels were bursting forth, but the Pescadors took us to a part of the sandy beach, some way from the hotels and tourists, where we spent a happy Sunday in and out of warm water. This included being dive-bombed by friendly brown pelicans in search of fish, which in-between attacks swam about among us. That night Pescador said that they would be driving back next day. He handed us two air-tickets, saying that as we had come so far to help them, he had arranged for us to stay for two more nights as their guests. As we were in a warm climate in December, we took full advantage of all the swimming we could. Sadly I later heard rumours that, sixteen years on, the waters in the bay had become too polluted for bathing. Such is progress.

Back in Mexico City the rest of the week was spent on work, mixed with sight-seeing. The work, as in other cases, included long discussions on tunnelling problems, with the added complexities of the peculiar soil and 6,000ft of altitude. I laid particular stress on the need for and type of intensive site investigation and the need for full scale tests. If such a long tunnel was to be driven, it was more important to explore along it than in the City where much was already known. As shafts and manholes would be needed, some of these should be carried out and used as trial pits in order to test out the behaviour of the ground by short lengths of full scale work. In other words, spend money in order to save it.

Christmas decorations were going up all along the streets, where we were taken to see the vast murals painted on the walls (and up the staircases) of several public buildings. These were painted by Diego Rivera and others. Suarez presented us with an outsize book *La Pintura Mural de la Revolution Mexicana, 1921-1960*, with all the paintings perfectly reproduced in full colour, but to see them in place is a startling experience. The skill is as terrific as the subjects are horrific. The communistically-inclined artists were concerned in many pictures to show the Spanish Conquistadors being beastly to the natives in full detail of torture and killing. Shades of the Gulag Archipelago. One of the curious aspects is the way in which the Spanish Mexicans have adopted the Aztec artefacts as their own culture. Incidentally, the artists did not spare the priests, who were shown as being as beastly as the soldiers. I doubt if these were shown to the Pope on his visit later in 1979.

There was much to interest us on our visit to the hill of Chapultepec, which terminates the apparently endless Avenue des Insurgentes. Here I turn once more to my favourite *Encyclopedia Britannica* to refresh my memory. Once, there was a gigantic figure of Montezuma carved on this hill of porphyry, but this, and the hanging gardens, were destroyed in the eighteenth century. In Montezuma's day and until they were drained, the city was surrounded by lakes of brackish water.

> For water supply the Aztecs used the main causeway through the city as a dam to separate the fresh water from the hills and the brackish water of Lake Texaco and obtained drinking water from a spring at the base of the hill of Chapultepec.

Then in 1848 the USA carried out some training for their Civil War by invading Mexico under General Scott.

> To enter the City by way of the Tacalaya Causeway it was necessary for the Americans to capture Chapultepec. The hill defended by about 4,000 Mexicans under General Nicola Bravo, was bombarded on September 12th and was carried by assault on the 13th and on the following day the City of Mexico surrendered".

This explains the striking memorial in the gardens of the palace on the hill top, which comemorates each of some forty boy-cadets who fought to the death, rather than surrender. After this battle, the Emperor Maximilian appeared on the scene and built the palace in its wonderful setting surrounded by mountains, but in 1867 it was once more captured by General Porfirio Diaz and in due course the Emperor was executed. From 1821 until the second election of Porfirio Diaz to the Presidency in 1884, the history of Mexico is one of almost continuous warfare. The history from 1884 to 1910 was then almost void of political strife. President Diaz's policy was to keep down disorders with a strong hand, to enforce the law, to foster railway development and economic progress….to introduce new industries. This explains my family connection with Mexico and why my visit was, in a sense, a pious pilgrimage. By way of explanation, what follows is a brief summary of my family connection.

John Body

Once upon a time in Cheshire there were two Mosley sisters, whom I never met. One was my maternal grandmother, who produced six daughters. The other married a Job Hamer, who was 'in cotton' and provided him with four sons and five daughters. The Mosley sisters must have had strong and lively personalities, as these persisted in both sets of daughters, who closely resembled each other. Around 1890, after his wife had died and the eldest daughter married, Job Hamer gathered his remaining brood of eight and set off for distant Mexico, where he helped Diaz in introducing new industries by starting the first cotton industry. The boys all stayed out there, made good and produced a number of my second cousins, some of whom I caught up with on the 1962 visit. One sister married a British civil engineer, Vivian Lister, who was building roads out there. Mary, the second daughter, married John Body, who was Lord Cowdray's right-hand man in Mexico and who contributed immensely to the country's progress. He built Vera Cruz harbour, the Tehuantepec railway across the narrow part of the isthmus (which rivalled canal schemes for many years), the drainage of Mexico and much else. But John Body's greatest work was his responsibility for calling Lord Cowdray's attention to oil deposits and then taking charge of the spending of Lord Cowdray's millions, which led to the formation of the Mexican Eagle Oil Company, in spite of hostility from the then monopoly of Standard Oil.

In 1911 John Body helped Diaz to leave the country, when he was overthrown in his old age by 'liberals'. In turn, these liberals found themselves unwillingly driven on by the bandit private armies of Villa, Zapata and other highly unpleasant types, which led to years of convulsion. After 1911, John made repeated trips to Mexico, during and after the 1914-18 war, to protect the interests of Mexican Eagle for the good of the United Kingdom. All the daughters also came back to England after the 1911 revolution and in 1914 John Body built a capacious house at Hindhead. I pay this tribute because of all the kindness that I received from him and his family, who provided me with many holidays throughout my impecunious youth. These included four month-long holidays in Scotland when I was at college. In 1920 John Body rented Hopeman Lodge, which later became the Prep Department of Gordonstoun, and in 1921 rented Gordonstoun itself from Sir William Gordon-Cumming. This came complete with dry and water dungeons and secret passages. One feature is the round square, built by Sir Robert Gordon in the 1700's. The traditional square of out-buildings in the form of a hollow circle was built because a witch had prophesied that this wicked man would be caught by the devil coming round a corner. This proved a conversational gambit when I met a famous old boy.[53] In 1922 followed the renting of Arndilly on the Spey and then in 1923 Beaufort Castle. I strongly recommend the biography of Lord Cowdray called *Member for Mexico* by Desmond Young (Cassell) which details what the British achieved and why they were still *persona grata* in that country when I visited.

Back to 1962 and we found Chapultepec Palace simple, with charming rooms, shady patios and with furniture intact. In the moat we found an exhibition, which was set up to glorify the revolution and its alleged heroic deeds and to set out the alleged misdeeds of the Diaz Regime, which seemed ungrateful. Our two hosts were far too young to have experienced the period, but I had listened to Mexican matters throughout my childhood. Our visit was crowned by an additional pleasure, as the Golders met us as they passed through from South America on their way to Toronto.

Mexicans in London

On the Sunday, December 9th, we flew on to Chesapeake, but it is worth mentioning the curious sequel, three years later in 1965, to the Mexican episode. We had exchanged Christmas cards, but had heard no more, until Wykstrom rang up from Stockholm. He said that Enrique Tamez was bringing a party of engineers to England after a visit to France and could I please arrange for them to visit tunnels and especially a factory for casting concrete segments for linings. In October, I went to their hotel, the Carlton Towers, to meet them. To my surprise, Tamez introduced their leader as Minister of Hydraulics and also President of the Mexican Institution of Engineers. He told me that they had wished to keep their visit private, but the British Embassy in Paris had got wind of it and the Minister was compelled to make some visits organised by the Ministry of Technology, through the then Mr C. P. Snow, Parliamentary Secretary. I invited all six of them to be my guests next evening at the monthly dinner of the Institution Council Dining Society. The Minister was delighted and he said, 'That will give me great pleasure. Now I can tell Mr Snow that I do not wish to hear Carmen sung in English at Sadler's Wells, as I shall be dining with the President of the Institution of Civil Engineers'.

---

53 This was almost certainly the Duke of Edinburgh, with whom HJBH was on several committees.

The Minister told us that first he had to go to meet people at the behest of our Parliamentary Secretary, who would then entertain all six to lunch at Lancaster House. The Minister went off with his aide, while I took the other four down a length of the Victoria Line, which was being driven under Green Park. Megaw, a partner of Mott, Hay and Anderson had kindly arranged the visit, so he and I walked the visitors down to Lancaster House for lunch. There the concierge insisted that only four names were on his list, so Megaw and I took the two outcasts to lunch at the United Services Club. Although they spoke no English they enjoyed this and we got by somehow. Next day the Minister was once again conscripted to make other visits. I had arranged for Kinnear and Moodie (K&M) to take two of the others to see the concrete segment casting yard, which had been the first to pioneer in 1937. I drove Tamez and another to Eltham, to take them into a tunnel, which was being driven under compressed air and to show them the plant and precautions required for success. That evening I took my party of six to the Council Dining Society dinner, where they received a very warm welcome. When I presented the Minister-President to Wynne-Edwards, who had succeeded me as ICE President, the Minister's secretary asked me, 'Is your Minister of Technology an engineer?'. I replied, 'No. Our Minister, Mr Frank Cousins, is a Trade Union Secretary' which surprised him. The Minister told us that the 'Min of Tech' had made him spend his time on visits that had no interest for him and had prevented him from seeing the tunnels, which had been the purpose of his visit. However, he was very pleased that his party had seen so much.

There were two further additions to the Mexican story. Firstly, I told Nigel Boxer of K&M that, if he wanted the work he should go out to Mexico to get it. As a result, they made many hundreds of segments after setting up in Mexico. Another offshoot was the building by Markhams of Chesterfield of three large tunnelling machines designed for the specific peculiar conditions. These were duly sent over and went to work. This ended my contribution to our invisible and visible exports.

Chesapeake Bay

On Sunday December 9th 1962, at the invitation of Raymond International Inc. we had flown off on our next assignment, which took us from Mexico to Norfolk, Virginia to see the building of the Chesapeake Bay bridge-tunnel. Their Mr Keating and his wife met us and installed us in a motel for two nights, where the bedroom had many knobs and switches for our comfort. One was connected to an alarm clock which, if this was set at ten minutes before reveille started a motor which caused the bed to vibrate violently up and down. The idea was to tone you up for the day, but other ideas can spring to an undisciplined mind. Next day, Mrs Keating carried off my wife to visit Williamsburg. The eighteenth-century colonial buildings of this former capital of Virginia had been carefully restored and were all lived in or otherwise in action. The shops and pubs were filled with period furniture and merchandise and period costumes were the order of the day. Thus visitors could have a taste of the past which, naturally, do not extend far back in the USA.

Keating took me first to the office, where the work was explained. The agent started off by saying 'That is the critical path but it is no good looking at it, because we have not worked for three weeks because of the appalling weather'. The Chesapeake bridge-tunnel was being built in what amounts to open ocean for about 17 miles across the mouth of the 180 mile long inland waterway. The bridge part consisted of a series of low level trestles made of three tubular concrete piles 54 inches in diameter. Precast deck units 75ft long, wide enough for two lanes of traffic, were being dropped in place by floating craft. There were two sections, with higher steel spans, to clear fishing craft, as the rest of the deck was only 30ft above the waves. There were two large gaps to provide shipping lanes by means of two immersed tubes, each about a mile long. Two pairs of islands had been made from sand dredged from the sea bed, each a third of a mile long, so that the roadway could slope down from the bridge into the mouth of the tunnels. These were made in the American style of steel cylinders with inner and outer skins. They were built on all slipways at Orange in Texas and towed for 1,500 miles all the way to the Bay. We went to see some of these in an old coal dock, where each unit, about 290ft long, was being filled with concrete between the two steel skins, to overcome most of the buoyancy before being towed out and sunk in a dredged channel.

52. Twenty-three mile mouth of Chesapeake Bay bridge-tunnel, rough sea

53. DeLong platform at work during construction of the Chesapeake Bay bridge-tunnel

When we turned to look along the south shore past the Naval Base, we beheld naval craft of every description as far as the eye could reach, from submarines to the Nuclear Aircraft Carrier *Enterprise*, the pride of the US Navy. This fleet had returned only a few weeks before from confronting Kruschev over the missile crisis off Cuba, so whoever now ruled the waves it did not seem to be Britannia. We then boarded a cabin cruiser, which Keating explained had been a bad buy for their purposes, as it was not intended for rough weather. This proved true enough, as we both suffered while we made the 20 mile trip from shore to shore along the bridge, which was three-quarters completed. It was far too rough to land on the islands, but some storm-tossed photographs were possible. We passed a DeLong platform with big pneumatic jacks for lifting the legs, it carried a derrick crane and pile frame and I was told that it took the place of a former platform, blown over in a gale. On the north bank there was an impressive and well ordered depot, in which the pre-stressed concrete units were being made and rapidly piling up. It was interesting to see the making of the 16ft long concrete tubes, 54 inches in diameter being cast by spinning in a mould at 450rpm. These would then be strung together by tension rods to make the piles up to 160ft long. We then came back, by choice, on one of the ferry ships, somewhat larger than Channel steamers. After a very long soak in a very hot bath, I was able to enjoy a pleasant dinner laid on for us by the Keatings in the

Navy Officers Club.

This was a most instructive visit from many points of view. In 1959 the Channel Tunnel Study Group had asked Parsons Brinkerhoff to report on a bridge-tunnel for the Channel and they had done so with advice against it. I could report back the same opinion, as the Chesapeake conditions were so very different to ours. Later, in 1968, my friend Professor A. L. Baker *et al* started to propagate another, but not fully designed, scheme for a bridge-tunnel. This led to a debate in Parliament and an inspired article in *The Times* by that worthy amateur engineer, Mrs René Short MP, with whom I joined in battle as referred to in the previous chapter. General Sverdrup, who was in charge of the Chesapeake work, gave a talk about it to the Civils and to Baker's chagrin Sverdrup declared it out of the question in the Channel conditions. Whilst on this subject, the Chesapeake Bridge, the Severn Railway Bridge, the Tasman Bridge and the Lake Maracaibo Bridge have all been damaged by ships getting out of lane or out of control, with many deaths in the last case.

The Bahamas

We left Norfolk early in the morning of December 11th for our final assignment, in the Bahamas, which proved to have an element of fantasy. After long delays and changes of plane at Atlanta and Miami, we arrived at Nassau in the evening and installed ourselves in the nice old-fashioned Royal Victoria Hotel. We had come in response to the cry for help from the Swede, Herr Lindrodt, who had asked for someone to assist on a tunnel problem, but he had not yet returned from New York, so we spent a happy day exploring Nassau and finding out more about the island of New Providence. On good advice, we lunched and swam at the Colonial Hotel beach on Paradise Island; on our map it was called Hog Island but it had been promoted – in several senses. The island was a long narrow strip which ran parallel to the Nassau coast and so formed a narrow waterway with the harbour at one end. To get across this waterway there had to be much haggling with the local boatmen who would then row visitors across.

The next afternoon we spent three hours in Lindrodt's office, where the mystery of his cry for help was revealed. Lindrodt had been the business manager for a Swedish millionaire who had bought most of Hog Island and on this estate he had built a fine mansion with superb gardens and then he had expired. Another slightly eccentric American millionaire had bought the estate, kept Lindrodt on as manager and turned it into a select and luxurious hotel for rich Americans. He rechristened the island 'Paradise Island' and called the hotel the Ocean Club. His name was Huntingdon Hartford. The hotel was over half a mile from the landing stage. Now, no rich American can contemplate walking that far. They would hire a Hertz car at the airport and be infuriated to find that they would then have to dismount, board a ferry-launch and have to walk or take the hotel taxi to the Hotel. Lindrodt advised that a bridge had been strictly forbidden because of shipping, so, please, would it

54. Courtyard of the hotel on Paradise Island

be possible to drive a tunnel under the waterway, so that rich Americans could be spared this intolerable hardship? This made a contrast to the heavy-weight matters that we had just visited.

On enquiry I found that the water supply was entirely from wells bored in the very permeable coral limestone, fed by plenty of rainfall which portends wet tunnelling conditions. After studying maps and other information, I devised a scheme for a small-sized immersed tunnel of concrete boxes, made from local materials. In order to inspect the site next day, Lindrodt conducted us to a beautifully equipped launch, which set out with much éclat and smart boat-hook drill by the uniformed bow-man, along the waterway to the hotel wharf. We were then driven by the hotel taxi along an avenue between densely planted vegetation, with lakes full of flamingoes and imported palm trees. These were intended to give a tropical atmosphere to a non-tropical island.

An annexe provided additional luxurious bedrooms on either side of an open-air Winter Garden that led into the mansion itself, which was furnished in beautiful taste. We were then taken down the half-mile of gardens, which gave a vista down to the waters edge. First we saw a swimming pool surrounded by every comfort for shady sitting and drinking *al fresco*, served by cheerful waiters from an adjacent bar. Then came sunken gardens of the highest standards. In one we found a beautiful little sculpture of a mother and child by Reid Dick, which we had seen in the Royal Academy some years before. The real surprise was to chance upon a gothic medieval cloister surrounding a square centre with sculptures, which was built on a small mound and ended the vista. This had come from an old French abbey where Randolph Hearst had bought it, had it packed in crates, shipped to his famous hide-out in California and like much else it had never been unpacked. Years later, Huntingdon Hartford had bought the cases and shipped them to Hog Island where it was realised that the parts had never been numbered, so the question arose as to who could solve the jig-saw puzzle of piecing it together. Hartford was advised that there was only one person, Monsieur X, a French archaeological architect who had been born in the town from which the cloisters had been ravished. So the call went out, 'Search the world. Find Monsieur X – no money to be spared'. The reply came back, 'Monsieur X retired. He is living in Nassau'! We met Monsieur X, who had taken the job, and found him most interesting. He showed us with pride how he was replacing broken and missing parts by making precise moulds and casting especially coloured concrete to match the stone.

55. *The cloister 'ruins' at Paradise Island Hotel*

Next day, Lindrodt took us on a cruise to see other parts of the estate, including a new golf course being watered from newly bored wells. It is lucky that the Bahamas are seldom stricken by drought. He then left us to make ourselves at home at the Ocean Club. We had the swimming bath all to ourselves as the hotel was empty, being prepared for the Christmas influx two weeks later. Later he and his wife crossed over to entertain us to dinner there.

We were to leave for home the next day, travelling via Bermuda, but when Lindrodt found that our plane did not leave until 9pm he insisted that we should again spend the day at the hotel, which we gladly accepted.

On my return home, I sent my report to Widmark and Platzer for them to follow it up. There are three sequels to this fantasy. First, a bridge was permitted after all, which settled that. Secondly, we read that the hotel had been burned down, which was a great pity. Finally, we went to a film and saw James Bond swimming up an inlet with a familiar little bridge across it, following which he was thrown into a swimming bath full of alligators, which had not been the case when we swam in it.[54]

Ceylon - 1963

The Mexican trip was not the end of the Swedish connection. In 1959, the largest contractor in Sweden, Skånska Cementgjuteriet, along with Widmark and Platzer as partners, had been awarded the contract for a hydro-electric scheme in Ceylon (Sri Lanka after 1972) to drive 4 miles of rock tunnel. As early as 1961, Platzer had been asking my opinion on a problem which had arisen and then, from January 1963, I was visited by a succession of Swedish engineers, bringing several inches of contract documents to study. In July, Sven Platzer asked me to find a British lawyer who could advise them on contract problems. My choice fell upon 'Tim' Singleton. Singleton was a partner in my solicitor friends Macfarlanes and I am pleased to say that the choice justified itself, as he was still advising Skånska 16 years later, after becoming Sir Edward and President of the Law Society. This was the start of a contractual struggle in which we were involved for over eight years and which ended with our client's success in a final arbitration. The Ceylon Government had taken unto themselves all the 'powers under the contract' which should have devolved upon the British consulting engineers who had designed the work. It was a peculiar contract, being based on the out-of-date Hudson Contract. In it, the contractor was responsible for everything under the sun, including any errors, misstatements, omissions etc in the documents... 'Unless the Government and consulting engineers shall deem otherwise'. As the errors proved numerous, the main struggle was to persuade the Government to deem otherwise.

This pear-shaped island is slightly smaller than Ireland and has a mountainous region towards the south of the centre of about 4,000 square miles. This rises to over 8,000ft and is made of a mix-up of pre-Cambrian rocks of the same age and composition as the Scottish Highlands, only the slopes grow tea and not heather. British engineers had built the Norton concrete dam across a river soon after the war and by 1963 a larger dam, the Castlereagh, had been built some miles upstream. Skånska's contract was to dredge the Norton Dam, build a power house at the upstream end of this and to drive 4 miles of rock tunnel parallel to the valley in the side of the flanking mountain. This ended 400ft or so from the upper dam, above the power house, and the water was to flow to the turbines down a steel flume on the surface of the slope, made by a Japanese contractor. I flew to the capital Colombo on a Thursday in September 1963, picking up Svensson of WP at Zurich and then on via Tel Aviv, Tehran and Bombay, bedding down at the Galle Face Hotel. As it was the tail of the monsoon, it was pouring with rain and Colombo, under Mrs Bandaranaike (who ruled in place of her husband who had been assassinated in 1959) seemed very tatty. The docks had been on strike so long that ships had to lie off the harbour for three months at a time. The

---

54 Coincidentally, it turns out that Sir Denis Wright was flown out in 1979 by the British Government to explain the denial of asylum to the by then deposed Shah of Iran, who was taking refuge on Paradise Island.

16. TRAVELLING BROADENS THE MIND

56. Pipeline and power station, Norton Dam

57. Pipe tunnel

ship carrying the crunching plant to start up the work gave up, turned around and unloaded it at Djibouti, so the build-up for the work had taken time.

The three major engineering institutions have joint groups in most of the Commonwealth countries so, as President-elect of the Civils, I took the opportunity to meet the Member of our Council for Ceylon, Professor Pereira, who was having a rough time with rebellious students, and his successor, Mr Wijeyesekera. The aim was to stiffen their professional morale. On the Saturday we were driven to the site, pausing on the way at a tea house on the banks of the river, where they had built the 'Bridge over the River Kwai' for the film. Then the best part of a week was spent in long sessions in the office, down the tunnel and around the site, chewing over the many problems and collecting details for future use. By then most of the tunnel, driven from several adits, had been lined with concrete made of crushed rock from the tunnel spoil. The most pressing problem was the need to complete the 600ft of tunnel round a ninety degree curve to join up with a short length driven from the valve chamber of the dam. This had been brought to a halt when it ran into bad ground. The rest of the tunnel had been in reasonably stable rock, but here they suddenly ran into a deposit of very soft clay. This was not a bed but a patch under a cleft 300ft above, where the rock had rotted in the coarse of time to kaolin. The face was a mixture of rock in the bottom, shattered rock in the rest of the face, mixed with very soft clay with water pouring in. The tunnel was lined with conventional steel arch ribs, imported with difficulty from Sweden. Two problems arose, the first was the method of tunnelling to be adopted and the second was whether the tunnel could be completed by Christmas, which would depend upon a forecast of what lay ahead.

The reason for this anxiety was simple. As rock-tunnel labour was almost non-existent in Ceylon, Platzer had imported a body of Filipinos who had been highly trained by WP on their hydro-electric contracts in their own country and their contract was due to expire in three months time. Before I left for Ceylon, I had called in Dr Gilbert Wilson, Reader in Geology at the Imperial College, who was working with me in protecting the Conservators of the Malvern Hills against the depredations of the quarry owners. From the plan supplied to him and together with the known conditions in the completed tunnel at the dam, Wilson had suggested locations for two more boreholes and he was able to forecast the probable length of troublesome ground. From this we deduced that the time limit could be beaten and so it was. Dr Wilson was to distinguish himself even further. Under the contract, the contractor had to have satisfied himself of the nature of the ground, without the modern clause 'so far as is practicable'. Certain borehole results had been included in the documents, but Dr Wilson got busy in the libraries of the Royal Geological Society and Imperial College and produced several papers published by geologists of the Ceylon Government. These papers referred to previous site investigations, which had not been revealed to the Swedish tenderer, who could hardly have had much chance of knowing about them. The papers were impeccable, so all the more reason to resent their concealment. One paper was very helpful, it described what sort of an investigation ought to be carried out, with emphasis on the need to bore over the tunnel at this particular troublesome spot. We then found out that the borings had been carried out alongside the road at the *bottom* of the valley to save the trouble of toiling up the mountainside. However, the dip of the strata was such that the bottom of the boreholes never penetrated into the beds in question by several hundred feet. This was valuable ammunition for the coming conflict.

Time was spent on the site showing the Swedes how they must prepare their records and the importance of making the engineer, and his staff, agree on fact, even if they did not agree with theory. Fifteen years later, one of them, Frederickson, said to Singleton at a meeting in Malmo 'I well remember Harding telling that it is no help being a virgin if nobody knows it'. In Sweden, Skånska had such a high reputation that, so they told me, they never had the day-to-day surveillance of consulting engineers. I think that I told him that they should blow their own trumpet, so I suppose that this remark was in the form of a parable. This assignment brought me a fresh number of Swedish friends. I became very attached to the two in charge on the site and we kept in fitful contact for a number of years. Kallstrand, the agent, who was young for the job, kept his cool in difficult circumstances and like his No 2, Frederickson, was intelligent and competent. The Kallstrands had a bungalow on the banks of the Castlereagh Dam, with a view across it to the Scottish-type mountains. It had belonged to a British tea planter, but this race of empire builders had been reduced from 3,000 to 300 under independence. Mrs Kallstrand invited me to stay with them, so I was in homely comfort, while Svensson lodged in the 'engineer's bunkhouse'.

58. Arch over the road to Columbo

At the end of the week, Frederickson drove me back to Colombo by way of Kandy, with a hairy moment involving a native lorry on a mountain road. Along the roadside were ornamental tombs. At Kandy we visited a remarkable arboretum made by the British, which included an avenue of every known type of palm and also a splendid orchid house, still kept in condition. After lunching by a stream where several elephants were being bathed by their drivers, we came to a place full of a milling crowd. We found that this was the birthplace of the lamented Mr Bandaranaike and the crowd were celebrating Bandaranaike Day. They seemed so cheerful that it was hard to guess whether it was a day of mourning or of celebration. This time my way back to London was via Karachi, Bahrain, Cairo and Rome. Many visits from my Swedish friends followed over the years while 'Tim' Singleton and I helped them with their case. It was three years later that Tim and I spent three days at Skånska's office in Malmo, where we tidied up all the details and the method of presentation of which, as an arbitrator, I had experience. Arbitration like constipation is the thief of time but in the end 'my side' won in front of a British arbitrator and even got paid!

The Episode of the San Francisco Subway – November 1966

The San Francisco Subway proved an unexpected, brisk and successful 'intrusion' before the Aberfan Tribunal (see Chapter Eighteen) really got underway. The morning of Wednesday November 2nd had been spent in dealing with letters for offers of evidence and discussions with the tribunal secretary, Lloyd Thomas. Then to complicate matters, at 5pm the telephone rang:

> 'This is Jack Bonny speaking from Boise, Idaho. You remember, President of Morrison-Knudsen. How are you Mr Harding? We are starting on a contract for part of the new San Francisco Underground and we would like to start it right for a change. Can you come out to us for a week and give us some advice?'.

This was rather a facer, Morrison-Knudsen were the largest contractors in the USA. They were involved with the Channel Tunnel and I had helped their Vancouver subsidiary the Northern Construction Co. The next day, I rang Lord Justice Edmund Davies, who said that he saw no reason why I should not go, as long as I was back for the resumed sessions, so I headed off to my bank to obtain dollars and to the US Embassy for a renewal of my visa. Just to add variety I had to hasten back to dress in order to attend a dinner that evening that the Duke of Edinburgh was giving at Buckingham Palace for the members of the Smeatonian Society of Civil Engineers and the Royal Society Dining Club! Sometime during this day I telephoned Jack Bonny to accept his invitation and to arrange dates. He then sent me drawings and details which I received on returning home after Day One at Merthyr Tydfil on November 8th, three days before flying out to the USA. The contract called for a large central ventilating shaft 93ft deep, from which twin tunnels 18ft in diameter were to be driven in either direction, using shields in compressed air or tunnelling machines if permitted. The ground water was only 10ft below road level in sands and was specified to be lowered by deep well pumping, to a depth such that the air pressure would never have to rise above 14lbs/square inch. This was to prevent workmen suffering from the bends (and reduce the cost of extra wages).

This project posed three major problems: one, the degree of success which might be achieved by the ground water lowering; two, the ability to tunnel in the sands with a rotary tunnelling machine; and three, complications in the shaft-sinking and the initiation of the work. I gathered together as much data on such matters as I could find and on Saturday morning November 12th, flew off by Pan-Am. I was met at the airport at 10pm by the project manager, but as this was really 6am GMT, I then stayed in bed in my suite in the Sheraton Hotel until 10am the following day. The Sheraton was in an old style Edwardian palace, so much nicer than the standardised Hiltons, with a vast dining room rivalling the Albert Hall and I would have to say that some of the inhabitants were equally remarkable. I then spent the whole of Sunday in my sitting room (ignoring about ten television channels) transferring the many borehole logs from two large sheets of drawings on to the longitudinal sections of the mile-long tunnels. This was to follow the advice of Sir Marc Brunel 'To make the plan to suit the ground and not the ground to suit the plan'. The 'plan' I had created revealed a number of points for improvement and discussion. That night I was joined by Noel Lambert of the Northern Construction Co. of Vancouver, who had driven the Highbury Sewer tunnel successfully in compressed air under my 'tuition' followed by a number of subsequent contacts. We were joined for supper by Mansell MacLean, a fellow member of mine on the World Bank Mission to the Litani Tunnel. He had wound up his firm after completing the Lincoln Tunnel under New York's Hudson River and in his advancing years had joined MK as an advisor. Next day we started five days of intensive labour from 8am to 6pm, with short breaks for a peculiar lunch in the local drug store. At breakfast and dinner at the Sheraton nothing was spoken except for 'shop' and no attempt was made to examine the city by night. How different from the days spent in Paris working with Soletanche, where work was hard but lunch and evenings were distinctly relaxed. I remember that the local San Francisco papers were getting worked up by the (then) novelty of topless waitresses which had struck the City but this did not appeal to me. What is good for the *Folie Bergère* is hardly suitable for handing round the soup.

MK had taken floor space in a tall building within walking distance of the Sheraton (it proved usual to walk in SF even if not in LA) where on the way we passed 'Wells Fargo' (which proved to be a bank and not a television series). A large board-room opened off a vast open office, complete with soft drink and coffee vending machines for the use of anyone at any time. In the board room, O'Dean Anderson, a MK Director whom I had met in Lisbon, took the chair and a dozen of us sat round all day in continuous session. His team were all in their thirties or forties, with varied experience, but whose tunnel experience had been in rock and not in compressed air. MacLean, Lambert and I embodied the experience which they lacked, with a lead of over twenty-five years. As the contractor was not being hustled by the engineers, MK very sensibly flogged away at every detail of what should, could or might be done or needed before rushing into the fray. After the Monday morning session, we walked the length of Mission and Market Streets to 'case the joint' and assess the situation. I was then taken to a warehouse to examine the many soil samples from the different depths, in order to assess the problems which the soils would cause. I see from my diary that I woke up next day at 4am still suffering from jet-lag and at once got down to an intensive study of the water lowering problem, before walking to the office by 8am to talk to the potential sub-contractor for this pumping and well boring work. This would involve boring wells at intervals along over 5,000ft of busy street.

We also discussed the choice of Robbins or Caldwell tunnelling machines, which were both capable of being converted to shields should the need arise. Then the project manager came in with an article from the *Engineering News–Record*, which gave details of astonishing progress of a Memco machine in Japan, so on an instant decision he flew off to Japan next day. The Memco machine cost nearly twice as much as the Robbins and I advised that it must pay for itself by greater progress (in the end the Memcos did the work but suffered from gremlins and did not match their Japanese performance - such is the gamble with new machines). I was able to quote the progress made on the Rotherhithe tunnel as far back as 1908, when the early Price machine drove steadily through very much the same ground in compressed air. Although I had not worked day after day behind a machine, I had visited a dozen machines in many countries on behalf of the Channel Tunnel Study Group, and in

16. TRAVELLING BROADENS THE MIND

1932 I had introduced deep well lowering into Britain, whereas it did not take on in the USA until after the war. I had also driven ten shields in air simultaneously at Bow-Leyton, so I found that I had something to offer.

Next day, after an evening of cogitation, I dropped my surprise idea into their laps, which had a dramatic effect on the work. Briefly, I proposed that the large square ventilation shaft should be sunk by the ICOS system, which ingeniously uses oil well techniques. The walls are dug out by grabbing through a continuous supply of heavy mud and water, sufficiently heavy to support the ground. Concrete is then poured into the trench from the bottom upwards, which displaces the mud. The idea was that the groundwater was to be lowered to 15ft below the shaft bottom, so that the shields or digger machines could be built in free air and driven far enough to build four air locks for the tunnel drives, uphill in both directions. I produced a progress plan and the idea that they should persuade the engineers to let them start at one end (16th Street Station) which could be done in free air while the shaft was being sunk, instead of waiting for it to be completed. I also said that I had grave doubts that the water could be lowered to 15ft below the shaft, due to the ground conditions, so the machines could be driven through the shaft walls before digging out inside the shaft. An air deck could be built in the shaft at a higher level and the bottom of the shaft dug out and concreted from the bottom of the tunnel. The shaft could then be dug out after the tunnel drive was over. This would save over six months and make many wells around the shaft unnecessary.

This idea really shook them, as they had naturally been bound by the plan of the contract. However an outside mind can sometimes be less involved. They then went off to try this out on the engineers, where again the value of contacts proved itself. There were two consulting firms involved. The first, Parsons Brinkerhoff, had been engaged by us to study a Channel bridge and an immersed tunnel, while the second, Bechtel, who in USA conditions could sometimes act as engineer on one contract while acting as a contractor elsewhere, had been one of the three firms, MK, Brown and Root and Bechtel who had been active on the Channel Tunnel on behalf of Technical Studies Inc. Together with Steve Bechtel, I had briefed both these firms with several contacts, so my scheme was put forward by someone whom they knew. Provisionally, they agreed on the spot, so I dictated a long report for Jack Bonny to a kind lady in diamanté spectacles who typed it out with skill and rapidity on the Thursday. That night I was taken out to dinner at a 'Steakery' and then in the dark we went by the fabled cable cars down the steep slope to Fishermans Quay, where we at once turned back and were wound up again, so my sight of this fabled portion of the city was brief. However, they were all very hospitable and friendly during my stay and I had the added pleasure of meeting old fellow campaigners, Mansell MacLean and Noel Lambert.

On the Friday morning I boarded Pan Am and flew home. The view of San Francisco Harbour is unforgettable, although I had seen less of the city than on my first visit. The Rockies seemed much more worn and smooth than in the Canadian range and, as there was no cloud, several vast dams were visible. We stopped at Winnipeg in the sunset with a temperature below zero and I was joined in the plane by a young mother with a charming thirteen-year-old daughter. They were going to London to join the American diplomat father and the daughter was desperate to visit Carnaby Street. However I lost much face by admitting that I did not know where it was in London. I later received a very warm letter from Jack Bonny, telling me that not only had my scheme been adopted, but it had caused the contract for the 16th Street Station to be given to them as well.

17

ARBITRATION UNDER ENGLISH LAW

In 1965, I was asked to start a series of lectures on arbitration for local associations of the ICE. The first one, on 9th February 1966, was at Leicester University. However, when I was appointed to the Aberfan Tribunal in October (see the following chapter) the rest had to be undertaken by others.

I opened with this sound advice:

> My text this evening is taken from the Gospel according to St Matthew, Chapter 5 verse 25. 'Agree with thine adversary quickly, while thou art on the way with him; lest at any time the adversary deliver thee to the judge and the judge deliver thee to the officers, and thou be cast into prison. Verily I say unto thee, thou shalt by no means come out thence til thou hast paid the uttermost farthing'.

Ninety per cent of our thousands of members never become involved in arbitrations and as these are conducted in private and not published, they remain a mystery to us until we ourselves become involved. Civil engineering contracts have adversarial positions built into them. If you consider that a car maker knows to a few shillings what the average cost of a car is when over 1,000 have been made, but the civil engineering contractor only sells personal services in advance, before venturing into the uncertainties of ground conditions, weather and labour behaviour, as well as the quirks of his employer or 'the engineer'. The standard contract sets out over 70 conditions, almost any of which could lead to a dispute if either or both sides behave unwisely or are overtaken by events. One cause of dispute is 'lack of diligence' leading to expulsion from the contract. Two others often occur:

- one, 'adverse physical conditions or artificial obstructions which could not have been reasonably foreseen by an experienced contractor' (a resident engineer was once cited as an 'artificial obstruction'!);
- another arises from 'alterations, additions and omissions'. In this case the engineer has powers (at the drop of a hat, though not so expressed) to increase or decrease the quantity of any work, omit or change the character, quality, kind, change levels positions and dimensions or execution and to include additional work of any kind necessary for the completion of the works. Where necessary, this should be corrected by 'Valuation of Variations'. The client or engineer shall determine the amount (if any) to be added or deducted at the rates defined in the contract, or if not applicable then prices shall be fined by the engineer.

In several cases in which I was involved, considerable variations were ordered at very short notice, while the client or engineer showed the utmost reluctance to value the cost, but insisted upon the contractor submitting claims (which were challenged) with the implication that the variation of the cost is all the fault of the contractor. This can take several years and may lead to arbitration or even the bankrupting of the contractor. A refreshing contrast to this was my own experience in settling several variations of an important nature on the Bow-Leyton tube extension. The resident engineer of Mott, Hay and Anderson and my deputy sat together (in calmer moments in the Blitz) to examine all the records and come to an agreed assessment. The standard ICE Form of Contract provides that 'when a dispute has arisen and if the parties cannot agree on an arbitrator he shall be nominated by the President of the Institution of Civil Engineers'. Thus, when I became an independent individual consulting engineer in 1956, I found myself nominated. The first was a small case, leading to several others. I sought advice from my old friend, Major W. Morgan. On retirement, he was much in demand as an arbitrator. He arranged with the parties for me to sit in on one of his cases, so that I could study the form and procedure. Once, at a luncheon for the Institution Honorary Members at the ICE, I mentioned arbitration to the famous law lord Viscount Radcliffe. He said 'Ah yes. Arbitration was intended for experienced men to make sensible decisions. Now, sadly, it has got into the hands of the lawyers'. This is very true. So let us now consider the procedure.

The late E. J. Rimmer QC worked in Mexico as a young civil engineer for Lord Cowdray's firm and then read for the bar. Due to his experience, he became the most called upon QC in civil engineering cases and for many years afterwards he trained up numerous young barristers, who were still referred to as 'coming from Rimmer's Chambers'. In April 1947 he wrote a paper to the Institution on 'The Civil Engineer as Arbitrator' and gave useful cribs at the end as to how the novice should compose the necessary notices and instructions. This was helpful for any beginner entering such quasi-legal waters and it included the warning that it is of vital importance always to write to the solicitors of both sides simultaneously, they being known as the Claimant and the Respondent. The first example is how to word the letter accepting the appointment and to ensure that it includes the section stating that 'My charges will be at the rate of £x per day (of 5 hours) during which I engage myself upon the duties of the reference together with all expenses and outgoings' this includes time spent in mental work as well as the actual sittings.

The arbitrator can make the parties appear before him so that he can draw up his 'Order of Directions'. This has to be strictly formal and to set out the dates for delivery of 'points of claim', 'points of defence', 'replies to defence' and counterclaim. 'Orders for Discovery' and the listing of documents for inspection, including figures and drawings, are to be agreed. This programme can be affected by requests for further and better particulars and replies thereto. One Respondent succeeded in wasting four years before the case could be heard and the Claimant asked me to undertake a medical examination at their expense as they wished to insure my life. The last important example in Rimmer's appendix is this.... 'I hereby give notice that I have made and published my award in writing on the matter and it may be obtained at my office between the hours of … and … on payment of my charges amounting to £x'. This is the point: sit tight on the award until one or other of the parties pays, otherwise you may never see your money. I like the word 'published' which really means that you are keeping it closely clutched to your chest. I have always insisted that the arbitrator shall be provided with some sort of room from which to emerge at the start and to retire to during intervals. As arbitrations are legal matters, it is important to keep a form of discipline to ensure the dignity of the matters, even though proceedings are comparatively informal – no wigs or gowns are worn and although all parties are seated it is usual for counsel and others to stand when the arbitrator enters and leaves.

This form of life brings one into contact with a quite different galaxy of personalities. Barristers who are retained for civil engineering cases make a small tribe of their own and those who are QCs are compelled by their 'union' to have a junior, much as a plumber must have his mate. This has an interesting effect on the ultimate costs. Most cases of any size are handled by a few top firms of solicitors and if one's performance has been reasonable, they often agree to nominate for another case, without going to the Institution for a nomination. Two small cases involved a small Urban District Council. As only solicitors appeared for each side, I was able to obtain all the necessary evidence and arguments for each case in a single day, but could claim my award for each case. The longest sitting was about a power station with a contractor vs. the Central Electricity Generating Board. This lasted for a total of 91 days and was spread over a year, with constant moves to fresh places as and when the solicitors for the Claimant could find them. These varied from several visits to the Arbitration Room in Grays Inn, a session in Holborn Town Hall (the mayoral chair was very uncomfortable) and a small court in the Law Courts among others. The mass of papers grew and the large files had to be numbered, as bundles contained years of correspondence and other evidence. Almost all letters exchanged between the parties over several years were dutifully read aloud and then the Respondent's counsel would say 'I must ask the learned arbitrator to refer to Bundle 6'. I would lean across the table and heave Bundle 6 into position, at which point counsel would say 'Oh, I am sorry, I mean Bundle 5'. So, two heaves were needed to return one Bundle and lift the other. I developed a new complaint 'Arbitrator's Elbow' – very painful! I have no need to dwell on this case. It essentially arose from variations where many buildings were redesigned, with little likeness to those tendered for. This brought in something called a 'Scots Schedule' in which many of the points were numbered and columns of figures attached to each point. In such cases, it is always surprising how the Respondent, who issued the variations and produced the new drawings, would fight to try to prove that there was little or no change when the changes were obvious at a glance to anyone of experience.

Relays of professional stenographers came and went, as in such cases words to a lawyer become sacred. The arbitrator must check the stenographers work each day, as well as take his own notes all through. This note-taking drives the points home (as well as keeping one awake) and if kept on one side of a suitable book, leaves the opposite page free for comments at the time for future notice. Only one example needs quoting: days 35-43 (or so) were spent by the Claimant proving that he had not the data to set out the railway sidings for the power station. When the Claimant's case closed, the Respondent spent a number of days trying to prove conclusively that the data was on the drawings.

> The arbitrator: 'How close were your two offices to the site?'.
> Answer: 'About 200 feet, Sir'.
> The arbitrator: 'I cannot understand why this was not settled over a cup of tea'.

Then on the last day:

> Counsel for the Respondent: 'Before making his award I must ask the learned arbitrator to read the whole of the transcript'.
> The arbitrator: 'I understand that it has as many words as the Bible, but the inspiration is different'.

The hearings then closed but this was not the end. The Respondent took time at the High Court, as he had asked for several 'points of law' to be stated and so some weeks were spent making an additional award, all accepted by the court. Having checked my diaries I can see that I was appointed to this case in April 1961, the first sitting was on June 26th 1962 and the final sitting on May 27th 1963. After the first award had been taken in the High Court, including reply to 'points of law' raised by Respondent's counsel, I then made my final award on April 3rd 1964. During this period I was dealing with many other clients, including numerous visits to Paris, North America and elsewhere for the Channel Tunnel Study Group. I see from the numbers on my file that I dealt with 100 cases between 1961 and 1964, and of these, 50 were during the period of June 1962 and May 1963. The fee is only payable for the final award and I worked for a total of 145 days over 3 years before receiving it. Arbitrators never grow rich – unlike learned Queen's Counsel!

In the case with the four year wait for the case to be heard alluded to above, the dispute was between a small contractor and a different Urban District Council. I accepted the nomination on 17th February 1962 and the final award was paid in December 1966. The sittings started at Grays Inn on 18th May 1965 and went on fitfully until the final speeches on 19th November – Day 24. The gaps were mostly to suit the lawyers and their sacred long vacation, with 24 sitting in all that time. No wonder the Claimant insured my life. This was a simple case of a water pipeline of only about 10 kilometres in length. Each party had their QC and juniors, and each had an expensive consultant witness who sat through every sitting but only gave evidence on one day. The engineer for the UDC was asked why he did not use his independent powers enshrined in the contract. He replied, 'Councillor Jones would not let me!'. A few years before, I had watched a British firm (backed by plenty of money) lay one mile of oil pipeline in a wild part of Persia in a day, with no human beings in sight. In this case in the west of England, the pipeline was to cross farmers' fields and village streets with bad traffic. The adverse physical conditions were small compared with Persia, but waxed large where there was little money to overcome them (as neither side had much), as well as long arguments as to whether hard chalk, unexpectedly encountered, ranked as rock under the Conditions. The artificial obstructions included irate farmers, prosperous villagers and a very young inexperienced resident engineer, anxious to show his virility by strong-man actions at every opportunity. The contractor's man on the site was older but no more sensible, and his chief wrote pompous letters that were more suitable to a contract of several millions, which showed a perfect example of unnecessary conflict by each party under pressure from their councils and companies. The award went to the High Court, as the QC for the Urban District Council asked me to state 35 points of law, divided into 7 headings, each with 5 sub-headings. This I composed with the help of my barrister, Mr Wordici and I was upheld in every point by Mr Justice Donaldson.

Looking back from my experiences of carrying out over seven 'unofficial' arbitrations where there are no lawyers but discussions and arguments with both sides (separately and together) and much studying of documents, this case could have been settled in a few days with only the cost of the fee of the arbitrator. However, this might not have been accepted by the District Auditor.

18

THE ABERFAN TRIBUNAL

In the past, natural disasters such as vast floods in China involving one hundred thousand people were beyond the capacity of distant newspaper readers to grasp. In contrast, with the almost instant reporting by television and radio, the man-made disaster at Aberfan, which included the loss of a significant number of young children, was of a dimension and horror that could be deeply and emotionally grasped all over the world.

Since memories fade, let me first recall part of our report (Part II).

> **What Happened at Aberfan?**
>
> 49. At about 9-15 a.m. on Friday, October 21st, 1966, many thousands of tons of colliery rubbish swept swiftly and with a jet-like roar down the side of the Merthyr Mountain which forms the western flank of the coal-mining village of Aberfan. This massive breakaway from a vast tip overwhelmed in its course the two Hafod-Tanglwys-Uchaf farm cottages on the mountainside and killed their occupants. It crossed the disused canal and surmounted the railway embankment. It engulfed and destroyed a school and eighteen houses and damaged another school and other dwellings in the village before its onward flow substantially ceased. Then, in the words of the Attorney-General:
>
>> "With commendable speed, the work of attacking this seemingly ever-moving slimy, wet mass began as people strove to release the afflicted. Essential services were brought to the village and there began the unprecedented and Herculean task of recovery. People came in their hundreds from far and wide to lend their hands, whilst from the local collieries there hurried the officials and the sturdy experienced colliers to use their strength and skill as never before".
>
> But despite the desperate and heroically sustained efforts of so many of all ages and occupations who rushed to Aberfan from far and wide, after 11 a.m. on that fateful day nobody buried by the slide was rescued alive. In the disaster no less than 144 men, women and children lost their lives. 116 of the victims were children, most of them between the ages of 7 and 10, 109 of them perishing inside the Pantglas Junior School. Of the 28 adults who died, 5 were teachers in that school. In addition, 29 children and 6 adults were injured, some of them seriously.... According to Professor Bishop, in the final slip some 140,000 cubic yards of rubbish were deposited on the lower slopes of the mountainside and in the village of Aberfan, whilst the amount actually crossing the embankment is estimated, very approximately, to have been about 50,000 cubic yards.

Parliament had acted quickly. On Tuesday 25th October 1966 resolutions were passed in both Houses of Parliament declaring that it was expedient that a tribunal be established for inquiring into a definite matter of public importance, namely, the causes of and all the circumstances relating to the disaster at Aberfan, Merthyr Tydfil, on Friday the twenty-first day of October, 1966. Early the next day, I had an urgent call from the Welsh Office, pressing me to be a member of the Tribunal and asking for an immediate decision. I accepted instantly, although the quoted statutory fee (taxable) was considerably less than my normal

18. THE ABERFAN TRIBUNAL

one. Our Introduction proceeds thus:

> 2. In pursuance of these resolutions a warrant was issued on Wednesday the twenty-sixth day of October 1966, over the hand of the Secretary of State for Wales, which, after reciting the terms of the resolutions set out above, proceeded as follows:-
>
> "NOW I, The Right Honourable Cledwyn Hughes, one of Her Majesty's Principal Secretaries of State, do hereby appoint Sir Herbert Edmund Davies, one of Her Majesty's Lords Justices of Appeal, Harold Harding, Esquire, and Vernon Lawrence, Esquire, C.B.E., to be a Tribunal for the purposes of the said Inquiry.
>
> AND I FURTHER APPOINT Sir Herbert Edmund Davies to be Chairman of the said Tribunal.
>
> IN VIRTUE of Section 1 of the Tribunals of Inquiry (Evidence) Act, 1921, I hereby declare that the Act shall apply to the Tribunal and that the said Tribunal is constituted as a Tribunal within the meaning of the said Section of the said Act."

The intense national emotion was shared by the MPs and led to this very brisk reaction and Mr Cledwyn Hughes' declaration was issued immediately after I had accepted. The final paragraph apparently is a protection to members of the Tribunal, giving them powers over witnesses to compel their attendance and the production of documents. The appointment was published in the evening papers and my telephone rang with many calls from journalists. The BBC got hold of a photograph from the Institution of Civil Engineers and we all three were shown on the news, with our descriptions. I was recorded as a Past-President of the Institution and consultant to the Channel Tunnel Study Group. Vernon Lawrence was a solicitor, a former clerk to the Monmouthshire County Council as well as a magistrate and had been chosen for his experience in local government.

Matters then became very brisk. I had a number of assignments in hand which had to be dealt with at the same time, and kept afloat somehow. The next days' agenda shows a good example of the pressure we were under, the morning was again spent dealing with calls from the press, but also receiving offers of help from firms and individuals, one in particular from Huntings, who specialised in aerial photography. After lunch, the members met for the first time in Lord Justice Edmund Davies' rooms in the Law Courts, where he briefed us on procedure and we were also introduced to Roger Lloyd Thomas, a leading civil servant in the Welsh Office, who was to become our invaluable secretary and organiser. To examine witnesses on its behalf, the Tribunal had the services of its own counsel and also the Treasury solicitor. We discussed who we should appoint as our own expert witnesses and considered who were likely to be produced by the many parties involved. I was asked to deal with this aspect. We also learned that the first hearing was to be at Merthyr Tydfil on the following Tuesday, after which we should adjourn for three weeks while the lawyers set about their duties of preparation.

We parted at 4:30pm and I at once made for Charing Cross Station bound for Chatham, as I was invited to stay the night with the Commandant of the Royal School of Military Engineering to attend the Annual Dinner of the Engineer and Railway Staff Corps, RET and AVR in which I was an Honorary Colonel. At drinks before dinner, the situation at Aberfan came up for much discussion. I expressed regret that more use was not made of sappers in civil works, compared with the use made of them in the USA. The Engineer-in-Chief then asked me to let him know at once in case there was any way in which he could help. We were soon to avail ourselves of his offer. By catching an early train next day, Friday October 28th, I was in time to attend a meeting of the Governors of the Imperial College and to ask the permission of the Rector to involve Professor Skempton and his colleague Professor Bishop as the Tribunal experts in the soil mechanics problems on which so much depended. I talked to Skempton and it was agreed that if he were appointed, Professor Bishop would take this on. Then the afternoon started with talks at the Welsh Office, after which I drove Lloyd Thomas to Huntings at Elstree, to see what photographs they could provide. Huntings then offered to make, free of charge, two models of portable size showing the slope before and after the disaster. These models were displayed throughout the hearings. I notice that my diary for this day also states: 'Channel Tunnel decision', possibly the Government allowing it to go ahead.

Monday October 31st was mostly spent at Imperial College, with Professor Bishop (Professor of Soil Mechanics), his senior lecturer Dr Hutchinson and A. D. M. Penman (Principal Scientific Officer at the Building Research Station) who all agreed to act for us, if appointed. We also drew in Dr Evans, a lecturer in Civil Engineering at University College, Swansea. These four formed the team that supervised the post-disaster site investigation that proved so successful. We spent until 5pm drafting a letter to the Treasury solicitor, setting out our scheme for the investigation. Next day, Roger Lloyd Thomas and I met with the members of the Treasury solicitor's staff and later with Sir Harvey Druitt, the Treasury solicitor himself, to discuss our proposals, including the wish to retain Dr Woodward, Assistant Director of the Institute of Geological Sciences.[55] He had written the definitive Geological Memoir of that area of South Wales and so was a valuable source of facts. After lunch we took the officials to meet Professor Bishop and his fellows and they were at once officially appointed. They were at work at Aberfan the very next day, having previously collected a formidable dossier of facts and photographs. A few days later, those on site asked me if I could find them a tracked vehicle, as it was almost impossible to get around the mountainside in the state it was in. I at once rang up the Engineer-in-Chief, Royal Engineers who replied, 'I have two articulated tracked vehicles which the Canadians had sent over for vigorous testing. They are at Brecon. I will send them over tomorrow'. This shows the value of personal contacts. The situation was saved and they were invaluable throughout the winter.

Just to keep up the pressure, following the surprise invitation to help out on the San Francisco Underground on 2nd November, the next morning, which was Friday, I flew to Newcastle with Wex of Freeman, Fox and Partners to meet the CEGS engineers to inspect a cofferdam and discuss other steps to prevent the vast pylon carrying the high tension lines across the Tyne from falling over. This assignment had been arranged some days before. I still had to earn my living and my diary shows that I spent all Saturday and Sunday at home dealing with six continuing cases, including letters to Malcor on the Channel Tunnel future which was still in active consideration. On Monday November 7th Lloyd Thomas and I took the 10am train to Cardiff and booked in at the Royal Hotel, then drove to Aberfan with Vernon Lawrence to meet our experts. We then went back to Cardiff to meet the judge and sat up discussing the details until midnight. On Tuesday 8th came our first ordeal. We set out on the road to Merthyr Tydfil piloted by a police car, pausing at the start to meet Sir Elwyn Jones who was to follow us to open the proceedings. On arrival at the large College Assembly Room, we had to line up by our car to face a battery of press men, TV and film cameras. The hall was designed to act as a theatre with a stage and a gallery at the end and we were assigned a small dressing room at the back. We entered the hall, to be confronted

---

55 The fancy title of what we used to know as the Geological Survey and is now called the British Geological Survey.

by a packed house of inhabitants of Aberfan, including relatives of victims. There was also a vast body of pressmen in a compound on one side and in the forefront a row of ten Queen's Counsel, representing the many bodies who were involved, with their juniors and instructing solicitors behind them.

In the process of advancing to the office of President of a learned institution, there is much experience gained in standing on platforms, but considering the reasons for our presence, this one was an emotional strain. The full stature of the Lord Justice could be realised by the calm, serious but courteous way in which he conducted the affair. For this preliminary sitting, he outlined how matters were to be conducted and each QC defined his reason for appearing. Then Sir Elwyn Jones, as Attorney-General, gave a fine introductory speech setting out the general pattern of events on which evidence was to be heard. The Tribunal hearings were next adjourned until 28th November. Lloyd Thomas, Vernon Lawrence, myself and our secretary walked to lunch in a local restaurant, where a room was reserved for us. All the way we were pursued by photographers, at times walking backwards ahead of us. One could sympathise with Royalty. We then drove to Aberfan for a general view of the disaster site and to see how the investigation was proceeding.

The next three days were spent dictating letters on tape, and dealing with six or seven other jobs, including drafting an arbitration award, as well as matters to do with Aberfan. Then on Saturday November 12th I was driven to Heathrow and after waiting the customary 2.5 hours 'regretted delay' I flew to San Francisco via Montreal. After I flew back from San Francisco on Saturday November 19th, there followed a busy week sorting out other jobs, including making an award on a long-drawn out arbitration, interspersed with many telephone conversations with Professor Bishop about his problem. My purpose is to recall what it is like to serve on a Tribunal, but not to pursue the sad story in detail. Once I was back, Lloyd Thomas took me to lunch at the Cabinet Office Mess in Whitehall, where I met many Permanent Secretaries 'at graze' which was one more experience. I then drove him to Elstree to inspect the models which Huntings were making and to brief ourselves on their aerial photographs and how they were able to contour them at five feet intervals, which became important in the investigation. Next came the first three weeks of hard labour.

On Monday November 28th we went to Cardiff. The judge had wisely decreed that we should stay at separate hotels, so that we would not get in each others hair, and I was allotted the Angel Hotel. This was comfortable, but as it was by the entrance to Cardiff Arms Park, it was vulnerable after rugger internationals. We then had a meeting with the Welsh Office officials for them to meet Professor Bishop's party and Major Farmer, RE of the Ordnance Survey, whom I had mobilised to provide large-scale maps and futher aerial photographs. On Tuesday 29th we started in earnest, with the first of 76 days of sittings. Each day a long and weary drive was made up the valley to Merthyr Tydfil, to listen to the evidence of bereaved parents and others to establish what actually happened. This was followed by a long drive back in the dark, during which I tried to instruct my colleagues on the general facts about the behaviour of loose soils in which the judge was an expert, but which have no connection with underground mining. A cutting from the *Sunday Times* of December 4th survives which shows how the opening appeared to a journalist:

> The early stages were like a low-key Edwardian drama in modern dress. Three sombre men in dark suits held the stage of the tiny, well-appointed assembly hall of Merthyr Tydfil's College of Further Education. In front of them similar men were humbly submitting, obtaining some apprehension of craving indulgences, and, occasionally

finding it pertinent to add.

> Nothing takes the passion out of an issue more quickly than attorney's English, and the Aberfan Tribunal which opened last Tuesday is no exception. Yet in a curious way the archaic circumlocutions seem appropriate for examining the cruel pattern of events leading up to the slide of Aberfan tip No. 7 on Friday October 21st – an avalanche which buried alive 116 children and 28 men and women.
>
> If spades were called spades the horror of the thing would give the impromptu courtroom an almost intolerable emotional charge. As it is, issues are emerging, politely, almost apologetically, but, nonetheless sharp.

The article goes on rather unkindly:

> The Welshness of the tribunal has been almost laboriously promoted. The Chairman Lord Justice Edmund Davies is a Merthyr man.[56] Sir Elwyn Jones was born in Llanelli and names like Evans, Rees and Williams crop up amongst the 17 Counsel who cram the four rows in front. But in a mining valley they still seem a race apart, men with hard Middle Temple voices and soft hands.

For the first three weeks until the Christmas break there certainly was an almost intolerable emotional charge on all of us. No doubt we seemed sombre, but with good reason, having to share with each of the witnesses the intense scrutiny of the assembled press and the public. Fortunately, after Christmas when most of the local witnesses had been heard, it was possible to continue at Cardiff in another College, this time of Food Technology. The college included a course on Hotel Management as well as cookery, so we then had not only our own private room but also a small dining room, where we were served by charming young students, who were being trained for all concerned at the College in waiting as well as cooking.

In Tribunal proceedings it is customary to have relays of trained stenographers who take down the proceedings verbatim. After twenty minutes, one stenographer is relieved by another, so that he or she can transcribe their notes before their next turn comes. When the court has risen, the whole transcript has to be collated and many copies made, which are distributed later in the evening to the members of the tribunal and also to the parties represented before it. The transcript must be read and studied, because the first matter next morning is to agree or correct what is in the transcript. So the members must do their homework. We were issued with large Stationery Office standard books much bigger than foolscap, in which we would make our notes throughout. It may seem unnecessary to do so when transcripts are made, but there are good reasons. While every word is not written, it is very useful to take down some verbatim questions and answers. If this is done on the right hand page, quick notes can be made on the opposite page or even rude comments. This all helps, when faced with the many pages of daily transcript, to turn quickly to any particular section and to underline important matters. The judge would say quietly to a witness, 'Watch our pens', this reduced the temptation to gabble and incidentally helped the witness to keep cool and not rush ahead of his thoughts when under pressure. Above all the taking of notes makes the members stay awake and maintain full attention, which is difficult when a witness is being asked the same question for the seventh time by the seventh QC, as often was the case.

According to my diary, the Christmas break was not wasted. There were meetings with the Stationery Office over aspects of our eventual report, there were discussions with our experts on findings up to date and drafting ideas for our report from a study of the notes.

---

[56] The journalist was mistaken, Lord Davies actually came from Mountain Ash, on the other side of the mountain.

There were telephone calls to Platzer in Sweden over his problems, in addition to getting up-to-date with cases that had been interrupted. There was also the final service in Westminster Abbey to celebrate the conclusion of their 1,000 year celebrations, at the end of which all of us who had served on the Exhibition Committee of the Royal Society were herded on to the dais before the High Altar among the Dean and Chapter, who were deployed in all their robes: 'Bottom, Thou art Translated!'

Sittings of the Tribunal were resumed in Cardiff on January 10th. The day before was spent in inspecting the almost completed Monmouth road tunnels, as I was advising the consulting engineer on the contractor's claims. He was able to drive me to the inspection and then take me on to Cardiff. I am happy to report that my settlement was accepted. We now started on a remorseless routine of sittings from Day 15 until Day 76, which was 28th April, with a welcome break while the College was closed for the Easter vacation. The line-up of counsel with their juniors behind them, and the solicitors who briefed them in another row, made a formidable array. At least five of them soon became judges, not to mention (Sir) Geoffrey Howe, who became Solicitor General and who, after an interlude in opposition, became Chancellor of the Exchequer in 1979.

Mr Tasker Watkins VC, QC, the Tribunal's counsel, sat in the centre with his juniors. It was his duty to introduce each witness and examine them, after we had let them know of any special points which seemed in need of clarification. Then each QC would take his turn on behalf of his clients. Mr Desmond Ackner, for the parents and residents and Mr Mars-Jones for the teachers, were concerned with the victims. The others were more on the defensive, representing the National Coal Board, managers and mineworkers, while the Borough of Merthyr Tydfil, Cardiff Corporation and Powell Duffrys came in on the fringe. Many days were spent working through the hierarchy of the Coal Board from lowly individuals up to the top, followed or interweaved with others who were put forward as witnesses. The whole story was so intricate that it took many pages to elucidate in our report (which is the only true record of the events, both before and up to that date) and no purpose would be served in trying to summarise it.

My admiration for our chairman, Lord Edmund Davies, increased as time went on. He was completely but quietly in charge and courteous to witnesses, which belied his reputation of being a hard man because he had sentanced the Train Robbers. We avoided any need to consult together in public by passing slips of paper to the chairman with questions to ask, if he thought them appropriate. He would then request each of us to ask any further questions in our own particular field. Vernon Lawrence lived locally and was an interesting and helpful colleague and Roger Lloyd-Thomas, our secretary, was invaluable especially as the Tribunal needed to remain aloof from outside contacts, so the amount of organisation behind the scenes was considerable. He also proved a stimulating and witty companion and became a close friend, kindly commenting on the script for this book.

The long winter sittings were a burden. It was desirable for us to remain aloof, so the only exercise was to walk the streets of Cardiff alone in the dark, while waiting for the transcript. Luckily the centre of Cardiff is slightly more attractive than that of many cities of its size, owing to the layout of the civic buildings in a park made available by the Marquess of Bute. My wife kept me company on two separate occaisions, where she was given a seat behind the lawyers and her observations were helpful. One evening she said, 'I notice how you three are imprisoned behind your long table with no relief while the QC's, unless they are talking, have nothing to do but go out for coffee or visit the gents!'. On the second occasion she visited, one of the very top men from the Coal Board in London was on the witness stand and he started on a high and mighty line, savouring his importance. She made a little 'comic strip' series as she watched, drawing him as he dwindled down in size under the relentless cross examination, which would not have been so severe but for the initial attitude adopted. To adapt early lessons on arithmetic, a witness who considers himself the 'highest common factor' is usually reduced to his 'lowest common denominator' as cross examination can reduce everyone to the same level. She produced another strip showing the tribunal members gradually disappearing under accumulated piles of transcripts and documents.

The speed with which coal can be got to the surface and processed is controlled by the speed with which the volume of waste can be removed. In the narrow Welsh valleys the collieries and their villages occupy the valley bottoms, so there is little space to tip waste except on the mountain sides, usually far up so as not to interfere with farming. Waste has to be moved for long distances up steep slopes and this involves haulage and lifting equipment of various kinds. This is the reason why the responsibility for all tips had been vested in the Board's mechanical engineers, who seemed to guard their domain jealously to keep their status. They would keep an eye on the efficiency of the mechanical devices on which the progress of spoil removal depended, but the actual tipping of the small trucks, laboriously hauled up the mountain-side, was carried out by simple labourers, who pushed the trucks or skips to the edge and tipped them down the face of the tip. This manual work was left unwatched, and far from the main scene of action. Out of sight equalled out of mind.

At Aberfan, when Tip No. 6 was ending its life, a search had been made for future sites for the next tip. In parallel to this search, the dip between Tip No. 3 and No. 4 proved a convenient space as an interim solution. However, the search ceased and what had become Tip No. 7 was left to extend month after month, out and down the mountain side. Much is summed up in a paragraph that I wrote into our report.

> It is true that flow-slides of the kind which produced the disaster are rare, but those who are minded to talk of 'unforeseeability' are confronted by the startling fact that three such flow-slides have occurred at Aberfan itself within a space of less than twenty-five years (that is in 1944, in 1963 and in 1966) and a fourth less than five miles away just a few years earlier.

Very large 'blow-ups' of aerial photographs were displayed on either side of our dais, showing all seven tips before and after the disaster. Tip No. 4 showed a vast concave cavity in its face, which to an expert eye showed the result of a previous slide in 1944, whose remains were still visible on the ground, although it just stopped short at the railway embankment above the school. The brief on which the Coal Board counsel had to work insisted that the slide could not have been foreseen and witness after witness became entangled in having to follow this line. By Day 30, I was driven to say to my colleagues, 'If only someone would do a "Perry Mason", we could soon go home'. Then, at Day 49, this stance was abandoned and the senior Coal Board Divisional Production Director did a 'Perry Mason' and confessed 'that the instability of Tip No. 7…could clearly have been foreseen' and that there was no tipping policy. The comment in the report on this volte-face was pungent, as we shall see.

The slide five miles away from Aberfan, which was mentioned in our report, occurred in 1939. It crossed a main road but no-one was hurt. Powell Duffryn, the then owners, produced a very sensible memorandum, *The Sliding of Colliery Rubbish Tips* with words on 'Precautions to prevent sliding'. However, on nationalisation it was put in a drawer and forgotten. When

18. THE ABERFAN TRIBUNAL

a scare arose in 1965 on another site, the divisional mechanical engineer remembered it and the divisional chief engineer sent copies to all concerned, with instructions to report on all tips, but the follow-up faded away. The area mechanical engineer, a man who proved a powerful personality, was told to inspect Tip No. 7 with his civil engineer opposite number. He so resented the intrusion in his preserves by a man of weaker personality (with whom he said that he was at loggerheads) that he failed to carry out the inspection, so good intentions were frustrated.

I gathered, rightly or wrongly, that before nationalisation each colliery was a self-contained unit run by a manager who was responsible only to the directors. When the Coal Board took over so many collieries, a vast hierarchy was bound to be set up. When we had been enlightened on the various sections involved, I made up a little 'table of Avourdupois' to remind me:

Four collieries make a Group.
Three Groups make an Area.
Two Areas make a Division.
Later I added *One Division makes a mess-up.*

Lack of communication between so many different talents, titles, ranks and responsibilities brought about failure to follow up hints and portent, or to follow up instructions. The handing down of accumulated experience, on the lines of father to son, does not flourish in such a setting. While we were hearing evidence, our team of experts were busy on the mountain side measuring, examining boreholes and digging. Other experts called in by bodies that were on the defensive (so to speak) co-operated with them, so that all the findings were agreed. Put over-simply, the Coal Board had postulated a burst of water which could not be foreseen, and that this had washed the waste down the mountain slope. Professor Bishop's party proved that this was not the case and that the water was released after the slide had come to rest. The rush of water greatly hampered rescue efforts, but it took place after the deaths had occurred. The experts produced a long report which is to be seen in most technical libraries.

Banks for railways and roads are consolidated by plant and so are made strong and resistant to most pressures. But heaps of colliery waste were either end-tipped from wagons or dropped in some cases from cableways, so remained in a loose condition. Again quoting from our report:

G. Soil Mechanics

267. Soil Mechanics has been defined as 'The scientific study of the behaviour of those materials that consist of an aggregation of discrete particles, and particularly of their interaction upon each other, and upon the fluid filling the voids between them'. A colliery spoil heap, tipped loosely from above and uncompacted, is an eminently proper subject for this study, though it appears that it has not often been considered in that light before this disaster.

268. Some further terms need to be defined in non-technical language.

Liquefaction This can occur in a heap of loose sand or in an uncompacted tip of mine rubbish. If the lower part of the tip contains water, filling the space between the particles, and if a sudden load or shock is applied (such as the slipping of the upper part) the water then supports the particles and the whole saturated body behaves as if it were a liquid.

Flow-slide If the liquifaction occurs in such a heap on a slope, the mass will rush down the slope as if it were a liquid, though actually a mass of wet solids. When it stops, it immediately reverts to a relatively dry heap.

This was borne out by the evidence of one or two who were first on the scene and who were able to walk on the heap on the side away from the following flow of water. Our report went into the findings of the experts at some length. The tip, with its steep slope of 1 in 5 of very loose waste, extended down the mountain side and over small water courses. The mountain, covered with a mantle of impervious boulder clay, consisted of layers of fissured pennant sandstone between seams of impervious clay. Professor Bishop and his team painstakingly dug gently down through the remains of the slide after the style of archaeologists, and uncovered a classic example of a slip plane through the boulder clay, which was nearly ten feet thick. The excavation showed conclusively that it had been gouged out by the moving mass and had released the water, pent up in the sandstone in normal times. Thus the flow slide was followed, but not accompanied by, what is known as a mud-run. This term is usually applied to incidents of common occurrence in mountainous districts. Torrents of water, rushing down a mountainside, collect and carry with them all available loose material in their path – in a 'run of mud'. This evidence finally demolished the submission of the Coal Board and was accepted by their counsel.

The last days were spent in hearing the technical evidence of the experts, our own and those called in by other parties.

> A pleasing feature of the scientific investigations was the spirit of harmony which prevailed and the mutual help which the experts retained by the different parties gave to each other. No less remarkable is the almost complete identity of their findings; such differences of opinion as were revealed seemed rather to lie in the field of semantics than in that of basic points of view.

By this time the tensions were slightly relaxed, as the matters were impersonal and only the press agencies remained, the journalists having departed as the evidence was not newsworthy. We then had our only (slightly) lighter moment, one of our counsel embarked upon a long discourse which I found hard to follow. From memory of the transcript:

> The Chairman: 'Mr Rees, Mr Harding has just muttered "What is the message?".'
> Counsel: 'My Lord, I am trying to prove (something or other).'
> Chairman: 'Mr Harding says "Message received".'

59. The Report of the Aberfan Tribunal and its accompanying documents were widely read for the insights on slope stability

Sir Edmund Davies had firmly refused to allow any photographers to take pictures of the Tribunal in session, but when it was all over he allowed photographs to be taken of us sitting at our long table, with our documents around us and the large blow-up plan of the area to which witnesses had been constantly referred as background behind us. So with great relief on Friday April 28th we left Cardiff for London, to start composing our report on Monday May 1st. We were allotted several rooms in the abandoned semi-basement of the old buildings of New Scotland Yard. Our material consisted of the evidence of 136 witnesses contained in 4,236 pages of foolscap in 76 daily volumes, 61 maps and

200 photographs, as well as 21 written reports submitted at our request or on behalf of the various parties involved. This is where our daily note-taking came to our aid as a form of indexing when needing to refer to a specific day.

In Cardiff we had had several ideas as to how the intricacies of 'who, when, where, why and what – not forgetting how' should be disentangled. We then agreed on a format and each sat down to write his own version of his own particular area. Vernon Lawrence sorted out the local government problems and I got down to the geological and engineering side. Then we had to turn attention to many other details, the whole work required much re-drafting and constant searching through the mountain of transcripts in order to make quite sure of every matter. The report was set out in six parts:

I. The physical features and history of tipping at Aberfan.
II. What happened on the day of the slide.
III. Tip policy and responsibility.
IV. Posed the question 'Should anyone be blamed?' This was the longest part. It began with an introductory passage on our concepts of blameworthiness, the history of other slides and events on the disaster tip which either could have a bearing on the causes of the disaster or serve to reveal the state of mind of those most intimately concerned. It went on to deal with the National Coal Board, the personal responsibility of various officials, the Merthyr Tydfil Borough Council and its officials and the National Union of Mineworkers.
V. Posed the question 'How and why did it happen?' which was answered by a summary of the views of the experts.
VI. Consisted of Lessons and Recommendations.

All four of us worked separately with constant discussions as we went along, while our devoted secretaries typed out draft after draft. The final report contained 70,000 words, but far more than these were written in our efforts to end up with a clear result. Lord Justice Edmund Davies then wove it all together in an admirable style, in which clear legal thinking was explained with Welsh eloquence, suitable to the subject. After careful proof-reading it was finally submitted and the House of Lords and the House of Commons 'Ordered (it) to be printed on 19th July 1967' and it was subsequently published on August 4th. When we had been appointed nine months earlier, more than one newspaper stated quite bluntly that this was to be a white-washing affair. When they had read the report, their headlines were of a very different character. For instance, the Evening News displayed a photograph captioned 'Three Just Men sit in Judgement' and their leading article was headed 'No speck of whitewash' and said:

> Today's report by the Aberfan Tribunal – surely one of the most competent, candid and lucid ever to result from a Government inquiry – has no spark of whitewash in it. Blame is certainly shown – and blame is cast without the mincing of a single word, upon the men whose 'ignorance, ineptitude and failure' allowed this dreadful thing to happen.

Part IV of the report went into considerable detail, with quotations from the evidence in order to make the story of 'eight years of folly and neglect' clear, both as lessons for the future and as a way to show how our conclusions as to blameworthiness were reached. When we had agreed on this, we each wrote our opinions independently, but found we were in close general agreement. In the final draft, all virtues and influences of those criticised were carefully set out, before balanced words summed up their responsibilities. The press reprinted these *in toto*, but the BBC showed photographs only with the harsh words and none of the carefully chosen words on their otherwise good characters. We felt that this nullified our intentions and was a form of trial by television. Most papers chose to quote a paragraph from our report, which will suffice:

> 47. ...there are no villains in this harrowing story. In one way, it might possibly be less alarming if there were, for villains are few and far between. But the Aberfan disaster is a terrifying tale of bungling ineptitude by many men charged with tasks for which they were totally unfitted, of failure to heed clear warnings, and of total lack of direction from above. Not villains, but decent men, led astray by foolishness or by ignorance or by both in combination, are responsible for what happened at Aberfan. That, in all conscience, is a burden heavy enough for them to bear without the additional brand of villainy.

Unlike some reports, ours was at once accepted by the Government, including our recommendations for changes in practice and also proposals requiring new legislation. The National Tip Safety Committee was duly set up and they had a hard task ahead of them. Several Professors of Engineering made our report required reading by their students and one told me that much could be learned on psychology, man-management and organisation, in addition to lessons on geology and behaviour of soils. The report was sold out almost on the first day, so must be sought in libraries. The House of Commons duly debated our report and my wife and I were given seats in the Distinguished Strangers Gallery. Prior to this, I had only heard a debate once before in 1926. During Question Time the House was packed and a good performance put on. It was amusing to find that Ted and Harold and Reggie and Barbara[57] were actually real and not merely the invention of television directors. Then, when Question Time ended, there was a rush to evacuate the Chamber, leaving only a handful of members scattered about the benches who wished to speak in the debate. The best speech was by a woman member of good appearance - we were told that she was a Mrs Thatcher. Richard Marsh as Minister of Power also spoke well. A large contingent from Aberfan had been given seats in the public gallery and I feel that they must have felt let down by the rush to evacuate.

Two years later there was an epilogue in the form of a conference organised by the Institution of Civil Engineers on 'Civil Engineering problems of the South Wales Valleys' (of which there are plenty). I was persuaded to chair the organising committee, which had to identify and persuade eleven authors from expert members of the South Wales Association of the Institution to make a contribution. Between 1960 and 1964 I had paid several visits to the Association at Cardiff as a Vice President and later as President of the 'Civils'. I had visited some of the valleys and as President I paid the customary call on the Lord Mayor at the Impressive City Hall where I was confronted by two paintings by my father-in-law, Edmund Blair Leighton, in place of honour. I also made a number of friends in South Wales, one of whom was Mervyn Jones, the head of the Wales Gas Board, naturally known as Jones the Gas. He kindly relieved the tedium of my spell in the Angel Hotel during the Tribunal's sittings by invitations to his home. Another was Norman Williams, who was at that time member of council for the South Wales Association. He had been resident engineer on the building of the steel rolling mills, which cover a vast area near Newport. They were built on a deep deposit of soft estuarine clay and every conceivable and available type of piling plant was mobilised, so I christened him 'Williams the Piles'. I am not sure whether he approved, but the name stuck.

I co-opted Roger Lloyd Thomas to our committee, as his connection with the Welsh Office

---

57 Edward 'Ted' Heath (CON), Harold Wilson (LAB), Reginald Maudling (CON) and Barbara Castle (LAB).

was valuable. The subjects covered came within the purview of a number of Ministries so 'Williams the Piles', who was a keen worker, asked if we could not organise a luncheon for all six Ministers involved in order to enlist their interest. Roger made a statement worth brooding over; he said 'Ministers never hunt in packs. Nothing infuriates a Minister more than to see the face of another Minister at the same dinner table'. His was the voice of experience.

When the time came I travelled down once more to Cardiff and as I approached and neared the Angel Hotel, I felt as deep a depression as if returning to Prep School after the holidays. However, as I entered the hotel, I was met by 'Jones the Gas', who had laid on a welcome dinner party for me with a dozen of the contacts which I had made in Wales. This was a moving and cheering evening. The papers and proceedings of the Conference provided a wealth of valuable information and creative suggestions and were much approved by the Welsh Office. I was enticed into the local BBC Studio for a few unrehearsed words on TV and ended by saying (without thinking) that we hoped that the results would be to prevent anyone from ignorance making a proper Charlie of himself. As this was just before the investiture of the Prince of Wales, I was surprised that they left this in when broadcasting it.

19

ARBITRATION UNDER INTERNATIONAL LAW

Oued Nebaana Dam, Tunisia - 1967

In July 1967 I was busy checking the last proofs of our Aberfan Disaster Report, when I was asked to meet a director of the Utah Construction and Mining Company. Utah was the plaintiff in a case over the contract for the Oued Nebaana Dam in Tunisia and asked me to accept their nomination as one of the three arbitrators necessary under the Conditions of the International Chamber of Commerce. The defendants were the Tunisian Government and an American consulting firm, with the French firm of Coyne et Bellier who were involved with them. Each side appointed a neutral arbitrator and then had to choose a chairman. Tunisia had nominated M. Tavernier, an engineer with Electricité de France at Lyons and we were able to persuade Professor Alfred Stucky of Lausanne to act as our chairman. He had designed and been responsible for over fifty dams of all sorts and was a 'top man' in the Congress des Grandes Barrages. Some sub-contractors were the hardest hit but could only be compensated if the main contractor succeeded, so after studying the usual mass of documents from both sides, we organised a preliminary session with all concerned in the Hilton Hotel, Tunis from 17th until the 22nd October 1967. My wife came along to keep me company, but was even more useful as she spoke French, which was the first language of the other arbitrators and the Tunisians. The Tunis Hilton, some way outside the city, had been recently built by a British firm. The beds were comfortable but the food was atrocious!

Unlike British arbitrations, each side did its own talking at the start, which our chairman had difficulty in keeping under control. Utah was led by their senior Vice President, A. L. Reeves, who fortunately was also a lawyer. The chief engineer of the Tunis department responsible for the dam was an experienced Jewish engineer, who led their side with great volubility and some heat. No doubt his position was not easy and he had to work even harder on behalf of Tunis. I soon gathered that this first sitting was an unusual experience for almost all except me. At this stage I had already sat seven times as an arbitrator, not to mention the Tribunal. As so often happens, the dispute was over major changes in design during construction, as well as unexpected physical conditions. The International, as well as the British, Conditions of Contract allow the engineer to alter the design as much as he likes, but the subsequent clause lays down that he shall assess the different cost, either up or down. The changes can be almost instant, but human nature being what it is, the delay in assessment and paying of increased expenses so incurred is put off as long as possible, to the extent that arbitration is sometimes the only means for the contractor to try to obtain what he feels are his just deserts. In this case, drilling and grouting for the cut-off curtain, as well as a reversal in the way that the earth bank was to be built, and a number of other changes, led to a considerable extension of time and overhead costs. Another factor was an attempt by the engineer to deny that the very involved new design for the overflow works was no more difficult than the simple original contract design. However, the difference would have been obvious to a second year student. Coyne et Bellier had designed the work, but under French custom, the consultant was only responsible for the design and had no concern over the execution. This contrasts with the British tradition where the engineer is responsible also for the way in which the contractor carries out his work and the engineer is supposed to be in a quasi-judicial position. Wives of some American engineers entertained my wife, including a visit to the Tunis Souk or market. My only glimpse of the city was a short courtesy visit to a Minister, but at least I was able to swim in the hotel pool. After four days of confused discussion, the three arbitrators adjourned and met for three days (November 8th to 11th) in Professor Stucky's large house and office in Ouchy, Lausanne, close to the lake. This was to sort out matters and educate our French colleague on etiquette and the need to let the chairman control the hearings. Professor Stucky and his son Professor J. F. Stucky were both charming men, especially the father, for whom we formed a deep affection. Sadly he died a year after we completed our work.

I had informed Leo d'Erlanger of my pending visit to Tunis, as I was still involved with the Channel Tunnel Study Group. He replied by inviting my wife and I to lunch at his villa in Sidibu-Said. Unfortunately, he was not going to be there at the time, but he had warned his Major Domo to be ready for us when we could set a time for the visit. Leo's father had built the villa before the first war and had designed it in exact Arab style, obtaining the leading craftsmen in all the necessary crafts to produce a small (or not so small) masterpiece. When Tunisia had become independent in 1956, the new Government seized all the surrounding acres of the property, but allowed d'Erlanger to retain the house. By coincidence one of the small sub-contractors on the dam was a Frenchman who was also the next door neighbour and tenant of Mr d'Erlanger, so by agreement with the defendant's counsel, he was allowed to drive us out to Sidibu-Said and back. It was an unforgettable experience to spend the afternoon in this beautiful house with terraced gardens, being well looked after and followed by a pleasant tea with M. and Mrs Brun in their smaller house. M. Brun then drove us around this resort beside the sea with its interesting white houses before taking us back to the unattractive Hilton. In due course, the main contractor won his case, so M. Brun was rescued financially from troubles not of his making.

From November 27th to December 6th all parties re-assembled in Lausanne for seven well-conducted sittings. We stayed at the Hotel d'Angleterre where Byron wrote *The Prisoner of Chillon* facing the Lake of Geneva and which was only a short walk from chez Stucky. The main city is on the hills above the lake, with a funicular railway to help pedestrians to reach the centre, which we were able to do in off-moments and evenings. One evening I was rung up in the hotel by my secretary, to advise that a confidential letter had arrived saying that the Prime Minister had it in mind to recommend me for the rank of Knight Bachelor and would I reply by return. This took Sophie and me by complete surprise and astonishment as it was so utterly unexpected. One is instructed to make no mention to others until it is announced in the London Gazette and the Press. This happened on January 1st 1968, when there was an announcement in the Times. My daughter and her husband were staying with us at the time and I left the paper on my son-in-law's chair with the heading prominent. Down he came to breakfast, took one look at the paper and swept it onto the floor with the words, 'New Years Honours! They give them to all the wrong people!'. Then he took up the Business section which contained a photo and said 'Good Lord!'. My son, Robert, and his wife were at that time living over our garages and were rung up by friends, while Edmund my elder son went to his office and wondered why people were congratulating him. As all the family were within reach, we were able to have a very cheerful dinner party. The announcement was 'for services to civil engineering' which pleased me. Knighthoods are rare for civil engineers and usually go to the heads of firms, but I had been working single-handed since 1956, so I was naturally very pleased. I see from my diary that I had to reply to 270 letters.

The next step in the Tunis arbitration was for M. Tavernier and I to return to Tunis and spend from January 7th to 12th at the Hilton adjudicating in individual prices. It was bitterly cold and Sardinia was deep in snow as we flew over. Sophie nobly supported me once more and we never stirred outside, the cold was so bitter. Friends wrote to say how lucky we were to have been in North Africa while they were so cold in England! Light relief and some confusion arose when two DC8 planes arrived. One discharged the USA Vice President Humphries and a vast retinue who occupied most of the hotel, while the second was full of the press. Security was so strict at the hotel that the Tunisian engineer was himself briefly detained as a suspicious character, which greatly pleased his Utah opponents. From January 27th to February 5th we were back in Lausanne to wind up the case. By then the story had been properly disentangled and in due course Utah got the award, but for far less that they had hoped. Their case was fully proven, but when the work started they had a weak man in charge and in the first year, which was during his time, the work was caught by flood and this person was later changed. To a dam engineer this is deplorable, so they lost out on overhead claims in that period.

When this session started, the Swiss, French, Tunisians and Americans were slightly baffled by my change in title, so I told them not to worry about it but address me as before. Once it was over, Mr Reeves gave a successful party for all concerned and we found that we had made new friends. One of Utah's engineers with worldwide experience was Paul Gadjeff, who was a charming polyglot of middle European descent and was the most active in the work. Paul continued to write to me every year to keep in touch. As usual, Sophie with her tact, fluent French and 'professional understanding' was a great help to all the parties, especially dear old Professor Stucky, and maintained the atmosphere of calm 'between rounds'. This experience had later repercussions.

'The Knight Bachelor' does not comprise a Royal Order but is the surviving representation of the ancient 'State Orders of Knighthood'. In 1968 no insignia was handed out, but one was urged to join the Imperial Society of Knights Bachelor who, for a sum, provided a badge to be worn on the left side parallel with the navel. In 1974 a neck badge and miniature were instituted and handed out at the investiture. New Knights are asked to report to the College of Arms and 'sign the book'. While I waited to do so, an elderly clergyman made an enquiry to the elegant young lady in charge. With a lordly gesture she waived her arm and said 'Garter is upstairs'. For the investiture, the various categories are first herded into different rooms in the Palace. About twenty K's were assembled, all of us no doubt worthy, but distinctly unromantic in appearance. We were drilled in the use of the stool on which to kneel; thoughtfully it has a tall pole at one side to help those in need of hoisting themselves to their feet. My wife and two sons sat in the 'stalls' for relatives with rows of empty seats behind and Ghurka officers added glamour. Candidates are lined up in an anteroom on the left side and let in one at a time where the Queen, in a simple green dress, looked charming and possibly more relaxed as no photographers were present and she was in her own home.

Jim Callaghan, as Home Secretary, called my name. I knelt on the stool and the Queen tapped me on each shoulder with a sword and then I rose to my feet (without the pole). The Queen shook my hand firmly with a friendly smile and then I took three steps backwards, right turned and exited out through the other anteroom and back to the empty seats behind the relatives. The military orchestra in the gallery played cheerful tunes, then they played a march which in my pseudo-military days we sang to the words 'You'd be far better off in a Home!' I turned to my neighbour to comment on this but one look showed that it would be lost on him. However, I think that Prince Charles would have appreciated it!

Lead up to Bombay Harbour Arbitration

The whole of 1968 was free from trips to Heathrow and was spent on matters for various clients too numerous to mention. The 100th letter to the XXI Club of April 1968 revives some memories:

> The Channel Tunnel is still awaiting decisions. It has presumably vacated Barbara's bosom for that of Richard Marsh or whoever took over last week.

The 101st letter of November 1968 has other news –

> Personally I continue to give good and bad advice to such as offer to pay for it, and also get let in for more charitable efforts. Last month I was persuaded to go to Cambridge to lecture to the University Engineering Society and tried to give them the facts of life below ground to counteract all that theory. It was funny to be given dinner in Queen's College, where I was billeted for seven months in 1918 in an (infantry) OCB. I also gave a lecture to the City and Guilds Third Year and Post Graduates on organising a Feasibility Study for Guess What. The latter is still in the doldrums.

The letter also describes being installed as a Fellow of Imperial College at their Commemoration Day celebrations in the Albert Hall. Luckily it does not carry 'letters' with it. The chief interest in personal affairs was when my elder son, Edmund, decided that after four years he had had enough of the ill-fated Dungeness B atomic power station and joined the consulting firm of Mott, Hay and Anderson, so following in my footsteps into tube tunnelling and briefly on the abortive Channel Tunnel. My younger son, Robert, spent six months as a Director on Ulster TV just before the troubles really burst and directed Rev. Ian Paisley and one of his sons in interviews.

In the so-called summer of 1968 we flew to Klagenfurt in Austria, in order to stay at Marisworth and swim in the alleged warm Worthersee. It sheeted with rain for half the two weeks, which meant we had to keep the central heating on. We then had ourselves driven in floods of rain through the Julian Pass to Rijeka, where the sun came out and we spent a pleasant twenty-four hours on a Jugoslav boat down the coast to Dubrovnik. An elderly Jugoslav gentleman with very good English asked if he could share our table and in conversation it emerged that he was the current President of the Jugoslav Institution of Engineers and a civil one at that. On finding that I was a Past President of the same ilk, he asked us in a future year to stay at his villa somewhere in the interior, but this never materialised.

About this time I was appointed as arbitrator on more than one case. However, in due course after giving my Order of Directions, peace was made without a hearing. I then slipped down the social scale from learned arbitrator to witness. The first time was due to a dispute between Kinnear Moodie and the West Kent Sewerage Board over a new sewer tunnel. The contractor had allowed for compressed air over half the length, but the engineer had said that it was not necessary and they were awarded the contract after deleting the allowance for air. In the end air had been essential for the whole length. I was asked to give my opinion on what should have been deduced from the Board's boring logs. Much of the Board's case rested on the idea that the engineer for the Board, although adept in the art of sewerage disposal work, had no experience in compressed air tunnelling. I was cross-examined by the formidable Sir Derek Walker-Smith QC, MP:

> QC: 'Sir Harold, should not the contractor have questioned the engineer as to whether he had proper experience?'
> HJBH: 'If he had asked that of Sir Harley Dalrymple Hay he would not have come out alive'.

My client won.

This was a prelude to a more protracted case: Bombay Harbour. Way back in 1957, Michael Gibb of Sir Alexander Gibb and Partners told me that Gibb had been so impressed by the way in which I had handled the settlement of their claim over Braehead Power Station, that they would be glad if I would study all the documents in another case and say exactly what I thought of the contractual behaviour of all parties. Gibbs had designed a new dockyard in the harbour area for the Indian Navy in Bombay, where an Indian contractor had won the contract in partnership with an Italian firm. The latter sent out a small armada of floating plant and then soon after the start, they fell out with the Indian firm and withdrew, but were not allowed to take away the plant. The Indian firm showed such lack of due diligence in doing almost nothing that, as contracts allow, they were sent off the site by Gibbs, the 'engineers'. The Indian firm then sued the Government, involving Gibbs. After reading all the matter and studying the drawings, I reported that in my opinion, Gibb and the resident engineer had performed perfectly correctly and agreed to give evidence if needed. An Indian arbitrator sat for some time and then suddenly died! Another was appointed and had to start all over again. The bones of the case were firstly, that the contractor flatly refused to use the Italian floating rock-breaker to reduce the bottom level, this craft worked off spud-legs by dropping a very heavy chisel. The Indian contractor said that the rock breaker could not do the work and had never even tried (a few years later the Indian Government carried out the work with Gibbs and direct labour and the rock breaker broke the rock). The other main point was refusal even to try to make the hollow concrete tubular piles that were to be set in holes in the rock to carry the deck. Colonel Ramakrishnan of the Indian Army was in charge of the work and I had meetings in London with him from year to year, while the case limped along. Then on February 19th 1969, Mr Hall of Sir Alexander Gibb and Partners, who was dealing with the case, organised for my wife and I to fly to Bombay and then to Delhi. So in 1969, twelve years after the start, I found myself off to India again, as a witness for the Government of India at their expense. While there, Hall looked after us, which I must admit was a welcome attention.

We were met at Bombay Airport by Colonel Ramakrishnan, who then controlled our every move. We spent two nights in the Taj Mahal Hotel and the wife of Gibb's Bombay man took my wife round the sights, including the vast municipal wash house which is not in a house, but in the open. The Colonel took me to inspect the completed harbour works on land, including boarding the famous rockbreaker by boat. We also crossed the bay to the far side to inspect a large quarry being run by a Jugoslav contractor, who was doing much new work. That night, the Colonel and his wife took us for cocktails with the Admiral in charge of the area and his family. These two days proved useful when facing cross examination about the work, after seeing that it had been completed with no difficulty.

Early on Saturday morning, we flew to Delhi and found that the Government had booked us in at the Ambassadors Hotel, on the outskirts of the newer part of the City. Although the hotel was distinctly two star, it was reasonably comfortable. We started work that afternoon and continued almost non-stop for the next four days. This meant working with the eminent Indian Barrister Mr Vyasa, who acted for the Government. He was a charming and efficient man, who had also practised in London. It also meant studying all the evidence to date and all the points in dispute and how to handle them. Then as a respite, Mr and Mrs Vyasa took us to lunch at the International Hotel, very different from the Ambassadors, with a large swimming pool. Vyasa's daughter-in-law then took my wife to visit cottage industries, which was her first break. We saw alot of the Vyasa's and their pleasant family and were entertained by them on several evenings. The daughters were highly educated and were elegant in beautiful saris. They had often visited Europe and complained that since there had been such an influx of Indian and Pakistan immigrants into England to seek work, it was no pleasure for them to come as visitors and be mistaken for such immigrants.

This trip was hardest for Sophie. While I was locked up with my lawyers there was little for her to do, as where we were staying was not the sort of area where she could walk out and paint. We were able to spend time in the Red Fort and a particularly pleasant morning was spent in the Ecological Gardens outside the walls of the Red Fort, with beautiful gardens and well-cared-for animals. We were able to pay several visits to the old and new city, where young British 'hippies' were hanging about and we were told that they sometimes scrounged for help from Indians. Seeing the beauty of these places was an antidote from seeing the odd shanty huts scattered in odd corners of Delhi where the poor were not so well cared for. The population explosion is no myth. Another relieving feature was the presence in the British High Commission of the nephew of my oldest college friend Grigor-Taylor. He invited us to a lunch of peacock (which seems to be a game bird in India and was delicious) and then for a welcome swim in the High Commission bath. Then, as he had to be away, he arranged for us to make use of the swimming pool when we could. On his return and after warning us that we might be shocked, he took us to a *Son et Lumière* at the Red Fort which was an impressive spectacle and, as expected, towards the end was strongly anti-British, immortalising Sen Dupta, who joined up with the Japanese in Burma in the hope of 'liberating' India. I do not remember Lord Attlee getting much credit.

A further surprise awaited us. Colonel Mcmullen had been to Delhi just before us and had informed the Institution of Engineering (India) of our impending arrival. The result of this was much unexpected hospitality. I was invited to the Institution's annual meeting and get-together and after being very kindly treated, asked to give a lecture to their members a few days later. In addition, the Engineer-in-Chief Lt General Lumba invited both Sophie and I to several private parties, where we met others and their wives. General Senrupta and Major General Kumer also had us to their homes for drinks. Sandhurst influence still lingered pleasantly.

Then on the evening of February 26th we had our first session with the arbitrator. He had been a High Court Judge, or its Indian equivalent, but was now practising as a barrister during the day in the courts and for this reason he held the sittings for two or three hours from 4 or 5pm at his house, which was some way outside of the city. He was a dear little man and I felt confidence in him. It was all much simpler than the London style. The arbitrator sat at one end of the table and on one side was our counsel, Mr Vyasa, with Colonel Ramakrishnan supporting him and Mr Hall. On the other side were three Indian gentlemen led by a barrister, Mr Das, who was an Indian MP (a graduate of Queen's College, Cambridge and Lincoln's Inn) who had a large chip on his shoulder. I took care to keep up the standard by standing up until the arbitrator had settled. Between February 26th and March 13th we had 14 sittings, during which I was examined by Vyasa and then cross-examined at length by Mr Das, who lost many points in the view of my supporters. I was able to give evidence of Sir Alexander Gibb and Partners from personal experience over contract behaviour as well as that of the resident engineer from personal knowledge. I quoted examples from my own experience as an arbitrator, on cases of 'lack of diligence' and my counsel took care also to mention the Tribunal to strengthen my *bona fide*. During this time, most mornings were taken up going over matters with our counsel, with occasional escapes. Mr Das attempted to pour scorn on my evidence by trying to prove that my only subject was soil mechanics. However, I had brought with me examples of much harbour and dock work on which I had performed, including photos of installing hollow concrete piles into holes bored in rock at Middlesbrough. As the contractor was claiming that this would not be possible, Mr Das was much embarrassed. Some years later, after many more sittings, I learned that my client had won.

While in India, we were able to entertain Major General Sir Harold Williams to lunch, after meeting him at some of the other occasions we had attended. He was very senior by then, as he had been the British Engineer-in-Chief of the Indian Army at partition. He was so well-liked that the Indians asked him to carry on in their new army for several years to help them over the change. He was so successful in his role that for some years he was invited to represent both India and Pakistan on the Council of the British Institution of Civil Engineers. He refused to leave India and eventually died there a few years later, a fine example of what both British and Indians have gained by our association in spite of all left-wing notions. We also visited areas around Delhi like Quitab Minor and the Indian Home Industries, where we spent many rupees on beautifully enamelled brass ware, being fashioned before our eyes. Then, as a final effort, we hired a car to drive us to Agra. On the way, in fairly bare country, we saw the inevitable procession of women walking miles for the privilege of filling large jars with water to carry home on their heads. The Taj Mahal was as beautiful as expected. Photographs do not do it justice, as they do not show the beautiful inlay work on the marble. We drove back via Fatehpur Sikri, which is also a necessity on such a visit. We were told that this palace built by Ackbar was never properly occupied because there was no water supply. Within its walls, it has a fascinating arrangement of halls, chambers and buildings, all in the best Oriental tradition. Our driver took us back by side roads in order to visit the Dinapore bird sanctuary and to see more of the villages. Again, groups of women were crouching around the village water tanks to fill their pitchers, which was picturesque, but that may not be much compensation for the hard life involved. We British do not know what we are missing!

On Monday March 17th – in the words of American commentators – we bid a long farewell to Delhi, flying over miles of Persian desert bound for Cairo. We were then told that, owing to a dense sandstorm, we were to be diverted to Beirut, an airport with which I was very familiar. Unfortunately, we found that our relief crew were waiting for us at Cairo and the emergency crew stationed at Beirut were all out shopping, so in order to help pass the four hours delay we were taken on a coach drive round Beirut, which I was glad for my wife to see. As usual, buildings around the coast had greatly extended, but now presumably many have been destroyed in the inexplicably complicated civil war. 1969 proved a busier year than expected and soon after our return I found myself once more bound for South Africa.

South Africa Again 1969-1970

Within two weeks of our return from India, I lunched with Sir Alexander Gibb and Partners to discuss the progress of the Delhi arbitration. Dr Oliver mentioned that there was a problem in which they were involved over the Hendrik Verwoerd (HV) Dam in South Africa.[58] They had given my name to the Union Corporation, who were part of a consortium building the dam, and this led to a meeting with Mr McWilliams at the Union Corporation London Office, where he explained their problems and asked if I would help them. We agreed terms for my visit, which included my wife accompanying me on the trip. On May 10th, after making the usual study of a bundle of papers, the contract and the drawings, we flew out to Johannesburg in that fine but neglected British plane, the VC10, this time in only 12 hours with one stop at Entebbe. Since my last trip in 1954 much had happened and the 'wind of change' had blown so strongly that the British Colonial Service was extinct. In 1960 South Africa had become a republic and come out of the Commonwealth, which had also greatly extended. President Obote, our former materials clerk, ruled Uganda and so did Kenyatta in Kenya, while Basutoland of childhood memory had become the independent Kingdom of Lesotho. In 1964 Zambia took over what had been Northern Rhodesia, while at this time in 1969, 'Smithy' had just declared UDI in Southern Rhodesia. However, the Portuguese still held their east and west colonies.

The HV Dam was the first of three dams intended to be built as part of the Orange River Project, at a cost of R450m (1965 values). The Orange River rises in Lesotho and, after the Caledon River has joined it, the river is diverted by the Drakensburg Range. It then trundles its way westward until, after forming the border between the fairly populous Orange Free State and Cape Provice, it wanders around a somewhat deserted area, finally reaching the Atlantic to separate Namibia from the Cape Province. Near Colesburg, the river wanders through a wide expanse of the stony infertile land of the Little Karoo and then has worn a course through a group of high rocky hills. This provided a natural site for a dam, which produced a reservoir about 50 miles long with little loss of fertile ground or human habitation. A vital sideline of the project was the driving of the Fish River Tunnel that took a large volume of water from the dam under the watershed, to discharge it into the Indian Ocean. On its way, it produced hydro-electric power and fertilised a considerable area.

Until the Project started, all dams in South Africa had been designed and built by the engineers of the Bureau of Water Affairs, so their hostility was aroused when the Government entrusted the HV Dam to two joint consortia of British, French and South African consulting engineers. Among these six firms were Sir Alexander Gibb and Partners and, once more, Coyne et Bellier from the Oued Nebaana Dam. The HV Dam is 280ft high

---
58 Called the Gariep Dam since 1996.

60. *Fish River intake, Hendrik Verwoerd (HV) Dam*

61. *View across the top of the HV Dam*

with a total length of 3,200ft, built of concrete whose aggregate was quarried from the particularly hard dolerite of the enclosing hills. Difficulties arose from its unusual design. The centre is an arch with a curved face, but instead of springing off the hills on either side the arch springs off massive 'gravity' blocks which then close the gap. These are riddled with many varied openings to allow water to pass for hydro-power, flood relief, compensation water and a supply for Bloemfontein. One consortium of consultants designed the arch and the other designed the mass concrete portion. Gold is not always on the up-and-up so, during a recession, the largely British-directed mining company Union Corporation decided to diversify into dams and tendered jointly with two French contractors: Dumez with a long list of successful dams, including the Dokan arch dam in Iraq; and Boris, to cope with the rock tunnelling below the work. They were successful against strong international competition. The work started in 1965 and by 1969 a number of problems had led to dispute, so the contractors wanted help in interpreting the contract.

The International Conditions of Contract had been stiffened and some items cancelled, in particular the reference to arbitration had been changed into the verdict of a Judge of the Orange Free State High Court in Bloemfontein. As in other cases the Variation Clauses were the chief cause of conflict. The engineers had made many variations but, in order to safeguard their client, were as reluctant as ever to come to a solution. It was thus becoming necessary for the contractor to put forward what are described in contracts as 'Claims'. The Union Corporation were used to being in complete charge of all their own work, from the initial construction of any new site as well as the mining work. It was a hard lesson for them to be under the domination of consultants, resident engineers and inspectors. Then, when the experienced French contractors began to formulate 'claims', they were shocked, as they had never needed to beg in all their life and it took some time for them to understand the situation.

We were met at Jan Smuts Airport with a warm welcome by Messrs Munroe and Gerricke, directors of Union Corporation, and installed on the 19th floor of the President Hotel to recuperate. It was early May and so into winter and the weather was cold although fine. Next day Mrs Munroe drove my wife around, while I spent all day in the Union Corporation

Office finding more of the facts of life. They had arranged for us to take the night train to Dankerpoort to be met there by a small 'reception committee'. During my early days, I had made at least five long night train journeys on their meter gauge lines, so it was no surprise when an attendant brought in large bags containing blankets and pillows, which he proceeded to make up on the seats for our comfort. Exactly as I so well remembered! The reception committee had duly assembled, led by Monsieur Belin, the project manager and his wife, with McNamara his joint second-in-command from the Union Corporation and others. Unfortunately, our carriage had proceeded beyond the end of the platform, so, in order to alight, Sophie tucked up her skirt and consumed with laughter took a flying leap to reach the ground with a graceful landing. This relaxed the tensions and gave a good start to our relationships.

The dam was to have a second value as a tourist and sports centre, so a motel had already been built on comfortable lines by an enterprising Boer Africana and we were duly installed next to M. Garbe and his wife. He was a director of Dumez, with whom I was to work on many days. A township called Oranjkrag had been built about a mile above the dam with a view to future use, after it had housed all the white personnel concerned in the construction. Further away was another township for the coloured work force and a third one a mile below the dam for the Bantu labour. 'Coloureds' are by no means all half-black and half-white, but are largely made up from the descendants of Malaysians and others from the old Dutch East Indian colonies and imported to South Africa as servants or indentured labour in the early Dutch days. Although poor, they came from countries which, at higher levels, had a considerable and old form of development and culture, far in advance of the Bantu who, at that time, were beginning to advance southwards to oust the Hottentots and Bushmen. In summary, most of the special craftsmen were coloured, with the Bantu providing the general labour. Next followed five weeks of intensive work, mixed with varied pleasant experiences. I worked in M. Belin's office for want of space, so was able to observe how he controlled his mixed bag of South African, French and English staff with the odd Italian and German to add variety. Macnamara worked next door in complete harmony with his French opposite number and naturally spent much of his time on the working site. This had been well set out and the plant was very good, as might be expected from a firm of Dumez' standard.

On such a site, the complexities are considerable, from the dolerite quarry up the hill, through the extensive train of conveyors carrying the raw material to stock piles and thence to the various crushers, including rod mills to reduce some of it to sand. Then there were the silos and the daily train-load of bulk cement to be transported from railhead to bulk storage, as well as the concrete mixers and the distribution plant. Most visible of this were the Blondins, or overhead cable ways, along which the skips of concrete could travel and be lowered to discharge at the appropriate place. These cables were fixed to towers on the far bank, but on the near side were carried on travelling towers, which could move laterally to allow for the curve of the dam. The drivers sat up aloft, with close-circuit television in the driving cabs, fed by cameras attached to the carriage, which travelled along the cables. The driver could then see how his load was behaving, helped by walkie-talkies used by the foreman inside the shutters. The young engineers thought that Blondin was the name of the maker, being too young to remember that Blondin crossed Niagara Falls on his tightrope.

It took some time to get to the bottom of affairs, including studying the contract and how it had been interpreted, as well as the problems, past and future. By the end of the first visit, we had got far enough to set further wheels in motion. By luck, before I had left the UK, I had read an article by Gibb's chief designer describing how, four years after the work was started, they were still trying to design the reinforcement by means of a computer. As no reinforcement had been shown on any of the drawings and as there was only one item of a few hundred tons of one class of steel in the Bill of Quantities, this proved an opening. The chief difficulty arose because, briefly, the design had to be altered constantly to try to accommodate all the extra steel needed, which included many changes in the internal structure and the many openings. Eventually, mass grids of heavy steel joists had to be used in places where the mass of steel rods would have been impossible to work in. I found that the Government Minister concerned had refused to admit that there had been any change in design, as the outside shape of the dam was the same. Professor J. F. Stucky had visited the site and made his report, with which I agreed. The country around the dam was dotted with curiously shaped rocky hills, quite different from the High Veldt kopjes familiar from the Boer War pictures and, although it was winter, there were great variations of colour from the curious vegetation and shrubs which covered the landscape. The dam itself had its own beauty with the high light-coloured walls rising out of the brown ground with blue sky above and water flowing by.

62. *Sophie, Mrs H and Mrs B at the HV Dam site*

I spent long hours on the site and the office overlooking the work, while Sophie had a well-deserved vacation. Freed from home life and after the frustrations of Delhi, she had plenty of scope to wander around and paint, in an air which rapidly dried the watercolours. The Macnamaras, with their charming adolescent son and daughter, took good care of our social life. In Oranjekrag, the South African families were bi-lingual, but only in English and Afrikaans, so there was a lack of social mix with the ex-patriate French, which sometimes caused a little tension, as the latter were short of English. Sophie was not only fluent in French, but all her life had studied French literature and had just finished reading all three tomes of *Les Lettres de Madame de Sevigné*, which René Malcor had given her. She had also spent much time in different parts of France, so with her ease of manner, she was made welcome in the French houses as well as the others and therefore helped to relieve some tensions. I also found that my years of working with French engineers, not only on the Channel Tunnel but also with Soletanche, Fougerolle and others made me more sympathetic with those on the site - to good effect.

Because the consultants were resolutely fighting off all their approaches, the upper echelon of Union Corporation was also under tension from anxiety over the financial burden. As often happens under such stresses, they began to doubt whether M. Belin was the right man to be in command. I soon settled that by saying that I had spent every working day in Belin's office, so as an experienced spectator I could vouch for the fine way in which he handled his mixed staff and every crisis. Much legal advice was being bandied about and the French raised the question of whether it would be better to cut further possible losses by 'pulling out' and abandoning the contract (this had been done by a French firm on the Kenya Sasamua Dam and they had lost the arbitration). I told them that it was a decision for them to take, but explained what this entailed under the contract, which was that the Government would take over all the plant and then negotiate for someone else to complete. This might be even more expensive, as their own troubles would be known, but the extra cost would be chargeable against the former contractors (unless they could successfully defend themselves, which might be difficult in a South African court). The French might get off well, as once they were back in France they might be able to put up a strong rearguard action and were out of reach. The Union Corporation were however much more vulnerable, as they might merely be told by their Government to go and dig out more gold, as there was plenty there. I am pleased to say that the work later carried on to a successful conclusion.

One Sunday we were lent a Volkswagen with a native driver, so that I could take Sophie to see the scene of my two-year never-forgotten and too-often-talked-about childhood in Jammersburg Drift, on the banks of the Caledon River. We climbed the local Kopje which still showed signs of the breastworks from the Boar War; she painted sitting beside the river where the old bridge and mill still stood. There were even more trees than there had been in 1954. By kind connivance of the local Border Police, we were allowed to drive briefly in Basutoland towards Mafeteng, where nothing seemed to have changed. We met the same Basutomen riding their ponies, with their wives walking patiently beside them among the flat-topped hills of Cape Town sandstone. Once the border was crossed, the roads were the same dirt tracks and the spruits or water courses had still to be crossed by drifts or fords. They were just the same as those over which we had driven for days as children, in a two-horse Cape Cart, passing the ox wagons with their teams of oxen. This was all in sharp contrast to 1969 South Africa, where the modern roads were beautifully constructed with all the new interval distance posts and never a drift, but always modern bridges over rivers and spruits and at times flying junctions, built where no buildings were in sight.

On another Sunday, Miss Murdock, the chief secretary on the dam, drove us for many miles to her family farm in the middle of the veldt. There her sister bred horses and we were much impressed by a magnificent prize-winning black stallion, which pranced around in a conceited manner as if he knew that he was beautiful. Her father, Pat Murdock, was known to be a considerable 'character' with a lively and varied past. As a young man in 1918 he had been in the South African Air Force in England and he became the first to fly a light plane from London to Cape Town. When I asked him about it, he said that at the time he had all his plans but no plane. It only became possible when he made friends with a girl named Babs who insisted that he should meet her father, John Body, who bought him a plane! When I said that Babs Body was my cousin and her father a sort of favourite uncle, there were drinks all round to celebrate such a curious conjunction in the middle of nowhere in particular! On another day, Macnamara took Sophie out to try a nine hole golf course that he had hewn out of the very rough land beyond Oranjskrag. It was built beside an ostrich farm and shortly before they arrived, a player found himself being chased by a bird which had escaped; ostriches can be quite ferocious and he just made it to his car in time.

The contract laid down minimum wages for coloureds and Bantu and also benefits which they were to receive in food, allowances, periodic travel vouchers and so on. Mr Prinsloo, the camp superintendent, took my wife all over the African Township and was very proud of the kitchens and the food provided. He said that as the men liked to eat when they were hungry and not at fixed times, food was available at any time for the day or night shift and soft drinks were available from several aluminium Rondavals (round huts) dotted about the site. One area of the day room was wired off, as some of the men wanted to be apart from the general crowd to be quiet and read. The men could have free passes back to their homes, sometimes on leave and sometimes because they wanted to consult their witch-doctor, who acted as their psychoanalyst. The men would walk the mile to the dam, but if they worked overtime they were run back by motor transport. On my several periods on the work, harmony reigned.

After several weeks we were driven to Bloemfontein and flown to Jo'burg and again bedded down high up in the Presidential Hotel. Next day, Mrs Munroe once again entertained Sophie, while I spent the day with the directors of Union Corporation, who were an effective but very pleasant body of men. They included the charming Lew Douglas, ex-USA Ambassador to Britain who had lost an eye when casting his fly on a Scottish salmon river, and was well-known at the time. On this long session, they had to be briefed on the peculiar rigmarole of the civil engineering Conditions of Contract and the ritual dances which had developed over the years, due to the adversarial nature of the contract, and over which I had become somewhat critical after seeing so much of them. It was hard for them to grasp that it had to be in the nature of a fifteen round contest, with no knock-out possible. Although their cause was obviously a just one to any intelligent observer, the proof required time and much detailed work. I explained that they must note every change of any sort and insist on the engineers' staff agreeing to every step as to fact and to be meticulous in giving every notice in good time. I was able to assure them that the whole work could not be faulted overall, in staff or plant or execution, but that they were the prisoners of an extremely rigorous contract, to which they had put their signatures, and which up to that point no quarter was being given as the engineers were themselves prisoners of what their Governmental client had evolved. They then surprised us by saying that our visit had been so helpful and had been such hard work, that they were sending us for a week's holiday to their own rest bungalow at Venue Timbers in the North Drakensburg. Mr Gericks lent his Mercedes with his skilled and gentle Bantu driver, to drive us for 250 miles to Graskop on the edge of the escarpment where the High Veldt drops from 6944ft to less than 1000ft in the Kruger National Park below. We approached through Long Tom Pass, where a monument reminded us that this was where the last of the Boer Army towed their remaining Long Tom cannon to refuge in Mozambique. We soon found ourselves in fascinating and beautiful country, with superb views over forests of trees planted over the steep sides of the many valleys and rifts in the mountainsides. This area is covered with vast plantations of Australian eucalyptus owned by different mining companies, as the demand for timber support in the gold mines seems insatiable.

We found our way to the rest house, which was a small bungalow set in a wide clearing. Nearby, in another one, lived one of the companies' Boer drivers who had married an African woman and she looked after us very well. The plantation was managed by Mr Herring and his son, of British origin, who made us welcome and from time to time were able to show us around. We were also able to explore the country by car. Tourists had not yet spoilt this area, although preparations for tourism were said to be underway as it is (for South Africa) historical. It is the scene of the first gold rush in 1870, before the days of Cecil Rhodes and the Rand. All round the little hamlet of Pilgrims Rest with its pioneer-style houses, there are abandoned headings where the ore had petered out. We admired the Mac Mac Falls, where a narrow stream drops hundreds of feet down a cleft to carry on as the Mac Mac

19. ARBITRATION UNDER INTERNATIONAL LAW

63. Mac Mac Bridge

River. Apparently, before the first Boer War the President of the Transvaal came to see what was going on and wanted to meet the miners. He was told, 'This is Macdonald, this is Macpherson, this is Mackenzie, this is Maciver....' and so on. The President had said, 'Why, they are all Mac Mac', and so the river was christened. The Sabie Falls were even more beautiful. This northern part of the Drakensburg is called the Great Escarpment and runs for 120 miles on a north south line. There were a number of places where we could drive and wonder at the expanse below from the precipitous edge. Another special place is called World's View. North of Pilgrims Rest is the Blyde Gorge, where the Blyde River has gashed a canyon over half a mile deep. The guidebooks claimed that it was the next deepest to the Grand Canyon of Colorado. It is certainly spectacular, although it did not seem possible to reach the edge to see the bottom, because of fencing. The opposite wall consisted of horizontal strata of contrasting colours and at one end there were three vertical eroded columns topped with conical grass slopes and naturally called the Rondavals.

We had to do our own shopping in the hamlet of Graskop, with its store and butchers shop. There were two doors to this, leading into one large room with a counter across the whole width. A single rope, three feet above the floor, divided the white from the black customers. This seemed to do no harm, as everyone, especially the blacks, seemed cheerful. In any case, the latter wanted quite different meat of different cuts from the whites, so the division of the counter was also helpful. It was all very friendly. Mr Herring took us on several trips to explore his widespread territory in his Land Rover, where he had cut down narrow earth roads along the steep slopes of the many valleys which his forest covered. As he had only one eye from the war and drove fast while waving to points of interest with one or both hands, there seemed a risk of cascading down among the stumps of the cleared areas. The eucalyptus in that climate grows ten feet in a year, so certain chosen areas are cut down after six years and the stumps are left in and fresh trees grow from the stumps after the surplus shoots are duly cut off. In one place, the trees had just been cut down by African timbermen with portable petrol-driven chainsaws and the white foremen were beside their tractors on the track above the slope, supervising. At the bottom were African wives, some with babies on their backs, cutting off the light branches. The trunks were then hauled up the slope close to the track, using the tractor winches where the white foremen marked off lengths with measuring rods and the African timbermen cut them to length with their chainsaws. These were then loaded onto trailers and towed to the sawmills.

On another beautiful sunny day, Mr Herring took us to a lovely wide valley between the masses of trees, through which the Mac Mac River ran among outcrops of rock, which formed pools and rapids. He showed us a patch of flattened grass and told us that a pride of lions had slept there on the previous night, but it did not matter as they had just eaten a donkey and would not be hungry for a day or two. He then proudly showed us two bridges of several short spans across the Mac Mac. He had read up something about concrete in a book and had then built these reinforced concrete bridges, using old rails for reinforcement. There was no shortage of timber for temporary supports and shuttering for the piers and deck. I would have to say they were impressive. That evening my wife produced a sheet of drawing paper and, at the top, painted a coat of arms, below this she wrote with a brush in proper style:

> This is to certify
> *(I then wrote in Herring's name in ink)*
> has been elected an Amateur Member of the Institution of Civil Engineers.

I signed it as Past President and she made a squiggle below as Secretary. There was a suitable spoof motto and, rather than the crane and beaver which the College of Arms had given us as supporters to the Institution crest, the supporters of the Coat of Arms were an ostrich proper and a lion regardant. Mr Herring was so pleased that he at once framed it and hung it in his office. This office was above the sawmill site which was open with a roof of galvanised iron, supported on timber posts to keep off the sun. Africans worked the circular saws, while totos or boys helped to assemble the timber cribs. These were made by laying logs in layers across each other to form cubes and spiked together. They were then taken by trucks, with a loading gang enjoying the ride, to a railhead some miles away to be transported to the mines for instant use.

On Sunday, we went past the village of brick bungalows that the Union Corporation had built for the Bantu families on the estate and passed the time of day with the men as they sat gambling beside the path. On our last day, the young Mr Herring drove us for a quick day-trip into the Kruger National Park. Here we saw hippo submerged in the Lower Sabie River, buffalo ambling across the road, troops of baboons behaving rather vulgarly, as is their custom, and from time to time all sorts of buck among the trees. The protective colour is so good that they are hard to discern in coloured photographs. We took some refreshment in one of the fenced-off enclaves for visitors and then saw more giraffe but no lions. These are more easily seen at Longleat! On the way out, we met a number of hyenas on the road and followed advice to keep the windows up, as they can leap and grab if the foolhardy leave theirs open. We had discussed visiting the Mala private reserve on the edge of the Park, which is much boasted in the guide books, but Herring agreed with our feeling that it should be avoided. One story we were told was that, for vast sums of money and having consumed a heavy meal, townsmen can sit inside the strong fence, while lions are driven towards the outside to show up in the floodlights. How vulgar can you get.

On Thursday June 12th with great reluctance, we let Michael drive us back to Jo'burg with younger Herring as a passenger. We had experienced a wonderful holiday. I spent the next day paying courtesy calls on some of the consultants, among them Mr Watermeyer, who I had received an introduction to on my 1954 trip, when in my opposite role as a contractor's director. We were seen off by Gericke on the Saturday and presented with the customary large spray of Protea. The VC10 took off at 9:30am and after stopping at Nairobi and Rome, landed us at Heathrow at midnight. In 1906, it had taken three days to reach the coast over land and then three weeks at sea by *RMS Armadale Castle*.

A month later, I spent three days in Paris with Professor Stucky for meetings with the directors of Boris and Dumez, while Gericke and Belin came from the site. Dumez had just built an ultra-modern office at 'Defense' north of the Seine opposite Neuilly, where dreary un-Parisian tower blocks were proliferating. In Lausanne, Professor Stucky and I had both seen Jacques Tatti's new film *Playtime*, which to satirise it had been shot in that complex. We shared the pleasure, unknown to our hosts, of noting how exactly the film-set had foreseen the modern office and its imposing vestibules. Steel armchairs on which we sat were heavily cushioned in black plastic and were the same as those on which M. Hulot sat and which emitted dubious noises when he (or we) sat down.

We were able to bring the French directors up-to-date and to organise an agreed approach to future actions, based on the independent reports from both Professor Stucky and me. As usual, we worked for long hours with no tea-intervals, but in the evening were revived by French hospitality, once a parking space had been found within three quarters of a mile of the restaurant. We also recommended Professor Schnitter of Zurich as a third advisor against the eight consultants ranged against us. Schnitter was rated just as highly as our opponents in the Congrès des Grandes Barrages. I had known his brother over the years, when he was the caisson expert in Pigeon House, Dublin and the Maas Tunnel. In August came the call for my second trip to Jo'burg from the 17th to the 27th, this time I went alone. The Sunday night VC10 flight touched down this time at Frankfurt and Nairobi and on arrival next morning, Gericke installed me once more in the President Hotel. He told me that in my absence there had been an exciting sex scandal in the hotel and someone suggested it should be renamed the Vice-President. I spent the rest of the day in Gericke's office and was then up at 5am on the Tuesday to join the shuttle flying to Bloemfontein, followed by a drive out to the site, this time with Monsieurs Garbe, Fouillade (who had been in charge of the Dokan Dam in Iraq) and Belin. All that day and the next two, we were in and out of the site office to get up-to-date with progress.

Progress was notable, in spite of the flow of new drawings, ever more newly-designed openings and reinforcement far beyond the contemplation of the contract. I spent time with some of the younger engineers and quantity surveyors, to save Belin in briefing them on the sort of records, drawings and times that they should be recording on their particular parts of the work. I also managed to get them to agree as to fact with their opposite numbers, so that the full story would be recorded beyond contradiction. Contractors are responsible for their own inefficiencies, so truth should be noted, even if at times slightly embarrassing. I gave suggestions for progress diagrams to a long scale, showing actual progress made and a companion one showing what progress would have been if the work had not been altered, based on true progress where this had not been affected. Nothing succeeds like true visual demonstration compared with too many wordy papers.

My evenings were not lonely, as I dined with the Macnamara family and also at a Belin barbecue. Then, after a few days, we made an evening trek to Jo'burg for two more days at Union Corporation office in long sessions with leading members of the Bar. Much legal advice was needed in drafting carefully worded letters to the Government, setting out the position and telling the client and consultants that, while continuing to press on with work, the matter would be fought to the end as the consortium's case was unassailable (or words to that effect). On the Saturday morning, when I came out of the hotel, which was near the railway station, I found myself entirely surrounded by Africans in cheerful mood as they had come in to shop in white shops. I met my cousin for lunch and then returned to work. Sunday was spent with others in the office working on draft reports. I had been due to stay for a few more days, but I developed a severe abscess in a tooth, which lowered my efficiency, so I flew direct to Jo'burg by a light four-seater plane which plied a private line nearby. That afternoon and next day were spent with Gericke working, with a brief interval when he had enough influence for his dentist to give me time to drain off the tooth.

It turned out to be a stroke of luck that I flew back by VC10 earlier than expected, as Sophie and I were able to drive at once to Tunbridge Wells in Kent to clinch the purchase of a house, which otherwise would have gone to a rival. We had decided to move, as we had one married son living at Brasted near Westerham (also in Kent), and the other was about to leave the house which I had built for them over my garages, to go to Mayfield in Sussex, so the move to Tunbridge Wells would place us between the two. My wife had always wanted a square white Georgian house and had found a listed white Regency property, with typical iron balcony, on the high ridge of Mount Ephraim, overlooking the common. It had a fine garden of almost an acre, with many valuable trees and shrubs. The semi-basement was dry and well lit, as it was surrounded by a brick-lined dry moat, so that the windows were above ground level. This gave me two rooms for my practice, as well as a workshop and a rumpus room for six visiting grandchildren (which soon had puppets in it). I decided to reduce my practice by declining unattractive propositions. For continued meetings and Institution activities, London was almost as easily reached from Tunbridge Wells as from the backwoods of Putney. The next few months were active with negotiations and disposing of 15 sacks of accumulated papers.

My third visit to South Africa was from December 2nd to December 13th 1969. The first two days were spent in Jo'burg with much discussion including lawyers and then on the third day I was up at 6am to fly straight to the dam in a small six-seater plane with one of Dumez's directors. Progress on the site had been so great that the full extent of the variations to the original design could no longer be denied. Much hard work right through the weekend was needed to make sure that all points were being properly recorded by the staff. By now it was high hot summer and the Oranjkrag Olympic-size swimming pool was open for welcome swims in the evenings after work. Again I was made welcome and entertained by the Macnamara family. On Tuesday December 9th I was up at 5:30am to drive with Belin to Bloemfontein and fly to Johannesburg, to spend the rest of the day with the Union Corporation and the French connection getting ready for the next day. This meant another dawn departure for Pretoria, for our first major confrontation with the massed consulting firms, under the Chairmanship of Mr Kriel, the Secretary of the Board of Water Affairs. Pretoria is the briefly historic capital of the Union of South Africa and is only a short drive from Johannesburg, but all I saw of it in several visits is the road to the Water Affairs Office and a nearby restaurant. On the way, the only remarkable scene other than the Veldt is the vast memorial to the Voortrekkers, on a hill some way from the road. Unfortunately, I never had time to visit it, but from photographs that I have seen of both the inside and outside, the British are not affectionately remembered.

On the Wednesday it was back again to Pretoria for another contest with the massed ranks of the eight consultants, where the secretary of the Water Affairs Board seemed more sympathetic than the others, who were very obdurate. After my clients had had their say, I took advantage of being an entirely independent person by being very outspoken, with chapter and verse from my many years of experience, including much arbitration. That evening we were back in Jo'burg for dinner at the Union Corporation flat, where M. Belin and Fouillade were staying. Munroe and Gericke joined us and we talked from 7pm until 11pm. The morning of the next day, Friday 12th, was spent in the office of one consulting group led by Mr Watermeyer, with a very young Frenchman from Coyne et Bellier whose self-confidence was in inverse ratio to his age, with all the theory from L'Ecole Polytechnique but scant experience. I was moved to tell him, 'Your trouble is, you are too highly educated'. The afternoon was spent with the other group of consultants, including Sir Alexander Gibb and Hawken. They were rather more sensible, but all were heavily defensive and passing all the need to prove a case to the contractors, although it was their own actions which had brought it about. I felt I had to remind them that earlier that year I had been a witness in Delhi on how well Gibb had behaved in another dispute.

On Saturday December 13th I concluded my early risings in order to be driven to Jan Smuts Airport to catch the daytime flight home, via Nairobi and Rome. I arrived home by midnight and again this was a lucky moment, as I found my young granddaughter with a temperature of 105, and my wife, son and daughter-in-law themselves about to take to bed with influenza. This left my six-year-old grandson and myself to nurse them, which was so successful that they recovered in time for their planned move to their house at Mayfield at the end of the week. The next few months were busy in preparing for our own move, but in the meantime much work had to be done as a result of the meetings over the dam, in preparing advice and forms of words for the pursuit of the claim in case it went to the courts.

I turned to the useful work of Max Abrahams on *Engineering Law and the ICE Contracts*, which also covers the International Forms. In the chapter on 'Delays, Disruption and Expedition Claims' I came across a paragraph on page 199 which so exactly fitted my case that I started to write it into my report to the clients. It started:

> In a recent unreported decision:
> The arbitrator in his findings of fact on a special case stated that 'the result, in terms of delay and disorganisation, of each of the matters referred to (causes of delay for some of which the employers and some of the contractors were responsible) was a continuing one. As each matter occurred its consequences were added to the cumulative consequences of the matters which had preceded it…It is therefore impracticable, if not impossible, to assess the additional expense caused by delay and disorganisation due to any one of these matters in isolation from the other matters.

I was writing down these wise words with enthusiasm, when I glanced at the footnote: 'Crosby v Portland UDC (1967) QBD, Donaldson J. I had been the arbitrator and these were my own words! In that case I had been asked to state 30 points of law by the defendant, but these did him no good, since in the Queen's Bench Division, Mr Justice Donaldson entirely upheld my award and in the end he held that:

> so long as there is no duplication…there is no reason why he should not recognise the realities of the situation and make individual awards in respect of those parts of individual items of the claim which can be dealt with in isolation and a supplementary award in respect if the remainder of these claims as a composite whole.

This proved valuable ammunition for the next round.

On February 24th 1970, after 37 years in the house in Putney (except for the war), came a four day move to Tunbridge Wells in frost and snow: *Sic transit gloria* Putney. In the middle of the move came a call to go as soon as possible back to South Africa, so three days later the Union Corporation sent a car to collect me from Tunbridge Wells for a meeting in London and then to Heathrow for the night flight, where I waited in the VIP lounge with two directors of the Union Corporation. A fireman's strike was in full swing, so no plane could leave after 8pm. Our take-off was to be at 7:45pm and just before 8pm we were strapped in, with engines running, when a voice said that there was an oil leak and the flight was cancelled. We were put to bed in Fortes airport hotel and called at a very early hour for take-off at 8am, when the firemen permitted it. There was a further slight delay, while three inches of snow were scraped off the wings of the VC10 and then we were off, via Frankfurt and Entebbe. Gericke and Monro unselfishly stayed up to collect me at the airport at 1:30 am, which was just as well, because when we reached my hotel (not the President this time) the porter tried to deny me my room, but they soon settled him.

After a brief rest, next day, March 5th was spent in the boardroom of the Union Corporation to battle with a row of eight consulting engineers. I was supported by Professors J. F. Stucky and Professor Schnitter. The war raged all morning and afternoon and we three did not mince our words and felt that we were making impressions. Both the Professors, unlike the British variety, spent much of their time in designing and actually supervising the construction of dams in different parts of the world and not chained to their Chairs in the Universities, so they had much to say. Professor Stucky had been involved in this problem almost from the start.

The contractors had already submitted a number of suggestions for solving the problem and I spoke out about this and said that whatever was put forward, 'You wipe aside without proper consideration'. My friend Angus Paton asked 'Who do?'. I waved my arm across all eight of them and retorted 'The whole lot of you!'. They did not seem to resent this unduly, especially as it was true. By that night, I was too tired to talk to anyone or even visit my cousin. Then next day, it was up very early as usual to drive with Gericke to Pretoria to meet all the same people in the office of Water Affairs. Here the battle raged again all day, but with a friendly interlude during lunch. In the late afternoon, Gericke drove me straight to the airport to catch the night plane, which left at 7:15pm. He spoke kindly of my efforts which was nice. I arrived in snow-covered England fourteen hours later, to find that the Union Corporation chairman had sent his Rolls to take me direct to Tunbridge Wells. I was very grateful after such a hectic two or three days. This trip was to be my last long trip abroad. I had begun to reduce my commitments, as I felt that at 70 I was entitled to do so. After three score years and ten, we can be considered to be 'playing in injury time' but there was still much to do.

After these meetings, the South African Government asked three other engineers of varied nationality to give their opinion. After a somewhat short visit, they reported that the variations in the work had been very considerable, as we had maintained. However, they went on to say that they thought that the payment could be arrived at from the existing priced items in the tender. This was manifestly impossible, as the whole time and overhead basis had been thrown into confusion. The battle now started in earnest. For two years my diary shows intermittent involvement in this case, while the actual work pressed on. Three days were spent preparing details and a dossier in Dumez's office with Gericke, who had flown from South Africa. At odd intervals, fresh questions were asked that required opinions. Then in March 1971, I was able to borrow an engineer from Mott, Hay and Anderson who, after I had spent a few days in briefing him, set out for the site. He was able to do much of the donkey work to help the site staff, who were also deeply involved in the construction work. He had been working on the site for nine months, when on January 4th 1972 I heard that sadly he had been killed in a car accident. The need for a substitute was acute and then on January 9th I heard that Dumez had found a French Canadian firm, Revay Associates, who had specialised in sorting claims on USA and Canadian dams. Two days later, I received a letter from Revay giving details of their engineer, a young Scot named Robb. I was then able to check his record with two British firms for whom he had worked and whom I knew. On January 13th Garbe and I met Robb in the Union Corporation London Office and we took a good view of him. We spent a long day briefing him, before he took off for the dam where he did yeoman work. Next followed ten days work discussing various dossiers, which were passing to and fro and also in consultations with solicitors and learned QC's in Lincoln's Inn and the Temple.

Too quickly came the last of my many visits to Paris. Gericke and Robb had flown in from the dam and we spent three days over the weekend from April 7th to the 9th in intensive work. Later in the year, there were other requests to vet company notes for their South African advocates and then the receipt of copies of the vast tomes that had been completed for the case. The advice to have a professional photographer constantly on the site had produced a mass of very explicit views of the course of the work, while Robb had been able to improve the many diagrams and records which had been insisted upon. Some vertical cross-sections showed on one side the work as tendered for and beside it, the work as carried out, with a drawing of a tiny man at the bottom which emphasised the scale. On August 25th 1972 came the news by telex that the Government had made an offer in the shape of an ultimatum. They offered many millions of Rands as settlement – take it or nothing. This did not cover the costs, which had been meticulously kept by the chartered accountants, Cooper Bros. and the only alternative to be considered was to fight a long case in the High Court of the Orange Free State in Bloemfontein. The result was dubious as, to judges, words are sometimes more sacred than facts. It would also lock up valuable staff of all the companies, who should be earning money elsewhere. The fact that such a large sum was offered is an admission that over more than four years, much additional work had been carried out with no proper payment. This is a curiosity of such contracts, even when it states 'if the engineers and contractor fail to agree prices the engineer shall fix such prices etc.' In this case no such attempt was made. On October 11th 1972, the consortium accepted the offer. The dam was completed on time and caught one of the largest floods ever on the Orange River, so that at the opening the dam was full and water poured over the face of the arch through the arched openings provided. Photographs showed the spectacular rush of water down the spillway under more than 200ft head, to be flung into the air by the 'ski jump' which dissipates the possible force of erosion.

20

SERVICES TO CIVIL ENGINEERING

The work of the civil engineer is only dimly comprehended by the general public and the expression 'services to civil engineering', which is often used in honours lists, may be equally mystifying. These can cover a multitude of good deeds beyond design or construction work or administration and are usually without financial regard, but involve the valuable contribution of the spending of time. A colleague gave his opinion that you only had to give the appearance of standing vertical, for others to come and lean on you. A less dogmatic approach might be to say that if someone is rash enough to say something useful or sensible, he can be in danger of being cajoled into becoming 'involved'. Such services have been spent almost entirely on behalf of my profession and the Institution of Civil Engineers. From my childhood reading, I had formed a respectful idea of the importance of the Institution, as well as a determination to try to join it, as a civil engineer. This attachment to an ideal is an occasional human characteristic, whether the object is a party, a church, a football club, a trade union or a freak religion. Those who become infected are those who work to keep such bodies in continued existence.

I first ventured to enter the sacred and imposing portals of the ICE as a first year college student, after applying to be enrolled as a student of the Institution. On this misty winter's day, the imposing Central Hall was empty when a small window was slid aside in a room near to the entrance, a face looked out – and shut it again. Footsteps rang on the stone floor at the far end and a man appeared, stared, and walked off again. My nerve failed, so I quickly retreated and went inside Westminster Abbey. It seemed much friendlier. Forty years later, I suggested that there should be an open reception area or counter in the Central Hall and that it should be occupied by a one-legged civil engineer! I explained that this meant a thoroughly experienced member, who had been prevented from active work. His function should be to give a proper welcome to members, young and old, to listen sympathetically to their questions and direct them to which of the many 'officers' they should go to to solve their problems. The result was 50% successful - the open reception area was provided, but staffed by a commissionaire. This was at least an improvement and provided a form of welcome. In my younger days, I had heard members complain that the Institution was not interested in them and did nothing for them. The idea of doing anything for the Institution does not occur to many, although it is only those who do so who keep it alive for other members.

When I offered our model of Piccadilly Circus Station to be shown at the 1927 Conversazione, I was fortunate to make a friend among the permanent staff. This was the man who was stage-managing the affair, named E. Graham-Clarke, who later became the secretary. A feeling of belonging was intensified, as I had just become an Associate Member

– AMICE. Years later this became MICE, while the older and more important MICE were inflated into Fellows or FICE, much to the confusion of more enlightened members of the outside world.

As my work was in or around London, I was able to attend many meetings at which papers were 'read'. In those days this was actually the case. They were solemnly read by the author, or, if he wished, by E. Graham-Clarke. This shortened the time for discussion, which in those days seemed pompous to us younger members. This improved later, although 'delivery' was sometimes poor. By good fortune, three contracts on which I worked became subjects of papers within my first ten years, the Camden-Euston Tube, Piccadilly Circus Station and Ford's Power House. It helps to have been connected with published works.

64. *HJBH as President of the Institution of Civil Engineers*

Then my real involvement began….

I have described earlier how Sir Henry Japp insisted upon my taking part in discussions where some of what I said, especially on the new 'processes' as well as practical work, led me to being invited onto committees. The first was on the one to advise the War Damage Council on earth movement caused by war damage. This soon led to helping to compose four of the then quite new Codes of Practice, as a member of the main committee. Site investigations, foundations, earth retaining structures and earthworks followed each other in quick succession. Another appointment was to a Regional Advisory Committee on technological education. I forget whom we were to advise, but I was able to meet the heads of a number of bodies, such as Woolwich and Northampton Polytechnics and the Brixton School of Building, which made mutually useful contacts. The lesson from this was that no-one takes the advice of an advisory committee and their only value is to increase one's circle of acquaintances.

This led to my being nominated, and soon after elected, by the Institution to become a Governor of Westminster Technical College. Graham-Clarke, now the ICE secretary, enjoyed being its chairman, in charge instead of 'serving' committees. This college had a side interest, as it included a Department of Hotel Management and so the Governors included the managing director of the Savoy Hotel and Frederick Hotels, as well as the famous architect, Easton Robertson. The college ran a very nice little restaurant where the students took spells as cooks, bottle-washers, waiters, wine-waiters and cashiers and so on, with very nice food which they had cooked. I was no longer a Governor when someone wrote an article about it in Punch, which produced long queues and discouraged visits for a quick lunch. At that time the Westminster Tech was run by the London County Council, which was dominated by Herbert Morrison and his fellow socialists, although they did not seem to object to the Hotel School, which catered for hotels with a higher number of stars. One Governor would turn up and snooze quietly through our meetings. He represented the TUC. The Head was completely under the thumb of the LCC, which meant that much of what the Governors proposed was quietly ignored if he felt this to be wiser. This in turn led to me being nominated as a Governor of Northampton Engineering College, part of Northampton Polytechnic. This was so called as it was sited in Northampton Square, near the Angel, Islington. Our meetings were held in the Drapers' Company Hall or that of the Dyers' Company, in more congenial surroundings than the room at the Westminster Tech.

During the war, the ICE were advised by its President, Sir Charles Inglis, to form a body of senior members who, if invited to do so, were willing to lecture to college engineering societies on subjects of their choosing. Most of these were heads of consulting firms of eminence, but by chance I caught Sir Charles' attention and he conscripted me onto his list, where I found myself giving a series of lectures 'free of charge' to each of the London colleges, as well as Cambridge, Southampton and Birmingham Universities. For these lectures, I chose the building of Ford's Power House or Piccadilly Circus Station as providing useful examples of complicated works far removed from their theoretical lectures. One of the most enthusiastic audiences was at the Northampton Poly, who asked for several lectures. I became an admirer of this college and chaired several special courses that they organised. The 'Northampton' eventually expanded and became City University and in 1970 they granted me the Degree of Doctor of Sciences *Honoris Causa*. To receive the degree, we attended a special service at St Paul's Cathedral and then sat on the dais of the Guildhall, arrayed in all the splendour of borrowed plumage. My wife, my son Edmund and his wife, Diana, sat in the Gallery. Diana is a professional artist and accomplished potter and when I stood up to be 'Doctored' by solemn speech, my wife said to her 'A job for you!' and in due course I was presented with a china penguin, complete with bonnet and hood and red robes!

Not long before the war, divisional boards were formed, where the members were chosen who were active on the subject and following this a works committee division was formed, on which I found myself an active member. We were able to cover types of construction and not merely formal papers or one isolated contract. Our meetings proved the liveliest of any of the boards, since the subjects were more exciting, as were the type of member, especially contractor's engineers. I was asked to write papers on the choice of expedients and on site investigation, this enabled me to invite some of my assistants to write, who otherwise might have been too diffident and it was a pleasure to watch their success. These meetings were so successful that they drained away the audiences from the formal papers, which were read after Council monthly meetings, with the President in the chair. Due to this, and while I was on a sabbatical year off the Council, the boards were dissolved and I was unable to fight for their survival.

In 1955 I was chosen by the City and Guilds Institute to represent them as Governor of the Imperial College and to do this I dropped out of my other Governorships. For twenty years we presided over the complete rebuilding of the new extended colleges, which was quite a business. Sir John Betjeman fought to preserve the Imperial Institute Tower. This entailed a complete re-design of the layout and £150,000 to prevent the Tower from falling over after the amusing but useless Imperial Institute had been demolished. The Queen came to open part of the new buildings, when the President of the Imperial College was a Trotskyite. He led the Queen and the Duke of Edinburgh among the assembly of the well-chosen 'great and good' important guests, in a scruffy beard and an open-necked grubby shirt. The Queen seemed unmoved and smiling, but the President ceased to be popular.

At the time of students' bloody-mindedness in the new universities, a debate arose

20. SERVICES TO CIVIL ENGINEERING

whether officers of the Students Union should sit in Governors' meetings. I expected much objection from the elderly and varied governors, but I said that we were urging students to learn more on management and it would be useful to have to talk with men of experience and not only to fellow students. To my surprise no-one objected, but one old Governor said, 'But suppose they create a scene?', to which I replied, 'It would certainly make a change'. So, amid laughter, this was passed and proved very successful with both alternate left and right wing Presidents. In 1974, the President was a left-wing West Indian, Trevor Phillips, but who was able and amiable. Five years later he was still President of the National Students Union. I attended most of the Commemoration or Graduate ceremonies at the Albert Hall in borrowed gowns and had the pleasure of being made a Fellow of Imperial College which carries no letters to follow a name.

Finally in 1975, after 20 years, I had to decline a request to serve for four more years. The Chairman spoke some kindly words of farewell and in reply I could only say, 'When I found that the Dean of the City and Guilds College became a student after my own son had graduated, I thought of the words of Oliver Cromwell to Parliament: 'You have sat here long enough – in the name of God, go!'

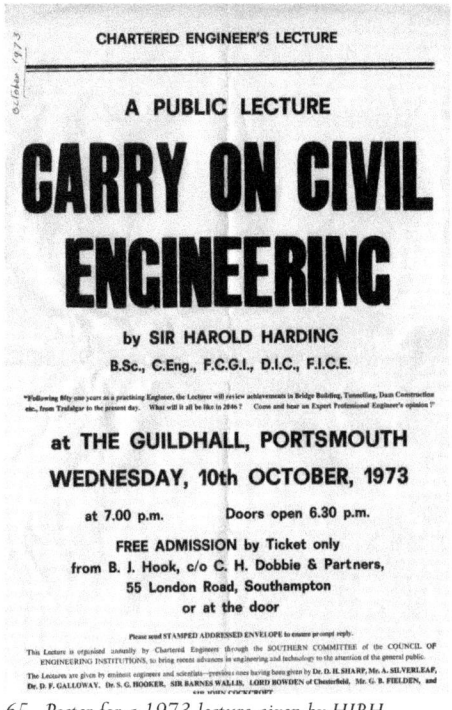

65. Poster for a 1973 lecture given by HJBH

PART 4
Amanda Davey

21

THE BRITISH TUNNELLING SOCIETY

The British Tunnelling Society was formed on the 18th February 1971, with the first meeting held on the 25th March and the constitution approved at the first Annual General Meeting on the 23rd September 1971. The founding chairman was HJBH. Unfortunately, due to a series of events, this planned and important chapter was never written. HJBH had made considerable leeway with the writing of the book and had started sending parts of the manuscript to the publishers that had been asking for him to write his life story. To his bemusement, their reaction was less than enthusiastic as they were looking for sensationalism, which isn't an intended civil engineering characteristic. By then, it was the late 1970s and early 1980s and the economy was not buoyant. He was also starting to suffer from compromised health, although he only shared this news with a few close friends. With the wind not gusting in his sails, having been proud of his previously robust constitution and formidable luck in the Blitz, he stopped working on the book. This means that I have had to stitch together this chapter from the available record. This is a shame, particularly since his pithy wit will be less present, although I hope not wholly absent. His insights would also have been very instructive. As the founding chairman of the society, there would have been many insights. In a letter published in *New Civil Engineer* on 7th September 1972 HJBH was commenting on some remarks made by a young engineer about the silver-haired gentlemen of Great George Street:

> Civil engineering is distinctly short of sex and of comedy and is of its nature, a slow-motion affair. What really interests the TV people is a disaster – which is not so good for Mr I.'s image.

The British Tunnelling Society

In the first Harding Lecture in 1996, on 'Risk: a tunneller's view,' Douglas Parkes (previously a chairman of the BTS himself) gave a brief introduction to the setting up of the Society:

> It seems appropriate that I spend a little time talking about our founding Chairman and about his part in the birth of this, our Society. The society itself was born out of the recommendation of the OECD Advisory Conference on Tunnelling of June 1970 that a national tunnelling focal point be set up in each of the 20 participating countries. The Institution agreed to provide support, and an ad hoc committee was established, comprised of five members of the Institution, including Sir Harold as Chairman, Alan, now Sir Alan, Muir Wood, the next Chairman, and another subsequently loyal member of the Society, Richard Triggs, then Managing Director of Edmund Nuttall. The two others were Sir Norman Rowntree and Roger le Geyt Hetherington.

The first meeting took place in this hall on 25 March, 1971, at the usual hour of 5.30 - preceded, of course, by tea: some things take a long time to change! An initial constitution had already been drawn up, very much under the guidance of Sir Harold, making membership open to all who were engaged or interested in tunnelling: the only restriction was the attainment of the age of eighteen, and there was the firm intention to try to make the Society relevant and of interest to all tunnellers, and not simply an academics' talking shop. We still keep trying, and it's always good to see some of our key foremen and inspectors here, at least the craic can't be bad. The first AGM was held the following September, when the first committee was elected and the Society has gone from strength to strength since then.

By the time the Society was formed, Sir Harold himself was in what he, in his Presidential address to the Institution some seven years previously, described as the fourth and final phase of his career, the advisory one.

HJBH's style was described in Colin MacKenzie's Harding Lecture in 2002:

> Harding was a formidable chairman, often calling directly on individuals to make a contribution when proceedings were a little sluggish. I recall one occasion, when the merits of rock TBMs were being discussed, the Managing Director of Nuttalls, Richard Triggs, sitting near the front, minding his own business, when out of the blue he was assailed by Harding with the following question - 'tell us Mr Triggs, do you feel that these new-fangled machines have a future, or would we be better advised to continue to drill and blast, as God intended?' To his great credit, Triggs rose and gave a very interesting dissertation on the current state of development of rock TBMs. I don't know whether Harding had a prior agreement with Triggs to call upon him, but he made it appear to be spontaneous, and Triggs didn't appear to know what was coming. Together they made it memorable, the proof of which is that I am telling you about it thirty years later.

The journal *Tunnels and Tunnelling* had been published from June 1969. From the start, it was considered of great importance to secure an informal link to allow the members to receive a copy as part of their membership. John King referred, in his Harding Lecture of 2000, to the habit in the first few years of papers published in advance in the journal, prior to the meeting at which they were to be discussed. This meant the author had twenty minutes to introduce the topic and time was allowed for in-depth discussion and contributions from the floor. 'This produced both more learned and more critical comment as well as more heated discussion...Harold Harding was probably the most compelling speaker.'

At the inaugural meeting in March, there was a great deal of discussion about the aims and intentions of the new society. The chairman was on characteristic form, with several illuminating anecdotes peppering his contribution. He referred to his ending of the James Forrest Lecture in 1939:

> I concluded my James Forrest lecture with a saying of Dr Johnson: 'Human experience, which is constantly contradicting theory, is the great test of truth.' Professor Pippard[59] then said afterwards that Dr Johnson said alot of silly things. The quotation went on, however: 'Experiences built on the discoveries of a great many minds are always of greater weight than the mere work of one mind, which can do little of itself.' We hope that this society will reveal the discoveries of a great many minds and so improve our environment.

The discussions were wide ranging, dealing with concerns for encouraging young tunnelling engineers into the society and making the discussions wide in their scope, encouraging contributions from people outside tunnelling but who were interested in the effects and arguments for their construction. HJBH agreed that this was important:

---

59 Professor of Civil Engineering at Imperial College, formerly the City and Guilds College 1933-1956.

> What to us is obvious is far from obvious to the public. Years ago, I was reading a novel by Dornford Yates in which the hero and a friend were digging a tunnel to get some buried treasure out of a castle where there were a lot of villains. So they drove a tunnel. First, they did a survey with a tape 100 yards long. Then they made the heading, which had to be done quickly, and all the muck was stored in a cellar. They made it 9ft high to give the men room to swing a pick. At that time I was doing a heading from Appendroke cellar, Piccadilly outfall to London Pavilion and we made it 4ft high because we had not room in the cellar to store all the muck.

In the BTS Minutes from the First Annual General Meeting, HJBH was able to report that the membership within that first year was already 60 corporate and 270 individual members. The March meeting had been very well attended. From that meeting it was clear that informality should be key... 'Immediate, uninhibited and challenging discussion of controversial topics seemed to be favoured by everyone - in passing, I find it difficult to think of many topics concerned with tunnelling which are not controversial, which I suppose is what makes it so fascinating a subject.'

The first committee contained the names of many subsequent chairmen of the society:

J. V. Bartlett (Chairman 1977-1979)
J. A. M. Clark
T. G. Hammond
Sir Harold Harding (Chairman 1971-1973)
D. A. Harries
H. D. W. Hoggard
J. R. J. King (Chairman 1976-1977)
Dr R. S. Millard
R. L. Triggs
A. M. Muir Wood (ICE nomination) (Chairman 1973-1974, 1975-1976)

At the second Annual General Meeting on the 26th September 1972, Douglas Parkes was elected onto the committee, nominated by Edmund Harding and one other. Douglas was later BTS chairman from 1974-1975 and 1979-1981.

At the third Annual General Meeting there was a report that only one meeting had had a poor attendance, with only 70 members present. This was for a meeting on the place of tunnelling in urban planning. The International Tunnelling Association was progressing slowly toward inauguration, expected to happen in the spring of 1974. *Tunnels and Tunnelling* had new publishers, but the co-operative arrangements continued. It was also announced that in spite of the increase in expenditure through the planned contribution to be made to the ITA, individual membership subscriptions were to be reduced from £5 to £3 for the financial year, reflecting the health of the accounts. The committee was also considering initiating an annual prize for which younger members only, probably up to the age of 30, would be eligible to compete.

> Mr Triggs proposed a vote of thanks to Sir Harold Harding, Mr Bartlett and Dr Millard for their services on the Committee, dwelling particularly on the inspiration and wise leadership which Sir Harold had given to the Society in its formative years; thanks to Sir Harold, it was already the largest and most active tunnelling society in the world. The vote of thanks was carried with acclamation, and the Chairman responded appropriately on behalf of the retiring members.

The British Tunnelling Society (BTS) was very important to HJBH. Of particular importance to him was the requirement, where possible, to ensure that the British society was free from Governmental control. He had spent a great deal of time lobbying behind the scenes to that end, not a foregone conclusion given that the early discussions were initiated from within Government circles. After their years working together in the Channel Tunnel Study Group years, (Sir) Alan Muir Wood had become a close friend and, judging from their letters to each other, shared a very similar view of the way the BTS should function. HJBH would often say that when speaking in public it was no good just reading from the script and keeping to the point, because the general attention-span of an audience is ten minutes. With this in mind it is necessary to make them laugh or raise their eyebrows approximately every ten minutes, in order to make sure that they stay awake. This will have been one of the many factors behind his intention that the meetings were informal and friendly. I know from the BTS meetings that I have attended that this informality and friendliness is in no way at the expense of content and there are some detailed and important matters that are aired. If these matters are discussed in a lighter atmosphere, then more of the audience will retain what they have learned and be in a position to use it themselves. That is a good legacy!

The Harding Prize – and a surprise

Shortly after the setting-up of the BTS, the committee acted upon the proposal that it should be important to encourage younger tunnelling engineers. A sum of money was set aside from the funds, to sponsor a prize named in honour of the founding chairman. The encouragement of young talent and recognition for their work was a subject dear to HJBH's heart and he was involved in many of the discussions. He was also understandably pleased to have the prize named in his honour. The prize is open to all under the age of 33 and given for an original paper relating to any aspect of tunnelling that the author considers to be of interest to those in the tunnelling industry. The first papers were submitted by the end of 1974, with the prize given in March 1975 to Trevor Orr, now Associate Professor at Trinity College, Dublin. Alumni of the prize include Gordon Masterton (1981) who was President of the Institution of Civil Engineers 2005-6 and Damian McGirr (2005) who was Chairman of the BTS from 2012-14 and who has been very helpful with bringing this book into publication.

In 1981, two years after staying in a Bristol hotel to go through the submissions for that year, a somewhat perplexing event took place. HJBH was visited by the police and the experience is outlined by him in a letter he sent to Douglas Parkes, asking for a character witness:

```
July 8th 1981

Dear Douglas

I have had the interesting experience of being briefly, I
hope, a murder suspect. A Sergeant of the Exeter CID and
his mate called on me on Monday evening over an enquiry
which they had from London over an acquaintance of mine.
   On arrival he said that it regarded one Henry Carr, to
which I briskly replied that was he not the one who had
been murdered? This took him aback slightly, saying 'That is
what I should have told you'. I replied that it said so in all
the papers and I had heard it discussed at the Institution
of Civil Engineers where he had been a member of staff.
   The next question was, had I stayed at the Unicorn Hotel
in Bristol two years ago? I found the date in the old diary
```

and read out this and the Unicorn Hotel and the street and explained that I was there for the Harding Prize meeting etc. He then said that in Carr's home they had found the hotel bill signed by him with three names as well, Irwin, someone and Sir HH. He had mentioned that Carr was thought to have homosexual tendencies. I could not place the other two and later remembered that they were two competitors. So I explained that Carr was the Societies secretary and as I was the guest of the Society he had to arrange and pay for the accommodation. They asked my age and were a little surprised to learn that I was now 81.

The next question surprised me. 'Was I in London in February this year'. I said that I had been up to London only once in twelve months and this was about then. I then found in my diary that I stayed for two nights at the Naval and Military Club and then corrected this to one night and one in Pangbourn with Muir Wood, February 16th and 17th to attend a meeting of the Fellowship of Engineering in St James's Palace with the Duke of Edinburgh in the chair. Then came the key question, where was I on Feb, I forget if this 21 or 28, but certainly a Saturday. So he wrote down quite happily, 'In Topsham with his wife'. So I duly signed a statement which he wrote down as is the custom which read quite sensibly. A curious episode for an ancient Past President. Will you be a witness to character? But a curious string of coincidences.

The Harding Prize has been given in most years since 1975. Tunnelling can be sensitive to the vagaries of the economy and in some years of tightened conditions there have not been the required number of submissions. For this reason it was decided in 1995 that from the following year, 1996, the Prize would alternate with another event in the memory of the founding chairman, the Harding Lecture, now called the Harding Memorial Lecture. This has been given every other year since then. The first lecture was given by Douglas Parkes on risk, as mentioned above. The lecture is given by an eminent speaker with significant experience of tunnelling related subjects. To date the lectures have been:

1996	Douglas Parkes 'Risk - A tunneller's view'
1998	Prof. Robert Mair 'Advances in geotechnical aspects of tunnelling in soft ground'
2000	John King 'A century of tunnelling and where we go now'
2002	Colin MacKenzie 'Reflections on 40 years in tunnel contracting'
2004	Sir Alan Muir Wood 'Ahead of the face'
2006	Colin Eddie 'Tunnelling into the future'
2008	Arnold Dix 'Sound engineering - The challenges of being professional in the 21st Century'
2010	Alastair Biggart 'The development and use of closed face tunnelling machines'
2012	Bill Grose 'Underground infrastructure - Meeting the future needs of society'
2014	Donald Lamont 'Health and Safety in Tunnelling: Evolution and Revolution.'

The Brunel Exhibition

In *New Civil Engineer* in 1975, in a special supplement to mark the event, HJBH wrote:

> In July last year, I accepted an invitation to become chairman of the Brunel Exhibition Project at Rotherhithe in East London. Formed by a group of Brunel enthusiasts from

many different occupations, the project's main aim is to renovate the Thames Tunnel pumphouse and access shaft near the southern bank of the river at Rotherhithe. With the area landscaped, the pumphouse will contain a permanent exhibition of the works of both Sir Marc Brunel and his son Isambard.

It is no coincidence that this year is the 150th anniversary of the start of the Thames Tunnel and I felt that the tunnelling society should not miss the opportunity to celebrate such an event and the subsequent development of British shield tunnelling which arose from Marc Brunel's creation.

A mainly pictorial display, held in London and open to the general public, seemed the most suitable way to show the tremendous contribution British engineers have made to the still developing art of soft ground tunnelling.

With financial help from many consultants, contractors, manufacturers and public bodies associated with shield tunnelling, the idea has developed into an extensive display of photographs, models and diagrams outlining not only the fascinating Thames Tunnel contract but also the professions' main achievements in soft ground tunnelling since those early days of subaqueous work.

The exhibition opens on 10 June at the Institution of Civil Engineers and will run until 1 July. Hopefully it will then move to London's Science Museum, for several months, before the section relating specifically to Brunel will be set up as a permanent display at Rotherhithe pumphouse.

A lot of effort has been put into creating an exhibition that will appeal to a large range of the general public. When you have looked at this special NCE supplement, which in part complements the exhibition displays, I hope you will find time to bring your family and friends to Great George Street to show them, and to remind yourself, what this branch of civil engineering really entails.

The Brunel exhibition in 1975 was arranged by the BTS to mark the 150th anniversary of the start of the construction of Brunel's Thames Tunnel. HJBH was chairman of the committee and as always heavily involved in the arrangements and behind-the-scenes negotiations, throwing his characteristic enthusiasm into the project. According to *London John,* the internal newspaper of Mowlem, the exhibition focussed on the development of shield tunnelling as first employed on Marc Brunel's tunnel. Once again, HJBH was collaborating with Mowlem for the exhibition work performed by the Basin South joiners under Bob Sinclair. Quoted in *New Civil Engineer* he said:

> As well as providing overhead protection, [the shield] helps to support the face, allows the lining segments to be erected easily and, in compressed air workings, it reduces the loss of pressure at the face. It can tow a platform ranging from Brunel's simple wooden staging to the modern 50 metre long materials handling sledge.
>
> A shield does not dig but it helps to steer a tunnel and trim it to exact size. Its main limitation is that it cannot turn sharp corners and, being expensive, should be used mainly in long tunnels where the extra speed and saving in face labour can recoup the additional costs.

The exhibition traced the development of shield tunnelling from Brunel onwards. It showed how James Greathead was inspired by Brunel's work to design and patent his circular iron shield and others with water traps and hydraulic segment erectors. Greathead's first tiny shield was used to build the subway beneath the Thames at Tower Bridge. This was repaired by Mowlem during the Blitz under the supervision of Sir Harold Harding.

Spon's Civil Engineering Price Book

In 1983 HJBH was asked to contribute a Foreword to the new Spon's Civil Engineering Price Book, which of course was a topic that appealed to him after all his contracting pricing experiences. The editor was delighted with what he wrote and said 'it is so obviously the product of a lifetime of experience, and it adds to the book a sparkle that is singularly lacking in the bulk of the contents'. Unfortunately the edits to the piece appear to me to have removed the sparkling bits, but the end result was probably more in tune with a book of its nature.

22

LAST YEARS AND A SIGNIFICANT LEGACY

While working on this book, three very striking characteristics to HJBH have become apparent that I hadn't been aware of as much as a youngster. First, he had an extremely well-developed ability to collaborate on almost all the projects with which he was connected, without ego but with great leadership.[60] Secondly, he could network in a way most business advisers of the modern era could only fantasise about, in a way that was warm and human and therefore a great deal more effective than the technique of the modern 'speed-networker'. The third characteristic was the ability to grab hold of a problem and keep going at it until it was solved or outmatched by what he would describe as the 'co-efficient of so-whatness'. A problem that he had been paid to grab hold of for many years was of course the Channel Tunnel and it is the mark of this extraordinary man that even in his last months he was writing letters and lobbying Government over what he believed to be the best solution, namely twin bored tunnels through the chalk. He was far from impressed by many of the bridge schemes.

Topsham Swimming Pool

In 1975, after gentle persuasion from my aunt Caroline, my grandparents moved from their house on Mount Ephraim in Tunbridge Wells to Monmouth Street in Topsham, just outside Exeter. The house was again three storeys high and just a little bit older, with a garden that was more manageable than the one in Tunbridge Wells which had become too large. With my aunt just down the road, they were able to spend time with her and her family. HJBHs 'retirement' in Topsham saw his attention turned to some local projects and the local bowls club. He often recounted a story about a letter my brother Gavin had sent him from boarding school, where he had said he hoped he was enjoying his bowels! He also spent time making his tapestry seat covers, firescreens and piano stool covers. One of the local projects came about because Caroline had started to collaborate with the Topsham community to bring the Topsham Swimming Pool to life and it was completely natural that she should go to her father for advice on the engineering elements.

> The saga of the Topsham Swimming Pool came to a head with the second official opening on the occasion of the 'swim-along with David Wilkie'. This character was touring the country giving exhibitions (at a price) and displaying his Olympic medals.
> The pool was first opened in July 1979, while the designer

---

[60] This is a rare, but thankfully not unique trait, since in the dangerous reaches of the earth it is an important survival strategy for the tunnelling engineer.

Caroline Oboussier was visiting the United States with her family, so she was not present. It at once became a great success all through the school holidays, with 300 and at times 400 paying visitors each day. The heating was kept up by gas as the solar heating from the nearby tennis court was out of action for want of a special valve. The cost of the heating was easily balanced by the increase in bathers. The Harding patriarchs were quickly into the water.

Before the David Wilkie visit, 200 children were needed to make sponsored swims to raise money to pay, but as these were falling short some adults on the committee decided to become sponsored. This led to the suggestion that Lady H might join in, though historians incline to the theory that it was her own idea. But once inspired, she went around to all the ladies twenty years younger than her but 'too old to swim' and got a long list of sponsors.

Two small boys in Monmouth Street who came to be sponsored heard of this and wanted to sponsor her. They insisted, as one said, I can pay as I have a paper round.

'Lady Harding' was the first name called, to her surprise and the announcer told the considerable body of spectators that she was the oldest swimmer and was the mother of Mrs Caroline Oboussier who had designed the whole pool complex free of charge as her gift to Topsham. But as unfortunately [this time] she was not well enough at the moment to be present, they sent their wishes and thanks through her mother.

Sophie then, after diving in with spectacular success, swam ten lengths. This was hard work as there were three children in each lane with some collisions. At the last moment many other children signed on and there was a line of 200 children waiting to go in. Sophie was much applauded and like the other competitors had to sit beside David Wilkie to have her photo taken. She raised £27.60 and was the subject of much praise and astonishment among the upper crust of Topsham as well as among all her many friends among the local children.

Tunnelling History and the autobiography

HJBH was persuaded by Hugh Golder to write his *Tunnelling History and My Own Involvement* which was published by Golders in 1981. He gave me stacks of proofs to use as scrap paper for my university work, all covered in his red pen edits. This 1981 book contained a great deal on the history of tunnels, a long-held passion as can be seen from the awareness of older and ancient schemes relevant to those mentioned throughout this book. It also contained some of the tunnelling projects he describes in this autobiography, but the scope of the book was tunnelling and therefore he was not free to deal in detail with the contributions that he had made in the field of soil mechanics, the war, the Aberfan Tribunal and many arbitrations. Almost as soon as *Tunnelling History* was out, his friends and several publishers started to ask for an autobiography with a wider scope. In 1979, in an article in *Tunnels and Tunnelling*, he had referred to the fact that only 40% of his time had been spent in tunnels, having done plenty of others things as well. He says in one letter to a keen publisher that he had started to write his autobiography in the 1960s and early 1970s but pressure of work had meant that he had put it to one side. This version was passed around to friends and his children and the once secretary for the Aberfan Tribunal, Roger Lloyd Thomas, who was particularly supportive in his comments and edits. My uncle wanted him to put in funny stories wherever possible, my aunt wanted him not to be too pompous and my father said that it was up to him. This is the script with which I have had the joy to work for the past few years.

Channel Tunnel again

66. *A partnership that lasted for nearly 60 years. HJBH and Sophie after repairs completed on the model of Piccadilly Circus that had cemented a pattern of collaboration*

As the Conservatives swept to power in 1979 with Margaret Thatcher as Prime Minister, the idea of a fixed link between the UK and France was again spoken of as a project that 'could happen'. To HJBH's consternation the political system required a new start in looking at the options and as Mrs Thatcher was not a railway fan she was therefore minded to look seriously at several of the bridge options. Out came the typewriter! He wrote letters to Government, to *The Times* (then the paper of choice for letters to be 'seen'), to the *New Civil Engineer* and to any bodies and individuals that were appropriate. He was interviewed many times by newspapers, both local and national and pieces were written on him in the professional press. As part of all this, the London Transport Museum relocated the Piccadilly Circus model and asked for advice on how to mend it, as one of the sections had come apart. In the end it had to be shipped to Topsham and my grandparents mended it themselves. This meant that we all got to see what HJBH always called 'their first child', as it was nine months in the making.

Lobbying for the Channel Tunnel – Phase 2

September 22nd 1981 - HJBH sends a letter to Margaret Thatcher as Prime Minister, lobbying for the twin tunnels option.

October 9th 1981 - letter received from the Department of Transport acknowledging receipt of the copy of the above letter that had been sent on to the Secretary of State for Transport. It can't have thrilled him to receive the comment:

> There is no doubt that full consideration must be given to the lessons learned last time. However it is only fair to those who responded to the previous Secretary of State's invitation to sponsors of privately funded schemes that the Department should thoroughly examine all the proposals that have been put forward.
>
> It is with this in mind that the Prime Minister and the French President recently announced new joint studies.

February 22nd 1982 - letter to *Tunnels and Tunnelling* referring to the idea for an immersed tube under the Channel waters. He pointed out several recurring themes and included reference to the great need for care with any ventilation shafts for an immersed tunnel, which would have to be resilient enough not only to withstand a hit from shipping, but also to avoid flooding the tunnel. He was also concerned that at this stage ideas were for a tunnel of 6m diameter, which he considered too small and likely to be a hostage to fortune.

February 4th 1984 - letter to my parents, Edmund and Diana

> …By the way (unloved by us) local vicar is the President of the OAPs Association and always ends with prayers before I take over [for the vote of thanks]. When he is away burying people I am expected to do so. He did not turn up [to a recent talk on catamarans], so instead of falling back on the Lord's Prayer I said that as our talk was all about sailors and as the great gales continued to rage, we should spend one minute in thought 'for those in peril on the sea'. My womenfolk think that this was artful.

May 28th 1984 - letter to *New Civil Engineer* suggesting a reprint of the 1961 paper on the work of the Channel Tunnel Study Group, to enable younger readers to have access to the 24-year-old material and decide for themselves which were 'new' suggestions for the fixed link.

October 21st 1985 - letter from Kevin Walton, a member of Council in the ICE requesting information about the history of the Channel Tunnel and the work of the Study Group for an initiative called 'Opening Windows on Engineering'. This request resulted in a short set of notes encouraging youngsters to review the Channel Tunnel situation in a journalistic way and to suggest their own assessment of the problems and proposed solution.

November 2nd 1985 - unpublished letter to *The Times* lobbying for the twin tunnels option.

November 25th 1985 - letter to J. Richard Graham, Director at the Ashridge Centre for Transport Management in Hertfordshire in reply to one he had written to *The Times* on November 23rd 1985:

> …the crossing would be a motorway. The false slogan 'You drive all the way' will misguide the paranoic motorist driver but you have touched the vital spot. In 1959 we consulted the Port Engineer of New York who controls all their cross-river tunnels. He extrapolated his statistics and found that for a road tunnel across the Channel, from the time that a car enters till the time that it emerges, there would be at least one accident. The combined solution would be an invitation to the quarterly pile-ups which we read about on our own motorways…. In a rail tunnel, the driver does not have to steer! Consider the record of the near hundred-year London Tubes with which I have had much experience….

December 9th 1985 - interview in the *Western Morning News*.

> Why Channel pioneer backs 1960s route: It is fully designed and no tunnel has ever been so exhaustively examined. There is no weather to contest with only the ground itself which being lower chalk is ideal for tunnelling machines, dry and solid.

January 6th 1986 - letter to *The Times*.

> The idea of driving over a bridge, down a helter-skelter along a long tunnel, up again and along another bridge is not so attractive as a quiet rest for half an hour in a train between long drives both sides of the Channel.

January 9th 1986 - a letter published in *The Times*:

> …I would wish the Minister of Transport to consider the views of French engineers, having worked with them for many years and found them highly intelligent and practical. Their view is the same as we expressed in our report in 1960. The cheapest form of construction yielding the quickest return on invested capital was a twin railway tunnel which could carry, with modern signalling, one train every five minutes and also carry 1,800 vehicles per hour in either direction… I fear that…enthusiasm has led to under-estimating in the hope of beating the rival solution. Among much else there are 100 ships in a peak hour passing the straits, with 143 known wrecks in the sea-bed among all the cables, with

extraordinary current fluctuations, an up-and-down sea-bed, and constant bad weather to interrupt the work, as we found in our work at sea in 1958-60 and in 1963-64. And we found that many insurance premiums were needed to cover the risks with shipping....

January 14th 1986 - a letter to J. H. Potter, Deputy Managing Director of the Dover Harbour Board:

...I am happy that you get my point that the ferries should continue with the rail tunnel and that the overblown Euroroute is the real threat to them (except that for periods if there is not a rail tunnel the bridge portion could be more vulnerable to weather than the ferries)......The Euroroute brochure shows two trucks and three motor cars on a short length of suspension bridge with open sides, a blue sky and calm. The slogan 'You can drive across' panders to the paranoia of all of us who drive and hate to stop. Mrs Thatcher never travels by rail as she has a chauffeur and it needs more thought than her opinion. We needed quite a fleet just for our marine borings and geophysical work with many marine insurance problems and days on end when too rough to work. I am sure the Euroroute has many unforeseeable expenses to rethink but it would take too long to enumerate. You will know well enough what the sea bed is really like".

January 20th 1986 - the Government announced that they were giving the go-ahead for a fixed cross-Channel rail link of two single track tunnels and a service tunnel bored through the chalk.

January 22nd 1986 - letter to the Editor of *The Times* about finance.

February 26th 1986 - a letter to *Tunnels and Tunnelling* explaining the relevance of the first site survey to the new scheme for the Channel Tunnel.

February 27th 1986 - an extremely fine photograph of HJBH appeared on the cover of the *New Civil Engineer* with the heading 'Harding: Tunneller of the century'. Inside was a celebratory article recognising the huge pleasure he felt with the news that the Channel Tunnel had been given the go-ahead after all those years. He was photographed holding the restored model of the Piccadilly Circus underground station.

In March 1986, only a few weeks after the announcement of the go-ahead for the Channel Tunnel, the illness that had marred his last few years finally took HJBH's life. It is such a Hollywood end, that he could have told some very funny stories about the timing of it. No-one can say if his interventions allowed common-sense to win through, but it is hard not to believe that he had a final influence on the decision. He had clarity of vision and acute understanding of his world that was much needed in this case.

67. The year that the Channel Tunnel was commissioned, the New Civil Engineer *celebrated a key figure in the history of the project with a place on the cover using this photograph*

Legacy

When the Channel Tunnel was built, opening in 1994 to much acclaim, it was (Sir) Alan Muir Wood who was the advisor, marking a crucial connection between the site investigation and feasibility stage of the project and the construction stage. The route and the scheme changed as time, techniques and ideas had moved on, but what had been the most significant project in HJBH's life had not been for nothing, although for a while both he and my father had believed this might have been the case. In a 1987 book on the Channel Tunnel[61] he was described as 'tunneller extraordinaire', which of course greatly pleased our family.

The British Tunnelling Society has continued to respect and honour their founding chairman. I have been invited to attend two Harding Lectures and two Harding Prize presentations where, at the latter event in 2014, I was encouraged to take 5 minutes (I managed to get it down to 7) to talk about this autobiography while the judges deliberated.

In 1993, my father and Sir Alan Muir Wood worked together to write the entry on HJBH in the Dictionary of National Biography. Sir Alan, in his letters to my father, noted his wish to 'emphasise [HJBH's] strong ethical code and his contempt for those who failed this test, especially the hypocritical'. There were several obituaries that were published in his memory. One that would have warmed him was written for the Institution of Royal Engineers at Chatham.

In the year 2000, *New Civil Engineer* published a special Millennium Supplement within which they gave the details of ten engineers that were voted as having been of national significance in the twentieth century. It was with great delight that we discovered that HJBH, fourteen years after his death, had been voted in at number 10, in a list that included Ove Arup, Alec Skempton and Karl Terzaghi. He would have been particularly delighted that the vote was cast by his own peers. At around this time I was introduced at a social function to an engineer who had been part of the team on the Channel Tunnel investigations in the 1950's. When he heard who I was, his immediate reaction was to grab me by the hand and pump it up and down with joy! That simple act told me a great deal.

Family Legacy

As a family we grew up surrounded by many Hardingisms. All of us carry guidance in our head that he gave to us in a gentle and calm manner. He did not value the concept of engineering dynasties. From experience he felt that they were the cause of varying shades of trouble, so he did not encourage any of his grandchildren down that path. My father came out of civil engineering and tunnelling at the age of 52 and became a versatile potter amongst other things, so still keeping a connection with clay! His sister, my aunt, was undoubtedly influenced by her father in many ways and became a well respected architect with a practice of her own, completing well over a hundred projects during her career. My uncle is retired from his TV directing career and has taken up writing, using his Navy experiences as inspiration. My brother Gavin did an analysis of the Channel Tunnel for his final year dissertation in his Economics degree at Heriot-Watt University, Edinburgh. For the rest of us, our world has been influenced as much by art as by engineering and the paths we have taken in life reflect this.

Another family member is HJBH's brother, Uncle Edward to us. Like their father he had led a simpler and less public life in insurance, that is until 1988 when he was suddenly catapulted to national fame as 'The Last Contemptible'. He was the last of the first 174,000 members of the British Expeditionary Force of 1914 (described by the Kaiser as 'that contemptible little Army'), to stand at The Cenotaph on the 11th November and say that at 92 he was pleased to be there. He had always been an enormous source of strength for his younger brother.

---

61 Jones and Ardill *The Tunnel: the Channel and beyond*.

Portrait of the President
Hugh Golder

Austere, severe,
Not bearded like the pard,
Beloved, revered,
By all who are not too hard
To see the virtue 'neath simplicity,
The worth in humble truth,
The wit that never in duplicity
Was cold or lacking ruth.
From careful hand
This portrait spreads on canvas,
From forebears grand
Evolved this skill tremendous.
Vaunting archbishop's baldness,
Smooth with never a wrinkle,
Cool avuncular blandness,
Eyes with icy twinkle
Beautifully rendered, spectacled, keen look,
Hands gently fondle undignified green book!

PEOPLE INVOLVED

While not exhaustive, this list is intended to give a bit of background to the people who have been mentioned and capture some extra anecdotes. There are some key figures in HJBH's world that he didn't manage to refer to in the text, but who have subsequently been of significance in the profession and also some characters who he never met but who were of great influence on him (these latter are shown in italics). Sources are wide and various, but include many obituaries. Any errors are unintentional and apologies are given for them.

O'Dean Anderson: Director of Morrison-Knudsen, involved with the tunnelling in San Francisco for the underground.

Louis Armand (1905-1971): French engineer, founder of the Societé du Tunnel sous la Manche in 1957. Head of SNCF from 1949, his role in revolutionising the French railways after the wartime destruction was in marked contrast to the negativity of Dr Beeching. Armand was an important member of the Channel Tunnel Study Group.

Peter Avery (1923-2008): As an expert on Iranian history, Avery interpreted for the Mowlem roads contract in Persia. He subsequently became a lecturer in Persian language at Cambridge University.

Sir Frank Baines (1877-1933): Architect of the ICI building and Thames House, Millbank, London. Had been head of the Ministry of Works in the second war. He was a demanding boss.

Sir Benjamin Baker (1840-1907): Famous British civil engineer and chief designer of the railway bridge over the Firth of Forth as well as consultant for London Underground bored tunnels from 1869. His writings were very influential on HJBH.

Professor A. L. Baker: Supporter and campaigner for a bridge-immersed tube scheme for the Channel crossing. This led to an exchange of thoughts between him and HJBH in the press. Professor at City and Guilds.

George Ball (1909-1994): Colleague of Al Davidson through Lend-Lease who had good French connections. Attended meetings of the Channel Tunnel Study Group. A US diplomat, he later became US Ambassador to the UN.

Robert Ogle Barnes (1895-1975): Was engineer-inspector for Dalrymple-Hay at Camden Town. He was godfather to Robert Harding. Although he had been a trainee engineer on the Welland Canal in Canada, the first war interrupted things, so although experienced and keeping up-to-date with American as well as British technical matters, he did not satisfy the ICE conditions. He joined his brother on his farm in Kenya, where he also did odd work for the Government. HJBH visited him in Kenya in 1954. He became Soil Erosion Officer for Kenya and a world authority on the art of contour ploughing and making of small dams. He later retired to Cambridge, from where he was sent by the British Government to Tehran to teach the Iranians the art of soil conservation, much needed after all the hungry goats had been at work.

John Vernon Bartlett (b. 1927): Worked under HJBH at Mowlem until the Persian episode, living nearby on the River Thames in the first years of his marriage. Later became senior partner at Mott, Hay and Anderson and heavily involved in the design of both the abortive 1974 Channel Tunnel works and the successful version in the 1980s and 1990s, (working in parallel with Sir Alan Muir Wood). President of the ICE 1983-84 and President of the Smeatonian Society 1993. Founder member of the BTS and Chairman 1977-9.

Steve Bechtel (b. 1925): Director of Bechtel, who were engaged to study the Channel Tunnel bored tunnel option by Technical Studies Inc. Also involved with the tunnel boring for the San Francisco Underground. Bechtel were later heavily involved with the construction of the Channel Tunnel (from 1987), with then Executive Vice-President, John Neerhout Jr.

named project chief executive of Eurotunnel in 1990.

Mr Edgar B. Beck (1877-1958): The first member of the board of Mowlem not to be a member of the family, brought in to look after the financial side of things in 1928.

Sir Edgar Charles Beck (1911-2000): Son of E. B. Beck, he later rose to run Mowlem on the retirement of Sir George Burt.

Dr Richard Beeching (1913-1985): His reports recommending closures in the mid 1960s were to have a devastating effect on the railways. This was at odds with events on the continent.

Curly Bennett: Chainman for HJBH, Edmund Harding and Douglas Parkes, among others. Before going to the City and Guilds College, Edmund Harding started as a pupil on the Brand's Kingston Power Station contract. In walked Curly Bennett:

> CB: 'Are you any relation to that chap in Mowlem?'
> ERH: 'He is my father.'
> CB: 'Blimey, he started under me! Do you know how long it took him to fix his first intersection point? Four bloody hours – but we never had to move it.'

Dennis Bertlin: Ran the consortium of contractors building the Owen Falls Dam (now called Nalubaale) in Uganda.

J. O. Bickel: With Parsons Brinkerhoff in New York when HJBH visited on the tour to get information for the Channel Tunnel Study Group on immersed tunnels.

Professor A. W. Bishop (1920-1988): Lecturer/Professor of Soil Mechanics at Imperial College and consultant to Mowlem. Studied under Skempton, later working as colleagues. Carried out the scientific assessments for the Aberfan Tribunal at HJBH's behest.

John Benjamin Body (1867-1940): Married to HJBH's mother's cousin, he was the right-hand-man of Lord Cowdray (Weetman Pearson) in particular in Mexico and a director of Pearsons. A civil engineer of talent, he wrote a paper on the drainage of the valley of Mexico which proved helpful on the Mexico visit. Provided much needed holidays when HJBH was a young man and it was in his house that Sophie and HJBH met.

Jack Bonny: Head of Morrison-Knudsen. Commissioned the work on the San Francisco subway.

Reginald Bosanquet (1932-1984): Journalist who carried out several interviews with HJBH in the Channel Tunnel feasibility study years. Later became a much-loved newscaster.

Nigel Boxer: Had worked under HJBH as a 'disciple' in Mowlem, he later became a director of Kinnear Moody and Co. There was a mutual involvement with Mexico.

John Brass: Managed the Mowlem Plant Department with McGibbon after Eric Burt's death in 1949. He had managed the Royal Engineers' plant in the Western Desert as a Lt. Colonel, after escaping from Calais with what was left of Tommy Burt's company of sappers. It turned out that Derek Hammett had served under Brass in the desert.

Professor John McGarva Bruckshaw (1906-1969): Professor of Applied Geophysics at Imperial College at the time of the Channel Tunnel Study Group. Interpreted the sonar readings and cores in particular.

John Buckley: Solicitor at Macfarlane's and family friend. Was a key source of support and guidance during the aftermath of the trip to Persia in 1956.

Sir George Mowlem Burt (III) (1884–1964): Grandson of the George Burt that was John Mowlem's partner and nephew by marriage. Managing Director of Mowlem for 48 years until retirement in 1961. Sat on Building Research Board for twenty years, chairman for ten. President of the Federation of Civil Engineering Contractors for seven years.

Eric Burt (1894-1949): Sir George Burt (III)'s much younger brother. Responsible for the Mowlem Plant department until his untimely death.

Edwin Burt (1875-1946): Son of Sir John Burt, cousin of Sir George Burt (III). A director of Mowlem.

Edward George Mowlem (Tommy) Burt (1904-1940): Son of Edwin Burt, brother of Kenneth Burt. Friend of HJBH. Was a Major in the Royal Engineers and died as a prisoner of war in France in 1940. His wife Edie was Edmund Harding's godmother.

Stewart Burt (1886-1963): Brother of Sir George Burt (III) and Eric Burt. With Kenneth Burt was on the committee in charge of plant following Eric's death.

Kenneth Rust Burt (1908-1988): Friend of HJBH and director of Mowlem. Son of Edwin Burt and brother of Tommy Burt.

Dr Carter (1922-2014): As a lecturer at Imperial College he contributed innovative biostratigraphy techniques to the Channel Tunnel Study Group investigations from 1958-61. Later work included site investigations contributing to the understanding of the geology in preparation for the Thames Barrier, 1970-2.

Brigadier Cavendish: Succeeded Brigadier Horsfield in running the road contract in Persia. Had also been at Christ's Hospital at the same time as HJBH.

Captain Clayton: HJBH met Captain Clayton in 1949 when he gave sailing lessons to Robert Harding at Shaldon, near Teignmouth. When a captain was needed for the Mowlem ships in the Solent in 1950, Clayton came to do the job. When he was 15, Robert had jumped at the invitation to be a temporary deck hand and this was described in a XXI Club letter of May 1952:

```
My junior son, Robert joined my crane-ship Ebury for 6
days before Easter as a deck-hand and had a great time,
testing the crane to 22 tons, swinging the compass and
then sailing from Portsmouth to Tilbury. This she did in 23
hours, between gales, and including a fog on the Goodwins.
She picked up a 20 ton diving bell of the Port of London
Authority at Tilbury and carried it to St Katherine's Dock,
where she is now using it to repair the lock cills and
causing considerable interest. We rescued Robert on Good
Friday, much to his regret. This old fashioned diving bell
differed little from the one used by the Brunels to inspect
the cavities over the Thames Tunnel, except that their air
pipe was made of leather. Rubber has simplified life in so
many ways since then.
```

Vincent Harvey Collingridge (1910-): Having joined Mowlem in 1936, he succeeded HJBH as works director, later becoming director of Soil Mechanics Ltd.

Leonard Frank Cooling (1903-1977): Leader of the soil mechanics section at the Building Research Station and close colleague of Skempton's.

Sir Edwin Cooper (1874-1942): Architect of the 1929 Lloyds Building and the National Provincial Bank on Cornhill. Designs had to be amended following the Commercial Union Building excavations collapse and the resulting raised awareness of health and safety principles.

Sir Harley H. Dalrymple-Hay (1861-1940): High profile engineer of underground railway construction, especially in London where he worked on the Bakerloo Line, The 'Hampstead tube' (now forming a major part of the Northern Line complex) and the Piccadilly Line. In *Engineering Wonders of the World* he wrote about shield tunnelling, inspiring the young HJBH. In 1922 he was the engineer on the City and South London Railway, HJBH's first job. A mutual respect developed.

Lord Justice Herbert Edmund Davies (1906-1992): British judge who tried the Great Train Robbery trial in 1963, before being appointed Lord Justice of Appeal and chairing the Aberfan Tribunal in 1966.

Frank P. Davidson (1918-2014) and **Alfred (Al) Edward Davidson (1911-2002):** Frank and his brother Al were American lawyers and visionaries who were involved with the Channel Tunnel Study Group and whose energy and enthusiasm were a major catalyst and contribution to the project. They ran Technical Studies Inc. one of the four component organisations of the Group.

Colonel DeLong: Inventor of types of jack for use in oil-well drilling platforms. Part of a consortium seeking to construct an immersed tube tunnel across the English Channel. Involved with the construction of the Chesapeake Bay Bridge-Tunnel.

Kingman Douglas: Partner of the bankers Dillon Read and part of the discussions on options for the Channel Tunnel Study Group.

Lew Douglas: Director of Union Corporation and ex-USA Ambassador to Britain. Involved with the Hendrik Verwoerd (HV) Dam, South Africa.

Sir William Arthur Harvey Druitt (1910-1973): H. M. Procurator-General and Treasury Solicitor at the time of the Aberfan Tribunal.

Prince Philip, Duke of Edinburgh (b. 1921): Gentleman member of the Smeatonian Society of Civil Engineers. He and HJBH were also co-members of the Fellowship of Engineering and HJBH received a warm thank you letter on receipt of a copy of *Tunnelling History, my own Involvement*.

Charles Putnam Dunn (1886-1966): Director at Morrison-Knudsen and another key player in the Channel Tunnel Study Group. Design team leader for the twin bored tunnel option.

Ebtehaj, Abolhassan (1899-1999): Had been a prominent banker in Persia and had been with the International Monetary Fund in Washington DC from 1952-4. On his return to Iran, he took the role of running the Plan Organisation with a programme of dam and road building. His behaviour was the catalyst for HJBH leaving Mowlem.

Dr Haydn Evans: Lecturer in Civil Engineering at University College, Swansea involved in the site investigations carried out for the Aberfan Tribunal in 1966.

Major T. W. G. Farmer: South-West Regional Officer of the Ordnance Survey who gave evidence to the Aberfan Tribunal, presenting surveys and aerial photographs with maps made from them.

Professor Wolmar Fellenius (1876-1957): Swedish engineer who developed the analysis of the stability of saturated clay slopes and the assumption that the critical surface of sliding is the arc of a circle. Key papers were in 1916 and 1926. He chaired a 1916 committee that originated the word 'geotechnique'.

Sir Francis Fox (1844-1927): High profile English civil engineer. HJBH found his autobiography very useful, particularly where he described encountering a spherical cavity tunnelling in sand and gravel due to a 'run of ground' from poor tunnelling for a sewer.

Sir Ralph Freeman (1880-1950): Structural engineer who joined Douglas Fox and Partners in 1901, later to become Freeman, Fox and Partners. Taught HJBH Structures in his final year of studying.

G. Friese-Greene: Son of the inventor of the first cinema camera, who had been immortalised later by Robert Donat in a film at the time of the Festival of Britain. Had been a friend of HJBH's in the London Rifle Brigade and managed Starkie-Gardner, the company that made the suspenders for the repairs to the Northern Outfall Sewer in 1941.

Rudolph Glossop (1902-1993): Had been a contemporary of HJBH while at the Royal School of Mines (part of Imperial College) and a cousin to XXI Club friend, Grigor-Taylor. HJBH arranged for him to join Mowlem, from which point they were close and collaborative colleagues. Following involvement at Chingford, he got to know Skempton, Golder and

Cooling at the Building Research Station. The founding of Soil Mechanics Ltd by HJBH, Wynne-Edwards, Glossop and Golder took place a little later. Until 1956 their careers were sometimes running in parallel, including both constructing Mulberry Harbours in the war, running Soil Mechanics Ltd and both becoming directors of Mowlem in 1950. Glossop was involved in the Persia contract but managed to keep out of the trouble and retain the friendship.

Jean Goguel: Head of the French Geological Survey, involved in the Channel Tunnel Study Group investigations.

Hugh Quintin Golder (1911-1990): Joined Mowlem from the Building Research Station in 1942 after working with HJBH, Wynne-Edwards and Glossop at Chingford. Helped establish Soil Mechanics Ltd as one of the three founding directors, later running the company largely by himself. Resigned in 1956 out of loyalty to his friend and colleague, after the problems in Persia and left England for Canada, where he started Golder Associates in 1960. Called in those days H. Q. Golder and Associates, it focussed on soil mechanics and ground engineering. In 1981, Golder Associates published HJBH's *Tunnelling History and my own involvement*. This book is still given as a prize to winners of the BTS's Harding Prize.

W. G. Gove (1910-): Chief accountant and later company secretary of Mowlem. Got uncomfortably tangled in the Persian troubles.

A. Grange: Founder of the firm Societé Generale Exploitations Industrielles, appointed by the French to the Channel Tunnel Study Group investigations. Worked closely with (Sir) Alan Muir Wood.

James Henry Greathead (1844-1896): British civil engineer who improved upon the tunnelling shield invented by Marc Brunel for the Thames Tunnel, when it had proved unwieldy. He later worked on the City and South London Railway (1886), using the shield and pioneering the use of compressed air.

Derek Hammett: Architect employed on the Mowlem Welham Green plant construction. It turned out that he had served under John Brass, who was running the plant, in the Western Desert.

Sir Frederick Handley Page (1885-1962): Chairman of the City and Guilds Institute. Influential British industrialist and aviation pioneer.

Gilbert Charles Harding (1907-1960)*:* English journalist and radio/television personality, appearing in particular on *What's My Line* as a panellist. The tabloids labelled him as 'the rudest man in Britain'. This was more entertaining on the screen than in real life. He was no relation.

William 'Bumper' Harris: Experienced and famous tunnel miner who ran the workings of the junction chambers near Mornington Crescent Station. Provided guidance to HJBH on tunnelling in running ground. Famous for the time in 1911 when, having a wooden leg as a result of an accident in the construction of Waterloo, he was employed on the first day of the escalators at Earl's Court, to go up and down to show members of the public that they were safe.

Joseph Curtis Raymond Head (1899-1988): Had been Willie Rowell's assistant and returned to Mowlem to help with tendering for the final contract at Piccadilly Circus following Rowell's death in 1927. He had joined as a trainee after the RAF, later rising to become second to Sir Henry Japp and became works director, with considerable influence and success in gaining work for the firm.

Arthur Montagu Holbein 'Bean' (1897-1970): Member of the XXI Club (see p. vii). A significant construction and design civil engineer, he was another involved in the building of the Mulberry harbours and Phoenix units so crucial for D-Day in the Second War. Was friendly with the actress Martita Hunt.

Brigadier Eric Horsfield: Had been Chief Engineer of the Fourteenth Army in Burma. First involved with Mowlem for the Ground Nuts affair of the late 1940's. Ran the road contract in Persia until this was taken over by Brigadier Cavendish.

Sir Geoffrey Howe (b. 1926): Was a barrister at the Aberfan Tribunal before becoming a high profile Conservative Minister in Margaret Thatcher's governments, including as Chancellor of the Exchequer and Foreign Secretary.

Donald Hunt: Public relations consultant at Whittaker, Hunt and Company. Brought in to help the Channel Tunnel Study Group after 1960. He continued to do the public relations into the Eurotunnel years and wrote a book on the history of the Channel Tunnel in 1994.

John Leigh Hunt (1901-1942): Contemporary of HJBH at Imperial College. On H. Dalrymple-Hay's staff at Piccadilly Circus. Assisted in the preparation for the model of Piccadilly Circus Underground Station by helping to smuggle Sophie into the workings. Went on to be an Admiralty Engineer and was lost in the second war to the Japanese at Singapore.

Martita Edith Hunt (1900-1969): Renowned English actress born in Argentina. She featured on stage and in films, notably playing Miss Havisham in David Lean's 1946 *Great Expectations*. She was a friend of Arthur Holbein 'Bean'.

Dr John Hutchinson: Colleague of Professor Bishop's at Imperial College. Part of the team investigating the Aberfan slip. Later became a Professor and Glossop lecturer at Imperial College, now Emeritus.

Roger Hutter (1911-1998): Ran the French railway company SNCF at the time of the Channel Tunnel Study Group.

Ernest Ischy (1905-): Head of Soletanche at the time of the Channel Tunnel Study Group.

J. Donovan Jacobs (1908-2000): Consulting engineer from San Francisco. Led the World Bank Mission to the Litani Project in the Lebanon in 1960.

Marc Jacquet: French Minister of Transport at the time of the Channel Tunnel Study Group.

R. T. (Jimmy) James: Member of the XXI Club. Built up a very successful structural consulting firm, R. T. James and Partners following a period in Malaya. Designed, with his partner Hausser, the structural work that underpinned the RIBA building in 1934.

Sir Henry Japp (1869-1939): Cousin of Ernest Moir, he had driven the four railway tunnels under the East River, New York for Pearsons, at very high pressure, in which he was a pioneer. By 1930 he had been persuaded to help Mowlem at Battersea Power Station. He was instrumental in improving HJBH's conditions at Ford's Power House, Dagenham. Later, he became a Christian Scientist and strong promoter of experience in public speaking.

Sir Elwyn Jones (1909-1989): Attorney-General at the time of the Aberfan Tribunal, he later became Lord Chancellor.

Isaac James Jones: Legendary tunnelling engineer first with Mowlem in 1928, following Willie Rowell's death. HJBH worked for him on and off for 20 years, including at Millbank, Battersea Cable Tunnel, Ford's Power House at Dagenham and Bow to Leyton tube. Inspired in HJBH the development of the 'Nelson Touch'.

Mervyn Jones (1910-1989): Head of the Wales Gas Board, who became a friend during the sittings of the Aberfan Tribunal.

Jack Kell: Symington's deputy on the City and South contract. Became senior partner at Mott, Hay and Anderson and was involved with the World Bank Mission to Litani project.

Sir Cyril Reginald Sutton Kirkpatrick (1872-1957): Founded Sir Cyril Kirkpatrick and Partners in 1925, specialising in heavy foundations, docks, harbours and sea defences. Was President of the ICE in 1931/2 and member of the Smeatonian Society among others. Designed the Ford Power House at Dagenham.

Sir Ivone Augustine Kirkpatrick (1897-1964): Had been Permanent Under-Secretary at the Foreign Office before retirement and then became chairman of the Channel Tunnel Study Group, alternating with M. Massigli. On his death he was succeeded by Lord Harcourt.

Noel Lambert: Friend of Wynne-Edwards. Head of the Northern Construction Company of Vancouver, a subsidiary of Morrison-Knudsen. HJBH gave advice on a soft ground tunnel in Vancouver after a collapse. This involved standing up to American soft ground gurus in the advice, which proved sound. Lambert was also involved with the San Franscisco Subway.

Herr Lindrodt: Swedish site manager faced with the problem of tunnelling on Paradise Island.

Roger Lloyd-Thomas (1919-2010): Secretary to the Aberfan Tribunal. Became a close friend and was heavily involved in commenting on early drafts of this book.

José Machado: Uncle to Philippe Oboussier. Lived in Connecticut and was a regular host to HJBH on his visits to the USA.

Mansell MacLean: Was part of the World Bank Mission to the Litani Project in 1960. Later joined Morrison-Knudsen as an advisor and was involved with the San Francisco subway.

René Malcor (1908-1997): Ingenieur en Chef des Ponts at Chaussées in France at the beginning of the Channel Tunnel Study Group. He directed the studies and asked for a British counterpart, who was to be HJBH. Their working relationship was close and lasting.

René Massigli (1888-1988): Former French ambassador to Great Britain and alternating chairman of the Channel Tunnel Study Group with Sir Ivone Kirkpatrick and later Lord Houghton.

Jean Mathieu: A contemporary of René Malcor from the French Ministry of Transport and opposite number to Colonel McMullen. He was the French chairman of the joint Franco-British Commission of Surveillance for the Channel Tunnel survey operation 1964-5.

McGibbon: Mechanical and aeronautical engineer who managed the Mowlem Plant Department with Brass after Eric Burt's death in 1949.

Colonel Dennis McMullen: Chief inspecting officer of railways for the Ministry of Transport, he was the British chairman of the joint Franco-British Commission of Surveillance for the Channel Tunnel survey operation 1964-5, opposite M. Mathieu.

Professor Cyril C. Means (1919-1992): Arbitration director of the New York Stock Exchange. Travelled to Europe for the Davidsons to explore the intentions and purpose of the advocates for the Channel Tunnel prior to the Channel Tunnel Study Group.

Major W. H. Morgan DSO: Received his DSO for work in a tunnelling company in the first war. Subsequently County Surveyor for Essex and Middlesex. Worked for Mowlem both before the first war and in the 1950s. Ran the Mowlem Persia Committee, after which he became an arbitrator and advised HJBH in his early forays into arbitration.

Sir Alan Muir Wood (1921-2009): Tunnelling engineer who joined Halcrow in 1952, becoming senior partner in 1979 and worked as a consultant to Halcrow from 1984 to his death. Founding President of the International Tunnelling Association in 1973, President of the ICE 1977-8 and second (and fourth) chairman of the BTS 1973-4, 1975-6. His involvement with the Channel Tunnel began in 1958 working on the feasibility report, with further input in 1964. He was part of the joint consultancy team for the design and construction of the abortive 1975 start, specialist adviser to the House of Commons select committees in 1981 and a member of the Anglo-French Disputes panel 1988-98. He wrote, with Edmund Harding's help, the DNB entry for HJBH in 1996.

Edward R. Murrow (1908-1965): American radio and television journalist who interviewed HJBH on the Channel Tunnel after the release of the feasibility report in 1960.

Roger Newport: HJBH's assistant at Mowlem at the outset of the Persia contract.

President Apollo Milton Obote (1924-2005): Was the materials clerk on a bridge in Uganda for Mowlem during HJBH's 1954 visit. He later led Uganda to independence in 1962 as Prime Minister (1962-70). President replacing Mutesa from 1966-71, when he was overthrown by Idi Amin. Obote deposed Amin in 1979. President again from 1980-85. His presidency struggled with internal fighting and he was again forced out of office in 1985.

Jack F. Pain: XXI Club member and long-term friend. Was in charge of the team that built the Sydney Harbour Bridge for Dorman Long and Co., later becoming a director. Was a proponent of a bridge solution to the Channel Tunnel crossing and several discussions took place.

Douglas Parkes: Friend and colleague of Edmund Harding's from their days working for Charles Brand. Became good friend of HJBH as a result of the work they did together in the early days of the BTS. Became the third chairman of the BTS in 1974-5 and again in 1979-81. At Ove Arup and Partners for many years. Carried out the risk assessment for the Channel Tunnel in 1987-8, following an assessment of safety provisions for the tunnel boring machines and preceding a safety audit on the Howden machines. Author of the investigation into the Heathrow tunnel collapse 1994. Gave the first Harding lecture on 'Risk' to the BTS in 1996. Has been very important in getting this book to print.

Professor Ralph B. Peck (1912-2008): Based at the University of Illinois at Urbana, he was a soil mechanics guru in the USA, having worked with Terzaghi, who was also his friend. He was advisor on the Morrison-Knudsen tunnel problems in Vancouver. Professor of Foundation Engineering at Illinois 1948-1974.

Arthur D. M. Penman (1922-2008): Was Principal Scientific Officer at the Building Research Station where he had worked under Skempton shortly before he left to return to Imperial College. A specialist in soil mechanics and slope stability he carried out analyses for the Aberfan investigations. He became a recognised authority in embankment dams.

Felipe Pescador: Key HJBH contact in Mexico, he worked for the contractor ECSA and provided entertainment and excursions while HJBH and Sophie were in Mexico City.

Jacques Georges-Picot (1900-1987): Head of the Suez Canal Company, he was a senior figure in the Channel Tunnel Study Group.

Professor Alfred John Sutton Pippard (1891-1969): Professor of Engineering at Imperial College from 1933. President of the ICE in 1958 and member of ICE Council for many years. He also lived in Putney, so HJBH will have come across him in several places.

Sven Platzer (1914-2014): Met HJBH during a Harding family holiday, with work related study, to Sweden in 1947. Head of Widmark and Platzer (WP), to become Platzer Construction after 1969. The two men worked together on many projects, notably in Mexico, Ceylon and the Paradise Island tunnel.

Commander Christopher Powell: Secretary of the Parliamentary Channel Tunnel Group in the 1970s and also secretary of the Parliamentary Scientific Group.

Hector Prudhomme (1903-1993): Advisor on the Seven Year Plan under which the Persia roads contract was undertaken. Had been employed by the World Bank from 1951, before heading the team of advisors on the plan while on secondment.

T. S. Rao (1898-1968): Chief engineer of the Bombay Electric Supply and Transport Undertaking. Mastermind of the Bombay Underground in the 1950's, the first section of which was to open in 2014.

A. E. (Sandy) Reid: Resident engineer for Dorman Long and Co. on the Ford Power House at Dagenham, he became a long-term friend. He went on to make a considerable professional reputation by building bridges and sinking caissons in Egypt and Iraq. He did much behind the scenes in Iraq in the second war when Iraq was on the verge of joining the Nazis. Sandy Reid, like Sir Henry Japp, came from Angus. Another Scottish engineer who appeared briefly on the jetty contract gave HJBH some uncalled-for advice. He came from Glasgow and his advice was: "Watch him. Reid comes from the wrang side 'o Scotland!" After his wife died and he had retired, Sophie and HJBH spent a week with Sandy in his home town of Kirriemuir, thirty years after Dagenham.

Lord Brian Hubert Robertson (1896-1974): Distinguished British Army General, he preceeded Lord Beeching as Chairman of the British Transport Commission (1953-61).

Vernon Robertson (1890-1971): President of the ICE 1949-50, he and HJBH had worked together on committees for several years before 1956. He was an early contact to gain introductions in the new life after leaving Mowlem. Provided the introduction for the Bombay Underground investigations.

William Rowell (c.1854-1927): Had been a contemporary of Greathead and he was HJBH's first boss at Mowlem. He ran the civil engineering work and was a seasoned tunneller under London. He ran the first part of the Piccadilly Underground Station job until his sudden death.

James Segrave: Was second in command to the agent on the Camden Town to Euston contract. He showed HJBH the ropes when he first started at Mowlem and they became close colleagues.

Sir Edward 'Tim' Singleton (1921-1992): Solicitor who worked for Macfarlanes from 1954. HJBH brought him in to advise Skånska and WP on contract problems in Ceylon. Became President of the Law Society in 1974, then the youngest to do so.

Professor Alec Westley Skempton (1914-2001): After studying at Imperial College, worked at the Building Research Station with Dr Leonard Cooling and Hugh Golder in the Soil Mechanics section. Involved with the investigations at Chingford. He returned to Imperial College in 1947 to work with Professor Pippard. Arranged for Bishop's involvement in the Aberfan research. President of the Smeatonian Society in 1981.

Professor Alfred Stucky (1892-1969): Swiss engineer who specialised in dams and hydraulics, he held the Chair of Teaching and Research in Hydraulics at the School of Engineers of Lausanne. For the Oued Nebaana Dam, Tunisia, he was chairman of the arbitration team for the Utah Construction and Mining Company, while HJBH was arbitrator. His son Professor J. F. Stucky, a hydraulic engineer, was also involved.

Professor J. F. Stucky: Hydraulic engineer and dam specialist who worked with his father, Professor Alfred Stucky in Lausanne, Switzerland. After his father's death, he carried out investigations on the Hendrik Verwoerd Dam in South Africa alongside HJBH.

Saturnino Suarez: Head of a firm of Mexican engineering contractors collaborating with WP in Mexico.

George E. Suderow (1913-1961): Chief engineer and inventor at DeLong when HJBH visited the Hyperion Outfall Sewer in Los Angeles in 1959.

Major-General Leif Sverdrup (1898-1976): Designed the bridge-tunnel-bridge link at Chesapeake, but in 1969 at the ICE he spoke in favour of a bored tunnel for the English Channel, declaring, 'the building of a bridge across the Channel would be a hazardous operation indeed.'

Enrique Tamez (1925-): Director General of Solum, a soil mechanics specialist firm at the time of HJBH's visit to Mexico City. Had studied under Terzaghi at Harvard.

Professor Karl Terzaghi (1883-1963): Austrian engineer whose research from 1925-29 at the Massachusetts Institute of Technology established the science of soil mechanics. He was back in Vienna from 1929-38 and it was at this time that Wynne-Edwards sought his support for the theories of Golder, Skempton and Cooling at Chingford. A significant guru and moral support to all who worked at Chingford and greatly valued by them. He was a staunch supporter of HJBH's early days working as a consultant. In 1938 he emigrated to the USA, teaching at Harvard University, where he became Professor of Civil Engineering in 1946.

From his ICE obituary:

> Terzaghi first visited this country in 1938. He was staying in France when he received a telegram asking if he would act as advisor on the construction of an earth dam in England. An interview was arranged and a few hours later the engineer-in-charge arrived in Paris. He was a worried man. Without a word he spread out plans and records, which Terzaghi examined.
>
> T: 'Where is the dam located?'
>
> V: 'North of London.'
>
> T: 'That dam must have been designed by an enemy of the British nation because it will fail, whereupon your Parliament and Westminster Abbey may be washed into the Thames.'
>
> V: (now smiling). 'It has failed already.'
>
> T: 'What instructions did you get from your boss?'
>
> V:: 'Show him the plans and watch his face. If he remains calm, take your hat and go. If he looks disturbed, bring him over on the next plane.'
>
> The same evening the two men flew to London together.

(The boss was almost certainly HJBH).

J. C. G. (Creed) Vowler: Member of the XXI Club and HJBH's best man. He acted as Hon Secretary to the XXI Club from 1926 to 1977. Worked for the Metropolitan Vickers Electrical Company, then Associated Electrical Industries Ltd, retiring in 1964.

Captain Henry Blythe Thornhill (Teddy) Wakelam (1893-1963): An old friend of Eric Burt's, during the General Strike of 1927 he ran an emergency milk depot for Mowlem before joining the newly formed sports commentary team at the BBC. He was the first sports radio commentator and then in 1938 became the first TV sports commentator. Wakelam told of how he had to climb a ladder to a little hut on the top of a scaffolding tower and found himself confronted by a notice: 'Watch our language – J. C. W. Reith'. Wakelam's name did not seem to appear in the BBC's self-congratulatory celebrations, but HJBH considered that he bore a heavy responsibility for the monstrous regiment of commentators which he released, although his style was very different. This success led him to take up journalism, so he and HJBH lost contact.

Sir Tasker Watkins VC (1918-2007): Counsel for the Aberfan Tribunal in 1966, he had received his VC for leadership that saved his men and influenced the course of a battle following the D-Day landings in the Second War. Made a judge in 1971, promoted to the Court of Appeal in 1980.

Brigadier Sidney Albert Westrop DSO: Had been chief salesman for Ruston-Bucyrus and then helped to extend the Maginot Line along the Belgian frontier and was involved in the Mowlem wartime work on trenches and pill-boxes as a result. Afterwards he became a Brigadier in India and for a time tried to get Mowlem to look for work out there, in spite of the complications of partition.

Styg Wikstrom: Worked for Widmark and Platzer and had played guitar for HJBH's rendition of 'Widdicombe Fair'.

George Wild (1913-): Had been HJBH's assistant at Mowlem, he accompanied Brigadier Horsfield to Persia to establish their base. Another who was sad to see HJBH's departure from Mowlem, he stayed and later became a civil engineering director.

Norman Williams: Had been a resident engineer on several Welsh steel rolling mill construction sites and was a member of the ICE South Wales Association. Became a friend of HJBH's at the time of the Aberfan Tribunal.

Professor Gilbert Wilson: Was Reader in Geology at Imperial College while working with HJBH in the Malvern Hills. His input to the Ceylon problems of WP was of great assistance, flagging up important information not passed on to the contractors. He later became Professor, renowned for his inspirational lectures.

Dr. Guthlac Wilson (1902-1953): Civil engineer who had studied under Terzaghi at Harvard in 1938, assisting him on roll-fill earth dams. From 1944 he was in private practice, joining William Scott in 1945 to form Scott and Wilson, known as Scott and Wilson, Kirkpatrick and Partners from 1951 following a merger with Sir Cyril Kirkpatrick. Killed in a plane crash near Dar Es Salaam in 1953.

Sir Denis Arthur Hepworth Wright (1911-2005): British diplomat who was at the Foreign Office at the time of the Persia contract, having been Chargé d'Affaires in Tehran in 1953. He was to become British Ambassador to Iran in 1963, where his in-depth knowledge of the country and the culture proved of significant benefit. Flown out by the Foreign Office to Paradise Island in 1979 to break the news to the sheltering Shah of Iran that he was not to be allowed asylum in Britain.

Sir Robert Meredydd Wynne-Edwards (1897-1947): Civil engineer whose early career was spent in Canada, where he first came across Karl Terzaghi's work on soil science. In 1935 he joined Mowlem and he and HJBH became close friends. His first major job was Chingford Reservoir, where after the 1937 slip he was the driving force that introduced Terzaghi to HJBH, Glossop, Golder and Skempton and led to the formation of Soil Mechanics Ltd during the war. Wynne-Edwards served as Director of Plant in the Ministry of Works with much success through the war and then became a Director of Richard Costain Ltd and founded the chemical-cum-civil engineering firm of Costain-John Brown. When this was sold to John Brown he went with it. HJBH's suggestion that they should call themselves John Brown's Bodies was not approved, so, to keep familiar initials of CJB they called it Constructors John Brown. This firm and others like William Press and Son were formed just in time to prevent the chemical engineering industry from falling under USA dominance. Wynne-Edwards went on to become the 100th President of the ICE in 1964, succeeding HJBH and to be elected first Chairman of the newly formed Council of Engineering Institutions. He emerged as Sir Robert Wynne-Edwards, CBE, DSO, MC, DSc, MA, FICE which HJBH suggested indicates his personal contribution in war and peace.

TABLE OF CONVERSIONS

The measurements given throughout the book are in Imperial units as those were the ones in use for HJBH's career. A 'conversion' between Imperial and Metric can be unreliable, but in order to make it easier to understand the measurements, the conversion information is given for lengths:

1 inch	=	2.540 cm
1 ft	=	0.305m
1 mile	=	1.609km

Examples used in the text (rounded to the nearest decimal place):

1.5 in	=	4cm
3 inch	=	7.6cm
2ft	=	0.61m
3ft	=	0.91m
4ft	=	1.22m
6ft	=	1.8m
7ft	=	2.1m
7ft 8in	=	2.3m
8ft	=	2.44m
10ft	=	3m
12ft	=	3.66m
14ft	=	4.27m
22ft 4in	=	6.8m
25ft	=	7.62m
26ft	=	7.9m
18ft	=	5.5m
20ft	=	6.1m
22ft 4in	=	6.8m
25ft	=	7.62m
26ft	=	7.9m
30ft	=	9.14m
50ft	=	15.2m
150ft	=	45.7m
6in x 3in	=	15 x 8cm

SELECTED BIBLIOGRAPHY

There is a collection of HJBH's professional papers in the Special Collections Archives at The University of Exeter. These were given by him to the Department of Engineering in the early 1980s for use in teaching and relocated on the closure of the Harrison Library in 2008.

Bruckshaw, J. M., Crognel, J., Harding, H. J. B., & Malcor, R., (1961) 'The Work of the Channel Tunnel Study Group 1958-1960,' *Proceedings of the Institute of Civil Engineers.* **18,** pp. 149-178.

Fetherston, D. (1997), *The Chunnel. The Amazing Story of the Undersea Crossing of the English Channel.* Times Books, USA.

Golder, H. Q., Harding, H. J. B. and Sefton Jenkins, R. A., (1961) 'An Unusual Case of Underpinning and Strutting for a Deep Excavation Adjacent to Existing Buildings'. *Proceedings of the Fifth International Conference on Soil Mechanics and Foundation Engineering*, Paris.

Harding, H. J. B., (1920) 'The Uniflow Steam Engine.' *City and Guilds Engineering Society*. 20th May.

Harding, H. J. B., (1934) 'Chemical Consolidation of Ground.' Manchester.

Harding, H. J. B., (1935) 'Ground-Water Lowering and Chemical Consolidation of Foundations at Bentalls, Kingston,' *Concrete & Constructional Engineering*.

Harding, H. J. B., (1936) 'Underpinning and Foundation Work in Loose and Water-logged Ground by Chemical Consolidation, Ground Water Lowering and Other Means.' *The Structural Engineer*. June, pp. 289-294.

Harding, H. J. B., (1939) 'Recent Applications of the Ground Water Lowering Process.' *The Engineer*. 28th April.

Harding, H. J. B. and Glossop, R., (1939) 'Modern Processes in the Support of Excavations.' *The Engineer*. 19th May.

Harding, H. J. B., (1939) 'Compressed air. Its use in Engineering and Tunnelling'. *Encyclopaedia of Civil Engineering*. Unpublished.

Harding, H. J. B. and Glossop, R., (1940) 'Chemical Consolidation of Ground in Railway Work'. *The Railway Gazette*. 2nd February.

Harding, H. J. B., (1941) 'The Civil Engineering Contractor and the War'. *The Central*, pp. 3-5.

Harding, H. J. B., (1945) 'Emergency Repairs to the Tower Subway after Air Raid Damage.' *Journal of the Institution of Civil Engineers*. **25** (1), pp. 73-79.

Harding, H. J. B., (1946) 'The Principles and Practice of Ground Water Lowering' presented to three Local Associations 1945-46.

Harding, H. J. B., (1946) 'The Choice of Expedients in Civil Engineering Construction'. *Works Construction Paper* No 6, pp. 3-31.

Harding, H. J. B., & Collingridge, V. H., (1947) 'Underpinning a church at Ealing,' *Concrete & Constructional Engineering*. May.

Harding, H. J. B., (1948) 'Some recorded observations of Ground Water levels related to adjacent Tidal Movements' Second International Conference on Soil Mechanics and Foundation Engineering – Rotterdam.

Harding, H. J. B., (1949) ' Site Investigations including Boring and other Methods of Sub-surface Exploration'. Works Construction Paper 12. *Journal of the Institution of Civil Engineers*. **32** (6), pp. 111-137.

Harding, H. J. B., & Golder, H.Q. (1950) 'The influence of soil mechanics on engineering' *Engineering*, Nov 24.

Harding, H. J. B., & Glossop, R. (1951) 'The influence of modern soil studies on the construction of foundations,' *Building Research Congress*, pp. 146-158.

Harding, H. J. B., (1951) 'The Practical Training of Civil Engineers.' *Joint Engineering Conference*.

Harding, H. J. B., (1951) 'The Civil Engineering Contractor at War.' *The Central*.

SELECTED BIBLIOGRAPHY

Harding, H. J. B., (1952) 'The Progress of the Science of Soil Mechanics in the last Decade.' James Forrest Lecture 1952. *Proceedings of the Institution of Civil Engineers*. **1** (6), pp. 658-681.

Harding, H. J. B., (1957) 'The Resources available for Extensive Highway Construction, and Methods of Handling a Large Programme.' Paper 7 *Conference on the Highway Needs of Great Britain*, Institution of Civil Engineers. 13-15 November.

Harding, H. J. B., (1960) 'Survey of Civil Engineers'. *Times Weekly Review*.

Harding, H. J. B., (1961) 'Channel Crossing Controversies: Review of a variety of schemes submitted,' *The Dock & Harbour Authority*, London.

Harding, H. J. B., (1963) 'Experience and the ICE'. *The Engineer*, 8th November.

Harding, H. J. B., (1964) 'Presidential Address'. *Proceedings of the Institution of Civil Engineers*, **27**, pp. 1-16.

Harding, H. J. B., (1966) 'Research & Progress in Tunnelling Techniques,' *The Financial Times*, 14th November.

Harding, H. J. B., (1968) 'Driving Tunnels - Sir Harold Harding on his work as a civil engineer'. *The Listener,* 19th September. From Scientifically Speaking. Talk given to BBC Radio 4.

Harding, H. J. B., (1971) 'My three favourite characters'. *Construction News*. 2nd September, pp. 52-3.

Harding, H. J. B., (1975) 'Mothballs for the Tunnel'. *The Consulting Engineer*, April p. 41.

Harding, H. J. B., (1978) 'Royal Engineer contacts of a Sapper Manqué'. *The Royal Engineers Journal*. **92** (3), pp. 158-165.

Harding, H. J. B., (1979) 'Tunnelling is still an art'. *Tunnels and Tunnelling*. June, pp. 29-32.

Harding, H. J. B., (1981) *Tunnelling History and my own Involvement*. Golder Associates, Toronto.

Harding, H. J. B., (No Date) 'A Link with the Continent?' Problems for the Consulting Engineer," Courtesy Esso Magazine pp. 28-9.

Hunt, D. (1994), *The Tunnel. The story of the Channel Tunnel 1802-1994*. Images Publishing, Upton-upon-Severn.

Malcor, R. & Harding, H. J. B., (1961) 'Technique de Recherches pour L'Étude du Tunnel sous la Manche', *Supplement aux Annales de L'Institut Technique du Bâtiment et des Travaux Publics*. **158,** pp. 240-257.

Malcor, R. & Harding, H. J. B., (1966) The Channel Tunnel Grouting & Injections in chalk. Report. March.

Watson, J. G., (1989) *The Smeatonians: The Society of Civil Engineers*. Thomas Telford, London.

Williams, R. E., (2011) *Rudolph Glossop and the Rise of Geotechnology*. Whittles, Dunbeath.

Speeches Addresses etc.

1953	Opening to Symposium on Engineering Education, City and Guilds College, March. 'How can the future needs of the country for Engineering and Scientific Technologies be met'.
1954	Old Centralians Dinner 'The Guests'
1956	Address to the Reunion of F.B.I. Overseas Scholarships holders
1960	Northampton College of Advance Technology Inaugural Lecture. 'The Civil Engineering Industry'.
1961	Northampton College of Advanced Technology, Course 1
1961	Northampton College of Advanced Technology Inaugural Lecture 2
1963	Presidential Address to the ICE
1963	Careers Masters' Conference 'What the Civil Engineer Does'
1964	President's speech Annual Dinner ICE, 28th April
1965	Societé des Ingénieurs Civils de France, Progrès Apportés a la Construction Métro de Londres depuis cent ans, 25th February
1966	Financial Times - Tunnels

1967	Chairman's Address Commonwealth Conference. Survey Officers. Cambridge
1968	Cambridge University Engineering Society Lecture. 'The Engineering Underworld: Practical Side of Civil Engineering'
1968	Imperial College of Science and Technology Commemoration Day
1969	Careers Masters Conference. 'The Overall Picture of Civil Engineering including Training'
1970	Concrete Society Lunch Address
1971	BBC Interview – Scientifically Speaking
1971	Dinner address to Sir William and Lady Glanville
1973	Chartered Engineers Lecture, Portsmouth. 'Carry on Civil Engineering'
1973	Introduction to International Symposium on the Aerodynamics and Ventilation of Vehicle Tunnels, University of Kent, Canterbury

INDEX

A

Aberfan Investigations 194, 247, 249, 250
Aberfan Tribunal iii, 136, 185, 188, 192, 193, 194, 195, 196, 197, 198, 200, 201, 202, 204, 210, 236, 237, 243, 244, 245, 247, 248, 251, 252
Admiralty 104, 105, 154
Air-raid shelter 76
Alpine Geophysical Services 142
Amin, Idi 117, 118, 119, 249
Anderson, O'Dean 186, 242
Arbitration 120, 136, 148, 150, 182, 188, 189, 191, 195, 204, 206, 211, 212, 215, 220, 236, 248, 250
Arbitrator 120, 124, 185, 189, 190, 191, 204, 208, 209, 210, 221, 248, 250
Armand, Louis 137, 147, 155, 242
Avery, Peter 124, 126, 127, 130, 131, 242

B

Baines, Sir Frank 41, 242
Baker, Prof. A. L. 180, 242
Baker, Sir Benjamin 38, 242
Balfour, John and Partners 113
Ball, George 149, 242
Bank of England 39
Barking Station 51
Barlow, Peter W. 13
Barnes, Robert Ogle 16, 119, 242
Bartlett, John Vernon 162, 230, 242
Basutoland (Lesotho) 2, 122, 123, 211, 215
Bath Oliver biscuit tin 1
Batignolles 91, 92
Battle of Britain 79
Bazalgette, Sir Joseph 19, 79
Bechtel Corporation 141, 144, 161, 187, 242
Bechtel, Steve 146, 187, 242
Beck, Edgar B. 33, 43, 100, 109, 131, 132, 243
Beck, R. I. 132
Beck, Sir Edgar Charles 117, 126, 131, 132, 243
Beeching, Dr Richard 137, 151, 242, 243
Belin, M. 213, 214, 218, 219, 220
Bennett, Curly 8, 10, 12, 99, 243
Bentalls Department Store 64
Bertlin, Dennis 118, 243
Bickel, J. O. 146, 243
Biggart, Alastair 232
Binney and Partners 142
Birkett, Sub-Lieutenant 83, 85, 90, 104
Bishop, Prof. A. W. 192, 193, 194, 195, 199, 200, 243, 250
Blair Leighton, Edmund 29, 202
Bleriot Plage 153
Blondin cable way 213
Body, Babs 22, 215

Body, John 16, 22, 43, 169, 173, 176, 177, 215, 243
Bombay Electric Supply and Transport Undertaking 135, 249
Bombay Sappers and Miners 100
Bomb Disposal Squad ii, 82
Bonny, Jack 168, 169, 185, 187, 243
Bored cores 158
Bored piles 78, 210
Bored wells 65, 181
Boreholes 44, 108, 139, 142, 154, 156, 157, 158, 184, 199
Boris (French contractors) 212, 218
Bosanquet, Reginald 147, 243
Bowler hat 7, 86
Boxer, Nigel 168, 178, 243
Boyer, Rev. James 1
Braehead Power Station 96, 208
Brand, Charles 99, 249
Brass, John 112, 243, 246, 248
Bridges 5, 54, 69, 76, 81, 91, 101, 118
    Bailey Bridge 76
    Callender-Hamilton 76, 81
British Association 110
British Channel Tunnel Company 137, 141, 147, 161
British Rail 150, 151, 161, 162
British Tunnelling Society i, iii, 161, 162, 228, 230, 231, 233, 240, 242, 248, 249
    Harding Lecture 228, 229, 232, 240, 249
    Harding Prize 231, 232, 240, 246
Brooke, Sir Alan 78
Brown and Root 141, 144, 187
Bruckshaw, Prof. J. M. 138, 140, 148, 156, 243
Brunel Exhibition 232, 233
Brunel, Henry 139
Brunel, Isambard Kingdom 13, 34, 139, 233
Brunel Museum 14
Brunel, Sir Marc Isambard 13, 139, 186, 233
Buckley, John 132, 133, 243
Building Research Station i, 58, 71, 72, 83, 89, 93, 95, 98, 194, 244, 246, 249, 250

Bunbury, Cheshire 2
Burt, Edward George Mowlem (Tommy) 33, 112, 243, 244
Burt, Edwin 9, 17, 33, 244
Burt, Eric 9, 28, 42, 107, 112, 243, 244, 248, 251
Burt, Kenneth 33, 112, 132, 244
Burt, Sir George Mowlem 9, 31, 33, 36, 42, 44, 48, 57, 71, 83, 89, 93, 100, 109, 129, 131, 132, 243, 244
Burt, Stewart 112, 244

C

Cambridge
    King's College Chapel 108

Canada
  Vancouver 70, 71, 165, 166, 167, 168, 185, 186, 248, 249
Cardiff 194, 195, 196, 197, 200, 201, 202, 203
Cardiff Corporation 197
Carter, Dr 141, 244
Castle, Barbara 161, 202
Cavendish, Brigadier 131, 244, 247
Ceremony of the Wires 12
Ceylon (Sri Lanka) 182, 183, 184, 249, 250, 252
  Castlereagh Dam 182, 184
  Colombo 182, 184
  Norton Dam 182, 183
Chainmen 8, 12, 26, 28, 90, 99
Channel Crossing
  Bridge option v, 138, 141, 149, 151, 180, 187, 235, 237, 238, 239, 242, 249, 250
  Euroroute option 239
  Immersed tunnel option v, 138, 140, 141, 143, 146, 147, 149, 151, 157, 187, 237, 242, 243, 245
Channel Tunnel iii, v, 148, 158, 160, 162, 164, 170, 172, 235, 237, 238, 239, 240, 242, 248, 249
Channel Tunnel Study Group iii, 137, 138, 140, 141, 147, 148, 149, 150, 151, 152, 153, 154, 155, 159, 160, 161, 162, 172, 180, 185, 186, 187, 191, 193, 194, 205, 207, 214, 231, 238, 240, 242, 243, 244, 245, 246, 247, 248, 249
  Feasibility study 138, 141, 146, 239, 248
Charles Brand and Son Ltd i, 168, 243
Chemical consolidation 57, 87
  Joosten process 57, 69
Chemical injection 23
Chesapeake Bay Bridge-Tunnel 172, 177, 178, 179, 180, 245, 250
Chief Leratholi 123
Chingford Reservoir slip failure i, 70, 71, 72, 89, 125, 134, 245, 246, 250, 251, 252
Christ's Hospital, Horsham 1, 2, 9, 131, 244
  Governor 2
City and Guilds College, Imperial College 2, 3, 4, 7, 91, 93, 101, 110, 112, 227, 229, 243
  Old Centralians 110
City and Guilds Institute 226, 246
  Associate 5
  Fellow 101, 113, 114
City University 109, 226
Civil Constabulary Reserve 27, 28
Civil Engineering Price Book 233
Clay-pocketing 9, 44
Clayton, Captain 104, 106, 244
Codes of Practice 96, 225
Collingridge, Vincent 168, 244
Commission of Surveillance 69, 83, 136, 137, 142, 150, 151, 152, 153, 154, 155, 156, 157, 158, 159, 160, 209, 248, 250
Compagnie Generale d'Equipment pour les Travaux Maritimes 156
Compressed air 9, 13, 14, 44, 50, 52, 53, 54, 58, 68, 73, 76, 96, 140, 144, 163, 165, 166, 178, 185, 186, 208, 233, 246
Conference on Large Dams 96
Congress des Grandes Barrages 204
Constructors John Brown 128, 252
Contracts 204, 212, 216
  Hudson Contract 182
  Standard ICE Form of Contract 189
Cooling, Leonard Frank i, 71, 72, 244, 246
Cooper, Sir Edwin 38, 40, 244
Coronation of Queen Elizabeth II 40, 105, 113, 114, 115, 116
Corporation of London 38
Costain 95, 121, 143, 252
Coyne et Bellier 204, 205, 211, 220
Crowborough, Sussex 78, 79
Croydon Airport 91

D

Daftarian, Mr 129
Dalrymple-Hay, Sir Harley H. 1, 6, 9, 14, 16, 23, 24, 26, 27, 30, 33, 35, 36, 39, 69, 242, 244, 247
Dam 29, 72, 84, 95, 98, 118, 122, 130, 145, 163, 176, 182, 183, 204, 205, 206, 211, 212, 213, 214, 215, 216, 220, 221, 222, 223, 245, 250, 251
Dam board 79
Davidson brothers 147, 248
  Davidson, Alfred Edward 245
  Davidson, Frank 137, 146, 172, 245
Davies, Lord Justice Edmund 185, 193, 196, 197, 200, 201, 244
Decca Navigation System 105, 139, 153
Deep well dewatering i, 63, 91, 185, 187
DeLong, Colonel 143, 146, 245
DeLong Corporation 144, 156, 179
Department of Transport 237
d'Erlanger, Leo 137, 147, 151, 153, 205
Dictionary of National Biography 240, 248
Dix, Arnold 232
Dorman Long and Co. 40, 41, 50, 52, 53, 118, 149, 249
Douglas, Kingman 146, 245
Douglas, Lew 216, 245
Dover Harbour Board 239
Dover, Kent 139, 141, 142, 153, 155, 157, 158, 159, 161
  Dover Castle 152, 153
  Shakespeare Cliff 140, 161
Druitt, Sir Harvey 194, 245
Dumez (French contractors) 212, 213, 218, 220, 222
Dungeness, Kent 105, 141, 153, 159, 207
Dunn, Charles Putnam 144, 145, 169, 245

E

Ebtehaj, Abolhassan 124, 126, 129, 131, 245
ECSA 171, 173, 249
Eddie, Colin 232

INDEX

Edmund Nuttalls Ltd 161, 228, 229
Egypt 104, 138, 249
    Suez Canal 104, 138
Encyclopedia Britannica 1911 iv, 117, 164, 175, 176
Engineer and Railway Staff Corps 111, 193
Engineering Wonders of the World 1, 79, 145, 244
Esso 105
Evans, Dr Haydn 194, 245
Evening News 24, 67, 201
Experience, the value of i, ii, 1, 22, 37, 38, 42, 48, 50, 51, 56, 57, 58, 73, 86, 89, 90, 93, 95, 99, 101, 104, 108, 109, 110, 111, 112, 124, 128, 134, 137, 160, 165, 185, 186, 189, 190, 193, 195, 199, 203, 204, 206, 208, 210, 220, 227, 229, 232, 234, 238, 240, 247

F

Faber, Oscar 5
Farmer, Major T. W. G. 195, 245
Federation of Civil Engineering Contractors 108, 109, 111, 243
Fellenius, Prof. 94, 245
Folkestone, Kent 140, 146
Ford Power House, Dagenham i, 46, 47, 48, 49, 50, 51, 52, 55, 56, 57, 68, 96, 225, 226, 247, 249
Forzky 154
Fougerolle 141, 214
Fouillade 219, 220
Foundation Engineering 95, 249
Foundations i, 9, 10, 34, 37, 39, 40, 43, 47, 51, 54, 64, 65, 72, 109, 149, 225, 247
Four decades 57
Fox, Sir Francis 18, 245
France
    Calais 112, 152, 153, 155, 157, 160, 243
    Dieppe 86, 92
    Dunkirk 77, 91
    Freyssinet docks 91
    Normandy 87
    Paris 71, 91, 92, 138, 140, 141, 142, 147, 149, 150, 151, 155, 156, 158, 160, 163, 177, 186, 191, 218, 222, 251
    Sangatte 140, 141
Francois Cementation Company 23, 48, 121
Frederickson 184
Freeman, Fox and Partners 194
Freeman, Sir Ralph 5, 245
Friese-Greene, G. 80, 245

G

General Strike 27, 28, 251
Geology
    Blackwall Rock 53
    Boyn Hill Terrace 23
    Chalk 139, 142, 158
    Clay 26, 27, 46, 48, 49, 50, 53, 71, 73, 82, 95, 125, 142, 166, 173, 183, 200, 202, 245
    Deccan trap 135, 136
    Glacial Clay 113, 166, 200
    Gravel 15, 18, 20, 23, 27, 28, 29, 34, 41, 44, 48, 49, 50, 53, 55, 65, 71, 75, 84, 94, 96, 113, 127, 166, 245
    Kaolin 183
    Laterite 136
    Limestone 142, 160, 164, 171, 181
    London Clay 8, 12, 13, 23, 39, 41, 44, 71, 83, 84
    Sussex Clay 78
    Taplow Terrace 23, 26, 28
    Woolwich and Reading Beds 13
        Shepherd's Plaid 13
Geophysics 139, 141, 153, 239
George Stephenson Gold Medal 149
Geotechnical processes i, 58, 89, 232
Geotechnical Society 95, 121, 142
Gericke 218, 219, 220, 221, 222
Germany 57, 58, 59, 68, 90
    Berlin 59, 62
    Nazis 61, 63, 250
Glasgow 42, 96, 101, 134, 168, 250
Glossop, Rudolph i, 33, 70, 71, 72, 74, 75, 78, 86, 87, 89, 90, 93, 96, 100, 107, 109, 110, 117, 124, 126, 129, 131, 245, 246, 252
Goguel, Jean 138, 148, 246
Golder, Hugh i, 71, 72, 87, 89, 90, 93, 107, 110, 121, 166, 167, 177, 236, 241, 245, 246, 252
Gove, W. G. 124, 131, 132, 246
Graham-Clarke, E. 224, 225
Grange, A. 141, 152, 155, 157, 246
Greater London Council 19
Greathead, James Henry i, 8, 13, 82, 233, 246
Grigor-Taylor, W. R. vii, 5, 70, 209, 245
Grose, Bill 232
Ground consolidation 58
Ground Nuts 99, 100, 117, 124, 247
Ground water 23, 63, 65, 87, 91, 92, 185

H

Hammett, Derek 113, 243, 246
Handley Page, Sir Frederick 113, 246
Harbour 43, 85, 94, 100, 103, 135, 136, 140, 154, 176, 180, 182, 208, 209, 210
Harcourt, Lord 137, 160, 248
Harding, Arthur Boyer 1
Harding, Diana 168, 226, 238
Harding, Edmund 3, i, 45, 73, 76, 86, 88, 94, 99, 102, 116, 161, 168, 206, 207, 226, 230, 238, 243, 244, 248, 249
Harding, Edward 2, 182, 202, 240, 244, 245, 248, 250
Harding, Gilbert 96, 246
Harding, Helen Clinton (née Lowe) 2
Harding, John Rudge (Uncle Jack) 5, 6
Harding, Lady Sophie (neé Blair Leighton) 3, i, 22, 27, 29, 30, 31, 32, 45, 48, 51, 54, 55, 59, 62, 76, 84, 88, 104, 114, 115, 116, 125, 126, 132, 133, 135, 147, 165, 169, 170, 171, 172, 174, 178, 197, 202, 204, 205, 206, 208, 209,

211, 212, 213, 214, 215, 216, 218, 219, 220, 221, 226, 236, 237, 250
Harding, Lelgarde 206, 221
Harding, Robert 3, 73, 76, 84, 104, 114, 115, 116, 119, 206, 207, 221, 242, 244
    Footmen 114, 115, 116
Harding, Robert (goldsmith) 2
Harding, Thomas 1
Harris, William 'Bumper' 15, 18, 246
Harry Sykes and Co. 75
Harvard Club 146, 172
Harvard University 72, 99, 137, 172, 250, 251, 252
Hausser, P. C. G. 64, 247
Hawkshaw, Sir John 139
Head, Raymond 31, 99, 100, 124, 125, 246
Hendrik Verwoerd Dam (Gariep Dam) 211, 212, 214, 245, 250
    Fish River Tunnel 211
Herring, Mr 216, 217, 218
Hindhead, Surrey 76, 77, 177
Hitler, Adolf 58, 59, 62, 63, 79, 108
Holbein, Arthur Montagu 'Bean' vii, 85, 110, 113, 114, 115, 116, 246, 247
Horsfield, Brigadier Eric 100, 117, 124, 126, 127, 128, 129, 130, 131, 244, 247
Houghton, Lord 248
House of Commons 40, 69, 109, 149, 158, 201, 202, 248
Houses of Parliament 101, 192
Howe, Sir Geoffrey 197, 247
H. R. H. the Duke of Edinburgh 111, 113, 114, 177, 185, 226, 232, 245
Hughes, Cledwyn 193
Hunt, Donald 148, 157, 158, 247
Huntingdon Hartford 180, 181
Huntings 193, 194, 195
Hunt, John Leigh 26, 29, 33, 247
Hunt, Martita 115, 116, 246, 247
Hurlingham Club 171
Hurst, B. L. 65
Hutchinson, Dr 194, 247
Hutter, Roger 151, 247
Hydro Electric 95, 121, 159, 163, 169, 182, 183, 211, 212

I

ICI (Imperial Chemical Industries) 40, 41, 43, 93, 242
ICOS system 187
Imperial College 3, 5, 62, 72, 93, 110, 138, 141, 183, 184, 194, 226, 229, 243, 244, 250
    Diploma 5
    Fellow 207, 227
    Governor 110, 132, 193, 226
    Imperial Institute 3, 226
    Royal School of Mines 66, 71, 93, 110, 245
India 26, 100, 124, 127, 136, 208, 209, 210, 211, 251
    Bombay (Mumbai) 131, 135, 136, 207, 208, 249, 250
    Delhi 208, 209, 210, 211, 214, 220
    Great Indian Peninsular Railway 136
    Mumbai (Bombay)
        Back Bay 135, 136
Institute of Geological Sciences 194
Institution of Civil Engineers (ICE) 13, 24, 30, 31, 40, 52, 62, 69, 71, 72, 80, 83, 92, 101, 110, 112, 118, 121, 124, 134, 138, 149, 152, 165, 188, 189, 193, 202, 210, 218, 220, 224, 225, 226, 228, 230, 231, 233, 238, 242, 247, 248, 249, 250, 251, 252
    Council (HJBH) 101, 111, 132
    Great George Street 228, 233
    James Forrest Lecture 72, 110, 229
    Parliamentary and Scientific Committee 148
    Past President (HJBH) 193, 207, 218, 232
    President (Gordon Masterton) 231
    President (HJBH) 57, 148, 152, 172, 177, 183, 195, 202, 225
    President (Muir Wood) 153
    President (Sir Charles Inglis) 226
    President (Sir Cyril Kirkpatrick) 51
    President (Vernon Robertson) 134
    President (Wynne-Edwards) 178
    Vice President (HJBH) 148, 149, 202
    Works Construction Division 95
International Tunnelling Association (ITA) 230, 248
Iran/Persia 109, 124, 126, 127, 128, 129, 131, 132, 135, 191, 242, 243, 244, 245, 246, 247, 248, 249, 251, 252
    Abadan 126, 127, 128
    Karun River Bridge 127
    Khorramabad 128
    Seven Year Plan 124, 126, 129, 130, 132, 245, 249
    Shush 127
    Shushan 128
    Tehran 124, 127, 128, 129, 130, 131, 182, 242, 252
Iraq Petroleum Co. 103, 104
Ireland
    Dublin 67, 68, 219, 231
Ischy, Ernest 141, 247
Isle of Sheppey, Kent 105
Israel 104, 163

J

Jack London (author) 143
Jacobs, J. Donovan 163, 165, 247
Jacquet, Marc 152, 153, 247
James, R. T. vii, 32, 63, 64, 247
Japp, Sir Henry 43, 44, 47, 50, 51, 69, 70, 71, 89, 99, 100, 225, 246, 247, 250
Jones, Isaac James 33, 35, 38, 40, 41, 42, 43, 45, 47, 50, 51, 73, 247
Jones, Mervyn 202, 203, 247
Jones, Sir Elwyn 194, 195, 196, 247

K

Kell, Jack 74, 163, 165, 247
Kent 76, 208, 219
Kenya 118, 119, 121, 211, 215, 242
    Nairobi 100, 117, 118, 119, 120, 121, 218, 219, 220
    Sasamua Dam 120, 215
King Farouk 117
King George V 4, 35
King, John 229, 230, 232
King of Buganda 117
Kinnear Moodie and Co. 168, 178, 208, 243
Kirk and Randall 43
Kirkpatrick, Sir Cyril 47, 48, 49, 51, 52, 68, 247, 252
Kirkpatrick, Sir Ivone 137, 147, 248
Knight Bachelor 205, 206

L

Lambert, Noel 165, 166, 167, 168, 169, 186, 187, 248
Lamont, Donald 232
Lawrence, Vernon 193, 194, 195, 197, 201
Lea, Dr 98, 99
Lebanon 163, 164, 247
    Beirut 131, 163, 211
    Litani Project 163, 165, 186, 247, 248
Lesotho 2, 122, 123, 211
Lindrodt, Herr 180, 181, 182, 248
Liss, Hampshire 76, 77, 79, 81, 84
Livesey and Henderson 141
Lloyds of London 154
Lloyd Thomas, Roger 185, 193, 194, 195, 197, 202, 237, 248
London 1, 2, 3, 7, 8, 9, 13, 19, 31, 37, 40, 41, 45, 46, 47, 59, 68, 69, 73, 76, 77, 79, 81, 83, 84, 85, 88, 91, 93, 101, 110, 115, 120, 121, 123, 124, 125, 126, 129, 130, 131, 136, 141, 147, 149, 156, 159, 165, 167, 168, 171, 177, 185, 187, 198, 200, 208, 209, 210, 211, 215, 220, 221, 225, 226, 231, 232, 233, 238, 242, 244, 250, 251
    Barking outfall 80
    Battersea cable tunnel 44
    Battersea Power Station 43, 44, 45, 51, 82, 103, 247
    Beckton outfall 19, 80
    Bridges
        Lambeth Bridge 40, 50, 116
        Lambeth Suspension Bridge 40
        London Bridge 44, 46, 69
        Waterloo Bridge 69
    Buildings
        Britannic House i
        Commercial Union Building 38, 244
        Cornhill 38, 244
        Lloyds 9, 38, 40, 109, 244
        Mansion House 38, 39
        National Provincial Bank 38, 40, 244
        Port of London HQ 40
        Thames House, Millbank 40, 41, 42, 242
        Tower of London 80
        Bush House 37
    Coventry Street 28
    Cromwell Road 45, 59, 81
    Dagenham i, 45, 46, 47, 48, 51, 54, 55, 58, 59, 70, 96, 103, 113, 250
    Dartford 163
    Docks
        East India Dock Road 20
        King George V Dock 41
        Port of London 50
        Surrey Dock 85, 86, 87
        West India Docks 20
    Ealing Common 7, 16
    Eros statue 24, 25, 34
    Euston i, 6, 7, 10, 37, 93, 225, 250
    Geological Museum 135
    Greenwich 17, 37
    High Holborn 37, 38
    Holborn
    Gas explosion 37
    Isle of Dogs 18, 19, 20, 21, 22, 34, 42, 109
    London Main Drainage 19
    London Transport Museum i
    Main sewers 19, 79
    Northern Outfall Sewer 79, 80, 81, 245
    Oxford Circus 10, 156
    Parks
        Hyde Park 28, 59, 84
        St James's Park
        Duck Island 84
    Pumping Station
        Abbey Mills 19, 79, 81
    Rivers
        Lea 73, 79
        Walbrook 39
    Rotherhithe 232, 233
    Rotherhithe Tunnel 186
    Tottenham Court Road 10, 37, 38
    Trafalgar Square 11, 79, 83
    Underground
        Bakerloo Line 15, 23, 24, 26, 244
        Bow to Leyton 73, 75, 76, 90, 163, 187, 189, 247
        Camberwell Tube Extension 107
        Central Line 10, 73
        Charing Cross Station 23, 24, 25, 48, 57, 83
        City and South London Line
        Malden 27
        City and South London Railway 6, 7, 9, 13, 14, 23, 27, 33, 74, 244, 246
        Camden Town 6, 7, 10, 11, 14, 15, 16, 22, 27, 29, 99, 242, 250
        District Line 19, 23, 40, 45
        Elephant and Castle 107, 116
        Hampstead and Highgate Line 14

Hampstead line 23
Hampstead Line 6, 7, 8
Inner Circle Line 135
Piccadilly Circus Underground Station i, 23, 24, 25, 26, 29, 30, 32, 36, 156, 224, 225, 226, 237, 239, 246
Piccadilly Line 30
South Kensington 11
Tower Subway 13, 80, 82
Victoria Line 156, 178
Waterloo and City 9, 13, 15, 16, 52, 69, 116, 162, 246
London and North Western Railway 7
London and South West Railway Co. 27
London and Tilbury Railway 46, 80
London County Council 9, 19, 20, 22, 24, 28, 35, 40, 46, 69, 76, 80, 81, 88, 101, 225
London Electric Railway Co. 7, 15, 17, 24, 32, 33
London Gazette 205
London Main Drainage 79
London Matriculation and Intermediate 5
London, Midland and Scottish Railway 93
London Passenger Transport Board 73
London Power Company 43
London Rifle Brigade 80, 83, 245
London Transport 108, 113, 135, 156
London Transport Museum 30, 237
London University 3, 5, 19
London University Officers Training Corps (LUOTC) 3
Long litter 21
Lowe, Rev. William 2
Luttrell Avenue, Putney 62, 81, 88, 90, 115, 132, 220, 221

M

Machado, José 146, 172, 248
MacKenzie, Colin 229, 232
MacLean, Mansell 163, 186, 187, 248
Macnamara 213, 214, 215, 219, 220
Mair, Prof. Robert 232
Malcor, René iii, 105, 137, 138, 141, 142, 147, 148, 149, 152, 154, 155, 156, 157, 159, 160, 161, 162, 194, 214, 248
Marchon Chemicals 108
Marples, Ernest 151, 152
Marquess of Bath 114, 115, 116
Massigli, René 137, 147, 149, 248
Mathematics ii, 3
Mathieu, Jean 152, 155, 248
Mau Mau 117, 119, 120, 121
McGibbon 112, 243, 248
McMullen, Colonel Dennis 151, 152, 155, 158, 209, 248
McNaughton 155
Means, Prof. 146, 248
Merthyr Tydfil 185, 192, 193, 194, 195, 197, 201
Meteorological Office 155
Metropolitan Railway 37

Metropolitan Water Board 70, 71
Mexican Eagle Oil Co. 176, 177
Mexico 22, 71, 169, 171, 172, 173, 174, 175, 176, 177, 178, 189, 243, 249, 250
Teotihuacan 173
Micro-fossils 141, 157
Middlesex Regiment 3
Ministry of Agriculture, Fisheries and Food 154
Ministry of Transport 76, 154, 159, 248
Mitchell, Sir Godfrey 95, 112
Moir, Sir Ernest 43, 247
Morgan, Major W. H. 124, 125, 189, 248
Morning Post 25, 32, 68
Mornington Crescent Station 7, 8, 15, 17, 246
Morrison-Knudsen 130, 141, 144, 145, 167, 168, 169, 185, 186, 187, 242, 243, 245, 248
Mott, Hay and Anderson 33, 38, 74, 161, 162, 163, 178, 189, 207, 222, 242
Mowlem i, iii, 3, 6, 7, 9, 15, 28, 33, 35, 38, 40, 43, 46, 49, 52, 65, 69, 70, 71, 73, 76, 84, 90, 91, 92, 94, 96, 99, 100, 104, 106, 107, 108, 109, 112, 114, 117, 118, 124, 128, 130, 131, 134, 135, 154, 162, 168, 233, 242, 243, 244, 245, 246, 247, 248, 250, 251
Basin South 41, 42, 43, 47, 48, 93, 104, 105, 109, 233
Depots
Marshgate Lane 70, 76, 77, 85, 89
London John 233
Mowlem Construction Company 117, 119, 135
Offices
Hampstead Road 7, 10, 25
Litchfield Street 25, 31, 74
Muir Wood, Sir Alan 152, 153, 155, 157, 162, 228, 230, 231, 232, 239, 240, 242, 246, 248
Mulberry Harbour 49, 84, 85, 86, 246
Murdoch, Dr 95, 142
Murrow, Ed 147, 248

N

Nassau, Bahamas 172, 180, 181
National Coal Board 121, 155, 197, 198, 199, 200, 201
National Physical Laboratory 82
National Tip Safety Committee 202
'Nelson Touch' 50, 74, 77, 247
New Civil Engineer 228, 232, 233, 237, 238, 239, 240
Newhaven, Sussex 77, 92
Newport, Roger 124, 249
Northampton Engineering College 109, 110, 225, 226
Governor 109, 226
Northern Construction Co. 165, 168, 185, 186, 248
North Sea 23, 46, 154, 156, 157, 160

O

Oban 108

INDEX

Obote, President Apollo Milton 118, 211, 249
Oboussier, Caroline (née Harding) 3, iv, 1, 31, 35, 62, 76, 88, 90, 99, 101, 102, 116, 159, 176, 187, 205, 214, 235, 236, 240
Oboussier, Philippe 172
OECD 228
Office of Works 3, 9
Oil 80, 103, 112, 128, 139, 143, 154, 155, 156, 158, 176, 187, 191, 245
Ordnance Survey 11, 142, 195, 245

P

Pain, Jack vii, 118, 149, 249
Paradise Island, Bahamas 172, 180, 181, 249
    Bridge 180, 182
    Cloister ruin 181
    Immersed tunnel option 181
Parkes, Douglas i, v, 162, 228, 230, 231, 232, 243, 249
Parliamentary and Scientific Committee 148, 149
    Vice President 149, 172
Parsons Brinkerhoff 141, 146, 180, 187, 243
Patent Office Library 5
'Paternoster' lifts 60
Pathé News 153, 157, 159
Paton, Angus 222
Paulings 91
Pearson, Weetman 1st Viscount Cowdray 43, 44, 51, 173, 176, 177, 189, 243
Peck, Professor Ralph B. 167, 249
Penman, A. D. M. 194, 249
Pescador, Felipe 171, 173, 174, 175, 249
Phoenix Houses 84, 87, 246
Piccadilly Circus model i, 29, 30, 31, 66, 114, 120, 130, 151, 155, 158, 224, 237, 239, 247
Pick, Frank 30
Picot, Georges 137, 138, 147, 249
Pippard, Prof. 110, 229, 249, 250
Platforms 141, 143, 153, 154, 155, 156, 157, 158, 160, 245
    Gem III 156, 157, 158
    Neptune 156, 157, 158
    Sea Gem 156
Platzer, Sven 94, 95, 97, 169, 170, 171, 172, 173, 182, 183, 197, 249
Ponts et Chaussées 137, 152, 161
Port of London Authority 9, 42, 44, 93, 103, 109, 244
Portsmouth, Hampshire 85, 104, 244
Portugal
    Lisbon Suspension Bridge 169
Posford and Pavry 155
Powell, Commander Christopher 148, 149, 249
Powell Duffryn 197, 198
Powell, Enoch 151
Practical training i, ii, 10, 110
Prentice, Dodie (née Machado) 146, 172
Press Conference 147, 148, 158, 159
Priestley, Gravesend 161, 168
Prime Minister v, 205, 237
Prix Croisseau 147
Professor Mitchell 5
Prudhomme, M. 129, 249
Public speaking 3, 69, 70, 247

Q

Queen Elizabeth II 120, 206, 207, 226
Queen's College, Cambridge 3, 207, 210
Queen Victoria 137

R

Rail tunnel 138, 151, 238, 239
Rao, Mr 135, 136, 249
Rattee and Kett 108
Raymond Company 65
Raymond International 124, 178
Reckless economy 73, 113
Red Cross 5, 84
Reid, A. E. (Sandy) 50, 53, 54, 249, 250
Rendel Palmer and Tritton 141
Reservoir i, 70, 71, 89, 211
Retaining wall 38, 39, 41, 42
Ribeyre, Charles 138
Rio Tinto Zinc 161
Risk i, iii, 37, 39, 174, 217, 232, 249
River Beam, Dagenham 46, 49
Roads Research Laboratory 72, 76, 93
Road tunnel 138, 141, 148, 149, 161, 171, 197, 238
Robb 222, 223
Robertson, Lord 151, 250
Robertson, Uncle Jack 2, 122
Robertson, Vernon 134, 250
Rochester Bridge, Kent 76
Rowell, Willie i, 7, 9, 14, 15, 25, 27, 31, 33, 99, 246, 247, 250
Royal Engineers 3, 90, 111, 112, 116, 127, 194, 240, 243, 244
Royal Geographical Society 147
Royal Geological Society 184
Royal Institute of British Architects 63, 64, 247
Royal Society 160, 185, 197
Ruston-Bucyrus 100, 251

S

Schnitter, Prof. 219, 221
Science Museum 14, 30, 233
Seaford, Sussex 77, 78
Second World War
    D-Day 85, 89, 246, 251
    Maginot Line 100, 251
    Pill boxes 77, 78
    Rabbit 84
    Second World War bomb ii, 79, 80, 81, 82, 83, 85, 87, 88, 90
Segrave, James 7, 10, 16, 17, 25, 31, 33, 250
Self-priming pump 21, 22, 58, 65

Setting out i, 10, 11, 28, 29, 36, 40, 70, 195
Sewer 18, 20, 21, 28, 29, 39, 67, 75, 82, 166, 173, 208, 245
Sheet piles
    Larssen Piles 41, 48, 68
    Steel 41, 50, 96
Shell 29, 99, 105
Shields, Wentworth 57
Shield tunnels 1, 12, 13, 233, 244
Ships
    Caulville 154, 155, 156, 158
    Ebury 40, 100, 104, 105, 106, 244
    Grosvenor 104, 105, 154, 155, 158
    GW14 155, 156, 158
    King John II 139
    Tank landing craft 103, 104, 105, 141, 153, 156
Siemens Bauunion 57, 58, 59, 60, 62, 63, 71, 72, 90, 91
Singleton, 'Tim' 182, 184, 185, 250
Sir Alexander Gibb and Partners 96, 118, 130, 135, 208, 210, 211, 214, 220
Sir Cyril Kirkpatrick and Partners 47, 247
Sir Robert McAlpine and Sons 42
Sir William Halcrow and Partners 108, 121, 135, 136, 141, 153, 161, 248
Site investigation 72, 85, 89, 90, 136, 138, 142, 152, 175, 194, 226, 239
Skånska Cementgjuteriet 182, 184, 185, 250
Skempton, Alec Westley i, 71, 72, 193, 194, 240, 243, 245, 250, 252
Slip failure i, 71, 72, 94, 134, 192, 200, 247, 252
Smeatonian Society 111, 168, 185, 242, 245, 247, 250
    President 111
SNCF 137, 149, 150, 151, 152, 242, 247
Societé des Ingenieurs Civils de France 147, 149, 156
Societé du Tunnel sous la Manche 242
Soft ground 53, 93, 108, 165, 170, 233, 248
Soil mechanics i, 48, 58, 71, 72, 83, 85, 87, 89, 90, 93, 94, 96, 99, 100, 103, 107, 109, 110, 116, 129, 134, 141, 167, 171, 173, 194, 199, 210, 236, 243, 244, 246, 249, 250, 252
Soil Mechanics Ltd i, 72, 85, 87, 89, 93, 96, 100, 103, 107, 109, 134, 244, 246, 252
Soletanche 141, 154, 186, 214, 247
Solum 171, 250
Sonar 140, 142, 243
South Africa 2, 71, 123, 124, 211, 213, 215, 216, 220, 221, 222, 245, 250
    Bloemfontein 212, 216, 219, 220, 223
    Jammersburg Drift 2, 122, 123, 215
    Johannesburg 121, 122, 211, 216, 218, 219, 220
    'Mac Mac' 217
    Mac Mac Bridge 217
    Mac Mac Falls 216
    Mac Mac River 216, 217, 218
    Orange River Project 211
    Wapener 2, 123

Southern Rhodesia (Zimbabwe) 122
    Salisbury 121, 122
Southsea, Hampshire 104, 105
Spain 125, 171
    Seville 125, 126
Spanish 124, 125, 126, 154, 155, 169, 171, 173, 174, 175
Sparker 139, 140, 142, 153
S. Pearson and Sons 43, 173, 243, 247
Spherical cavities 18
Spithead 104
Star and Garter Home 5, 40
Steam vii, 5, 19, 41, 42, 47, 48, 49, 50, 54, 55, 80, 94, 96, 105, 112, 121
Steel 23, 26, 27, 34, 39, 40, 41, 43, 45, 46, 47, 48, 49, 50, 52, 55, 61, 74, 89, 96, 101, 104, 105, 108, 113, 118, 127, 128, 138, 140, 143, 144, 145, 156, 165, 166, 170, 173, 179, 182, 183, 202, 214, 252
Storebælt 155
    Bridge option 155
    Immersed tunnel option 155
Straits of Dover 139, 154
Stucky, Prof. Alfred 204, 205, 206, 250
Stucky, Prof. J. F. 205, 214, 218, 219, 221, 222, 250
Suarez, Saturnino 169, 170, 171, 173, 174, 175, 250
Suderow, George E. 143, 250
Suez Canal Company 137, 141, 249
Surrey 76
Sussex 76, 219
Sverdrup. Major-General Leif 180, 250
Swan and Edgar 24, 26, 33, 35
Sweden 94, 96, 97, 99, 169, 170, 182, 183, 184, 197, 249
    Board of Waterfalls 97, 170
    Stockholm 94, 95, 97, 169, 170, 177
Switzerland
    Lausanne 204, 205, 206, 218, 250
Symington 74, 247
Syria
    Baniyas 103, 104, 141

T

Tactical handling 42, 69, 89, 100
Tamez, Enrique 171, 173, 177, 178, 250
Tate & Lyle 92
Tavernier, M 204, 206
Technical Studies Inc. 137, 141, 146, 172, 187, 242, 245
Territorial 27, 111
Terzaghi, Prof. Karl i, iii, 58, 71, 72, 99, 110, 134, 167, 172, 173, 240, 249, 250, 251, 252
Thatcher, Margaret v, 202, 237, 239, 247
The Netherlands
    Rotterdam 54
The Tunnel (movie) 66, 240
Timber 8, 10, 12, 15, 16, 18, 20, 21, 24, 26, 34, 38, 48, 49, 50, 52, 55, 68, 77, 78, 84, 101,

INDEX

104, 106, 118, 127, 166, 170, 172, 216, 218
Tin Lizzies 46
Toplis and Harding 105
Topsham, Devon 232, 235, 236, 237
Trans-Atlantic Telephone Cable 108
Triggs, Richard 228, 229, 230
Trinity House 142, 152, 154
Tunbridge Wells, Kent 162, 219, 220, 221, 222, 235
Tunisia 204, 205, 250
    Oued Nebaana Dam 204, 211, 250
    Tunis 204, 205, 206
Tunnel
    Bored 149, 150, 151, 239, 242, 250
    Twin bored iii, v, 148, 235, 245
    Twin tunnels 8, 52, 185, 237, 238
Tunnel liner plates 166
Tunnelling History and my own Involvement iv, 236, 245, 246
Tunnel machines 161, 185, 186, 238
    Beaumont, Colonel F. E. B. 139
    Caldwell 186
    Diamond drilling 78, 95, 141
    Howden 249
    Marietta 140
    Markhams 178
    Memco 186
    Priestleys 161
    Robbins 145, 156, 159, 160, 186
    Whitaker 140
Tunnel miner 9, 246
Tunnels
    Clyde Tunnel 168
    Thames Tunnel 13, 139, 233, 244, 246
    Tyne Tunnel 163
Tunnels and Tunnelling 229, 230, 236, 237, 239

U

Uganda 117, 118, 119, 120, 211, 243
    Entebbe 117, 118, 119, 211, 221
    Kampala 118, 119
    Owen Falls Dam (Nalubaale) 117, 118, 243
Uniflow Steam Engine 5
Union Castle Line 104
Union Corporation 211, 212, 213, 214, 215, 216, 218, 219, 220, 221, 222, 245
    Munroe 212, 220
USA
    Chicago 167
    Detroit 46, 47, 65, 71, 165
    Los Angeles 143, 144, 250
        Hyperion Outfall Sewer 143, 250
    New York
        East River 43, 247
        Subway 146
    Oahe Dam 145
    Salt Lake
        Lucin Cutoff Railroad Bridge 145
    San Francisco 143, 144, 146, 163, 185, 186, 187, 194, 195, 242, 243, 247, 248
        San Francisco Subway 248
    Seattle
        Deas (George Massey) Tunnel
        Immersed tunnel 167
    Utah 144, 163
Usherwood, T. S. 3
US Navy 125, 139, 179
Utah Construction and Mining Company 204, 206, 250

V

Vermuyden, Cornelius 49, 50
Voluntary work 109
Vowler, J. C. G. (Creed) vii, 28, 31, 32, 251
Vyasa, Mr 209, 210

W

Wakelam, H. B. T. (Teddy) 28, 251
Wales Gas Board 202, 247
War Office 77, 78, 111
Waste ii, iii, 80, 198, 199, 200
Watermeyer 218, 220
Waterworks River 73
Watkins, Tasker 197, 251
Welham Green 113, 168, 246
Welkom 123
Well-point 65, 75, 82, 83
Welsh Office 192, 193, 194, 195, 202, 203
Westacott 118, 119
West India Dock 84, 86
Westminster Abbey 40, 83, 108, 160, 197, 224, 251
Westminster Technical College 101, 225
    Governor 101, 225
Westrop, Major 100
White mice 53
Whittaker, Hunt and Company 148, 247
Widmark and Platzer 169, 171, 173, 182, 183, 250, 251, 252
Wikstrom, Styg 173, 174, 177, 251
Wild, George 124, 126, 127, 129, 130, 133, 251
Wilenski, Sir Roy 122
Williams, Norman 202, 203, 252
Williams, Sir Owen 149, 196, 202, 203, 210, 252
Wilson, David 161
Wilson, Dr. Guthlac 121, 252
Wilson, Prof. Gilbert 183, 184, 252
Wilton Castle, Middlesbrough 93
Wimpey iii, 105, 112, 141, 142, 154, 156
Wimpey-Forasol 154, 158
Woodward, Dr 194
World Bank 163, 186, 247, 248, 249
Wright, Sir Denis 131, 132, 182, 252
Wynne-Edwards, Sir Robert i, 70, 71, 72, 95, 101, 128, 134, 165, 166, 178, 246, 248, 252

X

XXI Club vii, 27, 31, 32, 35, 43, 44, 54, 58, 62, 70, 76, 90, 94, 103, 110, 114, 120, 143, 159,

207, 244, 245, 246, 247, 249, 251

Y

Young talent 231

Z

Zambia (Northern Rhodesia) 121, 211
 Lusaka 121

THE COMMON GROUND OF WILD AND CULTIVATED PLANTS

BOTANICAL SOCIETY OF THE BRITISH ISLES CONFERENCE REPORT NUMBER 22

The publication of this Report was partially funded
by the Botanical Society of the British Isles

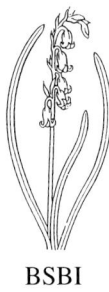

BSBI

Cover illustration: Mimulus × *maculosus*, Glen House, Peebles. (*Photo A.J. Silverside*)

The Common Ground of Wild and Cultivated Plants

Introductions, invasions, control and conservation

Edited by

A. Roy Perry and R. Gwynn Ellis

National Museum of Wales, Cardiff

AMGUEDDFA GENEDLAETHOL CYMRU
NATIONAL MUSEUM OF WALES

CARDIFF · 1994

© Department of Botany, National Museum of Wales 1994

Copyright of the photographs remains with the photographers

Published by

Department of Botany
National Museum of Wales
Cathays Park
Cardiff
CF1 3NP
U.K.

ISBN 0 7200 0408 X

Camera-ready copy produced in the
Department of Botany, National Museum of Wales

Printed in Wales by J.& P. Davison, Pontypridd

Contents

Preface   vii

Conference Programme   ix

1   A common ground for wild and cultivated plants
  Paul Evans   1

2   British wild species and varieties in gardens
  R.G. Woods   7

3   Ergasiophygophytes in the British Isles – plants that jumped the garden fence
  E. Charles Nelson   17

4   Some problems with introduced plants in the wild
  J.R. Akeroyd   31

5   A proposal for an alien register in the British Isles
  R. Gwynn Ellis   41

6   Ferns – a case history
  Martin Rickard   51

7   *Mimulus*: 180 years of confusion
  A.J. Silverside   59

8   *Symphytum* – Comfrey
  Franklyn H. Perring   65

9   *Hypericum* wild and cultivated – the common ground
  N. Robson   71

10   Proposal for a register of aberrant forms
  M. Cragg-Barber   77

11   The reintroduction of plants into the wild: an integrated approach to the conservation of native plants
  Mike Maunder and Margaret Ramsay   81

12   Principles of collecting plants as a conservation exercise
  R.A.W. Lowe   89

13 The identification of cultivated plants
    S.G. Knees and J.C.M. Alexander 95

14 Propagation not collection
    M.I. Read and B.A. Thomas 101

15 The introduction of trees to Roman and Medieval Britain
    J.H. Dickson 111

16 The possible effects of cultivated introductions on native willows (*Salix*) in Britain
    J.E.J. White 113

17 Conservation of rare temperate rainforest conifer tree species: a fast growing role for arboreta in Britain and Ireland
    Christopher N. Page and Martin F. Gardner 119

18 *Hedera*
    H.A. McAllister 145

19 The genus *Cotoneaster* in the British Isles
    Jeanette Fryer 151

Index 159

Preface

Gardening has long been a major preoccupation of many of the members of the Botanical Society of the British Isles. This is borne out by the findings of a recent survey of BSBI members which indicated that over 77% had enjoyed gardening in the last year. It seemed fitting, therefore, that the BSBI should consider organizing a conference to bring together botanists and gardeners to discuss common themes.

When it was announced that a Garden Festival would be held in Wales, at Ebbw Vale in the Gwent valleys in 1992, staff of the Department of Botany, National Museum of Wales, suggested that such a conference could be held in Cardiff. The suggestion was put to the Botanical Society of the British Isles and the Royal Horticultural Society, and approved.

The conference was held in the National Museum of Wales, Cardiff, 10-15 July, 1992. There were two days of lectures and discussions on wild plants in gardens and garden plants in the wild, with post conference tours of the Garden Festival Wales site, Bute Park, Cardiff and Dyffryn Gardens.

The Conference started on the Friday evening by Dr Max Walters giving a succinct lecture on the origin of our garden plants entitled 'Where do our garden plants come from?'. He pointed out that the relationship between wild and cultivated plants is a complicated subject which can be studied from many different angles. For the purpose of this Conference his talk considered only plants grown by man for amenity, our garden plants in a strict sense, and excluded those grown for food or for other human uses (on some of which our lives depend). He said that the very existence of the word 'Paradise', a Persian word meaning 'an enclosed garden', reminds us that our association with plants for pleasure is an ancient one. It was not therefore very surprising that we find the origin of many more familiar garden flowers obscure. Roses provide an excellent example: in detail we cannot hope to trace all stages in the ancestry of a modern rose, though we can learn something of the long and complicated history of garden roses. Dr Walters considered a range of garden plants, beginning with those wild plants we tolerate (or even welcome) in our gardens, through examples which can be presumed to be simple genetic variants of wild plants, to the more complicated situations where hybridization of original wild species has been involved. He concluded with some considerations on the effect of planned creation of garden plants and made some interesting speculations as to the possible future role of modern genetics in making the garden plants of the twenty-first century.

Dr Walters' paper is not published here as his thoughts on the subject have recently appeared in the Collins New Naturalist book *Wild and Garden Plants* (1993).

The further nineteen lectures were arranged under general themes. On Saturday the two topics 'The Common Ground' and 'Case Histories 1 – Herbaceous Plants' were covered, then the following day 'Identification and Conservation' and 'Case Histories 2 – Trees and Shrubs' completed the indoor part of the Conference. The lectures (except for that of Dr Walters as mentioned above, and that of Prof. J.H. Dickson which is being published elsewhere) are presented in this volume.

Two social events took place during the weekend. The first was on the Saturday evening when a reception was held in the National Museum of Wales with the kind permission of the Director. The second was a Conference Dinner held in the Traherne Hall of Residence, University of Wales College of Cardiff, on the Sunday evening.

The following day an excursion to the Garden Festival Wales at Ebbw Vale, Gwent, was made and delegates saw the development of an old industrial site. Then on Tuesday morning A.R. Perry led a walk round Bute Park Arboretum, Cardiff where there are some exceptionally fine specimens of trees such as *Pterocarya fraxinifolia,* many species rare in cultivation, and specialist collections of *Pinus, Quercus, Acer, Magnolia, Sorbus, Crataegus, Prunus, Malus, Pyrus, Carpinus, Alnus* and *Fraxinus.* In the afternoon a conducted tour round Dyffryn Gardens in the Vale of Glamorgan, was led by the Head Gardener, Mike Thurlow. This jewel of a garden, often referred to as the 'Garden of Wales', was created by Reginald Cory around the turn of this century and is now administered by a joint committee of South and Mid Glamorgan Councils.

The Conference was organized jointly by the National Museum of Wales and the Botanical Society of the British Isles and supported financially by both bodies. The programme was arranged by R.G. Ellis and B.A. Thomas and the excellent local organization was taken care of by the staff of the Botany Department of the National Museum of Wales. The publication, of this Report was partially funded by the Botanical Society of the British Isles and is BSBI Conference Report Number 22.

We thank the Keeper of Botany, National Museum of Wales, Professor B.A. Thomas, for allowing delegates free access to the National collections (NMW) during the course of this Conference.

Botanical and common names of plants are taken exclusively from Professor Clive Stace's *New Flora of the British Isles*, 1991. Authorities have mostly been omitted except for those species not in Stace.

A. Roy Perry
R. Gwynn Ellis

Conference Programme

THE COMMON GROUND OF WILD AND CULTIVATED PLANTS

National Museum of Wales, Cardiff: 10 - 15 July 1992

Friday 10 July

7.30 pm - **INTRODUCTORY LECTURE**
Where do our garden plants come from? – Dr S.M. Walters

Saturday 11 July

9.30 am - 12.30 pm – **THE COMMON GROUND**

    9.30   The common ground of wild and cultivated plants – P. Evans

    10.00 British wild species and varieties in gardens – R.G. Woods

    10.30 The ones that jumped the garden wall – Dr E.C. Nelson

    11.00 Tea/Coffee

    11.30 Problems with introduced plants in the wild – Dr J.R. Akeroyd

    12.00 A proposal for an alien register – R.G. Ellis

12.30 pm - 2.00 pm   LUNCH

2.00 pm - 5.15 pm – **CASE HISTORIES 1 – Herbaceous Plants**

    2.00 Ferns – M. Rickard

    2.30 *Mimulus*, 180 years of confusion – Dr A.J. Silverside

    3.00 *Symphytum* – Dr F.H. Perring

    3.30 Tea/Coffee

    4.00 *Hypericum* – Dr N.B. Robson

    4.30 A proposal for a register of aberrant forms – M. Cragg-Barber

    5.00 General Discussion

7.00 pm – RECEPTION in National Museum of Wales

Sunday 12 July

9.30 am - 12.30 pm – **IDENTIFICATION and CONSERVATION**

    9.30   The reintroduction of rare plants into the wild – M. Maunder

    10.10 Principles of collecting plants as a conservation exercise – A. Lowe

    10.50 Tea/Coffee

    11.15 The identification of cultivated plants – Ms S. Knees

    11.55 Propagation and not collection – Dr B.A. Thomas

12.30 pm - 2.00 pm – LUNCH

2.00 pm - 5.30 pm – **CASE HISTORIES 2 – Trees and Shrubs**

    2.00   The introduction of trees to Roman and Medieval Britain – Dr J.H. Dickson

    2.30   *Salix* – J. White

    3.00   Conifers – Dr C. Page

    3.30   Tea/Coffee

    4.00   *Hedera* – Dr H.A. McAllister

    4.30   *Cotoneaster* – Mrs J. Fryer

    5.00   General Discussion

7.00 pm – CONFERENCE DINNER

Monday 13 July

9.30 am - 5.30 pm – Tour of Garden Festival Wales site at Ebbw Vale, Gwent

Tuesday 14 July

9.30 am - 5.30 pm – Tour of Cardiff Parks and Dyffryn Gardens, South Glamorgan

1

A common ground for wild and cultivated plants

PAUL EVANS

Brambles, 69 Regent Street, Wellington, Telford, Shropshire, TF1 1PE

ABSTRACT. Plants are the basis of life on earth. We are here because plants matter. But they matter in different ways. I hope to show that the different ways in which plants matter to us reflects what we are doing to the environment and that finding a way of describing our relationship with wild and cultivated plants is important in deciding what we should do in the environment and how we should do it.

The common ground

A common ground between wild and cultivated plants exists in our experience of nature. This experience is of the world as it is lived and is expressed in the values that we hold. To understand this common ground between wild and cultivated plants is to 'unpack' our attitudes to nature.

As environmentalism, the emerging spirit of our age, reaches out towards horizons, to that which is not yet known, to a wisdom which will shape our future on earth, so we reach into nature, and our experience of nature, for meaning and value.

We do this because the science that supports botany and the ecology of plants is claimed to be value free. Although science tells us the story about how plants work in the world and what sort of things we can do, it cannot tell us what we ought to do. It might provide the means to manipulate vegetation; but it doesn't tell us how we should employ these means or what we should seek to bring about.

Because of the ecological crisis facing the biosphere the challenges facing plant conservation, from gardens to wild habitats, are challenges of our existence. And to understand what we do with plants is to understand the way we experience nature. This experience of nature is part of everyday life.

As a gardener and nature conservationist I am absorbed by the way our attitudes to wild and cultivated plants shape the world we live in and I am deeply concerned about the threats to plants in the world as it is lived. As an awareness grows that plants are more than the floral wallpapering over cracks in the environment, and that the threats to them from habitat destruction, pollution and climate change increase daily, so plant habitats are becoming the front line of nature conservation.

Although 40% of all the productivity of photosynthesis from plants on land is now directly consumed, diverted or wasted by human activity (Vitousek *et al*. 1986), we experience plants not collectively as a life-giving force, but as things – things with value dependent upon the way they matter to us; things of cultivation or things of the wild.

Perhaps I have hijacked Dylan Thomas' meaning when he wrote, 'The force that through the green fuse drives the flower / Drives my green age' but I feel that this intuition points to the essential connection between ourselves and plants.

We use them for food, fibre, fuel, drugs, building and we exploit their aesthetic and scientific values in every aspect of our daily lives. Look around you; how many plant-influenced designs can you see? In Britain 1.3% of the land's surface is used for horticulture and we spent £6bn on gardening in 1991. We are a gardening nation. In an opinion poll undertaken in 1982 on the factors contributing to the quality of life, the following preferences are listed (Coleman 1988):

safe streets	72%
attractive countryside	53%
unpolluted atmosphere	51%
good public transport	46%
wildlife	37%
gardens	37%
access to car	35%
sports facilities	33%
theatres	19%

The equality between gardening and wildlife and the importance of attractive countryside (and perhaps an unpolluted atmosphere), all point to the value of plants in our lives. But the essential nature of plants is concealed within the distinctions we have made according to the ways they matter to us. We have concealed the essential nature of plants within the dual phenomenon of 'cultivation' and 'wildness'. It is this distinction which must be the subject of a broader understanding of our life with plants on which to base strategies for conservation.

By conservation I mean 'saving' – not only sparing or safeguarding from harm but a positive action to 'free', in a proper sense, essential beings in nature to dwell in peace – and to do so as a fundamental character of our own existence.

It is necessary to review the philosophical foundations for nature conservation because the conventional approach has not prevented the loss in Britain of 98% of our lowland raised bogs, in the last 150 years (and still exploited for horticultural peat), almost 50% of old woodland including 32,000 hectares of ancient woodland in England and Wales since 1930, 97% of traditional meadow land (which is still being lost at 3% per year), and 100,000 miles of hedgerow (which is still being lost at 4000 miles per year). There are currently 93 plant species protected by the Wildlife and Countryside Act of 1981 with a proposal to raise this to

A common ground for wild and cultivated plants 3

Fig. 1. *a* (above), the enigma value: Scots Pine (*Pinus sylvestris*), the Wrekin, Shropshire; *b* (below), drawn from the wild – the herbaceous borders, Powis Castle, Wales. (*Photos P. Evans*).

131 but there are a possible 883 Red Data Book plants (rare, endangered and threatened species), that receive virtually no protection at all. Nature Reserve designations and particularly the Sites of Special Scientific Interest system have offered scant protection from habitat destruction and can offer no protection at all from climate change and sea level rise caused by global warming.

All this, in a culture where the plant-centred traditions of gardening and natural history are so important, points to the need to discover this common ground between cultivation and wildness on which to build a conservation ethic to save plants.

The conventional view of conservation – that put forward by government agencies and most voluntary organizations – is that we have a prime duty as stewards, holding the earth in trust for future generations: to preserve our heritage so that it may become theirs. This 'heritage' includes the lithosphere, the biosphere, hydrosphere and atmosphere. Secondly, conservation has direct and immediate benefits for humankind. The health of the natural world is inextricably linked with our own well-being and its resources underpin every aspect of our way of life. All of this may be true and even if nature has value as our heritage and our duty is to conserve nature as a resource for ourselves and future generations of ourselves, this does not preclude having respect or love for the inherent worth of other species, habitats, ecosystems or for nature as a whole.

But although many conservation agencies are prepared to add a policy statement referring to the value of wildlife for its own sake, they do not explain what this means. This is because we have yet to develop an environmental ethic to deal with the way values are distributed throughout the diversity of 'wildlife', from the dolphin to the equally enigmatic, though rare and unnoticed, Perfoliate Penny-cress.

Flower or weed, mushroom or toadstool, friend or foe, edible or poisonous – these prejudices stem from the conflicts between cultivation and wildness. I believe there is scope for exploring the commonality of plants beyond these prejudices (held by gardeners and scientists alike) to see whether an inter-species equity may be found in currents within our gardening and natural history traditions. If all the plants in a garden or habitat, have 'interest' as things in themselves and for their contribution to the whole, then our relationship to them is through their community. It is therefore the interests of the plants in the context of the community that concern us. As indeed they always have.

When we carved the leys, or clearings, from the great wild forests of Britain, to build our gardens all those thousands of years ago, that very forest and wildness was the context in which our civilization developed. Now that the wildness is split into fragments and winds through the hedges which bind the areas of cultivation together, this great garden we created with its agriculture and industry is the context of our civilization now.

The scientific line on the value of species as subjects of scientific study contributing to the resource of current and future knowledge does not necessarily suggest any responsibility to those species, only to future scientists. Though science underpins conservation, and ecology can help us arrive at an understanding of intrinsic value in nature, when science is used as the justification for conservation, important problems arise. Positivist science can only tell us part of the story about what nature does and nothing about what it means. No matter how high we raise the mountain of factual evidence for what something 'is' there may be no corresponding moral imperative for what *should* be done about it. Even though the facts are of such weight that they suggest action or restraint, only an environmental ethic can bridge the gulf between scientific fact and moral action. Such an ethic is necessary to remove the barriers between an anthropocentric view of the world which maintains cultivation and a biocentric view which attempts to preserve wildness and for plants to transcend the duality of cultivation and wildness.

The basis for such an ethic exists in our experience of nature. Dylan Thomas also wrote 'behold the wilderness banging in my blood' which conveys the intuition that wild nature is not a phenomenon 'out there' but an essential element of the lived world.

Cultivation and wildness occur together. Our landscapes, particularly in Europe, result from the interactions of countless generations of people with ecological systems. Grassland, woodland, heath and moorland are examples of 'cultivation' and contain many plant communities that exist because of human intervention.

So land, cultivated in its many senses, is a laminate or palimpsest overlaying land, the biotic community or wildness. Farming, forestry and gardening are ways that build our lived world. Land is not shaped as a by-product of our activity; to shape land, to organize our space in nature, to build with and into growing things is the very essence of our being. The intuitions that we gain from gardening may illustrate this.

In my own gardening experience – and this is true for many working with plants – a deep significance transcends the usefulness of plants as food, objects of curiosity or appeal to the senses. Gardening, in many forms, is a way of experiencing the integral or intrinsic values of plants and working them into our lives. The same can be said for my botanical experience, and is also true for many. Scientific study of plants generates a great sense of wonder, awe and respect for species and communities that goes beyond a fascination for their intricacies and beauty. It is the recognition of their intrinsic value that forms the common ground for wild and cultivated plants in our experience of nature.

There are few areas of wilderness where human activity has little influence. There are cracks and punctures in 'cultivation' which allow wild nature to be expressed. Threading through, beneath and beyond the woven heritage of the community of land, is the self-determination of landscapes and ecosystems. There is a constant colonization of wild nature to extend cultivation throughout the biosphere (and beyond?). Land is the tension between human earth-ownership and the self-determination of the biosphere.

Philosophers like Mary Midgely and Baird Callicot have described the way that domestic animals have always formed part of (otherwise) human communities and that we can reconcile obligations to them with regard to obligations to human groups, like the family. Our environmental obligations arise within communities which must be understood in relation to the wider ecological or biotic community of species of which we are a member.

I suggest that cultivated plants are also members of (otherwise) human communities and that obligations towards them are as those to domestic animals. Our environmental obligations arise within this community of people and plants and must be understood in relation to the communities of wild plants, of which we are also a member. This relationship between a community of people and the plants they cultivate within the ecological framework of the biotic community of wild plants, is what I call the 'community of land'. Because of this community of land we have responsibilities to plants in habitats which we have caused to exist – old parks and gardens, meadows, canals, coppiced woodland, orchards, churchyards, chalk downs, heather moorland, and so on – and to dwell in peace with wildness and to 'free' the self-determination of landscapes and ecosystems.

I feel that this is not unlike the duties and the obligations we have to future generations of people. Our actions, our policies, our technological innovations, actually influence who will exist and what sort of persons they will be. This gives us a certain responsibility to them as people. The same can be said of our conservation strategies and land use policies. If we build ourselves into a community of land (a community of cultivation and wildness), we have responsibilities to that community like those we have for each other and future people in the lived world.

References

Coleman, D. (1988). Social benefits of planned land use: recreation and public access in the lowlands. *In* Oxford Forestry Institute Occasional Paper no. 38.

Vitousek, P.M., Ehrlich, A.H. & Matson, P.A. (1986). Human appropriation of the products of photosynthesis. *Bioscience* **36**: 368-373.

2

British wild species and varieties in gardens

R.G. WOODS

Countryside Council for Wales, Ithon Road, Llandrindod Wells, Powys, LD1 6AA

ABSTRACT. Many wild plants are cultivated as forms which differ from the commonly occurring wild form. The extent of this variability is described by reference to particularly garden-worthy forms. It is not easy to quantify the part played by British wild plants in our gardens. Many important plants such as pansies and polyanthus are of hybrid origin. They carry genetical material from their wild ancestors, though we may no longer be certain which species contributed a particular character.

A minimum of 491 species of wild British plants were offered for sale in Britain in 1989. The most commonly offered species are described, the account concentrating on the species and cultivars still recognizably related to their wild ancestral species, an attempt being made to quantify the use we make of this wild resource in our gardens.

How many wild British native species are grown in gardens?

This is a difficult question to answer. Let us first define a wild native species as one which Prof. Clive Stace in his *New Flora of the British Isles* (Stace 1991) considers to be native or 'probably native'. But what do we mean by 'grown'? Many wild plants grow in gardens yet gardeners spend an immense amount of time trying to exterminate a number of them. In my garden, weeds such as *Galium aparine* (Cleavers), *Capsella bursa-pastoris* (Shepherd's-purse) and *Elymus repens* (Common Couch), are expelled, whilst others, for example *Bellis perennis* (Daisy) in lawns, and ferns in brickwork, are often suffered. A few such as *Epipactis helleborine* (Broad-leaved Helleborine), *Lonicera periclymenum* (Honeysuckle) and *Rosa arvensis* (Field-rose), may be actively encouraged whilst others like *Misopates orontium* (Lesser Snapdragon) and *Calystegia pulchra* (Hairy Bindweed) are considered to be weeds or 'flowers' depending on exactly where they appear. Finally, there is a large group of species which are planted, because I like them or want to know more about them. With such difficulties in classifying the part played by wild plants in gardens, a nation-wide survey of gardens to establish the use made of wild plants does not appear to be a profitable exercise, especially given the immense time and resources that would be required.

Wild plants purchased for planting in gardens might be more easily quantified. It seems likely that a relationship exists between those plants offered by the nursery trade and those grown in gardens. We are fortunate that Chris Philip of the Hardy Plant Society has compiled in *The Plant Finder* a catalogue or directory of over 40,000 plant 'varieties', together with the nurseries that offer them for sale (Philip 1992).

By searching through the 1989-90 directory it was possible to assess the numbers of British native species offered through the nursery catalogues submitted to Chris Philip for the 1989 season. Whilst this is clearly not exhaustive (the 1992 edition of *The Plant Finder* now lists 55,000 entries) it is probably as good a sampling method as could be devised. There may be some bias to perennial and woody species at the expense of annuals, but a search of leading seedsmen's catalogues revealed few species not included in *The Plant Finder*.

Lists in *The Plant Finder* show that 491 wild British plants were offered for sale by nurseries. Whilst many garden-worthy forms were listed, 470 are offered for sale apparently as the wild-type plant. Taking Gwynn Ellis's figures in *Flowering Plants of Wales* (Ellis 1983), of 1511 native British plant species, this represents almost one-third of the total species diversity (or in Wales with 1011 native species, almost one-half of the natural diversity). I have to confess a doubt as to whether all are really garden-worthy. You can only marvel at the sales pitch required to sell *Epilobium montanum* (Broad-leaved Willowherb), *Anthoxanthum odoratum* (Sweet Vernal-grass) or *Aphanes arvensis* (Parsley-piert). These species are, however, only offered by, at most, two specialist nurseries and it might be more instructive to look at the most widely available species.

Thirty-eight species were offered by 20 or more nurseries (see Table 1). Simply classifying them (and accepting that some could equally well go into more than one class) into trees and shrubs, alpine garden and herbaceous border subjects as would a gardener, the largest element with 17 taxa is made up of alpine garden plants. The remaining two groups contain a similar number of taxa. The trees and shrubs are, however, a notable proportion of the total number of British native species but the species in the herbaceous border element seem pitifully few.

A consideration of those species offered by between 10 and 19 nurseries adds a further 20 trees and shrubs to account, in total, for over three-quarters of our native woody species. Forty-nine herbaceous species are also added, including a large proportion of our wetland plants, such as *Caltha palustris* (Marsh-marigold), *Lythrum salicaria* (Purple-loosestrife), *Carex pendula* (Pendulous Sedge) and *Lychnis flos-cuculi* (Ragged-robin). The competition for garden space in this recently popular area of wetland and pool-side plantings from non-native species seems to be less than amongst other herbaceous plant species with a longer history of cultivation and a greater number of cultivars. Twenty-three additional plants for the alpine garden are offered by between 10 and 19 nurseries.

Combining the results from these two analyses, 10 or more nurseries offer 59 species of native British plants for the herbaceous border, 40 species for the alpine garden and 32 species of trees and shrubs (131 species in total). But this is only part of the story.

Table 1. British wild plants, classified horticulturally, offered as the wild-type form by more than 20 nurseries (extracted from Philip 1989).

Trees and Shrubs	Rock Garden	Herbaceous border
Acer campestris (Field Maple)	*Alchemilla alpina* (Alpine Lady's-mantle)	*Campanula glomerata* (Clustered Bellflower)
Betula pendula (Silver Birch)	*Arctostaphylos uva-ursi* (Bearberry)	*Convallaria majalis* (Lily-of-the-valley)
Buxus sempervirens (Box)	*Betula nana* (Dwarf Birch)	*Iris pseudacorus* (Yellow Iris)
Carpinus betulus (Hornbeam)	*Dianthus deltoides* (Maiden Pink)	*Linum perenne* (Perennial Flax)
Fagus sylvatica (Beech)	*Draba aizoides* (Yellow Whitlowgrass)	*Malva moschata* (Musk Mallow)
Fraxinus excelsior (Ash)	*Dryas octopetala* (Mountain Avens)	*Phyllitis scolopendrium* (Hart's-tongue)
Hippophae rhamnoides (Sea-buckthorn)	*Fritillaria meleagris* (Fritillary)	*Polemonium caeruleum* (Jacob's-ladder)
Pinus sylvestris (Scots Pine)	*Galium odoratum* (Woodruff)	*Primula veris* (Cowslip)
Prunus avium (Wild Cherry)	*Genista pilosa* (Hairy Greenweed)	*Primula vulgaris* (Primrose)
Quercus robur (Pedunculate Oak)	*Gentiana verna* (Spring Gentian)	
Taxus baccata (Yew)	*Lychnis alpina* (Alpine Catchfly)	
Viburnum opulus (Guelder-rose)	*Lysimachia nummularia* (Creeping-Jenny)	
	Primula elatior (Oxlip)	
	Primula scotica (Scottish Primrose)	
	Pulsatilla vulgaris (Pasqueflower)	
	Salix lanata (Woolly Willow)	
	Viola odorata (Sweet Violet)	

Variegated plants

Gardeners have seldom been satisfied with the wild-type species and have selected a range of forms, sometimes found originally in the wild and sometimes turning up as sports in gardens, which may differ significantly from the normal form of the wild species. A good example of such forms are the 55 species of British native plants with variegated leaves which are listed in *The Plant Finder*. The two commonest, and those which also occur in a bewildering array of forms, are *Ilex aquifolium* (Holly) and *Hedera helix s.l.* (Ivy). Eight other species are offered by 20 or more nurseries. The two species most widely grown are *Ajuga reptans*

(Bugle) which occurs in a number of forms such as the cream variegated 'Argentea' and 'Rainbow' with reddish and yellowish patches on the leaves, and *Phalaris arundinacea* (Reed Canary-grass) otherwise known as the infamous 'Gardeners Garters'.

In the wild, this latter species is a vigorous plant of river sides and marshes. It has probably seriously dissuaded gardeners from using variegated grasses in gardens, such is its reputation for spreading and engulfing its neighbours. There is, however, a number of other very worthwhile grasses such as the now widely offered *Holcus mollis* (Creeping Soft-grass) and *Molinia caerulea* (Purple Moor-grass) which are less invasive.

The remaining four widely offered variegated species are *Scrophularia auriculata* (Water Figwort), *Iris pseudacorus* (Yellow Iris), *Sambucus nigra* (Elder) and *Mentha suaveolens* (Round-leaved Mint). A further 23 species are offered by 10 or more nurseries, providing evidence of a high degree of popularity for those few plants sporting variegated forms. There still are, however, a significant number of garden-worthy forms only recently found in the wild or very rarely offered for sale. A silver variegated form of *Carex bigelowii* (Stiff Sedge) with curled leaves was recently found in Scotland by Dr Oliver Gilbert and is now being propagated. It will be a very fine scree bed plant, probably superior to the attractive *Carex ornithopoda* (Bird's-foot Sedge) which is at present widely grown.

The Aureo-variegatus form of *Alopecurus pratensis* (Meadow Foxtail) also deserves wider cultivation, as do variegated forms of *Arrhenatherum elatius* (False Oat-grass), *Dactylis glomerata* (Cock's-foot) (a variegated form with the young leaves curled, has recently entered the trade to provide a more vigorous form to that already on offer) and for dry shade, *Melica uniflora* (Wood Melick).

Perhaps not strictly variegations, the leaf colourings of *Lamiastrum galeobdolon* (Yellow Archangel) provide valuable winter and spring colour. The commonest form sold has been recently described as ssp. *argentatum*. Its origin is uncertain, but it is a robust plant now colonizing woods and roadsides even in mid and west Wales, having probably been dumped with garden spoil. Slugs don't seem to like it, in contrast to the very attractive and non-invasive form called 'Silver Carpet'.

Other leaf colour changes

The two commonest additional forms of leaf colour change favoured by gardeners are aurea forms in which the leaves become yellowish in colour, often by the loss of pigments, which leave the leaves more or less susceptible to damage from bright sunlight, and purpurea forms with often intensely purple-black leaves.

Aurea forms

Thirty-one species were offered for sale as aurea forms in 1989. The most commonly offered (by 20 or more nurseries) were *Calluna vulgaris* (Heather in a wide variety of forms), *Erica vagans* (Cornish Heath as the form 'Valerie Proudley'), *Fagus sylvatica* (Beech as the form 'Zlatia'), *Filipendula ulmaria* (Meadowsweet), *Hedera helix* (Ivy as the form 'Buttercup'), *Humulus lupulus* (Hop), *Lysimachia nummularia* (Creeping-Jenny), *Milium effusum* (Wood Millet as the form 'Bowles Golden Grass'), *Origanum vulgare* (Wild Marjoram), *Sambucus nigra* (Elder) and *Viburnum opulus* (Guelder-rose).

Plate 1

Silene dioica flore plena
(Photo RGW)

Bellis perennis 'Hen and Chicks'
(Photo RGW)

Ranunculus ficaria double-flowered
(Photo RGW)

Plate 2

Primula elatior 'Hose in Hose'
(Photo RGW)

Primula vulgaris 'Jack in the Green'
(Photo RGW)

Primula vulgaris 'Alba Plena'
(Photo RGW)

Purpurea forms
Sixteen species with purpurea forms were listed in *The Plant Finder*. Amongst the trees the most frequently offered for sale were beech, silver birch and elder, but also sold were purpurea forms of *Prunus padus* (Bird Cherry), *P. spinosa* (Blackthorn), *Quercus petraea* and *Q. robur* (Sessile and Pedunculate Oaks) and *Carpinus betulus* (Hornbeam). The commonest herbs were *Ajuga reptans* (Bugle, as the form 'Burgundy Glow'), *Plantago major* (Greater Plantain), *Euphorbia amygdaloides* (Wood Spurge), *Trifolium repens* (White Clover as the form 'Calvary Clover') and *Ranunculus ficaria* (Lesser Celandine as the form 'Brazen Hussy').

Other coloured leaf forms
Two species with coloured leaf veins are offered for sale. The Blood-veined Dock (*Rumex sanguineus*) is grown as a form with the major veins of the leaf picked out with broad blood-red stripes. It is this form that gives this plant its name. The form 'Susan Smith' of the Red Clover (*Trifolium pratense*) has leaf veins picked out in pale yellow.

Leaf shape changes

Finely divided or crisped leaves have a quality of form which attracts the gardener. Thus more of the cultivars of wild plants differ in leaf shape from their wild parents by an increase in the division or degree of folding of their leaves, than those with less divided or folded leaves. Twenty-three species are listed in *The Plant Finder* with laciniate or incised leaves. The majority are either ferns (dealt with elsewhere in this volume by Martin Rickard) or are trees or shrubs. So you can grace your garden with cut-leaved forms of *Alnus glutinosa* (Alder, Fig. 1a), *Betula pendula* (Silver Birch), *Carpinus betulus* (Hornbeam), *Corylus avellana* (Hazel), *Fagus sylvatica* (Beech, Fig. 1b), *Sambucus nigra* (Elder) and *Sorbus aucuparia* (Rowan). In contrast, simplification has taken place in *Fraxinus excelsior* f. *diversifolia*, the 'One-Leaved Ash', which has simple leaves with a jagged outline (Fig. 1c); in the form 'Cockleshell' of Beech, with small rounded leaves; and the form 'J.C. van Tol' of *Ilex aquifolium* (Holly) which is almost devoid of spines. Yet this latter species also produces leaf forms in the hedgehog holly ('Ferox') which are armed with spines on the leaf surface as well as the leaf edge. Other interesting leaf shape changes occur in ivy (*Hedera helix s.l.*), but this type of variability appears to be rarely found or perpetuated in herbaceous species. 'Crispum' forms of *Tanacetum vulgare* (Tansy), *Origanum vulgare* (Wild Marjoram) and *Teucrium scorodonia* (Wood Sage) are, however, widely grown.

Leaf tomentum changes

Hairs on the leaves and stems can impart a silver-grey hue. Ten species are grown as forms which are significantly more silver-grey than the wild-type forms. The most widely offered for sale are *Artemisia absinthium* (Wormwood as the form 'Lambrook Silver', amongst others), *Calluna vulgaris* (Heather as the forms 'Sister Anne' and 'Silver Queen'), *Erica tetralix* (Cross-leaved Heath as the form 'Alba Mollis'), *Salix alba* (White Willow as the form 'Splendens'), *S. repens* (Creeping Willow as the var. *argentea*), *Sorbus aria* (Common Whitebeam as the form 'Lutescens' with its young leaves creamy-white tomentose) and finally *Thymus polytrichus* (Wild Thyme as the var. *lanuginosus*).

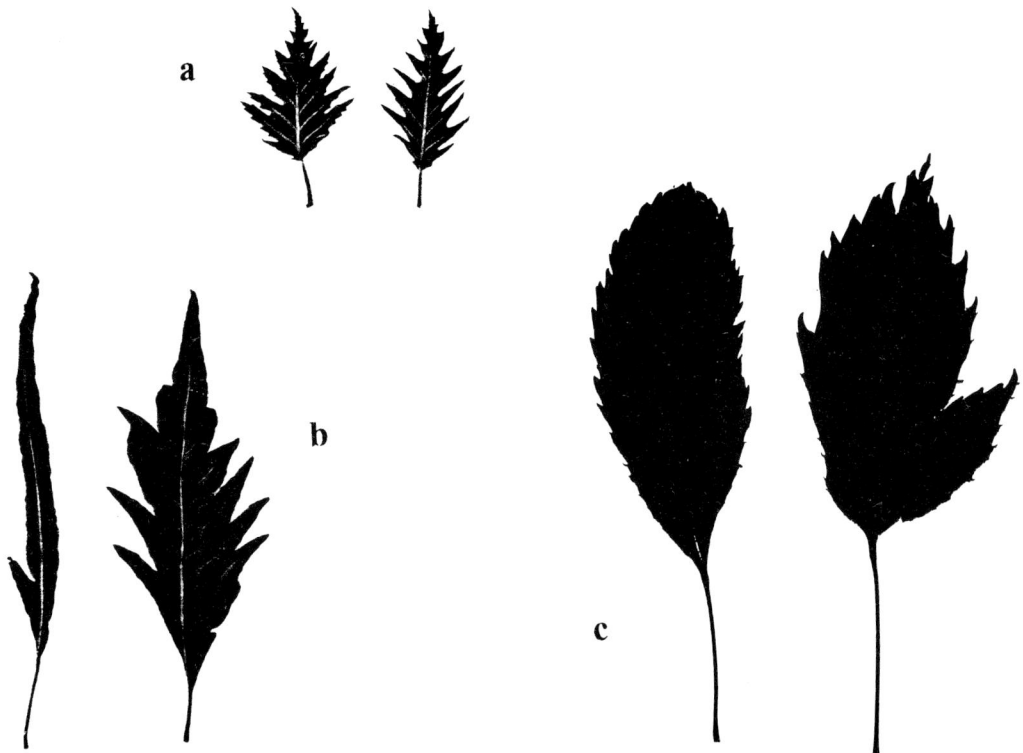

Fig. 1. Silhouettes of leaf shape changes. *a*, Cut-leaved Alder (*Alnus glutinosa*); *b*, Cut-leaved Beech (*Fagus sylvatica*); *c*, One-leaved Ash (*Fraxinus excelsior*).

Other growth form variants

Pendulous and prostrate forms

Weeping forms of trees and shrubs offer the gardener a restful shape seldom found in the wild, whilst prostrate forms offer the prospect of ground cover. Twenty-four species of normally upright native plants have pendulous and/or prostrate forms listed in *The Plant Finder*. I will consider them both together, since many prostrate forms can be converted to pendulous forms by the gardener staking them in youth or grafting them onto the normal form at a suitable height. Perhaps the most popular conversion of this nature is the 'Kilmarnock Willow', a prostrate/pendulous form of *Salix caprea* (Goat Willow). Other trees and shrubs with 'Pendula' or similar forms include *Betula pendula* (Silver Birch in the forms 'Tristis' and 'Youngii'), *Buxus sempervirens* (Box), *Carpinus betulus* (Hornbeam), *Corylus avellana* (Hazel), *Crataegus monogyna* (Hawthorn in the form 'Pendula Rosea'), *Fagus sylvatica* (Beech in the forms 'Pendula' and 'Purpurea Pendula'), *Fraxinus excelsior* (Ash), *Ilex aquifolium* (Holly), *Juniperus communis* (Juniper in a number of forms including 'Hornibrookii' and 'Repanda'), *Populus tremula* (Aspen), *Prunus avium* (Wild Cherry as the form 'Plena-Pendula'), *Quercus robur* (Pedunculate Oak), *Salix purpurea* (Purple Willow), *Salix repens* (Creeping Willow as the form 'Boyd's Pendulous') and *Taxus baccata* (Yew as the forms 'Dovastoniana' and 'Repandens'). Amongst shrubs, perhaps one of the more interesting forms to both botanist and gardener is the prostrate form of *Cytisus scoparius* (Broom) which occurs naturally on sea cliffs in Western Britain. A single nursery in *The Plant Finder* lists a prostrate form of *Ulex europaeus* (Gorse).

Fastigiate and pyramidal forms
These growth forms offer compact growth, and the architectural shapes thus created have earned a place in the garden for 16 wild species. Yet many of these fastigiate forms are surprisingly scarce, e.g. *Quercus robur* (Pedunculate Oak), *Populus tremula* (Aspen), *Betula pendula* (Silver Birch), *Carpinus betulus* (Hornbeam), *Alnus glutinosa* (Alder) and *Crataegus monogyna* (Hawthorn). In contrast, *Sorbus aucuparia* (Rowan, as the 'Sheerwater Seedling') *Juniperus communis* 'Hibernica' (Irish Juniper) and *Taxus baccata* (Yew as the form 'Fastigiata') are widely offered for sale.

Size changes
Neat and compact growth forms tend to please gardeners. Staking becomes unnecessary, weeds are suppressed more effectively and flowering subjects make a more intensely colourful display. It is not then surprising that out of 56 wild plant species offered for sale in some form that differs in size from the wild-type, 41 are offered in 'Nana', 'Minor' or other dwarf or more compact forms. *Lychnis flos-cuculi* (Ragged-robin in its 'dwarf form') is a good example, with flowering stems less than 20 cm high. There is a similar form of the related *Silene dioica* (Red Campion).

Larger forms than occur in the wild are less frequent, but include a number of species with 'Splendens', 'Superba' and other forms which produce generally larger flowers or flower in greater abundance. *Persicaria bistorta* 'Superbum' (Common Bistort), *Lythrum salicaria* 'Firecandle' (Purple-loosestrife) and *Ranunculus ficaria* var. *major* (Lesser Celandine) are significantly larger than normal wild-types. Forms of plants with larger than normal leaves are even more scarce. *Andromeda polifolia* 'Macrophylla' (Bog-rosemary) is an interesting example.

Flowers

Flower colour changes
The human being has an insatiable appetite for the unusual. This is perhaps why such a wide range of wild plants are cultivated which differ only in their flower colour from their relatives in the wild. They are also, perhaps, less obviously 'weeds' and so can be cultivated with less fear of censure from an ignorant audience. Seventy-eight species are listed in *The Plant Finder*, many with more than one colour form. Some forms such as *Ajuga reptans* 'Pink Elf' (Bugle) seem less attractive than the wild form. A personal interest in a genus may also introduce an element of 'stamp collecting' and lead to the cultivation of colour forms which if occurring as the only colour form might not be considered to be at all garden-worthy. Examples include species such as *Ranunculus ficaria* (Lesser Celandine) which can be obtained in flower colours ranging from white and lemon through yellow to a deep orange.

Uncommon flower colour changes in a familiar plant may shock, for example the red-flowered form of Kidney Vetch (*Anthyllis vulneraria*) or the red-flowered Cowslip (*Primula veris*), whilst some colour changes add a new dimension of enjoyment to a well-known species and can tease the botanist. The blue-flowered forms of Wood Anemone (*Anemone nemorosa*), e.g. 'Robinsoniana', may be mistaken at a casual glance for a non-native species such as *Anemone blanda*.

Double-flowered plants
Double flowers seem to provoke strong reactions in gardeners, with some enjoying them and some hating them. Thirty-nine species of native British plants with double flowers are listed in *The Plant Finder*. Seven of them are offered by 20 or more nurseries. They are *Aquilegia vulgaris* (Columbine) with a number of forms such as 'Nora Barlow', *Calluna vulgaris* (Heather) with forms such as 'H.E. Beale', *Caltha palustris* (Marsh-marigold), *Cardamine pratensis* (Cuckooflower), *Crataegus laevigata* (Midland Hawthorn in the form 'Paul's Scarlet'), *Filipendula vulgaris* (Dropwort) and *Silene uniflora* (Sea Campion).

A further 13 species are offered by 10 or more nurseries. This group includes such attractive plants as *Anemone nemorosa* (Wood Anemone), *Geranium pratense* (Meadow Crane's-bill), *Lychnis viscaria* (Sticky Catchfly), *Silene dioica* (Red Campion), *Ranunculus ficaria* (Lesser Celandine) and *R. acris* (Meadow Buttercup). Nine nurseries offer double-flowered *Ranunculus repens* (Creeping Buttercup), though consumer resistance to purchase when not in flower must be high. In flower, it is a singularly attractive plant. The double primroses, many of fairly recent origin have, through micro-propagation, become popular in the last few years. A few ancient varieties, closely related to the wild-type, still survive, for example 'Alba Plena' and 'Lilacina Plena'.

Perhaps best considered here are those composite flowers which normally have a single outer row of ray florets, but which have sported forms in which all the florets develop rays. The Daisy (*Bellis perennis*) is the best example with a whole range of forms from neat pompoms as flowers such as 'Dresden China' to the monstrous forms used for spring bedding. *Achillea ptarmica* (Sneezewort) is the other commonly grown composite which has undergone a similar change. Forms like 'The Pearl' make a good long-lasting cut flower.

Other teratologically changed forms
If double flowers disgust, this latter group may appal. Yet botanically they are of considerable interest, providing experimental material for the geneticist and physiologist studying origins and development of plant organs, and they can be of great value for the flower arranger in pursuit of an unusual shape or form. They can also be viewed as the funfair end of the plant world. Considering the modifications to flowers first, examples in the genus *Primula* include Jack-in-the-Green primroses (*Primula vulgaris*) with the calyx growing out to form a leafy ruff behind each flower, hose-in-hose Primrose and Oxlip (*P. elatior*), a form with a petaloid calyx, and Galligaskins and Jack-a-napes-on-horseback forms with an enlarged calyx, part petaloid and part leafy.

The modification of flower parts into petaloid or leafy structures can vary greatly. In Lesser Celandine (*Ranunculus ficaria*) the form 'Colarette' shows the stamen filaments expanded into small petal or nectary-like structures, producing a neat orange button centre to the flowers, whilst in 'Green Petal' the stamens and petals form dwarf, leaf-like structures, creating a tiny cabbage-like flower. The 'Tuberosa Monstrosa' form of Daisy (*Bellis perennis*) has green, unripe strawberry-like flowers. The 'Virescens' form of Wood Anemone (*Anemone nemorosa*) has the tepals and stamens converted to small leaves and much resembles the rose plantains. These latter forms have been cultivated for centuries. Each

Fig. 2. The 'Rose-flowered' form of Ribwort (*Plantago lanceolata*) with many flowers subtended by leaves (from Masters 1869). (×0.4)

---

plantain flower is subtended by a leaf to produce a green, double rose-like, flower spike. Greater Plantain (*Plantago major*) is widely offered for sale in this form. Ribwort (*P. lanceolata*, Fig. 2) is rarely grown and what might, perhaps, have been the most attractive of all the rose-flowered forms, the Hoary Plantain (*P. media*), may have now been completely lost in cultivation. One final 'fun plant' is the Hen and Chicks Daisy (*Bellis perennis*). In this form the main capitulum (the mother hen) is surrounded by a halo of smaller capitulae (the chicks).

Other teratological variants include contorted forms of Hazel (*Corylus avellana*), spiral forms of Soft-rush (*Juncus effusus*), viviparous forms, for example of Tufted Hair-grass (*Deschampsia cespitosa*), monoecious forms of usually dioecious species (thus assuring a regular berry crop), for example Butcher's-broom (*Ruscus aculeatus*). The Plant Finder lists 15 species which I have placed in this category of 'other teratological variants'. Perhaps best placed here are flowering time variants such as early Dutch Honeysuckle (*Lonicera periclymenum*) which can greatly extend the flowering season. This aberration is particularly notable in the Glastonbury Thorn, a winter-flowering variant of Hawthorn (*Crataegus monogyna*), a plant with a rich folk lore.

Conclusions

Table 2 summarizes the numbers of wild British plant species listed in *The Plant Finder* as offered for sale. It is a surprisingly diverse range of species with, perhaps, an even more surprising range of forms. Whilst some of these forms arose in gardens many were found in the wild. The loss of so many of our wild places and their wild plants must have reduced the chances of finding garden-worthy forms of wild plants. Yet enough remains to provide a valuable resource from which many new garden-worthy forms may arise. The debt which the gardener owes the British wild flora is a considerable and generally underestimated one. The

Table 2. Number of species of native British wild plants listed in *The Plant Finder* (Philip 1989).

	No. of species
Wild type	470
Wild type and cultivars	491
Variegated forms	55
Aurea forms	31
Purpurea forms	16
Leaf shape changes	31
Increase in hairiness	10
Pendulous/prostrate forms	24
Fastigiate forms	18
Size change	56
Flower colour change	78
Fruit colour change	7
Double-flowered forms	39
Other teratologically changed forms	15

wild flora undoubtedly has much more to contribute to our gardens. This probability provides one more reason for conserving what is left of our wild plants and why gardeners should press for their conservation. Conservationists should be alert to the possibility that new garden-worthy forms may spring up from time to time and should be pleased to draw them to the gardener's attention.

References

Ellis, R.G. (1983). *Flowering Plants of Wales*. Cardiff: National Museum of Wales.
Masters, M.T. (1869). *Vegetable Teratology, an Account of the Principal Deviations from the Usual Construction of Plants*. London: Robert Hardwicke (for the Ray Society).
Philip, C. (1992). *The Plant Finder* (6th ed). Whitbourne: Headmain Ltd. in association with the Hardy Plant Society.
Stace, C. (1991). *New Flora of the British Isles*. Cambridge: Cambridge University Press.

3

Ergasiophygophytes in the British Isles–plants that jumped the garden fence

E. CHARLES NELSON

National Botanic Gardens, Glasnevin, Dublin 9, Ireland

ABSTRACT. Gardens in the British Isles, with their rich exotic flora which has been expanding rapidly especially during the past four centuries, provide ample opportunities for not less than 50,000 alien plants to escape into the wild and become naturalized. The already-established adventive flora comprises *ca.* 588 genera, *ca.* 1280 species representing each continent; there is a preponderance of aliens from the 'Old World', and from temperate regions. Also numbered among the interlopers are cultivars (cultivated garden varieties) that have been more successful in escaping and surviving than their parental species. Reports of new alien plants are frequent, and some of the established ones require investigation to determine their correct identities and names. Gardens will always contain an abundant reservoir of potential invaders, and some of these may unpredictably become serious pests in parts of these islands.

Introduction

The richness of the garden flora of the British Isles is extraordinary. One of the best measures of the diversity is the listing of commercially available plants issued annually as *The Plant Finder* (Philip & Lord 1992); this 1992 database (which does not include annual and agricultural seeds) amounted to 55,000 names which, allowing each cultivar separate status, was more than fifteen times the total native and naturalized flora of these islands. Put another way, there are more than fifty thousand different flowering plants, gymnosperms and ferns growing in gardens and nurseries, each with the potential to become an ergasiophygophyte (an intentionally introduced plant that has escaped from cultivation (Schroeder 1969; Zizka 1985)), to invade wild spaces and mingle with and, in some instances, entirely oust the established flora.

The extraordinary size of the cultivated flora is not declining. Contrariwise, it is steadily increasing as new cultivars are selected or bred; modern techniques of manipulating plant tissues may yield yet more cultivars, more rapidly. It also expands annually as novel exotics are introduced from the wild; the ease and speed of air travel, coupled with the general frequency with which we travel outside our archipelago, enhance the opportunities for exotic species to reach these islands and become established in gardens and thence in the wild (cf. Wace 1979, 1986).

The established garden flora comprises plants with contrasting ecological propensities; these pose different threats in different places. Frost-sensitive African taxa are unlikely to become invasive on the upper slopes of Scottish mountains, and a cultivar that requires acid soil will not colonize limestone pavements. Plants from tropical and subtropical habitats are usually unable to survive the rigours of the northern European winters without artificial protection; such plants will be killed by cold every time they escape from cultivation. Even the hardiest plants (defined in terms of climate) are not always candidates for naturalization: any plant that is miffy in a garden will be even more miffy beyond the pampering hands and watchful eyes of a gardener.

While there is a greater and a still-increasing chance nowadays of cultivated plants 'jumping the garden fence', not every one can 'jump the fence' and become naturalized beyond the confines of the garden. Some cultigens (and some species) are effectively sterile and even incapable of vegetative reproduction without human assistance. With tolerable certitude these represent no threat. Some plants may succeed in 'jumping the fence' often but on 'the other side' many cannot find permanent niches; unable to compete with established plants these wither away as often as they escape and thus pose little problem in wild habitats. As a general rule, therefore, any plant that has to be treated with extra special care may be ruled out. However, having deleted such plants from the list of possible invaders, there is still a huge – incalculable – potential lucky-dip from which anywhere, any time, a cultivated plant may use a garden as a 'springboard', become established in the wild and begin to invade.

Deliberate introductions

Sometimes the 'springboard' has to be turned into a 'catapult' and a helping hand given to the potential invader. Fortunately there are relatively few people who set out deliberately to introduce exotics into the wild; embellishing the flora of these islands in the manner of Frederick Burbidge (Praeger 1949: 36) or the Rev. Thomas Gibson (Doogue 1984) is not tolerated politely nowadays (cf. Akeroyd 1994).

Most of the naturalized species and cultivars presently recorded in the British and Irish floras are accidental invaders of wild places. While they were deliberately introduced into gardens, they spread by natural means, thence into 'the wild'. There are celebrated examples of deliberate direct introductions: perhaps the planting in 1886 of *Sarracenia purpurea* into Irish midland bogs is the best example ([Burbidge] 1886; Nelson 1986). To the list can be added (although success or failure is not recorded) John Templeton prancing over Cave Hill at Belfast gleefully sprinkling rhododendron seeds (Nelson 1990c), and John Claudius Loudon's friend with his packets of *Heracleum mantegazzianum* (Nelson 1991)!

Transporting and introducing plants

There are many accounts of deliberate plant introductions by collectors (cf. Gorer 1978), and the dispersal of plants by natural agencies (wind, water, animals) is well documented (e.g.

Murray 1986). There is one aspect of plant transportation and introduction which is rarely discussed because it is impossible to quantify plants which are accidentally transported by horticulturists and gardeners, 'piggy-back'.

Few instances of accidental introductions are clearly documented. Two exotic examples (of Australian species) are of horticultural interest: a species of *Wahlenbergia* (named *Campanula gracilis*) was raised by William Curtis before 1800 from ' ... mould that came with the roots of other plants'; *Callicoma serratifolia* Andrew germinated in a Wardian case sent from Sydney in February 1834 (Nelson 1990a).

Before recent times when stringent phytosanitary requirements were imposed by national authorities in order to exclude harmful organisms, plants were transported from country to country in their native soil accompanied by many of the plants and animals that inhabited the same niche. In the eighteenth century and earlier, saltwater leaking into the plant containers while these were on board ships, killed many plants and obviously prevented innumerable successful deliberate and accidental introductions. After 1833 when Nathaniel Ward 'invented' his famous case, the crop of 'passenger' animals and plants was better protected from salt, so the rate of introduction must have increased dramatically. Each Wardian case was filled not just with living plants growing in unsterilized soil, but also with concealed seeds, bulbs and corms, and other living things including animals and microscopic organisms. The occurrence of exotic wood-lice (e.g. *Trichoniscus stebbingi* Patience, *Trichorina tomentosa* Budde-Lund from Venezuela, *Nagara nana* Budde-Lund from Madagascar (Pack-Beresford & Foster 1911; Foster 1911)) in such gardens as the National Botanic Gardens, Glasnevin, Dublin, and Belfast Botanic Gardens Park, has long been acknowledged as the result of importation with plants. Indeed the Palm House at Glasnevin is celebrated for its exotic wood-lice; Pack-Beresford & Foster (op.cit.) also listed *Haplophthalmus danicus* Budde-Lund and *Armadillidium nasatum* Budde-Lund, from the greenhouse.

The presence of an Australian amphipod (*Arcitalitrus dorrieni* (Hunt)) in Britain and Ireland has been highlighted by recent collections which extend the species' documented range. Known as the Woodland Hopper, this small crustacean has been collected in south-western Britain, the Inner Hebrides and in Ireland; its Irish habitats are invariably in or near gardens which have or had fine collections of exotic plants, particularly tree-ferns. The possibility that the amphipod was introduced from Australia in batches of rooted, growing plants is generally recognized (see e.g. O'Connor, O'Connor & Holmes 1991).

In the nineteenth century the extent of accidental piggy-back transportation of plants must have been prodigious. Just one example can be cited: hundreds of tree-fern trunks were brought to the British Isles for sale to gardeners, and the presence of *Tmesipteris* sp. in an Irish garden provides incontrovertible evidence that these still carried their substantial epiphyte floras (Nelson 1988, 1992). Even today illicit importation of plants in soil takes place and there is every chance of plants and animals being introduced 'accidentally'.

To extrapolate from the tree-fern example is easy, but there can never be definitive answers to such questions as these: could apparently inexplicable introductions (e.g. *Juncus planifolius* in Connemara) have resulted from the importation of other plants growing in unsterilized soil; did zealous gardeners and nurseryman in previous centuries introduce *Inula salicina* or *Sisyrinchium angustifolium* (see below)?

A brief history of gardening

The escapee phenomenon is not one of recent origin. When the first gardener/farmers turned the first sod and planted the first seed – an exotic plant – they opened unlimited opportunities for plants to arrive, thrive and escape. Cultivation commenced in Ireland and Britain about five thousand years ago (see e.g. Molloy & O'Connell 1987), and exotic species and cultigens have been establishing themselves in wild places since then, including those that came piggy-back. With mankind's assistance, the so-called weeds of cultivation have entered our flora not because they were beautiful but because they were in the right place at the right time to hitch a ride.

Ornamental gardening began in the mediaeval period; flower gardens were established by monastic communities and thereafter the quest for ornamental plants began in earnest (Harvey 1981; Nelson 1990b). At first the principal source of plants was the European mainland and western parts of Asia, the Old World. As European mariners extended their adventures, plants arrived from southern Africa, the Far East and then the Americas after 1492. Latterly, after 1770, Australasia began to contribute to our garden flora (Gorer 1978; Nelson 1983a, 1990a). Each of these regions has provided a cornucopia of garden plants and each has contributed also to the alien flora.

Established alien flora

All the continents have contributed to the alien flora of the British Isles. Approximately 1300 species and cultigens are listed by Stace (1991) as adventive members of the flora (i.e. 'Introduced'); these represent almost 600 genera (see Table 1). Forty percent of the exotic species and half the genera are native elsewhere in Europe. Australasia, Africa and South America have contributed many fewer species mainly because those land masses have relatively small floras adapted to temperate climates.

Europe
That half of the genera and *ca.* 40% of the species naturalized in the British Isles are of European origin is not unexpected. The first introductions date from the beginnings of agriculture – the weeds of cultivation and the cultivated plants. When ornamental gardening commenced more new species were introduced. Escapees of European origin include useful plants such as *Allium ampeloprasum* var. *babingtonii* (an onion which produces bulbils in place of florets; Stearn (1978)), antique medicinal herbs and garden flowers like *Erysimum* (= *Cheiranthus*) *cheiri* (Wallflower), *Galanthus nivalis* (Snowdrop) and *Dianthus caryophyllus* (Clove Pink), and shrubs and trees including *Aesculus hippocastanum* (Horse-chestnut), *Acer pseudoplatanus* (Sycamore) and *Malus domestica* (Apple).

Asia
Plant introductions into our gardens from Asia have occurred for many centuries; cereals for agriculture were the earliest examples. Tulips from Turkey, cedars from Lebanon, lilacs from Persia, have been grown in British gardens since the sixteenth century at least.

The major influx from Asia, resulting in the contingents that have had the most dramatic impact on the flora, occurred in the last half of the nineteenth century and the early decades of the present one. More than forty (micro-)species of *Cotoneaster* from China and the

Table 1. Approximate numbers of adventive taxa from separate continents represented in the flora of the British Isles compiled by extracting from Stace (1991) a list of taxa (species, hybrids, cultivars) designated as 'Introduced'.

* The total number of genera is not the sum of the numbers cited as *ca.* 130 genera are native in more than one region.

Continent	Number of Genera	Number of Species
Europe	292	535
Asia	127	236
Africa	59	75
North America	112	184
South America	61	94
Australasia	44	73
Cosmopolitan	10	14
[cultivars]	68	91
TOTALS	588*	1303

Himalayas are now naturalized, and all reached the British Isles as garden plants. *Rhododendron* species from Asia became fashionable in the mid nineteenth century. Notorious Asian adventives include such weeds as *Fallopia japonica* (= *Polygonum cuspidatum*) and its congeners, *Impatiens glandulifera*, *Heracleum mantegazzianum* and (perhaps) *Rhododendron ponticum*. The pretty lawn weed *Veronica filiformis*, first introduced *ca.* 1780 from the Caucasus, was being sold as a plant suitable for rock gardens as recently as the 1920s; Praeger (1939) commented:

> 'This pretty little creeping Speedwell from Asia Minor introduced into gardens all over the country, is proving a nuisance not only in beds, but by spreading exuberantly in lawns. I do not know of its being fully naturalized in wild ground but it seems likely we shall soon hear of that.'

Buddleja davidii (Butterfly-bush), from central China, came in batches of seeds sent by French missionaries shortly after 1886 (Père Soulié was perhaps the first to introduce this species as *B. variabilis*); now it is a plague on derelict city plots in Dublin (Jackson & Skeffington 1984) even sprouting like the archetypal wallflowers from cracked chimney stacks and crumbling brick and stone walls.

Africa
The flora of Africa as a whole contains relatively few species that are well adapted to temperate, frost-prone habitats, and thus the number of adventive species is proportionately small despite the popularity of South African species as garden plants since the late sixteenth

century. Coastal cliffs in south-western Britain and eastern Ireland have been colonized by Aizoaceae (*Carpobrotus*, *Lampranthus*, etc.) while the annual *Lobelia erinus* persists year after year by seeding. Stace (1991) records *Phygelius capensis* as naturalized in County Wicklow, but it has not escaped from gardens yet. Among the bulbous taxa a few have occupied precarious niches in the Channel Islands and the Isles of Scilly. *Chasmanthe* and *Crocosmia* are perhaps the most prolific and aggressive taxa although the plants occurring as adventives are likely to be cultivars rather than 'pure' species (see below).

The Canary Islands have in recent decades provided *Echium pininana*, which is marching round the southern slopes of Howth Head, County Dublin, with a host of other exotics.

North America
Importation of plants from temperate North America rapidly followed European settlement on the eastern coast; the heyday was the eighteenth century when there was a fashion for 'American' gardens stocked with such plants as *Kalmia* and *Magnolia*. Western species introduced unsuccessfully into Russia in the eighteenth century (cf. Steller 1988: 18-24), made no impact until the mid nineteenth century when the moist temperate areas were explored by such collectors as David Douglas. His introductions included *Ribes sanguineum*, *Gaultheria shallon*, *Spiraea douglasii* and *Rubus spectabilis* (Steller had gathered it in 1741), all now 'weeds'. Northern American adventives are both spectacular and serious pests: for example *Sarracenia purpurea* may be categorized as both a spectacle and a pest, while the water-weeds *Elodea canadensis* and *E. nuttallii* are nuisances. *E. canadensis* was naturalized in Ireland as early as 1836 according to John New (1854), who discovered it in County Down:

> 'About eighteen years ago, the pond at Waringstown was cleared of overhanging trees, when the Anacharis was immediately observed after the planting of some aquatics, making it necessary several times during the summer to clear it out.'

E. nuttallii has more recently (since 1966) been rapidly replacing *E. canadensis* as a serious weed of waterways (Simpson 1984).

The much-debated North American element of the Irish flora, although now disdained by plant geographers (cf. Perring 1962; Nelson 1979a), is of interest in the present context because it included *Sisyrinchium angustifolium* Miller (misnamed *S. bermudiana*, cf. Walsh & Nelson 1987: 66) and *Hypericum canadense* whose Irish populations may be of garden origin; equally they may represent endemic taxa similar to, but genetically and perhaps morphologically distinct from, the North American ones.

South America
Among the most spectacular and potentially most damaging plants in the alien ornamental flora are the giant 'rhubarbs', *Gunnera tinctoria* and *G. manicata,* both from South America. The common opinion is that *G. tinctoria* is the more widely naturalized; it reproduces by seed and is all-smothering. Recent studies of a unique symbiotic relationship between *Gunnera* and a cyanobacterium, *Nostoc punctiforme* L. (Osborne *et al.* 1991, 1992) which lives within the stems and rhizomes, suggest that in the right conditions (plenty of moisture and little frost) *G. tinctoria* is unstoppable. *G. tinctoria* was introduced as a garden subject to France before 1845 yet the first report of it being naturalized in western Ireland dates from 1939 (as *G. manicata*: Praeger 1939).

Other Chilean and Argentinian species naturalized in the British Isles are *Amomyrtus luma* (= *Myrtus apiculata*) and *Gaultheria* (= *Pernettya*) *mucronata. Margyricarpus pinnatus* (Lam.) Kuntze (= *M. setosus* Ruiz & Pavon), a native of the Ecuadorian Andes, seems to have disappeared; it is not listed as naturalized today although Praeger (1946) reported it growing 'on exposed moorland... near the western extremity of the wind-swept Dawros promontory' in County Donegal – '... there is no garden within miles where a South American plant like this would be grown... A mystery!', he exclaimed.

Also from South America, *Fascicularia bicolor* Ruiz & Pavon (Bromeliaceae: misnamed *F. pitcairniifolia*), a hardy and spectacular relative of the pineapple is indubitably a garden throw-out in the Isles of Scilly and Guernsey and behaves as if it might 'jump the fence' in County Donegal. *F. bicolor* does not reproduce by seed in gardens in these island, but by producing lateral rosettes can build up substantial colonies rapidly.

Australasia
The other antipodes have as yet contributed only about seventy species. In the mildest regions of the British Isles naturalized populations of shrubby genera such as *Hebe*, *Griselinia* and *Acacia* are established outside the gardens in which they were originally planted as shelter belts. *Phormium tenax* (New Zealand Flax), another useful shelter belt species tolerant of salt-laden winds, is also naturalized.

Included among the Australasian adventives is surely one of the most successful alien plants, *Epilobium brunnescens* (New Zealand Willowherb), which was being sold by Irish nurseries ('a very pretty little creeping plant which soon spreads over the ground') in the early 1900s for sixpence a plant or four shillings per dozen (Nelson 1989b). By 1939 Praeger reported *E. brunnescens* (as *E. nummulariaefolium* Cunningham) 'in fair quantity at 1500 feet in the Hag's Glen, Macgillicuddy's Reeks' (County Kerry), and 'at 1200 feet on Blackstairs [Mountains]' (counties Carlow and Wexford). *Acaena* spp., also from New Zealand, have been 'outlawed' in Northern Ireland: it is an offence 'to plant or cause to grow in the wild' any species of *Acaena*.

Other men's weeds

It is important to remember that adventive floras are not unique to the British Isles. Mankind has been transporting plants around the globe for many millennia. Even the most remote Pacific islands have alien species. One hundred and forty one species classified as anthropochore (plants whose occurrence in a region is due to the activities of mankind) are recorded on Easter Island (Zizka 1991); the adventives range from *Cirsium vulgare* and *Trifolium repens* to *Cocos nucifera* L. (Coconut Palm). As European settlements spread in the Americas, southern Africa and Australasia, garden plants were imported to stock colonial gardens and, as in these islands, species escaped into the wild. In New Zealand and also in Australia *Ulex europaeus* is a pest. *Hakea* spp. and *Acacia* spp. (e.g. *A. cyclops* A. Cunn.) have escaped from gardens in the Cape of Good Hope and have overwhelmed thousands of hectares of species-rich fynbos (e.g. Palgrave 1983). *Cortaderia selloana* is a weed in California. *Buddleja davidii* grows on Kauai, and *B. asiatica* Lour. on Hawaii (Carlquist 1980).

Cultivars: examples established in the wild

For various reasons, cultivars (artificial hybrids, selected clones and others of assorted origins) are not so abundant among our naturalized flora as might be expected. Many cultivars are specially selected variants which must be maintained very deliberately as clones or seed lineages, perpetuated artificially to ensure their garden-worthy characteristics, which often, but not always, break down or dissipate when plants escape and become naturalized.

When recording exotic plants botanists do not always identify the 'wild' plant with a named cultivar and thus exact equations are lost. Sometimes, however, identifiable cultivars do become established in wild places; examples are *Fuchsia* 'Riccartonii' and *Crocosmia* × *crocosmiiflora*.

Fuchsia 'Riccartonii' (Fig. 1)

To the casual observer, *Fuchsia* is a highly successful plant in the British Isles. To quote Praeger (1934) it had 'run wild in the west' and has bamboozled many visitors into thinking it is a true native: it has been a feature of western Ireland since the 1850s (Hind 1857; H. 1872). Two distinct variants occur in Ireland and Britain, one with fat buds ('Riccartonii') and the other with slim buds (*F.* cf. *magellanica* s.s., usually listed as *F. magellanica* var. *gracilis*) (Nelson 1983b); they rarely grow together, considerable areas being occupied by a single clone. *Fuchsia* is a useful shrub for hedging fields because cattle do not like eating the foliage and shoots.

'Wild' *Fuchsia* does not appear to spread by seeding nor does it 'sucker', although stems and stem fragments will root with alacrity. A famous shrub of *Fuchsia* 'Riccartonii' at Glanleam (Valentia Island, County Kerry) spread by the rooting of peripheral shoots, and eventually a huge colonial plant formed (Fitzgerald 1879).

In native American habitats *Fuchsia magellanica* and its relatives are pollinated by humming-birds. Observations in western Ireland suggest that there is a tendency to self-incompatibility in the naturalized *Fuchsia* and that a second clone must be available for abundant fruit production. The absence from western Europe of a suitable pollinator does not hinder seed production when different clones grow in close proximity, for example in gardens. As noted, in the British Isles the usual situation is that only a single clone is represented within a substantial area so seed production is relatively infrequent and fruits drop without swelling. However, when seeds are produced they are viable and will germinate. *Fuchsia* seedlings are occasionally reported but they are frost-sensitive and rarely survive winter except in very mild areas or exceptionally clement winters.

Crocosmia ×*crocosmiiflora* (Fig. 2)

The orange-blossomed, ditch-dwelling 'Montbretia' is an artificial hybrid, created when two southern African species, *C. pottsii* (small, red flowers) and *C. aurea* (pendent, golden flowers), were brought together in Europe as garden plants. The hybrid's history is well recorded: Victor Lemoine of Nancy raised the original seedlings before 1880 after deliberately fertilizing *C. pottsii* with pollen from *C. aurea* (Morren 1881), and other nurserymen followed suit, raising many hybrids of complicated parentage. *C.* × *crocosmiiflora* was undoubtedly a prize when gardeners discovered that the hybrid was salt-tolerant, hardy, and capable of rapid vegetative increase. No doubt 'Montbretia' was discarded on the compost heap once too often and thus 'jumped the garden fence'. The

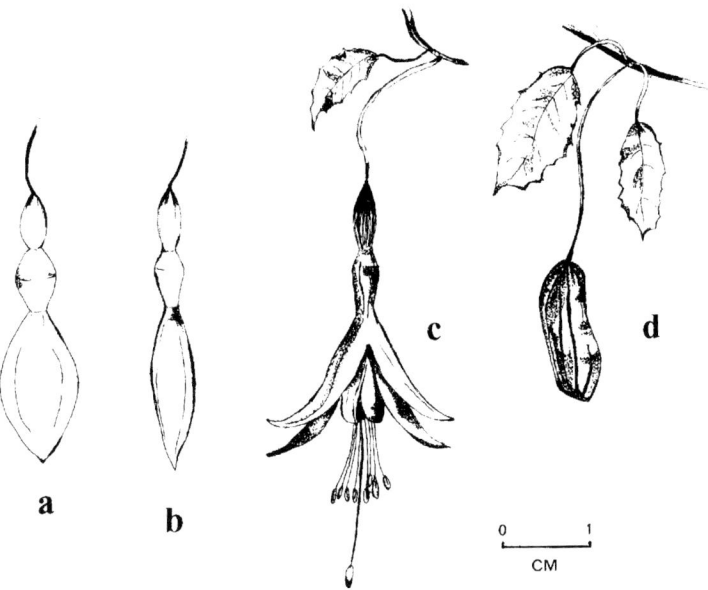

Fig. 1: *Fuchsia* naturalized in the British Isles. *a*, 'fat' bud of 'Riccartonii' (10 mm broad, sepals 7-9 mm wide); *b*, 'slender' bud of *F. magellanica* (this is the variant often labelled '*Fuchsia gracilis*') (6 mm broad, sepals 4-6 mm wide); *c*, flower of 'slender' (*b*) variant; *d*, fruit of *F. magellanica*. (Drawings by Bernie Shine, originally published in *BSBI News* **34**: 15 (1983).)

Fig. 2: *Crocosmia* naturalized in the British Isles (lateral views of flowers). *C. pottsii* (left) from County Donegal, the seed parent of *C.* × *crocosmiiflora* (centre) from County Clare; *C. aurea* (right), the pollen parent of *C.* × *crocosmiiflora*, which is *not* naturalized in the British Isles. (Drawings by Bernie Shine.)

consequences are as colourful as *Fuchsia,* indeed the two are not infrequent fellow denizens of Irish bohreens.

The commonly naturalized 'Montbretia' is most probably one of the original clones, but little attention has been paid to its taxonomy. One of its parents, *C. pottsii,* is naturalized in a few places (e.g. Lough Eske, County Donegal), but the other parent is certainly too tender to survive long out of doors in the British Isles and reports that it is also naturalized perhaps refer to yellow-flowered cultivars of *C.* × *crocosmiiflora.* Other *Crocosmia* taxa are also occasionally observed as adventives but their precise identity is by no means certain and some of the records cited by Stace (1991) are questionable.

Rhododendron cf. *ponticum*

Rhododendron ponticum is native in Bulgaria, northern Turkey, Lebanon and the Caucasus, with extraordinary disjunct populations in south-western Spain and Portugal (Chamberlain & Cullen 1982, map 94); subspecies have been described but are not maintained by Chamberlain & Cullen (op.cit.) because the differences are either 'too trivial' or are not geographically discontinuous. This plant is confidently reported as a fossil (both fruit and pollen) in interglacial peat deposits in Ireland (cf. Godwin 1975: 292-293; Mitchell & Watts 1970), and according to contemporary botanists (e.g. Cross 1975, 1982; Stace 1991) it is naturalized in many parts of Ireland and Britain.

But is the plant that has overwhelmed native oakwoods in County Kerry and invaded blanket peat in other places really *R. ponticum*? Could the invasive weed be a hybrid created unwittingly when veritable *R. ponticum* and, for example, the American species *R. maximum* L. (in European cultivation from 1734) were planted together in gardens about the end of the eighteenth century; other possible parents include *R. catawbiense* Michaux (also an American species, introduced 1809). *Rhododendron* is renowned for the ability of taxa to cross-pollinate with alacrity and produce fertile seedlings: there are countless named (and unnamed) hybrids grown in gardens throughout these islands.

The evidence that the 'wild rhododendron' is a hybrid is circumstantial but telling, although botanical, accounts (e.g. Cross 1975; Stace 1991) have not absorbed this. *R. ponticum* was introduced to British gardens about 1763 (Harvey 1988) perhaps from Gibraltar; by 1800 (at least) it was cultivated in Ireland. By 1806, hybrids were reported between the plants known to contemporary gardeners as *R. ponticum* and *R. maximum.* The major exodus of '*ponticum*' from gardens many not have occurred until about a century ago: evidence from Killarney (County Kerry) suggested to Cross (1975) that 'planting' did not take place there until *ca.* 1890.

Having seen native *R. ponticum* in Turkey, Cox & Hutchinson (1963) suggested the naturalized plant was a hybrid because it differed in its leaf morphology, and Clarke (1976: 743) noted that some naturalized plants in Surrey and Sussex showed clear evidence of hybridity in their floral characters. In his 'Biological Flora of the British Isles' study, Cross (1975) referred to Cox & Hutchinson's account but did not discuss the possibility that the plants he had studied in wild places were not *R. ponticum s.s.,* and most recently Stace (1991: 347) has stated simply that 'most or all our plants are ssp. *ponticum,* from SE Europe and SW Asia' without any reference to possible contamination by other species.

Other problem taxa

What are we left with: the misty, murky 'don't-knows' and 'can't-be-sures', plants like *Leucojum aestivum* which is an improbable native in Ireland, or the mysterious *Mycelis muralis* in places such as The Burren; these may have arrived with the aid of gardeners. *Erica vagans* in County Fermanagh may be included: Stace stated it was introduced and Nelson & Coker (1974) have argued that it is unlikely to have been deliberately planted! It is often impossible to be certain, and like trying to explain the presence of such plants as *Arbutus unedo* in Ireland, once claimed as a cultivated escapee (e.g. Lami 1958), it is impossible without unimpeachable historical evidence. In the case of *A. unedo* there is fossil pollen data stretching back almost six thousand years which suggests that it is a true native (Vokes 1967). Fossil evidence can be surprising: without exception, botanists supposed that *Erica erigena* was native in western Ireland (Nelson 1979b, 1989a), but its pollen does not appear in peat profiles suggesting that this species arrived in Ireland in the fifteenth century (500 years ago), perhaps as packing for Spanish wine casks (Foss & Doyle 1988).

Conclusion

We know much more about the flora of the world than perhaps the entire congregation of botanists who preceded us. We do not have to guess about the native habitats of *Gunnera tinctoria* or *Ribes sanguineum:* there are no doubts about the status of these plants in the British Isles nor that their presence in our flora is entirely due to the enthusiasms of gardeners and the great plant collectors. David Douglas, Augustine Henry and their fellows may have done horticulture an inestimable service and brightened our lives, but they also performed a great disservice by taking plants out of their native habitats and bringing them to localities where, freed from an overburden of pests and pathogens, they effervesced in their new milieu and 'jumped the garden fence'.

The future for exotic botanists (if that is their correct denomination) is exciting: one never knows what will turn up in a remote Connemara bog (recent arrivals have included *Juncus planifolius* and (before 1988) *Haloragis micrantha*), or in a New Forest pond (*Crassula helmsii*). We have no idea what delightful 'weed' a horticultural company will start distributing free of charge in its bales of compost, nor can we predict the potential of a new annual, a new herbaceous perennial, or a new tree for naturalizing.

References

Akeroyd, J.R. (1994). Some problems with introduced plants in the wild. *In* A.R. Perry & R.G. Ellis (eds), *The Common Ground of Wild and Cultivated Plants*, pp. 31-40. Cardiff: National Museum of Wales.

[Burbidge, F.W.] (1886). The Irish flora ... *The Garden* **30**: 239 [published under nom-de-plume 'Veronica'].

Carlquist, S. (1980). *Hawaii, a Natural History.* 2nd edition. Lawai: Pacific Tropical Botanical Garden.

Chamberlain, D. & Cullen, J. (1982). *Rhododendron ponticum. Notes from the Royal Botanic Garden, Edinburgh* **39**: 313-316.

Clarke, D.L. (1976). *Rhododendron ponticum. In* W.J. Bean, *Trees and Shrubs Hardy in the British Isles*. 8th edition. Vol. 3, pp. 741-744. London: John Murray.

Cox, P. & Hutchinson, P. (1963). Rhododendrons in north east Turkey. *Rhododendron Yearbook, London* **17**: 64-67.

Cross, J.R. (1975). Biological flora of the British Isles. *Rhododendron ponticum* L. *Journal of Ecology* **63**: 345-363.

Cross, J.R. (1982). The invasion and impact of *Rhododendron ponticum* L. in native Irish vegetation. *Journal of Life Sciences, Royal Dublin Society* **3**: 209-220.

Doogue, D. (1984). History of the flora. *In* P.W. Jackson & M.S. Skeffington, *Flora of Inner Dublin*, pp. 5-24. Dublin: Royal Dublin Society.

Fitzgerald, P. (1879). *Fuchsia ricartoni* as a hedge plant in Ireland. *Gardening Illustrated, 7 June*: 204-205. [See also *The Garden* **2**: 72.]

Foss, P.J. & Doyle, G.J. (1988). Why has *Erica erigena* (the Irish heather) such a markedly disjunct European distribution? *Plants Today* **1**: 161-168.

Foster, N.H. (1911). On two exotic species of woodlice found in Ireland. *Irish Naturalist* **20**: 154-156.

Godwin, H. (1975). *The History of the British Flora. A Factual Basis for Phytogeography*. 2nd edition. Cambridge: Cambridge University Press.

Gorer, R. (1978). *The Growth of Gardens*. London: Faber.

H. (1872). Flowering shrubs in the south of Ireland. *The Garden* **2**: 295.

Harvey, J.H. (1981). *Mediaeval Gardens*. London: Batsford.

Harvey, J.H. (1988). *The Availability of Hardy Plants of the Late Eighteenth Century*. [Sine loc.]: Garden History Society.

Hind, W.M. (1857). Dingle and its flora. *Phytologist* **2**: 97-100.

Jackson, P.W. & Skeffington, M.S. (1984). *Flora of Inner Dublin*. Dublin: Royal Dublin Society.

Lami, R. (1958). A propos du bois d'Arbousiers de Coat Hermit en Plourivo (Côtes-du-Nord). *Bulletin du Laboratoire Maritime du Dinard* **44**: 39-46.

Mitchell, G. F. & Watts, W.A. (1970). The history of the Ericaceae in Ireland during the Quaternary epoch. *In* D. Walker & R.G. West, *Studies in the Vegetational History of the British Isles*, pp.13-21. Cambridge: Cambridge University Press.

Molloy, K. & O'Connell, M. (1987). The nature of the vegetational changes at about 5,000 BP. with particular reference to the elm decline: fresh evidence from Connemara, western Ireland. *New Phytologist* **106**: 203-220.

Morren, E. (1881). Notice sur le Montbretia crocosmiaeflora (hybrida) de M.V. Lemoine. × Montbretia aureo-pottsi. *La Belgique Horticole* **31**: 299-303, tab.14.

Murray, D.R. (ed) (1986). *Seed Dispersal*. London: Academic Press.

Nelson, E.C. (1979a). Ireland's flora: its origins and composition. *In* E.C. Nelson & A. Brady, *Irish Gardening and Horticulture*, pp. 17-35. Dublin: Royal Horticultural Society of Ireland.

Nelson, E.C. (1979b). Historical records of the Irish Ericaceae, with particular reference of the discovery and naming of *Erica mackaiana*. *Journal of the Society for the Bibliography of Natural History* **9**: 289-299.

Nelson, E.C. (1983a). Australian plants cultivated in England before 1788. *Telopea* **2**: 347-353.

Nelson, E.C. (1983b). *Fuchsia magellanica*. *BSBI News* **34**: 15.

Nelson, E.C. (1986). Carnivorous plants in Ireland. 2. *Sarracenia purpurea*. *Carnivorous Plant Newsletter* **15**: 45-47.

Nelson, E.C. (1988). Some Australasian ferns in Irish gardens. *Kew Magazine* **5**: 129-136.
Nelson, E.C. (1989a). Heathers in Ireland. *Botanical Journal of the Linnean Society* **101**: 269-277.
Nelson, E.C. (l989b). An archaic duet: New Zealand's contribution to Ireland's garden heritage. *Annual Journal Royal New Zealand Institute of Horticulture* **16**: 4-11.
Nelson, E.C. (1990a). '... and flowers for our amusement': the early collecting and cultivation of Australian plants in Europe and the problems encountered by today's taxonomists. *In* P.S. Short (ed), *History of Systematic Botany in Australasia*, pp. 285-296. Melbourne: Australian Systematic Botany Society.
Nelson, E.C. (1990b). 'The garden to adorne with all varietie': the garden plants of Ireland in the centuries before 1700. *Moorea* **9**: 37-54.
Nelson, E.C. (1990c). 'It's a long way to Tipperary' – finding natural history archives in Ireland, with an appendix listing archives. *Archives of Natural History* **17**: 325-347.
Nelson, E.C. (1991). Small ad for a giant hogweed. *BSBI News* **57**: 26-27.
Nelson, E.C. (1992). Ferns in Ireland, cultivated and wild, through the ages. *In* J.M. Ide, A.C. Jermy & A.M. Paul (eds), *Fern Horticulture: Past, Present and Future Perspectives*. Proceedings of the International Symposium on the Cultivation and Propagation of Pteridophytes, London, 7-11 July 1991, pp. 57-86. Andover: Intercept.
Nelson, E.C. & Coker, P.D. (1974). Ecology and status of *Erica vagans* in County Fermanagh, Ireland. *Botanical Journal of the Linnean Society* **69**: 153-195.
New, J. (1854). [*Elodea canadensis*]. *Phytologist* **5**: 88.
O'Connor, J.P., O'Connor, M.A. & Holmes, J.M.C. (1991). Ornamental plants and the distribution of exotic amphipods (Crustacea) in Ireland. *Irish Naturalists' Journal* **23**: 490-492.
Osborne, B.A., Doris, F., Cullen, A., McDonald, R., Campbell, G.J. & Steer, M. (1991). *Gunnera tinctoria*: an unusual nitrogen-fixing invader. *Bioscience* **41**: 224-234.
Osborne, B.A., Cullen, A., Jones, P.W. & Campbell, G.J. (1992). Use of nitrogen by the *Nostoc-Gunnera tinctoria* (Molina) Mirbel symbiosis. *New Phytologist* **120**: 481-487.
Pack-Beresford, D.R. & Foster, N.H. (1911). The woodlice of Ireland. *Proceedings of the Royal Irish Academy* **29 B** (4): 165-190.
Palgrave, K.C. (1983). *Trees of Southern Africa*. 2nd edition. Cape Town: Struik.
Perring, F.H. (1962). The Irish problem. *Proceedings of the Bournemouth Natural Science Society* **52**: 36-48.
Philip, C. & Lord, A. (1992). *The Plant Finder* (6th ed). Whitbourne: Headmain Ltd. in association with the Hardy Plant Society.
Praeger, R.L. (1934). A contribution to the flora of Ireland. *Proceedings of the Royal Irish Academy* **42 B** (5): 55-86.
Praeger, R.L. (1939). A further contribution to the flora of Ireland. *Proceedings of the Royal Irish Academy* **45 B** (10): 231-254.
Praeger, R.L. (1946). Additions to the knowledge of the Irish flora, 1939-1945. *Proceedings of the Royal Irish Academy* **51 B** (3): 34.
Praeger, R.L. (1949). *Some Irish Naturalists*. Dundalk: Dundalgan Press.
Schroeder, F.G. (1969). Zur Klassifizierung der Anthropochoren. *Vegetatio* **16**: 225-238.
Simpson, D.A. (1984). A short history of the introduction and spread of *Elodea* Michx in the British lsles. *Watsonia* **15**: 1-9.
Stace, C. (1991). *New Flora of the British Isles*. Cambridge: Cambridge University Press.

Stearn, W.T. (1978). European species of *Allium* and allied genera of Alliaceae: a synonymic enumeration. *Annales Musei Goulandris* **4**: 83-198.
Steller, G.W. (1988). *Journal of a Voyage with Bering 1741-1742.* (ed O.W. Frost). Stanford: Stanford University Press.
Vokes, E. (1967). The late- and post-glacial vegetation development on sandstone and limestone at Killarney, Co. Kerry (abstract). *Journal of Ecology* **55**: 57P-58P.
Wace, N.M. (1979). Assessment of dispersal of plant species – the car-borne flora in Canberra. *Proceedings of the Ecological Society of Australia* **10**: 167-186.
Wace, N.M. (1986). Australia – the isolated continent. *In* A.J. Gibbs & H.R.C. Meischke (eds), *Pests and Parasites as migrants: an Australian Perspective*, pp. 3-22. Canberra: Australia Academy of Science.
Walsh, W.F. & Nelson, E.C. (1987). *An Irish Florilegium II*. London: Thames & Hudson.
Zizka, G. (1985). Anthropochore Pflanzenarten der Varangerhalbinsel und Sör-Varangers. *Dissertationes Botanicae* **85**: 3-102.
Zizka, G. (1991). Flowering plants of Easter Island. *Palmengarten Wissenschaftliche Berichte* **3**.

4

Some problems with introduced plants in the wild

J.R. AKEROYD

24 The Street, Hindolveston, Dereham, Norfolk, NR20 5BU

ABSTRACT. Much of our alien flora derives from garden escapes, but an increasing proportion has been introduced directly into the wild. People in Britain are belatedly aware of the destruction of our semi-natural grasslands, especially flower-rich meadows, which has led to the currently fashionable practice of re-creating grassland floras from commercially available wildflower seed-mixtures. However, much wild flower seed comes from agricultural and horticultural sources and is often not native in origin. Alien variants of widespread grassland species, together with a range of weed contaminants, have been widely reported in Britain. These alien plants are bound to have an ecological and genetic effect on the flora and vegetation of this country, although any long-term deleterious effects on native plants are still unclear. The use of wildflower seed should not be discouraged, but care needs to be taken in establishing or re-seeding stands of native or semi-native plants, and the aims of any project of this kind should be assessed.

Introduction

I have been interested for some years in the alien plants that turn up widely in town and countryside, not just species but intraspecific variants. Much of our adventive flora derives from garden escapes, but an increasing proportion has been introduced directly into the wild. It has become increasingly apparent that many adventive plants derive from the sowing of commercial wildflower seed. The plant conservation organization Plantlife therefore asked me to research the whole topic of wild flower seed in Britain (Akeroyd 1992b). My report to Plantlife is unpublished, but we are preparing a popular account for a wider readership. The present paper draws on material in the report. Here I shall concentrate on the more 'philosophical' aspects of the use of wild flower seed and make general comments about our countryside, its wild flowers and our attitudes towards them.

Introduced plants can cause various problems. Not only do many behave as injurious weeds of agriculture, ruderal and wetland communities, but they can also disrupt or replace native floras. Much of our alien flora derives from gardens but in some cases one wonders why a plant was cultivated in the first place. Common Blue-sow-thistle (*Cicerbita macrophylla*) is an example of one that is both invasive and scruffy. Another is Japanese Knotweed (*Fallopia japonica*), introduced as a garden plant and still spreading at a terrific rate throughout the British Isles, notably in derelict industrial land and cemeteries in South Wales. The alien weed problem, although of major economic importance, will not be discussed here. The theme of this paper will be the problems posed by the introduction of alien variants of native species, which in some cases have been given taxonomic recognition, into the countryside through amenity plantings of grassland species.

There is a great deal of variation in the British flora and these variants frequently exhibit distinct distributions and differing degrees of weediness. This variation usually has a genetic basis and the classic studies of Bradshaw and his co-workers (e.g. Bradshaw 1972), especially of heavy metal tolerance in grasses, together with the many examples cited by Briggs and Walters (1984) bear this out. This is certainly the case in many invasive taxa, as discussed by Baker (1974), but unfortunately there has been relatively little work carried out in this field in recent years.

The flora of Britain and Ireland already includes a rich alien element, although it is often difficult to be certain of a plant's native status. An example is provided by the well-known and well-loved Mediterranean element in our weed flora which probably invaded our shores during the Neolithic period and, ironically, is now disappearing at a rapid rate; it could be argued that this element is rarely native anywhere. Many alien plants used for amenity purposes have a special place in our affections. A group of aliens, of garden origin, that have become a feature of our countryside is the monkeyflowers (*Mimulus*), which brighten streamsides and river gravels, especially in the north. Probably the best example is the Horse-chestnut (*Aesculus hippocastanum*). So firmly rooted in our landscape, it is as much a part of our culture as the pub, the church or cricket on the village green: for more details I refer you to Akeroyd (1991b).

Planting wild flowers for amenity

In recent years there has been a drive supported by the BSBI, Plantlife and others, to plant native species for amenity, and a fashionable trend in the 1980s has been the planting of so-called traditional wild flowers, either individually or in packaged seed mixes, in order to replace the vanished grassland flora of the habitat loosely termed 'meadow'. A meadow is not just a flowery place: it is strictly grassland managed wholly or partly by mowing (the old song 'one man went to mow' gives the correct usage of the term). When managed along traditional lines, the meadows can be a rich carpet of colourful flowers and when we think of wild flowers in Britain we usually think of flower-rich meadows. It is a famous statistic that over 95% of this country's traditionally managed meadows have disappeared since 1945.

I can recollect, as a boy of twelve, being taken to watch badgers somewhere near Dorchester. To reach the woodland edge where the badgers foraged, we walked in the summer dusk through a meadow filled with flowers. I remember vividly the clovers and Yellow-rattle and the lovely feeling of the seed heads banging against the legs; I also remember the varied insect life in the same meadow. Twenty five years on from that distant summer evening, I stood not far from Truro in Cornwall, with fellow BSBI members Lady Rosemary FitzGerald and Keith Spurgen, looking over a gate into a meadow bounded by an ancient, overgrown hedge (which included the almost extinct Plymouth Pear, *Pyrus cordata*). The field, apparently forgotten, was a rich sward of mixed grasses interspersed with clovers and vetches, spotted-orchids and Yellow-rattle. It was a moving image of a lost countryside. We gazed across this colourful vista, witnessing a scene that would have been familiar to an earlier generation and one preserved by our writers and artists from Chaucer and Shakespeare to the lovingly crafted nature essays of Richard Jefferies and Henry Williamson. The bright green hues of rye-grass leys and suburban lawns fail to strike such a chord!

With the loss of these flower-rich meadows we have all suffered the terrible loss of a great slice of our culture, comparable to that represented by the great cathedrals. Much of this loss was inevitable as the horse gave way to the tractor and today's level of farm mechanization was reached, with farmers giving up mixed farming to specialize in livestock or arable. Above all, there are also too many instances of gratuitous destruction financed by the government, with the sanction of its frequently ill-informed and malevolent academic advisers. Nor are current plans for set-aside land likely to help – and yet more government cash to sustain the sort of management practices that took away our meadows, as the farming community seeks further ways to change the landscape. In 1991 a farmer in the little hamlet of Seifton in Shropshire where I then lived, dammed a brook running through his land to make an ornamental lake and trout fishery. Part of his plan involved creating a nature reserve, or so it was said. This did not materialize but he did cut down a fine old willow carr, banked his lake, sowed a mixture of coarse agricultural grasses and planted a couple of Garden Centre cypresses. Had he been advised about wildflower seed mixtures, it might have improved things a bit!

Change is inevitable in the countryside, but not on the scale that we have seen since 1945. Oliver Rackham has put the case with scholarship and eloquence (Rackham 1986). One piece of historical documentation in particular emphasizes the points that he makes when demonstrating the history of our countryside. In the August and September of 1940, the *Luftwaffe* exhaustively photographed southern and eastern England in a prelude to an invasion that never came. The landscape that they recorded in that distant summer would 'have been recognizable by Sir Thomas More' (Rackham 1986). For a glimpse of the atmosphere of that vanished world, have a look at the first chapter of *Daniel Martin* by John Fowles (himself a good botanist), describing a wartime harvest field.

Obviously with public awareness of this loss, and green matters having become fashionable (not least among academics and the chattering classes), there has grown up a movement to recreate our lost meadow flora. Television, radio, magazines, conservation organizations, including the BSBI and Plantlife, all urge us to go green and plant wildflowers. But, perhaps because systematic botany has been marginalized or eliminated, dismembered in both secondary and tertiary education in Britain, few of the pundits have thought through the consequences of their new enthusiasm.

Firstly, where is all the seed going to come from? We are already witnessing the pillaging of wild populations of bluebells, cowslips and snowdrops (though the native status of the snowdrop is arguable; but it is indeed a 'wild flower' and a most respectable member of our flora). Secondly, has enough thought been given to the impact on the landscape created by the establishment of these plants. More seriously, how will the meadows be managed in order to maintain a flowery sward? As our vanished meadows represent a product of sophisticated medieval and renaissance agrarian technology, all maintained by diligence, sweat and the hard labour of our forbears, is not the current fashion unlikely to yield long-term results? Above all (and few give this enough thought) are we sure that what we are sowing or planting really are native wild flowers? In other words, is the seed of wild native origin? This is my own main concern, as things currently appear to be going rather badly wrong. My fears arise both from my experience as a scientist and from my passionate love for the countryside of Britain and Ireland.

Let us consider some of these problems. The extensive sowing of wild flower seed is bound to have effects and these can be summarized in two categories, scientific and aesthetic. Not only will the study of plant distribution be made difficult and in the long run impossible (as we were warned at an AGM of this Society in 1989 by our late friend and colleague Dr John Dony) but there will undoubtedly be some ecological and genetic damage to the flora if seed is not of native origin. The greatest danger is to the natural and unique gene pool represented in the British flora, in other words our native biodiversity. This is especially significant as many of the species are grassland fodder plants; in other words they are of potential economic importance. Any damage will be complex to assess and we do not as yet have much evidence on how processes such as competition by, and hybridization with, these alien variants may proceed in natural populations. These events certainly do occur and a documented example is Sickle Medick (*Medicago sativa* ssp. *falcata*) hybridizing with the introduced crop plant Lucerne (*M. sativa* ssp. *sativa*) in the Breckland of East Anglia.

There is also the aesthetic and cultural problem; the use of wild flowers in gardening and landscaping, given the right management, has much aesthetic appeal and would certainly have improved the landscape such as the one in Shropshire mentioned above. Sowing of wild flower seed has much educational value, especially in urban areas, and can bring pleasure to those deprived of the open countryside, a very important but often overlooked point. Wildflower seed can also be used in the context of the conservation of individual species. However, we should be wary of encouraging wholesale attempts to recreate supposedly ancient landscapes of flowery meadows. Here we are treading on dangerous cultural ground and perhaps the word we are seeking is *authenticity*. Are we not in danger of creating merely a Rural Museum or Theme Park rather than a living landscape. True authenticity requires the people and the culture that made the meadows, but it certainly will not be achieved by foreign seed. In the Netherlands, where there is a small nation with a denser population, an even more despoiled flora than our own and large tracts of artificially created polder landscapes, the scientific problem has been recognized for a number of years. As in this country, public awareness of loss of habitat has led to a proliferation of the sale of seed and wild flower plants. Native wild plants have been reintroduced into the wild since the early 1950s and there is now an extensive network of wildlife gardens, especially on road verges in urban areas. From a scientific point of view, the sheer volume of alien plants has effectively brought about the effective 'end of plant geography in the Netherlands' (Mennema 1984). That may sound apocalyptic but the point is valid.

Mennema (op.cit.) praised the role of the wildlife parks and their help in education, but with a warning that not only were botanists losing track of real patterns of distribution, but that much of the plant material used was foreign and not of the same genetic provenance as native Dutch plants. He was especially concerned at the growth of the commercial sector which feeds the demand for wild flowers. On the basis of published catalogues, he concluded that much of the material was of commercial cultivar origin, cheaper and more accessible than native stocks. His final remarks are disturbing indeed: 'We hope that the cultivation of native plants will be restricted to the Netherlands, a relatively small part of Europe ... we sincerely hope that the warning [to other countries] will be superfluous'. Since that warning was issued, we in Britain have gone a long way down this road, and there is evidence from other countries such as Finland that landscapers have used seed from outside native regions. The philosophy behind the introduction of wildflower seed into the wild is sound enough, especially as so much damage has been done to grasslands that we often have no choice but to restore rather than repair if we are to see again anything like the rich carpets of flowers enjoyed by earlier generations. A wide range of accredited seed mixtures is now available and there seems to be no strong reason why widespread plants cannot be sown to provide pleasure and habitat enrichment for wildlife.

The landscaping of road verges involves a compromise between the practical and the aesthetic, but the current fashion has certainly put more pressure on suppliers to meet demand. Seed mixtures have always been obtained from the cheapest and most convenient sources, many of them foreign, and motorway verges provide an early example of alien variants of native species. A variant of Kidney Vetch (*Anthyllis vulneraria* ssp. *polyphylla*), from eastern and central Europe turned up widely on roadsides in the 1960s, including the embankments of an embryonic M4 near Maidenhead (Akeroyd 1991a). In recent years this variant has been replaced by *A. vulneraria* ssp. *carpatica* var. *pseudovulneraria*, a plant characteristic of alpine valleys in central Europe. I have several times seen this latter subspecies, which is similar in general appearance to the native plant, offered for sale as native Kidney Vetch on wildlife and market stalls all over the country.

When I was teaching genecology at Reading University in the mid 1980s, I became particularly interested in a group of similar foreign legumes on road verges and elsewhere – kidney vetches, medicks, clovers and Bird's-foot-trefoil. All are robust, tall and erect. These had presumably been selected for agricultural use, both for their vigour and for the erect habit which would facilitate cutting as a hay crop. In other words amenity plantings were being grown from readily obtainable commercial agricultural seed, much of it foreign. Their use was a practical move to improve fertility of the newly established landscapes through the fixation of nitrogen and probably also, to some extent, because of their attractive appearance. They had not been sold as wildflower seed as such but the Dutch experience, personal observations and those of the late Dr Dony, suggested that at least a proportion of the seed in this country was derived from similar agricultural or foreign sources. At least one well-known seed merchant, recommended by the BSBI, admits to using alien Bird's-foot-trefoil and Kidney Vetch in seed mixtures.

Many people still do not realize that there is a difference between native and agricultural variants of common plants. Consider Bird's-foot-trefoil (*Lotus corniculatus*), one of our most familiar and best-loved wildflowers. The grim scale of grassland destruction has meant that one is now pleased to see this charming, widespread and remarkably tenacious plant. The village green of my suburban childhood still retains not only a healthy population of Fiddle Dock (*Rumex pulcher*) but scattered stands of Bird's-foot-trefoil, and less intense, kinder management would probably encourage these to spread. This could also be achieved by scattering seed from adjacent heathy common land. The 'modern' way to encourage its spread would be to sow commercial wildflower seed, but perhaps without a thought as to its origin.

In the mid 1980s, the University of Reading sowed a bank with Bird's-foot-trefoil. Magnificent plants appeared, more or less erect, robust, half a metre or so tall, quite different from the prostrate to weakly ascending native variant in an adjacent old meadow. It would have been an easy matter to have consulted the staff of the University Herbarium and we could have banked a little seed from the meadow and sowed it with more positive results. Professor David Jones and co-workers (e.g. Jones 1977, Bonnemaison & Jones 1986) have examined some of the morphological differences between native and alien variants of this species as a preliminary study of the genetic effects of alien plants on native populations. The alien variant of Bird's-foot-trefoil has been widely planted in Britain and can be seen almost anywhere especially along road verges, which, ironically, are often the last refuges of the more compact native variant. A recent report of this alien from Ireland on new verges of the Bray-Newtown Butler bypass in Co. Wicklow, the main road south from Dublin, suggests further expansion in its use as amenity sowing. There are fewer reports of such plants from Ireland, presumably because that country has a less well-developed modern road system. Remember though that almost all wildflower seed in Ireland is foreign: it comes from Britain.

It is hard to believe that these plants are not having some sort of genetic or even ecological effect on our flora. For example, an interesting consequence of the Reading campus sowing was the appearance of plants of Wild Carrot (*Daucus carota*) apparently as a contaminant of the seed mixture, which also contained various grasses. The Wild Carrot is by no means an unusual species in Berkshire but these plants were apparently annual whereas the native form is usually biennial or a short lived perennial in Britain. In southern Europe and parts of the Mediterranean region, however, it is frequently annual. The Reading plants did not persist although their seed may still be present in the soil's seed bank. It cannot be a good thing for our native flora to have foreign seed dispersed so indiscriminately.

An extreme example of alien wild flower seed

The most extreme case I have come across was in the autumn of 1991 on the Gog Magog hills on the outskirts of Cambridge (Akeroyd 1992a). In 1989, The Magog Trust purchased designated set-aside land in order to establish new chalk grassland on former arable fields, with the intention of managing this as a public amenity. Some very reputable consultants gave advice, grants and donations were assembled, and the land was ploughed and sown with a mixture of grass and wild flower seed from very reputable suppliers. However, by September, Dr Max Walters and other Cambridge botanists were noting all sorts of unusual plants on the site. Some of these plants were indeed native, including an excellent suite of fumitories (*Fumaria* spp.), a dwarf variant of Fool's Parsley (*Aethusa cynapium* ssp. *agrestis*), locally common in stubble in Cambridgeshire, and the nationally scarce Cornfield Knotgrass (*Polygonum rurivagum*). However some of the other plants were a bit more exotic, although these would still be regarded, in the horticultural trade, as proper constituents of wildflower seed. It is important to remember that in the trade you are dealing with 'Pop Larkin' rather than the Council of the Royal Horticultural Society. There was evidence for the previous species composition of the site, since Chris Preston and I had surveyed the area during the BSBI's Monitoring Scheme in 1987. Notes on some of the interesting plants that turned up are given below:

- Yarrow (*Achillea millefolium*), a tall, robust variant, often with pink flowers, probably of garden rather than chalk grassland origin.
- Cornflower (*Centaurea cyanus*), many with pinkish-purple flowers, suggesting a garden origin.
- Corn Marigold (*Chrysanthemum segetum*), a plant of cultivated land on more acid soils; rare on chalk (A.J. Silverside, pers.comm.). It is a garden escape rather than a native plant in many areas.

- Bristly Hawk's-beard (*Crepis setosa*), an alien species, now quite widespread on roadsides in southern England, that has been spread with grass seed.
- Shasta Daisy (*Leucanthemum* × *superbum*), a variant often grown in gardens and probably a hybrid between *L. maximum* of gardens and the native Oxeye Daisy (*L. vulgare*) of grassland. It is more robust and branched and has bigger flowers than the native plant. This is my indicator species for the use of wildflower seed mixtures and has become a feature of new roads and grassland. It can be spotted easily from a moving car.
- Black Medick (*Medicago lupulina*), a robust agricultural variant, rather than the less vigorous and more prostrate plant of chalk grassland. The larger plant was cultivated in Cambridgeshire in the 1950s (P.D. Sell, pers.comm.); seed of another robust but less hairy plant is also found in some wildflower mixes. Both may be from southern Europe.
- Hawkweed Oxtongue (*Picris hieracioides* ssp. *grandiflora*), a robust southern European variant of a native Cambridgeshire plant.
- St Martin's Buttercup (*Ranunculus marginatus* var. *marginatus*), a buttercup, closely related to the native Hairy Buttercup (*R. sardous*). This undoubtedly alien plant is native to the Balkans and the Crimea to Iran and has been previously noted in Britain only as a rare casual; it was established for some years (as var. *trachycarpus*, with tuberculate fruits) on St Martin's in the Isles of Scilly, where it is apparently now extinct. This plant had been included as 'buttercup' in the seed mixture!
- Fodder Burnet (*Sanguisorba minor* ssp. *muricatus*), a robust fodder variant of Salad Burnet that can be recognized by its bigger, more winged, rugose fruits. It probably represents the 'salad burnet' of the seed mixture, as it is widely sold under this name for herb gardens. Salad Burnet (*S. minor* ssp. *minor*) grows nearby in a narrow strip of chalk grassland that has survived between a wood and an arable field.
- Red Clover (*Trifolium pratense*), a robust plant which has been widely grown as a fodder variant; this was probably previously established on the site.
- It was a little early for grasses and they had not flowered at the time the site was examined in 1991. However, in 1992 a very robust variant of Meadow Oat-grass (*Helictotrichon pratense*) covered a large part of the site. This plant, probably from the Balkans and south-eastern Europe, seems to be rather widely distributed in southern England.

It thus appears that the seed sources used by purveyors of our wildflower seed mixes require some scrutiny. I suspect we have given many the benefit of the doubt, not least because some seed mixtures are very good. For instance, both Emmersgate Seeds and John Chambers Seeds harvest mixed seed from the famous North Meadow at Cricklade in the Upper Thames Valley. However, it is not known how much bulking-up goes on and few individuals are able to recognize rogue genetic variants.

Even horticultural seed has its contaminants. My wife raised 50 plants from a packet of seed of Greater Selfheal (*Prunella grandiflora*). Of these, 48 were typical, one was the Spanish ssp. *pyrenaica* (possibly a separate species) and another was wild Selfheal (*Prunella vulgaris*). Fortunately, Selfheal sown on the Gog Magog hills turned out to be the native plant rather than one of these others.

A way forward?

Finally, let me set out a few thoughts on the possible way forward in this complex matter. Although I am deeply concerned about the effects of sowing wildflower seed in native plant communities, and I must emphasize that we need to think it through very carefully, I do not wish to discourage the use of this seed. I do, however, want people to be aware of the problem, to think about what they are doing and, if in doubt, to consult expert field botanists – apparently a rather infrequent event! It is essential that we do not confuse aesthetics with conservation. Fragments of native or semi-natural flower-rich grassland do remain on road verges, woodland margins, village greens, and especially in churchyards – Francesca Greenoak sums this up very well, showing how sacred ground is frequently the last vestige of grassland at a local level (Greenoak 1985). We would do well to imitate these sacred plots and to use them as a seed source with the approval of the appropriate authorities. Cowslip (*Primula veris*) is a great colonizer from such sites and will spread of its own accord.

A fascinating study of careful restoration of a wildflower-rich community was recently published by Dr Alison McDonald of Oxford University (McDonald 1992). In the Thames valley, for complex historical reasons, areas of ancient meadow survive on the flood plain around the city of Oxford. Somerford Mead is an ancient riverside meadow that had been improved with fertilizer during the 1950s and 1960s and ploughed for the first time in 1982. Six years ago, after a last crop of barley, grown without fertilizer, had been harvested, the land was sown with a seed-mix derived from nearby Oxey Mead. By 1989 a tall, mixed sward had been established that contained 80% of the species of Oxey Mead and, on a visit in June 1992, I was thrilled to walk into what was to all intents and purposes an old meadow; however, it had a more patchy sward without the characteristic lack of dominants of the true ancient grassland. It even had the early-flowering meadow ecotype, with ray florets, of Common Knapweed (*Centaurea nigra*). I had seen this variant the previous day in the old meadows at Reading University, slowly creeping back to their former glory.

Another heartening example was reported briefly a couple of years ago in *Country Life* (3 October 1991). After the construction of an underground reservoir near Bath, grassland was restored from local seed, collected before the work began. Also, in Sweden, successful experiments have been carried out to restore old meadow-land by a long-term mowing regime (Folkesson 1991). In these instances local seed and local practice have triumphed.

We must preserve such small areas of our native vegetation as remain and encourage those working towards this end. We should not be trying to recreate a vanished countryside, for we must accept that even when Richard Jefferies was writing in the 1880s, changes to the countryside were taking place. He certainly noted that farmers were locally improving meadows to the detriment of their flowers! The countryside has always existed in a state of dynamic, evolutionary change. Nevertheless, from time to time events move quickly, perhaps like in our own times, during one generation. There were still plenty of horses working the land during my childhood in the 1950s, contrasting with the early 1980s, when 65% of the Culm Measure grasslands of Devon were 'improved'. Although we can salvage little from the wreckage, we must remember that we have just caught the end of something glorious and have to be thankful for what remains. Let us cherish what little we have.

A cultural comparison

There is an interesting comparison with another aspect of our rural heritage, our folk music. By the turn of this century, the English folk-song and dance tradition had almost died, not least because of infiltration by material from the burgeoning pop music industry of the Music Hall. Ralph Vaughan Williams and others salvaged the last relics and many surviving folk tunes, for example 'Lovely Joan' and 'The Lark in the Morning', were to reappear in his own works, but so thoroughly did they permeate his music that it is frequently difficult to distinguish genuine folk melodies from those composed by himself.

Thanks to the great man and his colleagues, we still have our folk music today, if in a modified form. Just ten years after Vaughan Williams's death in 1958, a young and enthusiastic rock musician, Ashley Hutchings, began to cull songs and dances from the Vaughan Williams archives at Cecil Sharp House. He and his fellow musicians reworked them within the idiom of American rock 'n roll, itself a fusion of black man's blues and southern country music. Many of the dance tunes of Hutchings's present Albion Dance Band are difficult to distinguish as traditional or new. English country dance music has been conserved within a contemporary framework and is reaching more and more people. Perhaps here we have a model for our wild flowers.

But remember: despite the cultural value and aesthetic appeal of reconstructed traditional English dance music, it is still an artefact. In no way is it authentic. The social conditions that allowed folk music to evolve have gone, and probably my singing old songs as I cultivate my Norfolk allotment are closer to 'authenticity'! Similarly, with our wildflower meadows, we must be careful not to try to create a Museum of Rural Life, especially if we are creating it with foreign seed.

Both our folk music and our wildflowers represent but ragged remnants of a tradition. Perhaps the best way to conserve our wildflowers is in the garden, and this we must do at all costs. There are some good suppliers if you look and many of us can spare a patch of grass, an overgrown lawn, a wall or a dry bank for native plants However, to extend the concept of the garden into the countryside may have too many deleterious consequences. Present generations must hold the fort, for the biodiversity represented in our native flora is too precious to lose. Perhaps in the future the countryside will revert, through war, famine or plague, to something of its former anarchic state and floristic richness. Richard Jefferies' apocalyptic image of a London restored to nature after an obscure disaster in his strange novel *After London* is more in keeping with historical reality than the contrived, regulated utopia envisaged by conservationists, politicians and the chattering classes, but one seized upon by a public confined to rootless suburbia.

Acknowledgements

I should like to thank Plantlife for their financial support for my research on wildflower seed in Britain; Drs David Coombe and Philip Oswald for showing me round the Magog Trust's land near Cambridge, Dr Alison McDonald for demonstrating her experimental meadow at Somerford Mead near Oxford and Gwynn Ellis for his hospitality, much helpful discussion and the rescue of a lost draft of my manuscript.

References

Akeroyd, J.R. (1991a). *Anthyllis vulneraria* L. subsp. *polyphylla* (DC.) Nyman, an alien kidney-vetch in Britain. *Watsonia* **18**: 401-403.

Akeroyd, J.R. (1991b). Concerning conkers. *Countryman* **96**: 36-39.

Akeroyd, J.R. (1992a). A remarkable alien flora on the Gog Magog Hills. *Nature in Cambridgeshire* **34**: 35-42.

Akeroyd, J.R. (1992b). *Non-native Wildflower Seed in Britain*. Unpublished report, 20 pp. Plantlife.

Bonnemaison, F. & Jones, D.A. (1986). Variation in alien *Lotus corniculatus* L. 1. Morphological differences between alien and native British plants. *Heredity* **56**: 129-138.

Baker, H.G. (1974). The evolution of weeds. *Annual Review of Ecology and Systematics* **5**: 1-24.

Bradshaw, A.D. (1972). Some of the evolutionary consequences of being a plant. *Evolutionary Biology* **5**: 25-47.

Briggs, D. & Walters, S.M. (1984). *Plant Variation and Evolution*. 2nd edition. Cambridge: Cambridge University Press.

Folkesson, B. (1991). [The reappearance of meadow species on former arable land.] *Svensk Botanisk Tidskrift* **81**: 215-223.

Greenoak, F. (1985). *God's Acre: the Flowers and Animals of the Parish Churchyard*. London: Orbis.

Jones, D.A. (1977). On the polymorphism of cyanogenesis in *Lotus corniculatus* L. VII. The distribution of the cyanogenic form in western Europe. *Heredity* **39**: 27-44.

McDonald, A (1992). Succession in a three-year-old flood-meadow near Oxford. *Aspects of Applied Biology* **29**: 345-352.

Mennema, J. (1984). The end of plant geography in the Netherlands. *Norrlinia* **2**: 99-106.

Rackham, O. (1986). *The History of the Countryside*. London: Dent.

5

A proposal for an alien register in the British Isles

R. Gwynn Ellis

Department of Botany, National Museum of Wales, Cardiff, CF1 3NP

ABSTRACT. An alien is here accepted as a plant that has arrived in the British Isles directly or indirectly, as a result of human activity. The most recent estimate of the number of species of vascular plants growing outside cultivation in the British Isles is 3,400 and of these some 40% or about 1,360 are aliens. Many have become serious pests or noxious weeds but, because an interest in alien plants was unfashionable until comparatively recently, and their identification was often difficult or impossible, their spread into the countryside went largely unrecorded. The *New Flora of the British Isles* and the BSBI's new mapping scheme 'Atlas 2000', will increase our awareness of aliens, and allow their accurate identification, often for the first time. In addition, the possible change in our climate associated with global warming and the greenhouse effect, may not only trigger an increase in the rate of spread of already established although presently quiescent aliens, but may also allow new aliens to invade our countryside, posing further threats to our flora and landscape. Examples of some relatively recent invaders will be given and the results of today's more efficient monitoring will be compared with past experiences. The need to inaugurate an Alien Study Group and Alien Register will be explored and the anticipated benefits discussed.

Introduction

Contrary to popular belief at least among the more conservative members of the botanical community, aliens have been, until comparatively recently, a rather neglected component of the British flora. By 'comparatively recently' I mean the period from the end of the Second World War to the present day. An alien is here defined as a plant taxon that has been introduced or spread, either deliberately or accidentally, as a result of human activity. This definition covers not only the obvious aliens like garden escapes and some weeds but those plants that are native to just one area of Britain but have now spread much more widely, and also those genetic aliens that are mentioned elsewhere in this volume; however, this paper deals mainly with garden escapes and weedy species.

Aliens in the British Isles

There were two main reasons for this neglect of aliens. The first was a feeling, that still exists today, that there was something at best 'not quite right' and at worst definitely wrong about recording alien plants. They were not as important as native taxa and hence not worth recording in any detail. This is borne out by many of the local floras published in the 19th and the early part of the 20th centuries; if aliens were mentioned at all, it was often just to comment that they occurred in the county with no attempt at a detailed treatment. Just one example of this is provided by the two Floras of Glamorgan published completely independently in the early years of the present century. One of these, written by Revd H.J. Riddelsdell (a keen alien hunter) and published as a supplement to *Journal of Botany* (Riddelsdell 1907), includes *Impatiens glandulifera* (Indian Balsam) and *Matricaria discoidea* (Pineappleweed) in the section on aliens and gives two localities for each. The second *Flora of Glamorgan*, a supposedly more complete flora, edited by Prof. A.H. Trow and published by the Cardiff Naturalists' Society in parts between 1906 and 1910 (Trow 1911), makes no mention of Indian Balsam and although acknowledging that Pineappleweed had been recorded on numerous occasions, did not list the localities but dismissed it with the comment – 'Appears often as if introduced with seed used for feeding fowls'.

A second and perhaps more important reason for the neglect of alien plants was that without access to a good botanical library and herbarium, it was almost impossible to identify them. Most if not all *Floras* which dealt with the identification of British plants included only those aliens which were widely naturalized and were not obviously cultivated plants. Thus *Acer pseudoplatanus* (Sycamore) which looked like a native tree, grew in natural habitats, and was abundant over much of the British Isles, was always included. *Rhododendron ponticum* (Rhododendron) on the other hand, although quite widespread by the end of the last century, usually occurred in man-made habitats, especially parkland, looked like a garden plant and was ignored by *Flora* writers. This and other obvious garden escapes, together with more weedy aliens, although growing 'wild' in many areas were too exotic to be included in manuals concerned with British plants. The time had not yet arrived when it would become fashionable to record them; but this time was not that far into the future and the period following the Second World War was a momentous one for British botany.

A small group of University lecturers decided that the identification manuals available at that time were woefully inadequate. They had been written in the previous century and published in numerous editions many of which differed little from their predecessors, and all had inherited an obvious distaste for the more exotic alien plants. The two most notable were George Bentham's *Handbook of the British Flora*, first published in 1858 (Bentham 1858), and Joseph Hooker's *The Student's Flora of the British Islands*, first published in 1870 (Hooker 1870). These two are hereafter referred to as the 'pre-war *Floras*'. Both were admirable in their day but in the second half of the 20th century were very much out of date.

The time was ripe for a new *Flora of the British Isles* and the immortal initials CTW (standing for the authors Clapham, Tutin & Warburg), so well-known and treasured by all field botanists, came into being. Their *Flora* (Clapham, Tutin & Warburg 1952) contained for the first time detailed keys to, and descriptions of, all vascular plants (pteridophytes, gymnosperms and angiosperms), that were found growing wild in Britain. The authors wrote in the preface:

> 'There have also been changes in the flora itself, as well as in our knowledge of it, many of which will be apparent to every field botanist. A considerable number of introduced plants have become well established and some of them are now widespread. All those which persistently occur in natural or semi-natural communities

must be regarded as integral parts of the flora of the country and so should be included in any account of it. Others, which only maintain themselves by repeated reintroduction, are of frequent occurrence on rubbish tips, near ports and in railway sidings. These, though in a different category from the naturalized plants and of less importance to the ecologist, are of interest to the systematist and should also be included in a British flora.

This *Flora* was an immediate success; for the first time British field botanists could identify, or at least had a chance of identifying, most of the plants they were likely to meet with during a day's field work.

A second event that was to have tremendous consequences for the study of British plants took place shortly after. This was the decision taken in 1954 by the Botanical Society of the British Isles (BSBI) to press ahead with a survey to record the presence of all native and well-established introduced species of flowering plants and ferns in each of the *ca.* 3,500 10 km squares of the British Isles in which they occurred. Within the remarkably short time of five years the fieldwork was completed, and the *Atlas of the British Flora* was published in 1962 (Perring & Walters 1962). For the first time botanists were able to see at a glance the detailed distribution of any species within the British Isles. As far as alien species were concerned the maps were a revelation.

Rhododendron, introduced from southern Europe in 1763, which hardly rated a mention in the pre-war *Floras* or local Floras, was seen to be widespread throughout the British Isles (Fig. 1a), and Pineappleweed, similarly scantily treated in the pre-war *Floras*, could be seen to be present in almost all lowland squares and many of the upland ones as well (Fig. 1b), a quite remarkable rate of spread when one considers that it was not recorded in the wild until about 1871. These and many other examples of alien plants served to emphasize the fact that they were much more widespread than had been previously thought or assumed.

In the early 1990s a situation reminiscent of the early 1950s exists. Indeed one gets a distinct feeling of *déjà vu*, for a *New Flora of the British Isles* (Stace 1991) seems set to break the 40-year dominance of CTW, and the BSBI is embarking on a new mapping scheme. The new *Flora* includes many more alien taxa than any previous British *Flora*, reflecting, perhaps, the increasing interest being shown in these plants. In the introduction Stace writes:

> 'The coverage of alien taxa has been as thorough and consistent as possible. Many more aliens are included than in any other British Flora ... To merit inclusion, an alien must be either naturalized (i.e. permanent and competing with other vegetation, or self-perpetuating) or, if a casual, frequently recurrent so that it can be found in most years. All this applies as much to garden-escapes or throw-outs as to unintentionally introduced plants. Rarity, or the requirement of a highly specialized habitat, have not been taken into consideration (any more than is the case with natives). Cultivated species have been included if they are field-crops or forestry-crops or, in the case of trees only, ornamentals planted on a large scale. Exclusively garden plants, however abundant, whether crops or ornamentals, have not been covered, but most of the commoner taxa are included anyway because of their occurrence as escapes or throw-outs. Also excluded are non-tree ornamentals planted en masse on new roadsides or in parks, etc. The aim of this re-vamped and expanded set of criteria is to include all taxa that the plant-hunter might reasonably be able to find 'in the wild' in

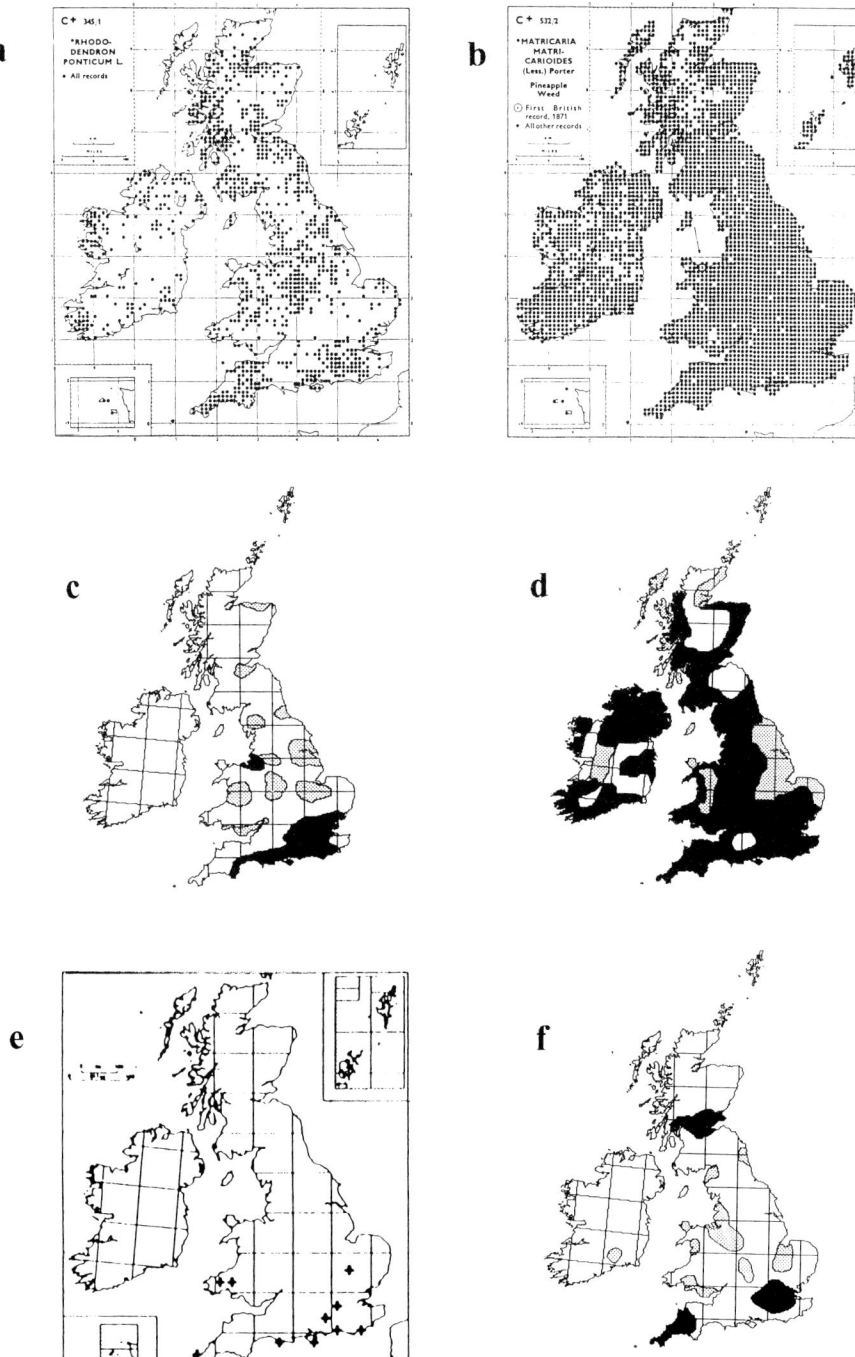

Fig. 1: Distribution in the British Isles. *a, Rhododendron ponticum*; *b, Matricaria discoidea*; *c, Crassula helmsii*; *d: Fallopia japonica*; *e, Campanula poscharskyana*; *f, Lamiastrum galeobdolon* ssp. *argentatum*. (*a,b,* from Perring & Walters (1962); *c,d,f,* from Ellis (1993); *e,* from Rich & Woodruff (1990))

any one year. Any such plant, whether native, accidentally introduced or planted, affects wild habitats and is part of the ecosystem, and botanists and others might be expected to need or want to identify it. Ornamental trees (but not shrubs or herbs) have been included because they are long-lived and frequently persist decades after all other signs of planting have disappeared from the area, so that the finder could not be expected to know that they were once planted.'

Stace went even further than this in his inclusion of alien plants:

'As well as the taxa treated 'fully' ... , other taxa that narrowly miss qualification, or fall far short of qualification despite their frequent inclusion in other Floras, or have been erroneously included in the past, are also briefly mentioned.'

The result of this is that in the *New Flora*, almost exactly 40% of the taxa treated 'fully', some 1,360 out of a total of 3,400, are aliens. This represents a tremendous increase over previous *Floras*, probably as great as CTW was over the pre-war *Floras*. Many of these alien taxa are rare but it is quite probable that some at least are 'rare' only because there has been, until now, no easy way of identifying them. An example is provided by *Cotoneaster*. Only one species is native to Britain and that is confined to limestone rocks on the Great Orme in North Wales. In the pre-war *Floras* this is the only species mentioned; CTW mentioned four others but only two were mapped by Perring & Walters (1962). Stace (1991) and Fryer (1994) list no fewer than 45 and this is probably a conservative figure which is sure to rise in the next few years. Of the 44 alien cotoneasters, many are 'rare', not necessarily because they *are* rare but, more likely, because they have just not been recognized, or have been confused with some other more abundant species. It is certain that records of these and other alien taxa will increase dramatically in the next few years, especially as there are two further factors to take into consideration.

Just as in the early 1950s, the BSBI is preparing to embark on an ambitious new scheme to resurvey the entire vascular plant flora of the British Isles and to produce a new *Atlas of the British Flora* by the year 2000. There has already been a partial resurvey, the BSBI Monitoring Scheme, which recorded the presence of species in approximately 11% of the 3,500 10 km squares in the British Isles. The BSBI, disastrously in my opinion, refused to publish the results of their survey, which included over 1900 up-to-date distribution maps; but these should eventually be published by the Joint Nature Conservancy Council (JNCC) in 1994, some 6 years after the survey was completed.

The taxa covered by *Atlas 2000*, as the new mapping scheme is commonly known, will include *all* that are fully treated in Stace (1991), just as the taxa covered by the original 1954 mapping scheme were most of those treated in the then newly published CTW. In this way 10 km square distribution maps will be available for a large number of previously unmapped alien taxa and although maps of all aliens recorded will not necessarily be published in *Atlas 2000*, they will be available on computer.

A third factor which could have a dramatic effect on the native and alien flora of the British Isles is the potential change in our climate associated with Global Warming and the Greenhouse Effect. If our climate does get warmer, and all the signs are that it will, even if only by 1° or 2° C, some of our native species will disappear or migrate northwards and new ones will arrive. We may find that some aliens already lurking in our parks and gardens, but insufficiently well adapted to our present climate to be able to spread far from them, will suddenly explode into the countryside. Similarly, gardeners are bound to grow new species and cultivars that need the new warmer climate. Some of these will undoubtedly escape, especially in the south.

There will also be some natural invaders. The number of different species native to Britain is very much lower than in comparable areas on the Continental land mass. The two main reasons for this are our cooler climate and the fact that the land bridge connecting the British Isles to the Continent (and Ireland to Scotland) was submerged beneath the rising seas (caused by melting ice-sheets and glaciers) *before* many of the species of southern Europe had had a chance to migrate here at the end of the last Ice Age. A number of these species now present along the opposite shores of the English Channel and elsewhere could invade us at any time and get a foothold in our countryside. In a recent paper, Rose (1988) discussed some of the plants that he considered worth looking for in the British Isles, since they were present in northern France and might just be found here as natives. Of the 20 or so species that he listed, six are treated in Stace (1991) but only as aliens. Whether they could be called *alien* invaders if they got here under their own steam is a moot point; perhaps it depends on how they cross. Some species with small or plumed seeds could be carried through the Channel Tunnel in the draught produced by high speed trains. A lot of effort is going into preventing rabid animals from using the tunnel as a migration route but it is doubtful if anyone has given serious consideration to the potential problems that could be caused by plants using the same route.

Cirsium oleraceum (Cabbage Thistle), is a prime example of a potential invader. Rose (op.cit.) writes of this ' ... a tall yellow thistle with broad cabbagy leaves and yellow-green bracts. This is common everywhere in northern France and Belgium, etc., in such habitats as roadside ditches, meadows etc.' It has just the sort of plumed seeds that might be sucked through the Channel Tunnel. When these arrive at the English end they will find a whole countryside in which to germinate and spread.

In the 1970s, the British countryside was transformed by the introduction of bright yellow into the normal agricultural patchwork of varying shades of greens and browns: Oil-seed Rape, *Brassica napus* ssp. *oleifera,* because of EC price subsidies and a directive that Europe become self-sufficient in edible oils, has in only 20 years become a major part of the spring scene. In the 1990s more and more fields are turning pale blue in summer as farmers turn to growing *Linum usitatissimum* (Flax). The ameliorating climate and increasing length of the growing season has made it economically viable to grow this crop which is cultivated here for the linseed oil from the seeds rather than the flax or linen from the stem fibres. It may not be very long before fields of Sunflower, *Helianthus annuus,* and Hemp, *Cannabis sativa*, are added to the increasing lists of crops that can be grown economically in Britain. Experiments at Birmingham University have indicated that: '... areas suitable for cultivating maize as a grain crop [could] migrate northwards at the rate of 10 kilometres a year as the growing season lengthens' (Gates 1992).

All of these crop plants already occur as casuals, but their spread into natural habitats can only be accelerated if conditions here become more suitable for their survival and growth. Whatever the means of entry, it seems inevitable that more and more alien species will invade the British and Irish countryside in the next few years and we must be ready to meet this challenge. Consider these words taken from Gates (1992):

'... within the Botanical Society of the British Isles a specialist group of botanists still exists which records the appearance of any new interlopers that appear on rubbish tips, railway lines and around ports. But for the most part these studies have been mainly for the pleasure of recording and identifying unusual plants. There has been very little detailed work on the dynamics of introduced species, attempting to measure their rate of increase or competitive ability compared with native species. The few studies of this type that do exist have been conducted only after introduced species have become well-established and troublesome weeds.'

This is almost exactly the point that I wish to make here. I consider it vital that we as botanists take steps to monitor the spread of alien plants. Let us consider two examples. The first where an aggressive alien species has been closely monitored not long after its first appearance in the wild and the results obtained used to good effect in attempting to check the spread of the invader. And the second, the opposite case, where the aggressive invader was more or less ignored for years and only recently has any attempt been made to find methods of control.

The classic example of the first approach is provided by *Crassula helmsii* (New Zealand Pigmyweed). Introduced by aquarists in the early part of this century, it was first found naturalized in the 1950s and has since become a serious pest of aquatic habitats in southern England, especially in the New Forest area (Fig. 1c). This alien was 'adopted' by a professional botanist, Hugh Dawson, of the Institute of Freshwater Ecology in the 1970s. He has since published many articles on the species, charted its spread in Britain since the 1950s, and written a leaflet on its control (NERC [1991]). Although the species is still spreading at an alarming rate and doubling its known sites every two years, the leaflet does give hope that it may prove possible at least to limit its advance into new and in existing areas.

The classic example of the second approach is provided by that most notorious of all alien invaders, *Fallopia japonica* (Japanese Knotweed). Introduced as a garden plant in 1825, it was ignored by botanists for many years and did not even rate a mention in the pre-war *Floras*. It was first recorded in the wild in about 1885 on cinder tips in the Maesteg area of Glamorgan, but virtually ignored until it started taking over parts of the countryside. It is now abundant in much of lowland Britain (Fig. 4) and has such a hold in many areas that nothing appears to shift it. It is now such a serious problem, especially in parts of South Wales, that, rather belatedly, the Welsh Development Agency initiated an investigation into possible methods of control. The result is a thirty-page document, *Guidelines for the Control of Japanese Knotweed*, which is a brave attempt to try and rescue something out of a seemingly hopeless situation (WDA 1991). The basic advice offered is to spray with 2,4-D herbicide several times a year for as many years as it takes to kill it.

Alien Study Group and Register

It is vital that we take steps to monitor the spread of alien plants when they first appear to be invading the countryside in any numbers, and to my mind the Society best suited to take the lead in this is the BSBI. This has the organization and the membership to take it on, and, with the help of institutions such as the National Museum of Wales, and other Societies such as Plantlife, there is no reason why there should not be a successful outcome in many cases.

There are already many botanists who take an interest in alien plants and it is quite likely that more could be attracted if some sort of Study Group were to be set up. I am thinking here along the lines of an informal group, rather like the BSBI's *Hieracium* Study Group. In such a group botanists could discuss the benefits and problems of closely monitoring new invaders. It may be an advantage if individual members of the group were to adopt a species or even a genus and take over the responsibility for monitoring its spread. This would provide a central reference point to which other members of the group or members of the BSBI could send records of the alien in question. The establishment of an Alien Register could be a further development. This would list those alien species considered to be spreading aggressively into the British and Irish countryside. For each species a short description, with ways of identifying it in the field, an illustration (line drawing or photograph), details of habitat, present distribution and any other relevant details, could be included together with a brief history of the alien in Britain. This report would be distributed to all members of the study group or even published in *BSBI News* with a request for members to keep a look out for it and report any sightings.

Whatever happens, it is vitally important that the changes to our climate and any associated changes in our vegetation are fully and completely documented.

There are already a few candidates for the Alien Register: species which appear to be spreading into the countryside, some after quite a long period of inactivity in gardens, others complete newcomers. Examples of the former are *Campanula portenschlagiana* (Adria Bellflower) and *C. poscharskyana* (Trailing Bellflower). Both were introduced from Yugoslavia and are much grown in garden rockeries. Neither was included in CTW or Perring & Walters (1962), but distribution maps were presented in Rich & Woodruff (1990) and Stace (1991) gives descriptions and illustrations of both. It appears likely that it is the Trailing Bellflower (Fig. 1e) that will cause most problems in the countryside. Both used to grow in my garden but it was only the Trailing Bellflower that spread rampantly over walls and rockeries. The Adria Bellflower formed nice discrete clumps until it was completely smothered by its aggressive cousin.

One of our most recent newcomers is *Lamiastrum galeobdolon* ssp. *argentatum* (Variegated Yellow Archangel). Variously treated as a species, a subspecies or a variety, this close relative of our native Yellow Archangel is a recently described plant of unknown, but probable garden, origin (Smejkal 1975). Its date of introduction to the British Isles is also unknown but was probably in the 1960s or 1970s (Rutherford & Stirling 1987). It is often offered for sale in Garden Centres as a ground-cover plant and it is certainly very effective at that. It suddenly turned up, uninvited, in my garden a few years ago and rampaged through my only flower bed. It has spread into the countryside in various parts of Britain (Fig. 1f) and is likely to prove much more common than present records indicate as it has been very much confused in the past with a variegated form of our native Yellow Archangel. This illustrates quite well the fact that records of many aliens are going to increase as botanists find out about them and how to identify them.

It is therefore proposed that an Alien Study Group be inaugurated and an Alien Register set up to monitor the spread of aliens in the British Isles, and, it is hoped, to find solutions to some of the very real problems that this spread may create.

I must confess to having a bee in my bonnet about alien plants, some colleagues complaining (with some justification) that I have a one-track mind on the subject. Perhaps I am over-reacting to a threat that might never take place. It is not, and never has been, my intention that a study of aliens should take priority over native plants, but I do feel very strongly that some action must be taken *now* before any new influx of aliens finds their way into our countryside.

Colour photographs and distribution maps of most of the alien plants mentioned in this paper are published in Ellis (1993).

References

Bentham, G. (1858). *Handbook of the British Flora*. London: Reeve.
Clapham, A.R., Tutin, T.G. & Warburg, E.F. (1952). *Flora of the British Isles*. Cambridge: Cambridge University Press.
Ellis, R.G. (1993). *Aliens in the British Flora*. Cardiff: National Museum of Wales.
Fryer, J. (1994). The genus *Cotoneaster* in the British Isles. *In* A.R. Perry & R.G. Ellis (eds), *The Common Ground of Wild and Cultivated Plants*, pp. 151-157. Cardiff: National Museum of Wales.
Gates, P. (1992). *Spring Fever. The Precarious Future of Britain's Flora and Fauna*. London: Harper Collins.
Hooker, J.D. (1870). *The Student's Flora of the British Islands*. London: Macmillan.
NERC ([1991]). *Crassula helmsii*. Focus on control. Peterborough: National Environment Research Council.
Perring, F.H. & Walters, S.M. (eds) (1962). *Atlas of the British Flora*. London and Edinburgh: Thomas Nelson.
Rich, T.C.G. & Woodruff, E.R. (1990). BSBI Monitoring Scheme 1987-1988. Report to the Nature Conservancy Council. Vols. 1 and 2.
Riddelsdell, H.J. (1907). A Flora of Glamorganshire. *Journal of Botany* **45** (Suppl.): 1-88.
Rose, F. (1988). Plants to look for in the British Isles some of which might be expected to occur as natives. *BSBI News* **49**: 11-12.
Rutherford, A. & Stirling, A.McG. (1987). Variegated archangels. *BSBI News* **46**: 9-11.
Smejkal, M. (1975). *Galeobdolon argentatum* sp.nov., ein neuer Vertreter der Kollektivart *Galeobdolon luteum* (Lamiaceae). *Preslia* **47**: 241-248.
Stace, C. (1991). *New Flora of the British Isles*. Cambridge: Cambridge University Press.
Trow, A.H. (1911). *The Flora of Glamorgan*. Cardiff: Cardiff Naturalists' Society.
WDA (1991). *Guidelines for the Control of Japanese Knotweed*. Cardiff: Welsh Development Agency.

6

Ferns – a case history

MARTIN RICKARD

The Old Rectory, Leinthall Starkes, Ludlow, Shropshire, SY8 2HP

ABSTRACT. Although ferns have been cultivated in Britain for perhaps 400 years they did not become popular as garden plants until the middle of the 19th century. At this time it was fashionable to grow ferns and to be seen to be collecting them from the wild. The emerging railway network allowed fern hunters to range far and wide in search of the rarer species listed in the contemporary fern books flooding the market. Eventually it became difficult to find new species but renewed interest in fern hunting was stimulated by the rediscovery of their natural tendency to sport (already documented in some 15 and 16th century herbals). Today it is usually the descendants of these sports, collected as wild plants, that we see in our gardens. The Victorian craze for collecting ferns from the wild in Britain has been accused of doing a lot of irreparable damage to our native fern flora. Undoubtedly some damage was done but perhaps in the long term it is less than generally supposed. British ferns, and an ever-increasing number of exotic species have usually done well in cultivation, and many have become established as first-class garden plants. Few have become naturalized away from their native haunts or have regenerated themselves in cultivation to become troublesome. Exceptions are wall ferns which have adapted to man-made habitats very successfully – but they have not become a problem.

Fern cultivation

For many centuries ferns have been cultivated in Britain. Common species have been used by herbalists for the treatment of many conditions, for example *Polypodium vulgare s.l.* (for the intestine) and *Asplenium ruta-muraria*. As far back as the late 16th century some cultivars were already being grown, for example *Asplenium scolopendrium* L. 'Multifidum' is mentioned in Gerard's *Herbal* of 1597, and over the following 250 years there were occasional records of other cultivars being brought into cultivation here and in Europe (Britten 1881). Notable examples are:

- *Asplenium scolopendrium* 'Crispum' – described as *Phyllitis crispa* in 1650-51 (Bauhin & Cherler 1650-51).
- *Polypodium cambricum* 'Cambricum' found at Dinas Powis near Cardiff in 1668 by Richard Kayse of Bristol (Dyce 1988), and presumably taken into cultivation at around the same time (Fig. 1).
- *Athyrium filix-femina* 'Kalothrix' found in the mountains of Mourne by Sherard about the end of the 17th century. This variety has been raised again recently (Dyce 1983).

Fig. 1. *Polypodium cambricum* 'Cambricum' at Dinas Powis, Glamorgan, 1978 (*Photo M. Wood*).

Fern collection

Despite the activity of botanists and herbalists, ferns did not become popular as garden plants for many years and it was not until the Victorian period that various social factors combined to make ferns fashionable (see Allen 1969). At the beginning of the so-called Victorian Fern Craze in the 1840s, interest was channelled towards the study and collection of as many of the British species as possible, whole plants being collected for cultivation or for herbaria. As the British pteridophyte flora is rather limited, for many collectors the commoner species were soon in the bag, and the determined pteridologist had to get further afield into the wilds to find anything new. Transport was still quite difficult but as the railway system grew the more remote parts of these islands became ever more accessible. Nevertheless a lot of walking was unavoidable if one was to explore the wildest corners: witness the botanical hikes of 1881 documented later by Druce (1929) and others by Praeger (1947). The railway companies were not slow to realize there was a potential here to fill trains into the middle of nowhere! The Corris Railway and Furness Railway even advertised in *The Fern Paradise* by F.G. Heath (1908), although by that time they may have missed the boat!

About ten or fifteen years after the initial rush to collect all the species, once again interesting cultivars of our British ferns were being found in the wild. Probably the first find in this renewed spate of activity was the discovery of *Polystichum setiferum* 'Divisilobum Wollaston' in a hedge bank near Ottery St Mary in East Devon in 1852 (Druery 1910). Other fascinating finds soon followed and the sport of variety hunting was born. So much so that less than 40 years later a total of 1861 cultivars of British ferns was described by E.J. Lowe (Lowe 1890); the majority of these were wild finds but with a large number of additional selections bred horticulturally.

Over the past 150 years, but particularly during the Victorian period, fern collecting reached such proportions that populations of some of the less common fern species near our large cities were seriously over-collected, as were the nationally rare ferns, whose habitats were well documented in the many emerging fern books (e.g. Newman 1854). Today the Victorians are roundly condemned for the destruction they caused. Even the *Radio Times* for 23 November 1991 claimed that 'several species of fern actually became extinct' (by inference in Britain). However, I think this is misleading. So far as I am aware no fern species reliably recorded as native to the British Isles since recording began (fossil records excepted) is extinct today; even on a world-wide basis very few, if any, species have disappeared as a result of over-collecting.

The Victorians did collect far too heavily, but I can think of very few sites where any of our rarer species are now extinct. One unfortunate example is the disappearance of *Woodsia ilvensis* from Falcon Clints in Co. Durham in about 1910. There are other sites where the *Woodsia* is no longer known, but, for example, who is to say it no longer grows on Cader Idris? Was it ever common there in the first place? *If* it is now extinct on Cader Idris was it lost through environmental changes or collection? This species was thought extinct in the Moffat hills but now at least two distinct colonies of it are known there (Rickard 1972).

Away from the grand rarities what about the more attractive very garden-worthy species such as *Osmunda regalis*? This was supposedly collected to extinction from the south-east of England, but was it? The prospect of seeing *O. regalis* wild in Essex seemed fanciful to me, as it was even rare in the county in 1862 (Gibson 1862); but in the years 1971 and 1972 I was shown three separate colonies, and others were known at the time (Jermyn 1974). I am therefore forced to wonder if even an extremely garden-worthy species like *O. regalis* is just as common in Essex today as it was in 1862 in spite of the collecting of the Victorians and the massive urban sprawl from London. How can this be? Perhaps it is due to environmental conditions triggering germination of spores from 'spore banks'. Even short-lived spores of species like *O. regalis* can live for a very long time in suitable conditions, perhaps 70-100 years in peat (Hainsworth 1992). It is likely, therefore, that other species with longer lived spores, could reappear by the same mechanism. Work is currently under way in Scotland to investigate the value of wild spore banks as a means of supplying new plants of the local strains of rare ferns for restocking suitable habitats. It would be marvellous if *Woodsia ilvensis* could be stimulated to recolonize Falcon Clints and thus negate the damage done by plant collectors mainly in the last century.

I don't think anyone would dispute that the Victorians did do some damage to our wild fern populations, but were all their activities on the debit side? What are the merits of collecting cultivars from the wild? Clearly it is against current codes of conduct for any plants to be dug up in the wild, but had this been the position in the past, our losses to horticulture and science would have been great. For example:

1) The cultivar *Polystichum setiferum* 'Plumosum Bevis' would have gone unnoticed and never been seen in a garden. This fern is sterile in most seasons and it was not until several years after its discovery in 1876 that any sporangia were noticed. The few spores produced subsequently gave rise to some of the most wonderful hardy British ferns in cultivation today: these are 'Plumosum Drueryi', 'Gracillimum' and the cream of the crop 'Plumosum Green'. Yet the parent of these marvellous plants was only ever found once, in a lane bank at Hawkchurch on the Devon/Dorset border. It was discovered by a labourer, Jno Bevis, who recognized it as different and pulled it from the hedge and delivered it to a local fern enthusiast, a Dr Wills. Would that hedge bank and that fern still be there today? Possibly, but I doubt it.

2) *Athyrium filix-femina* 'Victoriae' was originally noted near Drymen on Loch Lomondside in 1861 by a student, James Cosh. It was found at a most fortuitous time as the fern was about to be cut back, but Cosh organized its collection. The original clone of this fern is still in cultivation. No spore progeny of this cultivar is ever as good as the parent, so if Cosh had contented himself with spores the true plant would never have reached cultivation, and the original wild aberration would have long since disappeared.

3) *Asplenium scolopendrium* 'Bolton's Nobile' is a very handsome broad form of 'Crispum', a sterile variant of the Hart's-tongue. It was found on Warton Crag in North Lancashire early this century. Approximately 90 years later the original collection still thrives in the finder's garden (the finder is gone but his granddaughter still lives there; she is over 90 years old herself!). Fortunately it is not the only plant known and today it is a quite common cultivar because it can propagated vegetatively from frond bases. However, to do this it is necessary to dig the plant up and maul it about quite harshly, not an operation likely to be practicable on a plant growing in limestone rocks. Since all crispums are sterile it is only a matter of time before individuals left in the wild die out leaving no progeny.

4) *Asplenium trichomanes* (Incisum group) 'Percy Greenfield' was found in 1961 on a wall in Somerset. It too is sterile and quite the most beautiful form of *A. trichomanes* in cultivation today. It is still extremely rare, there being only one plant at Reginald Kaye's nursery in North Lancashire. Left on its original wall its chances of survival would have been remote. As it is still in cultivation there is a chance that it might become a little more widely distributed, and of course it is available for scientific research if needed.

5) *Asplenium scolopendrium* (Ramosum group) 'Fred Jackson' is an interesting example of the timely collection of a wild plant. This handsome cultivar grew in Brigham quarry in Cumberland. One plant of many was collected. A few years afterwards the site was revisited but the quarry had been filled in. Fortunately the plant was safe in Fred Jackson's garden.

From the above five examples of a much, much larger number it can be seen that some of the best new material for our gardens is often only available from wild sources. They also illustrate how insecure the wild environment is, with habitat destruction happening quite innocently all the time. In these circumstances is collection of 'one-off' cultivars, through proper channels, still justifiable today?

Had it not been for the Victorians' love of ferns and their collection, losses to science would also be significant. For example:

1) Many of the nation's herbaria would be much the poorer and indeed Britain would not be the well-botanized country it is today. It is hard, however, to condone over-collection for herbaria: I remember once seeing a sheet of *Woodsia alpina* in the herbarium of the Royal Botanic Garden, Edinburgh, collected on Beinn Heasgarnich by F.C. Crawford, containing 12 complete plants. (This was a 1902 collection, just outside the Victorian era.)

2) Again on the credit side, the Victorians must be applauded for their keen eye for anything different.

2a) *Polypodium cambricum* was recognized as something distinct about 100 years before chromosome counts uncovered the true story (see Moore 1855).

2b) Similarly *Asplenium trichomanes* ssp. *pachyrachis* was known as distinct from the Wye Valley (as var. *subequale*) (Lowe, 1890) as long ago as 1855.

2c) In 1859 Clowes collected a mysterious *Dryopteris* from a wood on the side of Lake Windermere. This has since been determined as the only known example from anywhere in the world of *Dryopteris* × *brathaica* (*D. filix-mas* × *D. carthusiana*) (Fraser-Jenkins & Reichstein 1977). This plant is still in cultivation and I believe that all interested in unravelling the intricacies of the British flora should be grateful to Mr Clowes.

3) There are also examples of collections made in the last century that stimulate scientific activity today. In 1857 a Mrs Bolland found the strange cultivar *Dryopteris filix-mas* 'Bollandiae'. This is sterile and recently Fraser-Jenkins (pers.comm.) has conjectured that it could be a hybrid between *D. filix-mas* and *D. aemula*. It certainly looks about right and geographically it is right, too, as it was found at Ashurst Park, Tunbridge Wells (Moore 1859). It is fortunate that Mrs Bolland collected this plant which, despite its sterility, is still in cultivation, so scientific work can go ahead on it 135 years later.

Today many ferns collected more than 50 years ago and their progeny are still in cultivation. But these stocks have been supplemented by the introduction of an ever-increasing number of hardy exotic species so that there are now in cultivation in Britain perhaps 500 of these, many of them only recently introduced. Perhaps one day one of these may become established in the wild!

Fern escapes

But with this very wide range of ferns in cultivation it is perhaps surprising that there are very few examples of ferns becoming established away from their native haunts. Clearly, many of the wall ferns so common throughout these islands would be rare were it not for man's activity in building suitable habitats; but these escapees are very rarely, if ever, from a horticultural source. It is recorded (Benoit & Richards 1963) that *Asplenium ruta-muraria* was not native to Merioneth as it was not known on natural rocks. However, I have seen it in the county on natural rock in company with *A. septentrionale*; I like to think it is native there and not a colonist from a neighbouring wall! Occasionally cultivated ferns do escape from gardens, usually as individual plants. I have seen a small paddock in Cumbria near the site of an old fern nursery, where there are hundreds of crested Lady-ferns. These have taken 60 or 70 years to become established and may now be a permanent part of the local flora.

Exotic species are also only rarely established in the wild. House fern species sometimes do get established for a short term until a cold winter wipes them out; for example, I have seen *Pteris cretica* on a wall in Saffron Walden, Essex over several seasons. But a more common and longer lived species is *Adiantum capillus-veneris*, established in many sites away from its native maritime habitats; a good example is the long-established colony on a railway bridge at Ledbury in Herefordshire (Whitehead 1976).

The escape of exotic species from gardens occurs from time to time. For example:

1) *Onoclea sensibilis* is established near Rydal Water in Westmorland – a garden outcast?

2) *Matteuccia struthiopteris* was established as a fine clump by the roadside at Kishorn in Wester Ross until destroyed by road widening (should it have been collected?).

3) *Cyrtomium falcatum* is known above high water mark on beaches in the Scilly Isles (Lousley 1971).

These are but a few examples, none of which threaten the indigenous flora. In fact I can think of only one escapee likely to be of any significance to British botanists: that is *Azolla filiculoides* and it only seems to be able to survive in each habitat for a few years before dying out.

Fern introductions

Up to now the whole emphasis of this paper has been on the collection of ferns from the wild and their very occasional escape. One other aspect of significance here is the intentional introduction of rare native species. It is on record that Herbert Stansfield planted Killarney Fern (*Trichomanes speciosum*) in 'some wild and congenial places' in North Wales (Hawkins 1928). For at least two reasons I think this action was misguided. The material planted out would almost certainly not have been of Welsh origin, and the chosen sites were not recorded, or at least not recorded publicly, so we do not now know if extant colonies of this, one of our two rarest British ferns, are natural or introduced in these sites.

Conclusion

In conclusion I am aware of no problems associated with the establishment of exotic species in the wild in Britain. Problems have been much greater in the reverse direction, that is the collection of native species for garden cultivation. This is regrettable but with the passage of time it can be seen that the long-term damage caused by collecting, mainly 100 years or more ago, is small. Today the only outstanding problem seems to be the question of whether or not to collect monstrosities (cultivars to fern lovers!) from the wild, but as long as the current laws covering the collection of plants are followed I cannot see gardeners again posing a significant threat to our native fern flora.

References

Allen, D.E. (1969). *The Victorian Fern Craze. A History of Pteridomania*. London: Hutchinson.
Bauhin, J. & Cherler, J.H. (1650-51). *Historia Plantarum Universalis*. Ebroduni: Fr.Lud. a Graffenreid.
Benoit, P. & Richards, M. (1963). *A Contribution to the Flora of Merioneth*, 2nd edition. Haverfordwest: West Wales Naturalists' Trust.
Britten, J. (1881). *European Ferns*. London: Cassell & Co.
Druce, G.C. (1929). The Flora of West Ross. Supplement to *Report of Botanical Society and Exchange Club of the British Isles*, 1928.
Druery, C.T. (1910). *British Ferns and their Varieties*. London: George Routledge & Sons, Ltd.
Dyce, J.W. (1983). Athyrium filix-femina 'Kalothrix'. *British Pteridological Society Bulletin* **2**: 243-247.
Dyce, J.W. (1988). Polypodium australe 'Cambricum'. *Pteridologist* **1**: 217-220.
Fraser-Jenkins, C.R. & Reichstein, T. (1977). Dryopteris × brathaica Fraser-Jenkins & Reichstein Hybr.Nov., the putative hybrid of D. carthusiana × D. filix-mas. *Fern Gazette* **11**: 337.
Gibson, G.S. (1862). *The Flora of Essex*. London: Wm. Pamplin.
Hainsworth, P.H. (1992). More adventures with spores. *Pteridologist* **2**: 118-121.
Hawkins, E.H. (1928). Obituary: H. Stansfield. *British Fern Gazette* **5**: 218-221.
Heath, F.G. (1908). *The Fern Paradise*, 8th edition. London: Hodder & Stoughton.
Jermyn, S.T. (1974). *Flora of Essex*. Colchester: Essex Naturalists' Trust Ltd..
Lousley, J.E. (1971). *The Flora of the Isles of Scilly*. Newton Abbot: David & Charles.
Lowe, E.J. (1890). *British Ferns and Where Found*. London.
Moore, T. (1855). *The Ferns of Great Britain and Ireland*, ed. by J. Lindley. London: Bradbury & Evans.
Moore, T. (1859). *The Nature-Printed British Ferns*. London: Bradbury & Evans.
Newman, E. (1854). *Popular History of British Ferns*. London: John van Voorst.
Praeger, R.L. (1947). *The Way that I Went*. London: Methuen.
Rickard, M.H. (1972). The Distribution of Woodsia ilvensis and W. alpina in Britain. *British Fern Gazette* **10**: 269-280.
Whitehead, L.E. (1976). *Plants of Herefordshire, a Handlist*. Hereford: Herefordshire Botanical Society.

7

Mimulus: 180 years of confusion

A.J. SILVERSIDE

Department of Biological Sciences, University of Paisley, Paisley, Renfrewshire, PA1 2BE

ABSTRACT. A taxonomic and nomenclatural revision of *Mimulus* in Britain is planned for publication in the near future. The account here necessarily anticipates some of the names to be used in the intended revision, but no new name or combination used here is yet to be regarded as validly published under the requirements of the *International Code of Botanical Nomenclature*. The extent of the revision needed, especially in the interpretation of *Mimulus luteus*, indicates the confusion that currently exists. As will be shown, the name *Mimulus luteus* has been applied to at least seven naturalized taxa in Britain, in no case correctly.

Discoveries in Chile and Alaska

The story begins with Father Feuillée, a French priest travelling in South America. While in Chile he discovered a yellow-flowered plant that he named '*Gratiola foliis subrotundis, nervosis, floribus, luteis*'. A description and illustration of this plant was included in his published journal (Feuillée 1714). Linnaeus published the name '*Mimulus luteus*' for this plant in 1763, his name being based solely on Feuillée's account, although he incorrectly stated that the plant came from Peru.

Around 1808 a Dr Langsdorff collected herbarium specimens, including mature seed, of a yellow-flowered *Mimulus* from Alaska. Plants were first grown from this material at Gorenk in Russia, and the plant was then distributed to France and Britain. Fischer, at Gorenk in 1812, called it '*Mimulus guttatus*', a name also used and validly published by De Candolle at Montpellier a year later. Also in France, it was subsequently described and illustrated by Loiseleur-Deslongchamps as '*M. punctatus*'. Donn, at Cambridge, called it '*M. langsdorfii*', but like Fischer published no validating description. However, J. Sims, in the influential *Curtis's Botanical Magazine* in 1812, declared this North American plant to be the same as that collected in South America by Feuillée a century before. By insisting that the Alaskan plant was *M. luteus*, Sims initiated the confusion that still persists, 180 years later.

The Chilean plant rediscovered?

This was a time when agents of the major plant nurseries were scouring the world for new plants of horticultural merit. In 1826 James M'Rae collected seed of a *Mimulus* in Chile. This differed most obviously from Feuillée's original Chilean plant in having a blood-red blotch in the centre of the lower lip of the flower. In 1827 Lindley published a brief description and illustrated the plant as *M. luteus* var. *rivularis*. In 1830 an excellent colour plate (by George Cooke) was published by Loddiges under the name '*M. rivularis*', but with a specific reference to its being the plant of Feuillée. There appears to have been no dispute that this was Feuillée's plant rediscovered, i.e. *M. luteus*, but throughout the rest of the century the majority of authors also applied the name to Langsdorff's North American plant.

The Alaskan and Chilean plants distinguished, but confusion continues

Green (1895) clearly differentiated between the North and South American plants, though using the name '*langsdorffii*' for the North American plant. Grant (1924) published a monographic revision of the genus, correctly referring the North American plant to *M. guttatus*. This should have cleared the confusion of these plants, but the names themselves continued the confusion. There was, on one hand, the Alaskan plant with yellow flowers, albeit with red dots in the throat of the corolla. Then there was the Chilean plant, by now subject to horticultural selection and hybridization, with heavily blotched corollas. There were two names in primary use: '*luteus*' meaning yellow, and '*guttatus*' meaning spotted. It was probably inevitable that the use of the names would be reversed, both in horticulture and in the naming of the plants that were, by then, well naturalized along British streams and rivers. Some botanists used the names correctly, others did not. Consequently, literature records of wild plants cannot be trusted, while the aggressive and horticulturally inferior *M. guttatus* is still sold as '*M. luteus*' in garden centres.

Even when the names were used correctly, there were still problems in naming naturalized plants. Miss M.N. Hamilton carried out important, though regrettably unpublished, work in the early 1960s, recognizing the widespread occurrence of sterile clones of what seemed to be *M. guttatus*. Meanwhile, Mr R.H. Roberts was realizing that almost all colonies of what seemed to be *M. luteus* were sterile hybrids with *M. guttatus* (Roberts 1964). In fact Miss Hamilton's plain-flowered plants and the majority of Mr Roberts's blotched plants are indeed a hybrid between *M. guttatus* and M'Rae's Chilean plant, and are correctly named as *M.* × *robertsii*.

However, the assumption that James M'Rae's plant is the same as that first described by Father Feuillée needs to be challenged. There are serious discrepancies between Feuillée's plate and the plant subsequently assumed to be *M. luteus*. This matter will be discussed elsewhere, but examination of a range of material has shown that there are good taxonomic reasons to recognize two quite distinct species. True *M. luteus* is a tall, yellow-flowered plant, superficially very like *M. guttatus*. It has apparently never been cultivated in Europe and is probably of limited horticultural merit. M'Rae's plant has attracted a number of names while in cultivation, but unfortunately none can be validly applied at specific level, if at all. It is my intention to take up a previous varietal name and call it '*M. nummularius*'. It is now rare in cultivation, but is naturalized in scattered localities in northern Britain and at one site in SW Ireland (Fig. 1).

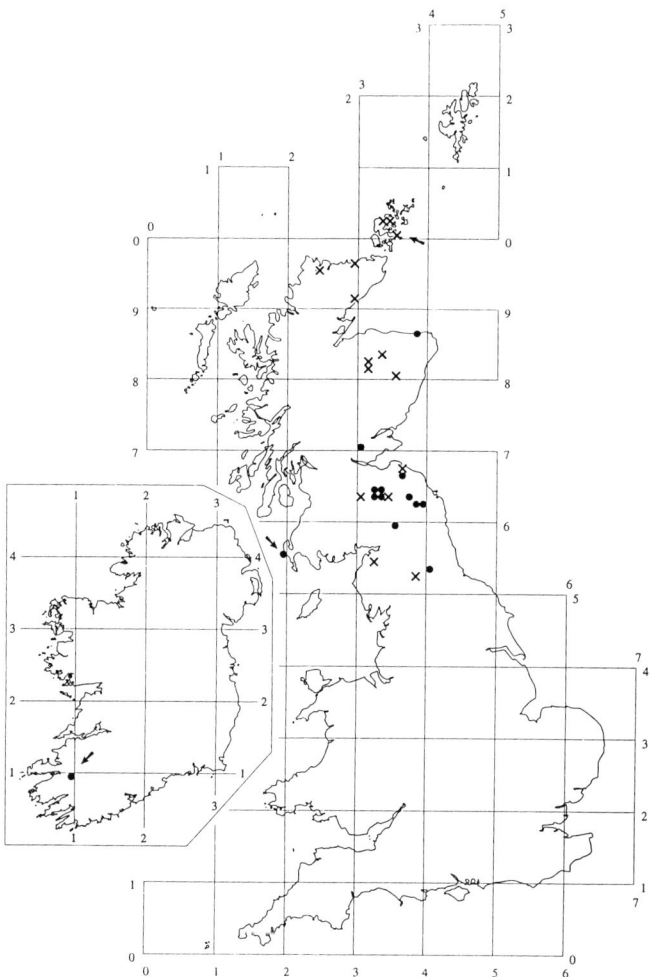

Fig 1. 10 km square records of *Mimulus nummularius* and *M.* × *smithii* as naturalized plants. (Closed circles = *M. nummularius*; crosses = *M.* × *smithii*.)

More South American introductions

There have, of course, been other introductions of *Mimulus* species into cultivation. *M. variegatus*, a plant with cream flowers with solidly purple-blotched lobes, was introduced from Chile to France and then to Britain, being first described and illustrated by Jaume Saint-Hilaire in 1832. This was quickly hybridized with *M. nummularius*, producing *M.* × *smithii*, a larger and more reliably perennial plant that evidently soon replaced *M. variegatus* in cultivation. *M.* × *smithii* is still sometimes grown and is also naturalized in scattered British localities (Fig. 1).

Table 1. *Mimulus* L. in Britain.

A summary of the main species in cultivation, infrageneric classification adapted from Grant (1924).

Subgenus *Schizoplacus* Grant
   Section *Diplacus* Gray
      Small shrubs, often of dry places, half-hardy in Britain. South-western N. America.
      M. aurantiacus Curtis (= *Diplacus aurantiacus* (Curtis) Jeps.; *D. glutinosus* Wendl.)
         South-western U.S.A.
      M. puniceus (Nutt.) Steud. (= *Diplacus puniceus* Nutt.)
         California, Mexico

Subgenus *Mimulus* (= subgenus *Synplacus* Grant)
   Section *Mimulus* (= section *Eumimulus* Gray)
      A few species, mostly perennials of wet places, in N. America, Africa, Asia and Australasia.
      M. ringens L.
         Widespread species in N. America. Frequently grown in bog gardens and as a semi-aquatic.

   Section *Erythranthe* Greene
      Perennials characterized by their clothing of long, sticky hairs and pink or red (rarely yellow) flowers, mostly (always?) in wet places, Mexico and western N. America.
      M. cardinalis Dougl.
         South-western U.S.A.
      M. lewisii Pursh
         By streams in the mountains of western N. America. Formerly placed in section *Paradanthus*.

   Section *Paradanthus* Grant
      Herbs, mostly perennial, of wet places. N. & S. America, Asia, Australasia. Grown as rockery plants, especially in damp spots.
      M. moschatus Dougl.
         Mountains of western N. America, established escape in ditches and wet places in Britain and Europe.
      M. primuloides Benth.
         Mountains of western N. America.

   Section *Simiolus* Greene
      Annual and perennial herbs, usually of wet places, in western N. and S. America.
      M. caespitosus Greene (= *M. tilingii* Regel var. *caespitosus* (Greene) Grant)
         Wet areas in the high mountains of western N. America. Cultivated as a rockery plant.
      M. tilingii Regel
         By streams in the high mountains of western N. America. Extent of current cultivation uncertain, since name may be used to include *M. caespitosus*.
      '*M. luteus* complex'
         Summarized in Table 2.

Table 2. Species of the *'Mimulus luteus* complex' in Britain.

North America

M. guttatus DC. (= *M. langsdorffii* Donn ex Greene)
Vigorous stoloniferous perennial, widespread along the western coast of N. America. Taxonomically complex, with populations showing incipient speciation. Collected by Dr Langsdorff (*ca.* 1808), introduced to Russia, then to France and Britain (by 1812). Cultivated as a waterside plant and a widely established escape in Britain and northern and western Europe.

South America (Chilean and Argentinian Andes)

M. nummularius (Clos) comb. ined. (= *M. luteus* L. var. *rivularis* Lindl.)
Creeping, mat-forming perennial of high altitude wet meadows ('vega'), introduced to cultivation in 1826. Now rarely grown, but very locally established in N. England, Scotland and SW Ireland.

M. variegatus hort. ex J. St.-Hil.
Introduced to France from Chile by 1832, then to Britain. Perhaps not perennial and quickly replaced by more vigorous hybrids with *M. nummularius*. Important source of garden cultivars but apparently no longer in cultivation.

M. cupreus hort. ex Dombr.
Clump-forming or shortly creeping perennial. Collected in Chile and introduced to Britain, first shown in June 1861. Widely grown as a rockery and bog garden plant. Taxonomically complex, with red- and purplish-pink-flowered variants, e.g. cultivars 'Whitecroft Scarlet', 'Roter Kaiser'. Recent introductions, in cultivation as 'Andean Nymph' and 'Inca Sunset', probably belong here. Claimed as naturalized in Britain, but in error for similar hybrids.

[*M. luteus* L. Not in cultivation, name misapplied to various cultivated or naturalized species and hybrids.]

Table 3. Hybrids of the *'Mimulus luteus* complex' in Britain.

N.B.: Crosses between the S. American taxa (*M. cupreus*, *M. nummularius* and *M. variegatus*) are usually at least partially fertile; crosses between this species complex and the N. American *M. guttatus* are usually sterile.

M. × *maculosus* T. Moore (= *M. cupreus* × *M. nummularius*)
First shown by W. Bull in 1863, now rarely cultivated. An established escape in three areas in Scotland.

M. × *smithii* Paxt. (= *M. nummularius* × *M. variegatus*)
In cultivation by 1834, known as an escape by *ca.* 1850 at Reay, Caithness. Now rarely grown, but established in a number of localities in N. England and Scotland.

M. × *hybridus* hort. ex Sieb. & Voss (= *M. cupreus* × *M. nummularius* × *M. variegatus*)
Usually annual or short-lived, much grown as a bedding plant. A rare and transient escape.

M. variegatus was regarded as a variety of *M. luteus* (including *M. nummularius*), defined simply by corolla patterning by Grant (1924) who saw only a few collections of these South American plants. Her treatment has been followed by subsequent authors, but this poorly known plant is certainly distinct from both *M. luteus* and *M. nummularius*. Its affinities may well be with the next species of the complex to be introduced from Chile, *M. cupreus*. This species was collected by a Mr Pearce for the nurserymen J. Veitch & Son of Chelsea, and was first shown at the Royal Horticultural Society's Grand Exhibition in June 1861. It is likely that Veitch was the originator of the name. Although best known as a plant with brilliant, copper-coloured flowers, *M. cupreus* is a complex and perhaps aggregate species, including yellow, red and purple-flowered variants. It probably includes a taxon with attractive straw-yellow flowers variably marked with pink, collected in Chile on three or more occasions and recently introduced to cultivation as the cultivar 'Andean Nymph'.

The three Chilean species that have been cultivated (i.e. *M. cupreus*, *M. nummularius* and *M. variegatus*) form a complex in which hybrids are usually at least partly fertile. They or their hybrids form typically sterile hybrids with *M. guttatus*. The result is a further array of taxa in cultivation, some also occurring as transient escapes or becoming naturalized. In some cases it appears that hybrids have also arisen in the wild as crosses between naturalized taxa, and have then been re-selected for cultivation. Names of hybrids, especially the name '*M. × hybridus*', are regularly misapplied, adding further to the difficulties in naming either cultivated or wild plants.

The current position

A summary of the more commonly cultivated species of *Mimulus* in Britain is given in Table 1. The cultivated species of what may loosely be termed the '*M. luteus* complex' are listed in Table 2, while their hybrids are summarized in Table 3. The name '*M. luteus*' has, at various times, been applied to naturalized populations of at least the following: *M. × caledonicus* ined., *M. guttatus*, *M. × maculosus*, *M. nummularius*, *M. × polymaculus* ined., *M. × robertsii* and *M. × smithii*.

References

Feuillée, R.P.L. (1714). *Journal des observations physiques, mathematiques et botaniques*. 2. Paris.
Grant, A.L. (1924). A monograph of the genus *Mimulus*. *Annals of Missouri Botanical Garden* **11**: 99-388.
Greene, E.L. (1895). *Mimulus luteus* and some of its allies. *Journal of Botany, London* **33**: 4-8.
Roberts, R.H. (1964). *Mimulus* hybrids in Britain. *Watsonia* **6**: 70-75.

8

Symphytum – Comfrey

Franklyn H. Perring

Green Acre, Wood Lane, Oundle, Peterborough, PE8 5TP

ABSTRACT. Stace's *New Flora of the British Isles* (Stace 1991) includes 11 species or hybrids in the genus *Symphytum*: *S. officinale*, *S.* × *uplandicum*, *S.* × *uplandicum* × *S. tuberosum*, *S. asperum*, *S. tuberosum*, *S.* × 'Hidcote Blue', *S. grandiflorum*, *S. tauricum*, *S. orientale*, *S. caucasicum* and *S. bulbosum*. Of these, only two, *S. officinale* and *S. tuberosum*, are native: the rest are introductions largely from SE Europe and adjacent parts of Asia. Most of the introductions are distinct species which cause no particular taxonomic or identification problems in this country. However, the *S. officinale* complex, which involves *S. officinale*, *S.* × *uplandicum* and *S. asperum*, comprises 3 native taxa and 3 introduced taxa together with hybrids between many of them, and this has long led to confusion. An attempt to unravel this complex is the main subject of the paper.

Introduction

The *S. officinale* complex, which involves *S. officinale*, *S.* × *uplandicum* and *S. asperum*, comprises 3 native taxa and 3 introduced taxa together with hybrids between many of them, and this has long led to confusion (see Fig. 1). This confusion was apparent during the preparation of the *Atlas of the British Flora* (Perring & Walters 1962) in which we could publish only a map of *S. officinale s.l.* and even the map of *S.* × *uplandicum* in the *Critical Supplement* (Perring 1968) had a text which finished with the words 'This map ... must be regarded as provisional'.

Symphytum in the British Isles

As a result the Botanical Society of the British Isles conducted a Network Research Project in the early 1970s – the *Symphytum* Survey. About that time two Dutch botanists, Messrs Gadella and Kliphuis, came to Britain as part of a European study of the complex and investigated 122 British plants cytologically. Some were plants which I had collected and grown in Cambridge University Botanic Garden. Many were later transferred to my own garden in Oundle, Northamptonshire where I lived with them for 12 years. The observations which follow are based on these three sources of information. The 6 taxa are:

1. *S. officinale* 2n=24. Short plants up to 1 m. Leaf decurrence exceeding one internode, buds always greenish yellow, flowers creamy white, open with corolla tube *ca.* 14

mm long and anthers longer than the filaments. It seems to be confined to fens in Cambridgeshire, Huntingdonshire and S. Lincolnshire (v.-cs. 29, 31 & 53). The two plants for which chromosome counts were made came from Wicken Fen and Welches Dam in Cambridgeshire.

2. *S. officinale* 2n=48. Plants taller, up to 1.5 m. Otherwise identical in leaf decurrence, flower colour and anther/filament relationship to the diploids but with the corolla *ca.* 18 mm long. Plants of river banks and ditches throughout lowland England, with scattered localities in E. Wales, SE Scotland and SE Ireland.

3. *S. officinale* 2n=48. Identical to 2. above but flowers have carmine coloured buds (59A in the RHS Colour Chart) and open corollas which are red/purple (72A). There is not a hint of blue in the colour. Also in ditches and on river banks, for example on the River Severn in Shropshire, and can occur in the absence of 2. above.

However, 2. and 3. do grow together and mixed populations with some creamy yellow and some reddish/purple flowers occur. This phenomenon is confined mainly to SW England particularly in v.-cs. 9, 11-13, 22-24, 33-37. Occasionally within these mixed populations there are plants which appear to be hybrids with corollas with alternating vertical, dark and light bands charmingly described by the late Ted Ellis as his 'peppermint stripe' form. Plants of this kind also have 2n=48.

In Holland there is a third cytotype in which 2n=40. This nearly always has purple flowers, and rough leaves with prickly hairs which are not as strongly decurrent as those of the diploids and tetraploids already discussed. It is apparently abundant in the Netherlands where it grows on peat under very moist conditions, but has not been recognized elsewhere. It should be looked for in Britain.

4. *S. asperum*. A tall plant up to 1.5 m. The stems and the midrib of the underside of the leaves are covered in stiff bristles. At least the lower and middle leaves are cordate and shortly petiolate whilst leaf decurrence is completely absent. In bud the flowers have a very small calyx, usually only 1/5 as long as the corolla but this enlarges rapidly in fruit. The corolla is red in bud but changes to a clear blue when open. The limb of the corolla gradually widens towards the top. The chromosomes of material from Croombe D'Abitot in Worcestershire were counted and were 2n=32. The same number has been found consistently in material from elsewhere in Europe examined by Gadella (1972). I have seen morphologically indistinguishable specimens in the garden of the late Dr G. Nelson which he was sent from Russia, and in Cambridge University Botanic Garden. In the 19th century it was grown for ornament and, in the 1954 edition of the *RHS Dictionary of Gardening*, it was recommended as a plant for the wild garden. It is now a very rare escape from cultivation that was probably much over-recorded in the past through confusion with one form of *S.* × *uplandicum* (5. below).

S. × *uplandicum*. This is a range of hybrids between *S. asperum* and *S. officinale*. They generally lack the broadly decurrent leaves of *S. officinale* but the leaves are never cordate and petiolate as in *S. asperum*. The calyx is similar in length to that of *S. officinale*, with acute segments; the corolla is long, up to 20 mm, with filaments equalling or longer than the anthers, and the corolla does not widen towards the top as in *S. asperum*. Corolla colour

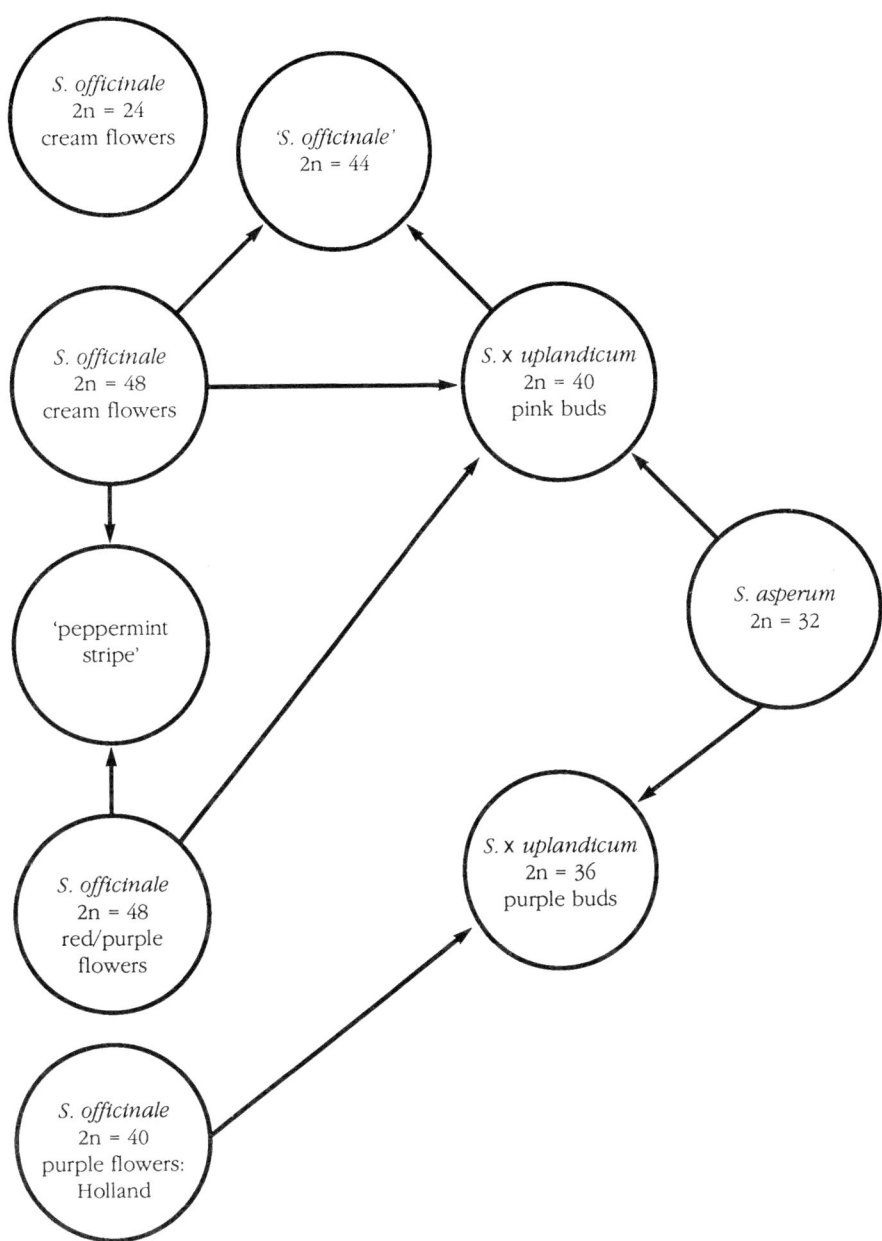

Fig. 1. *Symphytum 'officinale'* Complex.

varies from reddish/purple to violet, but is never clear blue as in *S. asperum* or creamy white as in some forms of *S. officinale*. Chromosome counts and morphological studies indicate that two cytotypes are found in the British Isles.

5. *S.* × *uplandicum* 2n=36. This is a plant with leaves which are not or only very slightly decurrent. The stems and leaves have prickly hairs and the whole plant is rough to the touch. The flower buds are generally deep purple (79A) or occasionally with a violet tinge but, on opening, the colour of the corolla changes to violet or violet blue (84A & 90A) or to a purple violet (77B, 80A or B & 82A). There is often a change to a bluer colour with age, but never to the clear blue of *S. asperum*. The general impression of a population is its uniformity in decurrence and flower colour. This hybrid almost certainly originated as follows: *S. asperum* (2n=32) × *S. officinale* (2n=40) giving the hybrid (2n=36). This has been duplicated experimentally by Gadella and Kliphuis (1971) in the Netherlands. It seems likely that this cytotype was introduced into the British Isles in view of the fact that hybridization between the 2n=40 form of *S. officinale* and *S. asperum* is impossible since the parental forms are either absent or extremely rare. Once introduced it does not appear to have introgressed with native forms of *S. officinale* or other introduced taxa within the complex; hence the uniformity of the populations noted above. It is widespread in the British Isles but is generally less common than the 2n=40 forms. The majority of records of *S. asperum* in the past were probably this taxon.

6. *S.* × *uplandicum* 2n=40. This plant is much more robust than the purple-budded form. Some leaves are usually slightly decurrent. The stems and leaves are not usually prickly hairy and the plant is soft to the touch. The flower buds vary from pink magenta (centred round 64A) but change on opening to a bewildering range of colour forms from pink (62B) to violet (88A). These variations often exist within a single population and it is these plants, with somewhat decurrent leaves, which must have been recorded as *S. officinale* in the past. This variable hybrid arose from crosses between *S. asperum* (2n=32) and the various colour forms of the tetraploid *S. officinale* (2n=48). Artificial hybrids produced by Gadella and Kliphuis (1971) were indistinguishable from these pink magenta budded forms of *S.* × *uplandicum*. This is the commonest taxon within the *S.* × *uplandicum* complex with about three times as many records as for the 2n=36 form. This is hardly surprising as it is clear that this is the hybrid received from Sweden by Henry Doubleday in 1871 and grown at Coggeshall in Essex and marketed by him and another nurseryman, Thomas Christy of Sydenham, as Russian Comfrey (though it came from Sweden) to distinguish it from *S. asperum* which was known as 'Prickly Comfrey'. This was grown as a forage crop and once introduced is extremely hard to eradicate. There is also a variegated form of this comfrey which is quite widely grown in gardens and marketed as 'variegatum'. It is not so vigorous as the normal form but exhibits the same characteristic of flowering twice in a season and remains attractive from late spring through to the autumn. That it does belong to this taxon is clear from the slight decurrence of the leaves and the pink buds.

As if the variation within *S.* × *uplandicum* (2n=40) were not enough, it became apparent during the survey of British material by Gadella and Kliphuis that some back-crossing has occurred between *S.* × *uplandicum* (2n=40) and *S. officinale* (2n=48). Two populations, one from Claydon in E. Suffolk and another from

Twyford Forest in S. Lincolnshire, have 2n=44 and experimental work showed that this hybrid can be easily produced artificially and that it is fertile.

In addition to *S. asperum* and the variegated form of *S.* × *uplandicum* (2n=40) there are a number of other taxa within the genus which may be found in, or perhaps should be considered for, British gardens.

S. grandiflorum, Creeping Comfrey, is a prostrate plant which spreads by leafy stolons. It has unbranched stems with leaves which are mostly petiolate and not or hardly decurrent. The flowers arise from pinkish buds which develop pale yellow corollas. They have deeply divided calyces and persistent exserted styles. The species is a native of the Caucasus but is widely grown in gardens where it is a useful ground cover plant for semi-shade with a long flowering period, often starting in February, reaching a maximum in April but continuing sporadically until mid summer. It is not infrequently found as a garden escape in central and southern Britain as far north as S. Northumberland.

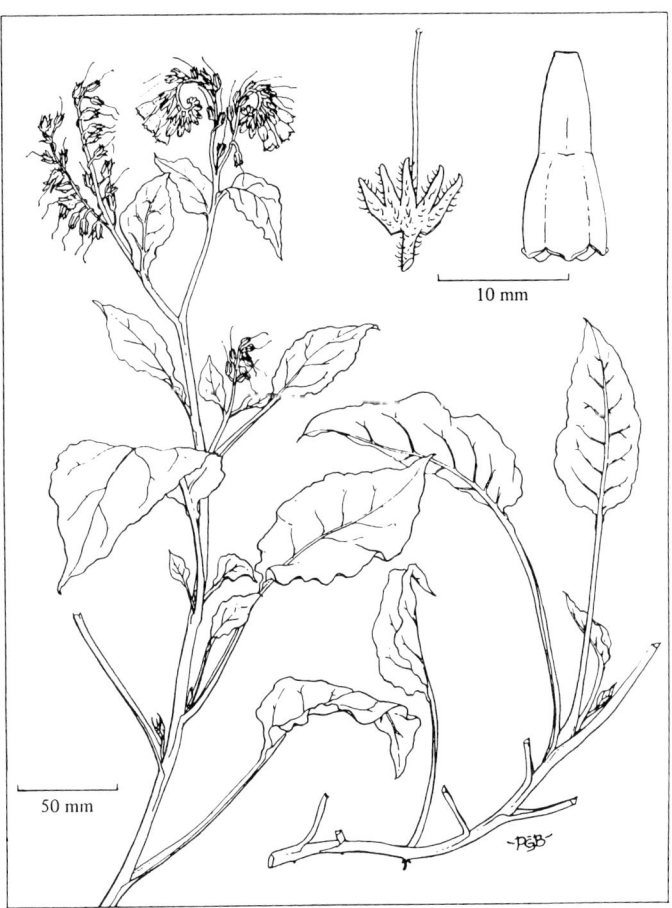

Fig. 2. *Symphytum* × 'Hidcote Blue' (from Leslie 1982).

S. × 'Hidcote Blue' (Fig. 2), is similar to *S. grandiflorum* but is a taller plant with ascending stems up to *ca.* 0.6 m. The flower buds are red (47A) but the corolla when open is two-coloured, almost white above and a delicate blue (101B) below making a pretty contrast. It has similar shade tolerance and flowering period to *S. grandiflorum* and is becoming increasingly available from nurserymen. It was first reported as a garden escape from Shropshire in 1980 (Leslie 1982) and has since been found in Guernsey and Surrey. It is almost certainly a hybrid between *S. grandiflorum* and perhaps one of the forms of *S.* × *uplandicum*.

S. orientale, White Comfrey, is well named as it is the only species likely to be found growing in the British Isles which has pure white flowers. It is an erect, softly hairy, little-branched plant up to 0.7 m. The leaves are sessile or shortly petiolate but not decurrent. The flowers are distinguished not only by their colour but by the exserted style and by a calyx divided only about 1/5.

This native of W. Turkey and SW Ukraine (it grows in Europe around Istanbul) was probably the first 'foreign' comfrey to be introduced into gardens in Britain; edition 13 of Donn's *Hortus Cantabrigiensis* reports it as in cultivation in 1752, 47 years before *S. asperum* and 63 before *S.* × *uplandicum*. In the ensuing 240 years or more it has frequently escaped from gardens and is well naturalized in town streets, hedge bottoms and other shady places, often self-sown, throughout S. & E. England and in scattered places elsewhere including Scotland and Ireland.

S. caucasicum, Caucasian Comfrey, is a softly hairy plant up to 0.4 m tall which spreads by rhizomes. The stem leaves are sessile, ovate-lanceolate and shortly decurrent. The calyx is divided up to 1/4 into obtuse teeth whilst the corolla is funnel-shaped and blue (101B), not unlike *S. asperum*. This species, known in the wild only from wet meadows and stream sides in the Caucasus, was grown in Cambridge as early as 1816: it occurs only rarely as a garden escape in SE England and has been reported from Flint in Wales (Wynne 1993) but, despite its rhizomatous habit, is often not persistent. As an attractive non-vigorous plant it deserves to be more widely grown in the herbaceous border. Happily it is now available from a good number of nurserymen.

References

Gadella, T.W.J. (1972). Cytological and hybridization studies in the genus *Symphytum*. *Symp. Biol. Hung.* **12**: 189-199.

Gadella, T.W.J. & Kliphuis, E. (1971). Cytotaxonomic studies in the genus *Symphytum* III. Some *Symphytum* hybrids in Belgium and the Netherlands. *Biol. Jaarb. Dodonaea* **39**: 97-107.

Leslie, A.C. (1982). A new alien *Symphytum*. *BSBI News* **30**: 16-17.

Perring, F.H. (ed) (1968). *Critical Supplement to the Atlas of the British Flora*. London: Thomas Nelson.

Perring, F.H. & Walters, S.M. (eds) (1962). *Atlas of the British Flora*. London and Edinburgh: Thomas Nelson.

Stace, C. (1991). *New Flora of the British Isles*. Cambridge: Cambridge University Press.

Wynne, G. (1993). *Flora of Flintshire*. Denbigh: Gee & Son.

9

Hypericum wild and cultivated – the common ground

N. ROBSON

Burn Edge, 48 Granville Road, Limpsfield, Oxted, Surrey, RH8 0DA

ABSTRACT. *Hypericum*, a genus of about 420 species with an almost world-wide distribution, has been cultivated in British gardens since medieval times. It can provide numerous examples of ways in which wild and cultivated plants interact:-
　1) Garden escapes. These include *H. calycinum* as well as the variable *H. hircinum*, which crossed with the native *H. androsaemum*.
　2) Spontaneous garden introductions from the wild. Apart from unwelcome incomers (weeds), one or two species (such as *H. humifusum*) have found an appropriate niche in gardens.
　3) Garden hybridization 'blurring' distinct wild species. Examples of this phenomenon are taken from the mainly North American section *Myriandra*.
　4) Differential introduction emphasizing differences that are 'blurred' in the wild. *H. olympicum* L. provides a good example.
　5) The influence of nomenclature on introductions. *H. patulum* Thunb. retained a hold on the horticultural imagination long after it became a rarity in cultivation.
　6) Hybridization in the wild followed by hybridization in cultivation. The story of *H.* × 'Hidcote'.

Introduction

Hypericum is a genus of about 420 species that occur naturally in most parts of the world where the climate is not extremely hot, extremely cold or extremely dry. It has been cultivated in British gardens since medieval times and can provide numerous examples of ways in which wild and cultivated plants interact.

Garden escapes

The best known of these is undoubtedly *H. calycinum*, a member of the ground flora of the broad-leaved woods of southern Bulgaria and along the south side of the Black Sea as far as Trabzon in north-eastern Turkey. It was introduced to Britain from near Constantinople in 1676 by the British Envoy to the Ottoman Sultan, Sir George Wheeler, and became widespread in gardens because of its large flowers and its ability to survive in dry shade. It soon escaped (Britten 1915) and became established in semi-natural habitats, largely as a result of vegetative spread, because our summers are usually too cool and our autumns too damp to allow the seed to ripen (Salisbury 1963).

An earlier introduction from southern Europe was *H. hircinum*, a variable species that has been divided into five subspecies. The largest and most widespread, both in the wild and in cultivation, is ssp. *majus*, which occurs from south-western France to south-western Saudi Arabia; but the smaller-flowered, obtuse-leaved ssp. *hircinum*, from Corsica and Sardinia, was also grown here by the 18th century. Later introductions were the relatively dwarf, acute-leaved ssp. *cambessedesii* (Cosson ex Barceló) Sauvage from the Balearic Islands and, recently, ssp. *albimontanum* (Greuter) N. Robson from Crete. The large flowers and broad undulate leaves of this subspecies make it a striking garden plant. The tall ssp. *majus* and dwarfer ssp. *hircinum* crossed naturally in gardens with our native *H. androsaemum*, producing the variable *H.* × *inodorum*; both *H. hircinum* ssp. *majus* and the taller form of the hybrid *H.* × *inodorum* became naturalized members of the British flora, the hybrid under the name *H. elatum*; and a selected form or back-cross to *H. androsaemum* has become popular as the cultivar 'Elstead'.

Other escapes have been less well established, such as *H. richeri* Vill. ssp. *richeri* from the western Alps, which escaped from a rock garden in Ledbury, Herefordshire early this century, or *H. barbatum* Jacq., a member of the same section, *Drosocarpium*. This native of south-eastern Europe (from Austria to Crete) was recorded from Strathearn, Perthshire by George Don (senior), and it was included in *English Botany* (Smith & Sowerby 1809) with an illustration by Sowerby; but it soon disappeared and has not been recorded in Britain since then.

More recently recorded escapes in the British Isles have included *H. xylosteifolium*, a systematically and geographically isolated shrub of the south-eastern Black Sea area, which turned up in West Lancashire (1978) and near Ripon, Yorkshire the next year (1979). It forms dense thickets by suckering. *H. olympicum* L. (from Greece and W. Turkey) and *H. nummularium* (from the Pyrenees and W. Alps) are both represented in the British Herbarium of the Natural History Museum (**BM**), respectively from South Devon (1943) and North Yorkshire (1972).

Finally, the recent influx of shrubby *Hypericums* from China and the Himalayas seems to have had practically no effect on the natural flora. It is rather early, however, to conclude that none is likely to become established in the wild; indeed, *H. pseudohenryi*, which seeds abundantly in our garden, has already been recorded from Glengarriff, Co. Cork, in 1983.

Spontaneous garden introductions from the wild

The garden incomers are often unwelcome and are then termed 'weeds', but they may find a place where they are a horticultural asset. For example, the short-lived *H. humifusum* can produce a fine effect between stones in a path, while *H. elodes* can flower brightly in a not-too-acid bog garden or shallow edge of a pool. The extremely 'Atlantic' *H. undulatum*, which is confined in Britain to south-west England and south-west Wales, occurs naturally in acid marshes; but it seems to thrive under much drier conditions in gardens.

Plate 3

Hypericum hircinum ssp. *majus*
(Photo NR)

Hypericum x *inodorum*
(Photo NR)

Hypericum xylosteifolium
(Photo NR)

Hypericum forrestii
(Photo NR)

Plate 4

Hypericum x *'Hidcote'*
(Photo NR)

Hypericum calycinum
(Photo NR)

Hypericum x *cyathiflorum 'Gold Cup'*
(Photo NR)

Hypericum x *dummeri*
(Photo NR)

Garden hybridization 'blurring' distinct wild species

The best example of this phenomenon is to be found in section *Myriandra*, which is centred in the south-eastern United States. In its species the stamen bundles that are a feature of most cultivated and wild Eurasian Hypericums have become completely merged to form an uninterrupted ring; and, particularly in the flowers of the southern Appalachian *H. frondosum* Michaux, they produce a powder-puff-like effect. This species has few large flowers (usually from the terminal node only) and relatively broad leaves. It is apparently always distinguishable in the wild from its nearest relative, *H. prolificum* L., which is more widely distributed in the eastern United States. *H. prolificum* has a cylindrical inflorescence of somewhat smaller flowers, from three stem nodes, and rather narrower leaves. Where these species are grown together, however, hybrid intermediates frequently occur. These were named *H.* × *vanfleetii* by Alfred Rehder and are now found in gardens in the absence of one or both parents. A similar situation exists between *H. prolificum* and *H. densiflorum* Pursh, which has even smaller flowers and narrower leaves. A low compact shrub grown in British gardens as *H.* × *arnoldianum* Rehder could be of this parentage, that is *H. prolificum* × *H. densiflorum*, but it is more likely to be a hybrid between *H. densiflorum* and either *H. lobocarpum* Gattinger or *H. kalmianum* L.

Differential introduction emphasizing differences that are 'blurred' in the wild

Variation in natural populations often makes it necessary to subdivide species into subspecies, in which the distinctions between the populations are incomplete, or into varieties, in which they are erratic and not associated with, for instance, geography or ecology. If introductions to cultivation are made from different parts of the range of these variable populations, distinct garden plants may result, especially if each introduction is propagated vegetatively. These distinct plants may be, and frequently are, recognized as cultivars, even though their distinctness is wholly due to the circumstances of cultivation and is blurred in their natural habitat by intermediates. A good example of this is *H. olympicum*, a plant of the northern Aegean region and W. Turkey (Robson 1980). Variation in this species is geographical, but it is almost impossible to classify because of numerous intermediates. The introduction of plants respectively from 1) northern Greece, 2) north-eastern Greece and north-western Turkey and 3) southern Greece has, however, produced three distinguishable garden plants. Although these do not appear to hybridize, they reproduce by seed and show a certain amount of variability. It is not therefore possible to treat them as cultivars, and I have classified them as botanical forms. Forma *olympicum* (from northern Greece and north- western Turkey) has large flowers and long or narrow elliptic leaves and is usually erect; f. *uniflorum* D. Jord. & Koz. (from north-eastern Greece and Bulgaria) has smaller flowers and relatively broad ovate leaves and is invariably erect except in shaded habitats; f. *minus* Hausskn. (from southern Greece) is a small-flowered and small-leaved variant of f. *olympicum* that varies from erect to prostrate. The erect variant has been confused in cultivation with the closely related south Turkish species *H. polyphyllum* Boiss. & Bal., and straggling variants have been misnamed *H. fragile*, a name correctly applied to a very small Greek relative of the British *H. pulchrum*. Both f. *uniflorum* and f. *minus* have given rise in gardens to pale-yellow-flowered variants, and these have correctly been named as cultivars, respectively 'Citrinum' and 'Sulphureum'.

Where species are dioecious or heterostylous, the introduction of one sex or one style length only will result in cultivated populations with no genetic variability. Thus the Mediterranean shrub *H. aegypticum* L. has been represented in gardens until recently by the short-styled form only; but since the long-styled form was introduced from Malta a few years ago, we may expect to see greater variability if the two forms cross.

The influence of nomenclature on introductions

Differential introduction sometimes leads to conservation problems and even nomenclatural confusion. After 1862, when Richard Oldham brought back *H. patulum* Thunb. from Nagasaki, that species became cultivated widely in the British Isles (Robson 1970). Although well known in the gardens of Japan and lowland China, it is a native of the lower regions of the mountains of central China, mostly between 450 and 1750 m. Coming from such low altitudes, it is not surprising that, by the end of last century, *H. patulum* appears to have succumbed to cold winters everywhere in the British Isles except in Ireland and possibly south-western England By then, however, a hardier plant from the lower regions of Yunnan (1500-2400 m) had been introduced by Augustine Henry and was given the name *H. patulum* var. *henryi* Veitch ex Bean in 1905. This, in turn, was replaced in the favour of gardeners by a plant from even higher altitudes in Yunnan, Sichuan and Burma (up to possibly 4000 m) introduced by George Forrest. This also was treated as a variant of *H. patulum* and named *H. patulum* var. *forrestii* Chittenden in 1923. In due course it, too, was replaced as the commonly planted tall shrubby Hypericum by a double hybrid *H.* × 'Hidcote', which, being sterile and having the hardy *H. calycinum* as a parent, flowers on until the autumn frosts. Even in a recent (1992) issue of *The Plantsman*, this is still wrongly listed under *H. patulum*, with which it has no connection.

Both the plants named as varieties of *H. patulum* are in fact good species (respectively *H. beanii* N. Robson and *H. forrestii* (Chittenden) N. Robson), but the nomenclatural influence of the originally introduced species persisted for a long time after *H. patulum* itself became a garden rarity. Fortunately, recent introductions, by Roy Lancaster and also by Chris Brickell and Alan Leslie, have made this desirable species once more available in the West. It is not yet sufficiently widespread to spoil my 'game' of guessing when a garden was laid out, or principally stocked, by seeing which shrubby Hypericums are represented in it.

Hybridization in the wild followed by hybridization in cultivation

I have already mentioned *H.* × 'Hidcote', which was introduced to the gardening public in this country at a Royal Horticultural Society show in 1950 and soon became popular on account of its hardiness and continuous flowering; but where and how did it originate? The name 'Hidcote Variety', given to it by the exhibitor T. Hilling, suggested an association with Lawrence Johnston's garden at Hidcote Manor, Gloucestershire; but the relevant records are apparently lost. The hardiness, as well as certain other floral and foliage characters, suggested that one parent of this hybrid was *H. calycinum*; but what was the other?

In an attempt to reproduce 'Hidcote', *H. calycinum* was twice crossed with *H. forrestii*, which seemed to have most of the characters of the missing parent. The result, both at Chelsea

Physic Garden and at Hillier's Nurseries, was a different plant, much lower than 'Hidcote' and with a tendency to creep like *H. calycinum*. This is now established in gardens as *H.* × *dummeri* N. Robson, named after Peter Dummer, who made the cross at Hillier's.

'The penny dropped' for me only after a visit to Hidcote Manor in 1979 or 1980, where I found only two shrubby Hypericums growing besides 'Hidcote'; one was *H. calycinum*, the other a rather uncommon garden hybrid known as *H.* × 'Gold Cup' (Robson 1985). This plant provided all the characters of *H.* × 'Hidcote' that were absent from *H. calycinum*, and so I became convinced that it was the missing parent. Like 'Hidcote', it could not be matched by wild-collected material, and so a hybrid origin for it was likely, too. *H.* × 'Gold Cup' combines the characters of two species: *H. addingtonii* N. Robson, which is endemic to north-west Yunnan, and *H. hookerianum* Wight & Arnott, which occurs there. I therefore gave it a specific hybrid name – *H.* ×*cyathiflorum* N. Robson 'Gold Cup'. It has $n=15$ chromosomes, which, when combined with the $n=10$ of *H. calycinum*, produces the $n=$ *ca.* 25 of *H.* × 'Hidcote'. *H.* ×*cyathiflorum* 'Gold Cup' probably originated in the wild and was brought back to this country, possibly by Lawrence Johnston himself (the 'Gold Cup' plants at Hidcote were being grown under the name 'Lawrence Johnston'); and the evidence of the name suggests that the cross with *H. calycinum* occurred, spontaneously or intentionally, at Hidcote Manor.

Conclusions

The interactions between wild and cultivated plants are multivarious, and *Hypericum* could provide other examples; but I hope that I have been able to demonstrate some ways in which cultivation can alter the state – or status – of some members of a rather large, almost world-wide genus.

References

Britten, J. (1915). Notes on *Hypericum calycinum* L. *Journal of Botany, London* **53**: 68.
Robson, N.K.B. (1970). Shrubby Asiatic *Hypericum* species in cultivation. *Journal of the Royal Horticultural Society* **95**: 482-497.
Robson, N.K.B. (1980). *Hypericum olympicum* – wild and cultivated. *Plantsman* **1**: 193-200 + errata.
Robson, N.K.B. (1985). Studies in the genus *Hypericum* L. (Guttiferae). 3. Sections 1. *Campylosporus* to 6a. *Umbraculoides. Bulletin of the British Museum (Natural History), Bot.* **12**: 163-325.
Salisbury, E.G. (1963). Fertile seed production and self-incompatibility of *Hypericum calycinum* in England. *Watsonia* **5**: 368-376.
Smith, E.J. & Sowerby, J. (1809). *Hypericum barbatum* Jacq. *In English Botany* **28**: t. 1986.

10

Proposal for a register of aberrant forms

M. CRAGG-BARBER

1 Station Cottages, Hullavington, Chippenham, Wiltshire, SN14 6ET

ABSTRACT. The topics covered include: (1) outline of teratology, touching on historical context, phyllody, floral and foliar proliferation, aerial tubers, laciniation, peloric forms, possible fasciation; (2) selected information as to causes of double flowers/single flowers; (3) use or interest in updating Masters' lists of 1869; (4) review of teratological forms (with one or two pigment aberrations) from selected corners of Wiltshire.

Introduction

The word 'teratology' has been generally used for aberrations of form rather than of pigment. There are interesting examples where both forms of aberration coincide. However, 'teratology' is already so diverse that to attempt to include pigment aberrations is perhaps a mistake. In the following review of Wiltshire *terata* observed in 1992 one pigment aberration which arose from a generally monstrous population will be described.

Review of recent aberrant forms from Wiltshire

In 1991, in one particular garden, the vast majority of the Foxgloves, *Digitalis purpurea*, that were due to flower failed to do so; their leaves became curled and contorted and though petals were discernible in some plants the flowers failed to open. The plants had become monstrous and deformed without giving rise to any specific teratological form. A foxglove virus was suspected. Two or three plants had escaped the supposed virus and in one of these plants a thin white line was noted up the stem of a subsidiary spike and continuing into the capsule. Seed from this capsule germinated to produce a majority of green seedlings but amongst them were a few variegated plants.

The yellow foliage form of Feverfew, *Tanacetum parthenium*, is noticeably consistent in its flower form. One particular plant in the same garden displayed a proliferation of leafy bracts beneath the flower on a few stems. Propagated material from these stems went on to produce plants with entirely normal flowers. In 1992 there were no flowers with leafy bracts on the original plant: instead, the flowers were fasciated in curves. A more constant aberration from the same garden is the form of Hart's-tongue, *Phyllitis scolopendrium*, with sori on the upper surface of the frond, a phenomenon noted by Masters. Double-flowered Creeping Buttercup, *Ranunculus repens*, also occurred there in 1992.

A local wood became prominent for the appearance of double-flowered and hose-in-hose flowers of Water Avens, *Geum rivale*. However, propagated material when grown in my garden produced only single flowers and two years later the only *terata* observed were specimens with the occasional leafy bract. Masters mentions that *G. rivale* is a plant very liable to teratological change. A small garden near Devizes produced an interesting form which last year had double red flowers. In the same place, for two consecutive autumns, a number of plants of Groundsel, *Senecio vulgaris*, produced perfect funnels of a leafy texture instead of flowers. Masters (1869: 313) supplies us with an illustration of the same phenomenon occurring in a lettuce plant (Fig. 1a). A specimen of *Philadelphus* discovered in July 1992 was also of interest. Its leaves formed funnels without terminating the shoot and had small holes as if produced by a gall-forming insect. Another interesting point was an increase in reports to Wisley of bracteate bluebells in 1991 (Fig 1b).

The susceptibility of the Toadflax family (Scrophulariaceae) to peloria may be demonstrated using an example from another Wiltshire garden (Fig. 1c). Most of the flowers were normal but peloric flowers appeared in the axils of the flowers at the bottom of the stem. Their initiation might well have coincided with the extremely wet weather at the end of May.[1]

In July 1992, another garden threw up an interesting form of *Lavatera olbia* where the normal single flowers were replaced by one large double flower which also terminated the shoot. It is possible to speculate that the dry weather through June 1992 played a part in this *teras*, as if the strong flowering impulse could no longer be paralleled by normal shoot extension.

Concluding suggestions

Teratological aberrations demonstrate the extremes of plasticity of each species. Part of the total picture of this plasticity is contained in knowledge accumulated in the past and is to be found in works on this subject which have largely petered out since 1950.[2] It can also be found in specimens dotted through herbarium collections and in the memories of field botanists and gardeners. The other part of this picture is outside, in the field, still awaiting discovery and it is to this that some sort of register or collation of reports is aimed. A group of enthusiasts could not only bring up to date the information contained in the literature and discover new *terata* but could also perhaps monitor simultaneous occurrences through the country. Masters (op.cit.) goes into little detail (other than in his appendix on double-flowered forms) as to whether *terata* were related to specific or aberrant conditions, so closely observed plants that have given rise to terata should enable us to add contributory factors such as weather conditions to our understanding of their development.

---

[1] In 1993 the same garden again produced a number of peloric flowers including one with eight sepals, four spurs and the normal mouth.

[2] This comment was a little premature. Since writing it a number of teratological references which include more recent items have been kindly sent to me by Dr Pierre Binggeli and Dr Brian Spooner.

Fig. 1. *a*, leaf of lettuce (*Lactuca sativa*) bearing on the back a stalked cup, arising from the dilatation of the stalk (?) (from Masters 1869); *b*, bracteate Bluebell (*Hyacinthoides non-scripta*); *c*, peloric form of Common Toadflax (*Linaria vulgaris*). (*b* and *c* drawn by Sue Herbert).

In order to identify and monitor the occurrence of aberrant forms of native plants, it is proposed, therefore, to set up a Register of such forms. Aberrant forms include both abnormalities of pigmentation (unusual colour of flower, foliage, stem or fruit) and morphological abnormalities (for example, double flowers, apetaloid forms, bracteate forms and fasciations). The aims of the register would be:

- To contribute to a historical record of the plasticity of each native species.
- To monitor the occurrence of such forms over space and time. Some years seem to be good for certain aberrations (e.g. bracteate bluebells). Other forms seem to occur simultaneously at great distances from each other.
- To retain the background information concerning the introduction into cultivation of forms of native plants.

Reference

Masters, M.T. (1869). *Vegetable Teratology, an Account of the Principal Deviations from the Usual Construction of Plants*. London: Robert Hardwicke (for the Ray Society).

11

The reintroduction of plants into the wild: an integrated approach to the conservation of native plants

MIKE MAUNDER[1] and MARGARET RAMSAY[2]

Conservation Unit [1] and Sainsbury Orchid Conservation Project [2], Living Collections Department, Royal Botanic Gardens, Kew, Richmond, Surrey, TW9 3AB

ABSTRACT. The creation of a protected area may not in itself preserve existing biodiversity; the effects of isolation, disturbance and biological invasion may necessitate an interventionist approach to species and reserve management. Additionally the effects of internal habitat dynamics and external influences such as climatic change, will influence the ability of an area to support species. An integrated approach to species conservation has the potential to operate at the levels of the species, habitat and landscape.

The reintroduction of rare plants is becoming an increasingly utilized technique. Reintroductions and associated activities are discussed with examples from the British Isles reviewed.

It could be argued that nature conservation in Britain is failing since plant extinctions are still occurring and protected areas such as SSSIs continue to be degraded. The British Isles are not immune from the global crisis of expanding rates of species extinction. The landscapes and habitats of the British Isles are changing dramatically both in terms of area and in the capacity to support the native biota. As such areas are subject to increasing degradation and isolation the traditional emphasis on obtaining protected areas for conservation will have to be supplemented by the active management of plant populations.

The conservation of biodiversity requires action at different levels. Soule (1991) has identified categories in the 'biospatial hierarchy' as follows: (1) whole systems at the landscape or ecosystem level, (2) assemblages (associations or communities), (3) species, (4) populations and (5) genes. Different techniques are required according to the level in the hierarchy. Whilst a protected area will operate on many of these levels a specific action plan for an endangered species will aim at saving one hierarchy, that of the species. The species cannot be saved in isolation from other levels of the hierarchy, each level subject to differing types of interference all potentially influencing the population of the target species. Soule (op.cit.) identifies six broad categories of interference: (1) the loss of habitat, (2) fragmentation of habitat producing deleterious area, edge, demographic and genetic effects, (3) over exploitation, (4) spread of exotic species and diseases, (5) air, soil and water pollution and (6) climatic change. A number of the factors listed above can at best be only partly dealt with through the action of protected area acquisition.

Accordingly a multi-disciplinary approach is required to counter these diverse threats. Such a strategy meshes the traditionally isolated approaches of *in situ* or habitat-based conservation with the more intensive *ex situ* approach. The reintroduction of extirpated populations is just one technique available to the conservation manager. It is a topic that has generated a considerable amount of debate. For some it possesses the attraction of reinstating extirpated populations; for others it poses inherent dangers through the potential to introduce inappropriate or diseased stock and the devaluing of original populations and habitats.

A number of guidelines for reintroduction procedure have been developed. The Reintroductions Specialist Group of the Species Survival Commission has been charged with updating the IUCN (1987) guidelines on species translocation. Reintroductions have been defined as follows (Stanley Price 1989) and merit comparison with the definitions established by the Botanical Society of the British Isles:

REINTRODUCTION: 'the deliberate release of individuals of a species into an area from which it has been lost, with the aim of establishing a self-sustaining and viable population.' This is synonymous with the BSBI's RE-ESTABLISHMENT defined as 'the deliberate release and encouragement of a plant in an area in which it formerly occurred but where it can now reasonably be held to be extinct; or, a plant so treated'. This 're-establishment' must be within 1 km of a previous record for a natural population of the species. This criterion of the BSBI will effectively reduce the number of strict reintroductions *sensu* Stanley Price and is only feasible in an exceptionally well-surveyed country such as the UK.

RESTOCKING: 'the release of individuals to reinforce an existing population with the aim of increasing population viability.' This is synonymous with the BSBI's REINFORCEMENT defined as 'an attempt to increase population size by adding further individuals to an existing population'.

INTRODUCTION: defined by the IUCN (1987) as 'the intentional or accidental dispersal by human agency of a living organism outside of its historically known native range'. The BSBI definition is far more geographically specific, defined as 'an attempt to establish a plant in an area in which it is not known either to occur now or to have occurred in historical times; or, a plant so treated'. According to the BSBI definitions an introduction takes place at a distance greater than 1 km from any previous record. RE-INTRODUCTION is defined by the BSBI as an 'an attempt to establish a plant in an area to which it has been introduced but where the introduction proved unsuccessful; or, a plant so treated'. It is apparent that the definitions developed by the BSBI are dependent upon a detailed knowledge of the distribution of the indigenous floras, but the UK is uniquely privileged in possessing possibly the world's most studied flora. The definitions do not take into account the dynamic nature of plant distribution and habitat change. Placing a defining constraint of one kilometre for re-establishment (synonymous with reintroduction *sensu* Stanley Price) would place a stranglehold on conservation practice. Reintroducing an extirpated population on a site more than 1 km from a recorded natural population but in a piece of contiguous habitat should not be termed as an introduction, a term that should be reserved for the establishment of exotic or non-native species outside their natural range or biotope. During a period of 20 years a habitat can change dramatically and replacing a population to the right geographical site will be futile if the habitat has changed and can no longer support the species involved.

Reintroduction is an operation distinct from amenity or aesthetic inspired projects, such as the seeding of roadsides with mixed-provenance wild flower mixes. Reintroduction projects aim to increase the viability of an endangered species through promoting existing populations and habitats, and through establishing new populations. The projects are long term, require secure land tenure and place great emphasis on the correct provenance of material and the undertaking of supporting research. Projects for amenity or aesthetic purposes, termed political habitat creation by Baines (1989), do not need to comply with any particular habitat model; an attractive display is the priority. Such areas have a great role in urban areas through encouraging public interest in conservation.

A different protocol is required for projects where the aim is to promote conservation of species and habitats and to restore habitats in rural areas. Baines (op.cit.) refers to this as 'ecological habitat creation'. Howell and Jordan (1991) have defined restoration (a preferable term since 'creation' suggests a biblical connotation, whilst restoration recognizes the need to use any remnant habitat components) as follows:

COMPLETE RESTORATION: ' the establishment of a group of species in abundance and proportions similar to those in natural communities such that natural processes occur.'

FUNCTIONAL RESTORATION: 'the use of community-like groupings of plants to perform general processes similar to those characteristic of the natural communities, e.g. nutrient cycles, or erosion control. In these plantings close adherence to the structure and composition of natural communities may not be important: these may be simplifications or may emphasize some species at the expense of others. . . . To be considered restorations . . . the plantings must be made up of native species in diverse groupings. We would consider functional plantings of non natives or low diversity to be "rehabilitation" or "revegetation" efforts but not "restorations".'

EXPERIMENTAL RESTORATIONS: 'the establishment of the visual or emotional essence of natural communities, often with a simplified array of native species.' This can in part be equated with Baines' political habitat creation. It may be argued that the visual portrayal of natural communities is not strictly necessary in an urban context where a large proportion of the populace will not have experienced water meadows or chalk grassland.

Species reintroduction and the restoration of degraded ecosystems are often viewed as distinct processes with reintroduction operating at the species level and habitat restoration operating with the reinstatement of large scale habitat components to a recognized archetype. In fact both operations can serve the same fundamental purposes, namely the restoration of an area's original biodiversity and ecological function (Maunder 1992a). The maintenance of functioning natural ecosystems should remain as the ultimate priority in conservation. The movement of species for reintroduction is only a useful tool when it promotes the survival of that particular population or the species as a whole.

Recent reviews of plant reintroductions (Maunder 1992a; Falk & McMahan 1988; Birkinshaw 1991; Falk 1987) have outlined the reasons for plant reintroductions. They are summarized as follows:

1. improving the status of an endangered species or population;
2. reinstatement of an extinct or extirpated wild population;
3. research into the conservation biology and autecology of the species or taxa concerned;
4. as mitigation against the destruction of populations and habitats by land development;
5. establishment of populations as part of a habitat restoration programme;
6. establishment of new populations to divert potentially damaging public attention from a vulnerable wild population;
7. the re-establishment of a keystone species important for ecosystem function;
8. the re-establishment of an economic or aesthetic resource;
9. the reinforcement of isolated and small populations to overcome the deleterious effects of small population size.

The reintroduction of plants can pose a number of threats to conservation:

1. pathogens/disease can be introduced to wild populations (there is an urgent need to research this topic particularly with regard to developing accurate tests for viral, bacterial and mycoplasmic infection);
2. through the introduction of inappropriate stock locally advantageous gene combinations could be disrupted;
3. the value of wild and original populations and habitats could be undermined;
4. the scientific value of original populations and habitats could be reduced.

Reintroductions should only be undertaken when the following criteria are satisfied (derived from the draft report of the workshop on the reintroduction of extirpated species held at the IV World Congress on National Parks and Protected Areas, Caracas, 1992 (IUCN 1993)):

1. the project has the full permission and involvement of the governmental agencies responsible for the area;
2. each reintroduction should be fully integrated with, and supported by, local communities, except where the security of the reintroduced population is at risk;
3. reintroductions should only take place where the original causes of extinction have been identified and either eliminated or controlled;
4. reintroductions should only take place when the habitat and landscape requirements of the species are satisfied, and likely to be sustained in the foreseeable future;
5. reintroduction projects should follow a recognized protocol, involving an initial feasibility study, a reintroduction phase and subsequent monitoring;
6. reintroductions of critically endangered species should be undertaken by agencies with a proven ability to plan, undertake and follow up projects.

Plants have been moved for centuries, but it is only within the present century that conservation motives have overtaken those of acclimatization for aesthetic or curiosity value. The Rare Plant Translocation Panel of the BSBI manages a database of translocations in the UK. The first recorded translocation is that of *Paeonia mascula* between 1824 and 1835 (Birkinshaw 1991). Since then a number of rare British species have been translocated or

reintroduced, these projects have been reviewed by Birkinshaw (op.cit.).
Traditionally reintroductions in Britain have been undertaken by a variety of agencies, with private individuals often initiating projects. The Botanic Gardens of the University of Cambridge in conjunction with the then Nature Conservancy Council undertook a number of reintroduction projects particularly developing expertise with the Breckland flora.

Until recently there had been no overview of Britain's threatened flora with management recommendations. This has been rectified through the production by the Nature Conservancy Council (Whitten 1990) of 'Recovery: a proposed programme for Britain's protected species', a document that reviews the conservation status and required actions for all species listed under Schedules 5 and 8 of the Wildlife and Countryside Act 1981; 86 higher plants, 6 ferns and fern allies and 1 non-vascular plant are reviewed. The document 'suggests means by which each of the scheduled species would become a secure, self-sustaining member of its ecosystem, and thus be considered for removal from the schedules'.

The Recovery plans recognize the need for an integrated approach for plant conservation and the need to employ a variety of approaches. Botanic gardens are one part of this strategy where the traditional skills of cultivation are meshed with modern advances in conservation biology. The modern botanic garden with a diverse range of facilities including glass-houses and open trials grounds, micro-propagation laboratories and seed-banks, is well equipped for holding and propagating valuable accessions of rare species.

Royal Botanic Gardens, Kew, and the conservation of the British Flora

Historically the Royal Botanic Gardens, Kew, have not played a major role in conservation of native plants. This is in part due to the institute's traditional responsibility for tropical and colonial botany, and to the fact that conservation has only recently, e.g. post-1970, become a major issue for botanic gardens. In Britain the University of Cambridge Botanic Garden started the Eastern England Rare Plant Project (Walters 1979). Concentrating on the rare plants of the east of England, cultivated collections were initiated and where appropriate reintroductions/translocations undertaken. This project showed the important role that botanic gardens can play in the collecting and multiplication of rare plants and played an important role in convincing conservation professionals that botanic gardens were more than places for afternoon strolls and order beds full of hybrid swarms.

In much the same vein the Sainsbury Orchid Conservation Project was established at the Royal Botanic Gardens, Kew, in 1983 with the aim of growing threatened orchids from seed in order to reintroduce some of the rarer species to safe sites. Orchids in Britain are threatened through the degradation and destruction of habitats and the destructive influence of the illegal collector. They are charismatic plants that can generate considerable public interest so are ideal candidates as a flagship group to promote plant conservation.

Orchids produce minute seeds, which under natural conditions form a symbiotic partnership with a fungus which aids germination. The emphasis of the project has been to imitate this process in the laboratory. A range of fungi has been isolated from the roots of mature orchids and cultured in the laboratory. Orchid seeds are sown with samples of these fungi on plates of specialized media (Mitchell 1989). The seedlings raised using this method are considerably more vigorous than those grown by conventional means and in addition the relationship between the orchid and the fungus may prove to be vital for the success of any subsequent reintroduction. Although progress has been slow with a number of the threatened species a number of notable successes have been obtained.

Initial trials were undertaken with the Loose-flowered Orchid, *Orchis laxiflora*. The species had been successfully raised with a symbiotic fungus and weaned from laboratory culture for cultivation in the glass-house. Seedlings were planted out in the autumn of 1987 at Wakehurst Place in West Sussex in grounds managed by the Royal Botanic Gardens, Kew. It was not considered appropriate to introduce a non-native species to a wild situation, so accordingly the population was established on an informal wildflower area within managed gardens. In May 1988 there were seven specimens in flower, and by 1989 35 were in flower. By the end of the 1989 season 350 plants were recorded indicating the rapid establishment of the colony, although hand pollination may prove necessary to ensure seed set. Such trials have proved useful in preparing for a planned reintroduction into a restored habitat on the island of Jersey.

A large proportion of the orchids protected by the Wildlife and Countryside Act are on the edge of their natural distribution and accordingly are vulnerable to poor pollination and fluctuations in season. Thus to ensure pollination of the wild plants, volunteers artificially pollinate a proportion of the population. Some of the seed produced is used to generate cultivated populations at Kew in addition to seed being scattered on site.

A research project has been initiated examining the population dynamics of an artificially supplemented wild population of *Dactylorhiza praetermissa*. In addition, the utilization of genetic fingerprinting for population management is the subject of a major Kew research programme. Stocks of wild-collected *Cypripedium calceolus* exist in a number of private collections but were these plants originally collected in Britain or from the Continent? Using the advancing techniques of molecular biology it should be possible to identify British material that could be used for bolstering the surviving wild stocks. For instance it would be possible to compare the genetics of the surviving two populations of *Orchis militaris* and compare these with Continental populations.

The Sainsbury Orchid Conservation Project has been able to advance the cause of orchid conservation by working on a sound scientific base and linking its work directly with the problems of *in situ* orchid conservation. The work has been greatly aided through the support and co-operation of NGO groups such as the County Naturalists' Trusts, the Nature Conservancy Council (now English Nature) and other land management agencies.

The work of the Royal Botanic Gardens (RBG), Kew, on the conservation of British natives, has expanded through involvement with the English Nature Recovery Programme. Under contract with English Nature, RBG Kew is expanding its conservation programmes to specifically deal with a number of very rare native species. In addition the long established programme for native plants at the Wakehurst Seed-Bank is receiving English Nature support. In the seed-bank samples of seed are first dried to equilibrium with 15°C and 15% Relative Humidity followed by storage at -20°C. At present over 576 taxa of the British flora are stored, some 40% of the species that can be stored in a seed-bank. Such batches of seed can be stored potentially for centuries, an invaluable and relatively inexpensive insurance against the loss of increasingly vulnerable wild populations. In addition the seed of orchids can now be stored in liquid nitrogen using cryo-preservation techniques. The aim is that most populations of England's rare plants will be securely held in the bank as an insurance strategy, a resource for research and a source of propagules for reintroduction projects.

As part of a three-year programme Kew is now propagating and researching the biology of the Plymouth Pear, *Pyrus cordata*, for reintroduction. It is a species restricted to the hedges and green lanes of Devon and Cornwall, where habitat destruction has reduced its sites to a mere handful. Cuttings and seed from all the known wild plants are being collected for propagation and subsequent reintroduction. The University of Reading Botany Department is undertaking a programme of genetic fingerprinting to assess the relationship between British populations and those on the Continent.

Conclusion

It is often assumed that habitat-orientated conservation will carry an area's full quota of botanical diversity. Although this approach will carry the largest number of species in the most cost-effective manner a number of species will leak through the coarse filter of habitat conservation (Hunter *et al.* 1988). In Britain there are 40 plant taxa now restricted to one or two sites. The future of these vulnerable populations within a changing pattern of landscape and land use will increasingly demand an interventionist approach to conservation. This will involve a variety of agencies coming together to improve the status of species and landscapes. This will be termed as 'meddling' by some groups but in actuality will be management for survival.

Acknowledgements

The authors would like to express their gratitude to English Nature and the Sainsbury Trust for supporting and facilitating conservation work at Kew. In addition the comments and help of Simon Linnington, Mike Fay, Andrew Jackson, Matt Ford and Joyce Stewart are greatly appreciated.

References

Baines, J.C. (1989). Choices in habitat re-creation. *In* G.P. Buckley (ed), *Biological Habitat Reconstruction*, pp. 5-8. London: Belhaven Press.
Birkinshaw, C.R. (1991). Guidance notes for translocating plants as part of recovery plans. Nature Conservancy Council, CSD Report No.1225.
Falk, D.A. (1987). Integrated conservation strategies for endangered plants. *Nat. Areas J.* **7**: 118-123.
Falk, D.A. & McMahan, L.R. (1988). Endangered plant conservation: managing for biodiversity. *Nat. Areas J.* **8**: 91-99.
Howell, E.A. & Jordan, W.R. (1991). Tall grass prairie restoration in the North American Midwest. *In* I.F. Spellerberg, F.B. Goldsmith & M.G. Morris (eds), *The Scientific Management of Temperate Communities for Conservation*. Oxford: Blackwell.
Hunter, M.L., Jacobsen, G.L. & Webb, T. (1988). Paleoecology and the coarse filter

approach to maintaining biological diversity. *Conservation Biology* **2**(4): 375-384.
IUCN (1987). The IUCN position statement on translocation of living organisms: introductions, reintroductions and restocking. Gland: IUCN.
IUCN (1993). Parks for life. Report on the IV World Congress on National Parks and Protected Areas, pp. 117-118. Gland: IUCN.
Maunder, M. (1992a) . Plant reintroduction: an overview. *Biodiversity and Conservation* **1**: 51-61.
Maunder, M. (1992b). Reintroductions: species conservation and restoration. Plenary Sessions and Symposium papers, pp. 208-220, IV World Congress on National Parks and Protected Areas, Caracas. Gland: IUCN.
Mitchell, R.B. (1989). Growing hardy orchids from seed at Kew. *Plantsman* **11**(3): 152-169.
Soule, M.E. (1991). Conservation: tactics for a constant crisis. *Science* **253**: 744-750.
Stanley Price, M. (1989). *Animal Reintroductions: the Arabian Oryx in Oman*. Cambridge: Cambridge University Press.
Walters, S.M. (1979). The eastern England Rare Plant Project in the University Botanic Garden, Cambridge. *In* H. Synge & H. Townsend (eds), *Survival or Extinction*, pp. 37-46. Kew: Bentham-Moxon Trust, Royal Botanic Gardens.
Whitten, A.J. (1990). Recovery: a proposed programme for Britain's protected species. Nature Conservancy Council, CSD Report, No. 1089.

12

Principles of collecting plants as a conservation exercise

R.A.W. LOWE

National Council for the Conservation of Plants and Gardens, The Pines, Wisley Garden, Woking, Surrey, GU23 6QB[1]

ABSTRACT. The principles of collecting plants as a conservation exercise are not readily available to the layman. It was not until 1978 when the Royal Horticultural Society decided to set up National Plant Collections, with its primary aim of conserving our unique plant heritage, that an organization was created to conserve garden plants. The body given the task of doing this was the National Council for the Conservation of Plants and Gardens (NCCPG). The work of the NCCPG started at the end of 1981 when my predecessor, Duncan Donald, now Gardens Advisor to the National Trust for Scotland, was appointed as the first paid full time officer of the NCCPG. Ten years ago there was only a handful of National Collections. Today there are 585 in the British Isles, over 30 in both France and Australia, and some 60 in New Zealand. America started setting up Collections to the same remit early in 1992.

The botanic gardens have of course been collecting plants for many years but when our forefathers collected plants their aims were not the conservation aims we espouse today. However, we must be grateful to these collectors, for whatever their motives, they have left us with a invaluable legacy of living plant collections to exploit in our efforts to save what remains of the world's diversity of plants.

It is not surprising that it is difficult to establish the principles for collecting plants as a conservation exercise, and in particular garden plants. It requires a new philosophy and it takes time to develop a new set of principles. NCCPG is the first in this field and has established a substantial lead. When the National Collections Scheme was first set up there was doubt about whether it would succeed. That it has says much for our character and passion for gardening. Those responsible for the idea of the National Collections were quite clear what they wanted to achieve but not how. Today we are clearer but the rules are still evolving.

In broad terms there are two main ways in which we can help conserve plants outside their natural habitat: through gene banks and through *ex situ* plant collections.

---

[1] Present address: 36 Sycamore Road, Reading, RG2 7LY

Living plant collections suffer a number of disadvantages in comparison with seed-banks. Many of the living plants we grow originate from very limited genetic stock – maybe only a single plant. The source of seed of *Leycesteria crocothyrsos* Airy Shaw, for example, was collected by Kingdom Ward from a single plant. Was this lonely plant a typical specimen? What genetic erosion has taken place since it has been adapted to cultivation? And so on. Seed collections will normally have a wider genetic variation within the species having been collected from a range of plants in a variety of habitats. Whilst it is true where practicable that we should conserve plants through seed-banks because it is efficient so to do it does not work for recalcitrant plants which represent 20% of the world's total; nor, of course, does it work for horticultural cultivars. These reasons alone are sufficient justification for the need to establish living plant collections. One should not ignore the aesthetic appeal of living collections and the pleasures that they can give. By comparison I cannot think that one's courtship would be much advanced by an invitation to stroll through my seed-bank!

My experience lies in the work of the NCCPG and its campaign to conserve our unique heritage of garden plants. Many gardeners do not realize that in these islands we can grow a wider diversity of plants than almost anywhere else in the world. In most gardens there is greater diversity of species than one would find in a similar area in the wild and most of the plants will have their origins in lands far distant. Our unique passion for gardening, emulated by few other countries, makes us the natural leader in this field of conservation of garden plants. Keeping any living thing alive is sufficient cause in itself but these plants may have in addition to their aesthetic appeal scientific and medical importance. In the past we found that *Salix* yielded aspirin and *Digitalis* yielded drugs beneficial to those suffering from heart problems; the Greeks overcollected silphium, a fennel used as a contraceptive, to extinction. More recently the Madagascar Periwinkle, *Catharanthus roseus* (L.) G. Don, has been found to contain compounds effective against leukaemia and *Castanospermum australe* Cunn. & Fraser may contain a possible treatment for AIDS. It is not known what other useful chemicals plants may yield. Even cultivars have their part to play. Much as some cultivars have particularly striking flowers or foliage so then perhaps some cultivars may have higher concentrations of useful chemicals than others. In Holland crossbreeding and rigorous selection of Afghan, US and Moroccan cannabis has raised the hallucinogenic content of tetrahydrocannabinol (THC) from 0.5% in 1960 to between 9 and 27% today. Knowing which garden plants to conserve, in common with *in situ* conservation, is a problem which we have yet to satisfactorily solve.

Currently there are some 55,000 different garden plants available commercially. This commercial availability depends on demand and this in turn depends on current fashion. A nurseryman will not stay in business if he tries to sell plants which no one wants to buy. Many garden plants have already been lost owing to changes in fashion. Not long ago one could choose from a range of 800 peonies; today only about 100 are commercially available. In 1985 there were only some 30 cultivars of *Crocosmia* being sold whereas in the 1920s when this genus was last fashionable there were over 100. The fashion cycle for plants is not dissimilar to those for music and millinery but when plant cultivars go out of fashion they disappear for good; once lost they cannot be recreated.

NCCPG also has a role to play in species conservation because there are a number of endangered species which are good garden plants. A specific example is the flora of New Zealand; 25% of their endangered plants are commercially available in British nurseries. However, returning plants to the wild is not part of NCCPG's remit.

Conservation organizations must be aware of the main thrust of the work being carried out by other organizations and societies. The conservation of the world's plant biodiversity is a mosaic, with each organization contributing a small piece to the eventual picture. We must collaborate and co-operate to ensure that any resource-consuming overlap is avoided.

NCCPG's commitment is to conserve the unique garden plant heritage of the British Isles (these include Ireland). The Deed of Trust states:

> 'For the benefit of the nation and the public to encourage the conservation of endangered or rare plants of species hybrids and cultivars which can be grown in the British Isles which have characteristics worthy of preservation in the interests of public education and in particular horticultural botany and the other natural sciences, medicine and environmental studies to ensure the availability of such plants for any purpose beneficial to the community as a whole.'

With such a wide remit some overlap is unavoidable but this can be reduced by modifying one's aim. For instance NCCPG does not maintain seed-banks or store tissue or cell cultures. Seed-banks are inappropriate to an organization whose main work is in conserving cultivars. These must be propagated by vegetative means. I am aware that it is said that some cultivars come true from seed but those that do so with 100% reliability are rare. NCCPG follows the lead from the Royal Botanic Gardens, Kew where species are concerned. The chances of seed collected from cultivated plants coming true are so slim as to invalidate the process particularly when genetic erosion and chance hybridization in cultivation are taken into consideration. However, the general gardener can enjoy splendid variety from seed cultivation and though this may not have a place in our conservation work this should not be denied gardeners who are seeking a spectacular display. We must avoid the elitist tag which says that only those who do what we do, are doing something worthwhile.

Another variation in NCCPG's policy is where culinary vegetables are concerned. As an administrative measure vegetables as garden plants are excluded from our scheme though we do maintain two comprehensive collections of rhubarb containing over 100 cultivars between them. You may be surprised by such diversity in the humble rhubarb because to many there appears to be little difference between cultivars. Recently, however, a supermarket sought the help of the Rhubarb Collection at Harlow Carr. They required a rhubarb which could reliably be expected to produce good-coloured stems 45 cm long – for why? – because this the width of a supermarket shelf. In general we rely on HYDRA (the Henry Doubleday Research Association, with whom we collaborate closely) to advise us on vegetable conservation matters.

Most agree that *in situ* conservation is the ideal but because of limitations on space the opportunity is limited. Many of the environments where it could be carried out are in countries where the principles and needs of conservation are not accepted or where there are inadequate resources. Until this changes we must rely on *ex situ* conservation.

For many years the majority of permanent *ex situ* conservation collections were within the botanic gardens. However, to quote Professor Vernon Heywood, 'The expansion of living collections in botanic gardens was often motivated by a generalized pretension for comprehensiveness rather that in response to any clearly articulated scientific policy. In particular the botanic gardens as a major force in conservation have been surprisingly neglected.' This shortcoming as we can see it from our present-day vantage point is being addressed by individual botanic gardens and the Botanic Gardens Conservation International (BGCI). Under their auspices botanic gardens are being asked to identify those plants within their collections which are endangered, making a special effort to conserve these, preferably in the most suitable location and with the minimum duplication and consequent depletion of scarce resources.

The botanic gardens initially viewed the work of the NCCPG as irrelevant but this attitude has significantly changed. To quote Professor Heywood again when he writes about the significant number of rare or endangered plants cultivated outside the botanic gardens, in public parks, school and college gardens and arboreta, private gardens, scientific and non-scientific institutions (in others words the typical locations where one finds National Collections), '....the holdings of conservation-worthy material from these sources is enormous and must not be ignored simply because it does not form part of the botanic garden system. Clearly botanic gardens must play a lead, in association with other bodies in cataloguing, assessing and monitoring such material.' However, when talking about the National Collections specifically he adds a warning: 'The National Collections function very well as a reference library, and are ideal for the conserving of horticultural cultivars. They are not, however, so well suited for the genetic conservation of wild species, because of the small number of individuals per species and the very high risk of hybridization. Also, few genera are composed of species all of which thrive in the same environment.' These limitations are recognized by the NCCPG.

Some duplication between the work of the NCCPG and the botanic gardens is unavoidable. In this country in particular the botanic gardens concentrate on species of which many are, of course, good garden plants and therefore fall within the remit of the NCCPG. Collaboration is essential and the realization that the collections held by both bodies are complementary is important.

Essentially the NCCPG follows the principles which the botanic gardens apply to their collections except perhaps that the NCCPG Collection Holders can often devote more resources to the study of their chosen genus. There are Collections in botanic gardens, for example, 285 species and cultivars of *Begonia* in Glasgow, 84 species and cultivars of *Betula* at Wakehurst, 28 *Skimmia* at Leicester University, but do the botanic gardens have the resources to devote to Collections such as the 2,500 *Narcissus* held by Martin Harwood in Surrey or 2,000 pelargoniums held by Fibrex Nurseries?

It was Professor Stern who pointed out that the National Collections and the work done by their custodians, in particular those in private hands, were in the very best traditions of botany. There have always been very knowledgeable and single-minded amateurs who have devoted their leisure time to the study of a specific genus. So far as National Collection Holders are concerned, where their study of their chosen genus coincides with studies of the species within that genus by a professional botanist then the combination of the two is particularly advantageous. A number of Collections have already benefited from this symbiosis: Pat Davies holds the *Alliums* containing 165 species and cultivars and has discovered that the cellular structure of the seed coat is unique within a species; Dr John Twibell holds the *Artemisias* which contain over 100 species and cultivars and has identified specific uniqueness by gas chromatography. The work between the professional and amateur is also exemplified by the research done by John Burwell and Margaret McHendrick who combined a study of nursery catalogues and other historical data with photographs, herbarium specimens and pollen analysis to enable them to show that a number of cultivars of the Japanese *Anemones* carried the wrong specific epithet. In time the National Collections will become the authorities for accurate naming within their Collections particularly in regard to cultivars. The National Collection Holder of *Lonicera* is already asked by his colleagues in the trade to correctly name stock that they wish to sell. Whether we can persuade commerce in general, and gardeners, to change the bad habits of a lifetime and adopt the correct name is another matter.

The work being done currently by Collection Holders emphasizes the main principle of collecting plants as a conservation exercise. One must first put together the most comprehensive collection of plants within the selected genus regardless of what name a particular plant may carry and regardless of the accuracy of that name. By doing this one hopes to ensure the survival of a wide representation of that genus and one can then at a later date carry out detailed research work.

You may have noticed that we have already strayed from the concept laid down by the founders of the NCCPG which you will recall was to conserve garden-worthy plants. As Collections mature they may come to include plants that you may think are of little or no garden merit; who is to judge anyway?

To conclude. The principles of collecting plants as a conservation exercise is first to amass the most comprehensive collection so that at a later date when time and money permit these plants may be exploited for scientific and horticultural study. In the meantime they are maintained in cultivation for the benefit of future generations. The resource thus created can be used for education in its widest form, of the public and government servants as well as students; it reduces the necessity to collect plants in the wild; it maintains the world's plant diversity, preventing the total extinction of parts of that diversity. Finally, that a plant exists is sufficient cause that it should be conserved though I do not think the general public would accept this view.

13

The identification of cultivated plants

S.G. KNEES and J.C.M. ALEXANDER

Royal Botanic Garden, Edinburgh, EH3 5LR

ABSTRACT. During the last ten years the British horticultural industry has doubled its earnings, from £240 million per annum in 1984 to approximately £600 million per annum in 1992. It employs well over half a million people, yet the plants on which this vast industry is based are frequently wrongly named. Evidence of this is easily found in catalogues and publicity literature in nurseries and gardens centres and, surprisingly, often in botanic gardens and private collections. The identification and nomenclature of cultivated plants has always been difficult and unreliable and the following paper attempts to analyse the reasons behind these problems and suggests how the situation can be improved.

Introduction

Problems in identification and nomenclature arise from several sources:

a) Plants in cultivation have originated from all over the world, often from areas whose wild flora is not well documented. Many have been introduced, lost, and then reintroduced under different names. This problem is particularly acute in large and widespread genera.
b) Plants in cultivation often represent only a very narrow selection of the natural gene base of a species. These have been propagated, thus maintaining a large group of genetically very similar plants, giving a false impression of the natural species variability. Floras will not always detail minor floral distinctions, such as colour and shape ranges and uncertain identifications based on scant descriptions always beg the question 'Is this a new species?'.
c) Cultivated plants often grow in conditions very different from their native habitats, and may show morphological and physiological differences from the original wild stock.
d) Plants established in cultivation have been subjected to both conscious and unconscious selection, giving rise to populations very different from those in the wild, for example double flowered forms, early flowering forms or those expressing exceptional hybrid vigour.
e) Many garden plants are the result of deliberate or accidental hybridization known only from cultivation and have no equivalent in the wild.

The lack of genetic diversity in many cultivated plants is largely the result of propagative material often being derived from one or two original introductions, in many cases made fifty or more years ago. This has frequently lead to the wrong assumption that several species are present when in fact all that is being represented are extremes of variation of the natural species.

Identification difficulties of cultivated plants

Examples of the confusion caused by taxonomic changes as a result of additional knowledge are more prevalent in some genera than others:

Rhododendron

The genus *Rhododendron* in the Ericaceae, with over 850 species, provides several examples, largely due to the history of separate approaches to classification being taken by herbarium and horticulturally based systems. One of the most recent approaches (Cullen 1980) combines many taxa previously thought to belong in separate species, where the taxonomy was formerly based on a few living individuals from a very narrow geographic range, or even from just one individual. The type specimen of *R. edgeworthii* Hooker, Subsection *Edgeworthia* (Hutchinson) Sleumer, an attractive species with fragrant white flowers was collected from the Sikkimese Himalaya in 1849. Nearly forty years later the type specimen of *R. bullatum* Franchet was collected in Yunnan, western China and in 1914 Kingdon Ward collected what was later designated the type of *R. sciaphilum* Balfour f. & Kingdon Ward from eastern Burma. Each was maintained as a separate species until Cullen's revision in which all are thought to represent variation of a single species, a factor confirmed by new collections made from across the entire range of the species in the last thirty years. The range of variation is quite considerable, with flower size ranging from 5.5 to 10.5 cm, colour ranging from pure white to pale pink, with or without a yellow blotch at the base of the corolla; indumentum also varies from dark orange to pale brown and leaf shape varies from oblong ovate to narrowly elliptic. Not surprisingly, vegetatively propagated material maintained these distinct entities until new material from recent expeditions became available. Now these plants are flowering and all the intermediates between the extremes are available to the horticultural as well as the botanical community, Cullen's revision no longer seems so radical in its approach and this variable species is now correctly known as *R. edgeworthii* with *R. bullatum* and *R. sciaphilum* as synonyms.

Another example of confusion within *Rhododendron* can be seen in Subsection *Cinnabarina* (Hutchinson) Sleumer, which contains two closely related species characterized by their fleshy waxy corollas containing copious nectar: *R. cinnabarinum* Hooker and *R. keysii* Nuttall. *R. cinnabarinum* now includes 3 subspecies of taxa formerly recognized at the specific level: ssp. *cinnabarinum* with *R. roylei* Hooker and *R. blandfordiiflorum* W.J. Hooker in synonymy at specific and varietal rank, from Nepal through India and Bhutan to China is recognized by its tubular flowers varying from red through orange to yellow; ssp. *xanthocodon* (Hutchinson) Cullen with *R. cinnabarinum* var. *pallidum* W.J. Hooker, var. *purpurellum* Cowan and *R. concatenans* Hutchinson in synonymy, from India, Bhutan and China has campanulate flowers varying from yellow through apricot to purple; ssp. *tamaense* (Davidian) Cullen from north Burma has purple flowers and a range of characters intermediate between the other two subspecies. In both cases artificial hybrids made between the taxa as represented in cultivation have further clouded the issue making the process of

Fig. 1. *The European Garden Flora*, volumes 1-3.

identification extremely difficult for all except the specialists. The result is a very confused nomenclature saturated with additional names.

Nomenclatural difficulties of cultivated plants

In general, many garden plants have arisen through hybridization in cultivation and have no equivalent in the wild, for example many of the complex, multigeneric orchid hybrids. Although there are various international bodies to control the naming of horticulturally raised hybrids and clones, problems still arise as it is impossible to have one body overseeing all horticultural developments throughout the world. The inevitable result is that cultivars and clones of certain groups, for example florists' Chrysanthemums, are raised independently in different parts of the world and are superficially identical yet have different names.

Development of a cultivated flora

Unlike wild plants with a known distribution, and often a published Flora to aid identification, cultivated plants are usually of unknown origin and hence a regional Flora is of little use. Cultivated plants have originated from all over the world, often from areas whose wild flora is not well documented. Many have been introduced, lost and then reintroduced under different names. This problem is particularly acute in large genera, especially those with a widespread natural distribution; for example, *Saxifraga* contains about 440 species from Europe, North Africa, Asia, North and South America (Webb &

Gornall 1989). Many genera in the Rosaceae present further problems, for example *Crataegus* contains about 280 species, but over 1000 names have been proposed, owing to the high number of apomictic or microspecies (Mabberley 1987). The problems are confounded by the fact that taxonomists have usually concentrated their efforts on wild plants, often ignoring garden material, while horticulturists are often unwilling to learn the technical procedures involved and are understandably reluctant to grapple with the taxonomic complexities.

Traditionally, the main reference works on which many horticultural identifications have been based are encyclopaedic or monographic rather than taxonomic in their approach; an example is *The New Royal Horticultural Society Dictionary of Gardening* (Huxley 1992). If the origin is not known and the inquirer has no idea of the family or genus of the plant in question, there is little hope of making an accurate identification without consulting a professional identification service.

Reliable identifications can be made only by using works that have five important attributes. They must:

a) be accurate and taxonomically up-to-date, i.e., give the currently accepted names for the plants included, together with important synonyms,
b) be comprehensive, i.e., include all the plants likely to be encountered,
c) give full descriptions,
d) be systematically (not alphabetically) arranged,
e) provide a logical and systematic means of finding correct names, i.e., through some sort of key.

The European Garden Flora (Fig. 1), a projected six-volume work arranged in taxonomic order, with keys to families, genera and species is partly published and will greatly aid the identification of about 17,000 species of plants commonly cultivated in Europe. An editorial committee was formed in 1977 and resolved to produce a work with the format of a conventional Flora, which would include all species commercially available or likely to be found in general collections and gardens in Europe. Although the *Flora* deals with all the botanical categories of family, genus, species, subspecies, and variety, no attempt is made to cover the vast plethora of cultivars found in many genera. However, commonly grown cultivars and representatives of important cultivar groups are mentioned. The technical language is as simple as possible without leading to lack of precision. Observations are given on the size, wild distribution and cultivation requirements of each genus, and on the wild distribution, hardiness and flowering time of each species. The *Flora* is illustrated with line drawings. Vernacular names are not included.

The problem with including vernacular names is knowing where to draw the line, and the editorial committee eventually decided that rather than producing a far from complete list of names that would always be open to question, it would be better to omit vernacular names altogether. An example of the difficulties likely to be encountered is illustrated by the following example where the listed totally unrelated taxa are all commonly referred to as bluebells in some parts of their range:

Campanula rotundifolia, a British native in the Campanulaceae, known as the Bluebell in some parts of Scotland but the Harebell in England; *Hyacinthoides non-scripta*, a familiar

European species in the Liliaceae (Dony, Jury & Perring 1986; Stearn 1992); *Mertensia paniculata* (Ait.) G. Don, from North America in the Boraginaceae (Stearn 1992); *Phacelia campanularia* A. Gray, from California in the Hydrophyllaceae (Bailey & Bailey 1976; Stearn 1992); *Sollya heterophylla* Lindley, an Australian species in the Pittosporaceae (Brickell 1989) and *Wahlenbergia polytrichifolia,* from Natal, South Africa in the Campanulaceae (Stearn 1992).

Progress to date

Three volumes have already been published covering pteridophytes, gymnosperms, all monocotyledons and part of the dicotyledons (Walters *et al.* 1984, 1986, 1989). Volume 4 is in press and includes many important horticultural families including the Crassulaceae, Leguminosae, Rosaceae and Saxifragaceae. Most of the text for volume five is with the editors and part of the last volume (6) is already written. This will provide an excellent aid to the identification of plants in cultivation and will also assist those wishing to name garden escapes.

References

Bailey, L.H & Bailey, E.Z, (1976). *Hortus Third. A Concise Dictionary of Plants Cultivated in the United States and Canada.* New York: Macmillan Publishing Co.

Brickell, C. (ed) (1989). *The Royal Horticultural Society Gardeners' Encyclopedia of Plants and Flowers.* London: Dorling Kindersley.

Cullen, J. (1980). A revision of *Rhododendron* 1. Subgenus Rhododendron & Pogonanthum. *Notes from the Royal Botanic Garden Edinburgh* **39**(1): 1-207

Dony, J.G., Jury, S.L. & Perring, F.H. (1986). *English Names of Wild Flowers*, 2nd edition. Botanical Society of the British Isles.

Huxley, A.J. (ed.) (1992). *The New Royal Horticultural Society Dictionary of Gardening*, 4 vols. London: MacMillan Press.

Mabberley, D.J. (1987). *The Plant-Book.* Cambridge: Cambridge University Press.

Stearn, W.T. (1992). *Stearn's Dictionary of Plants Names for Gardeners.* London: Cassell.

Walters, S.M. *et al.* (eds) (1984). *The European Garden Flora.* Vol. II. Monocotyledons (Part II). Cambridge: Cambridge University Press.

Walters, S.M. *et al.* (eds) (1986). *The European Garden Flora.* Vol. I. Pteridophyta, Gymnospermae, Angiospermae–Monocotyledons (Part I). Cambridge: Cambridge University Press.

Walters, S.M. *et al.* (eds) (1989). *The European Garden Flora.* Vol. III. Dicotyledons (Part I). Cambridge: Cambridge University Press.

Webb, D.A. & Gornall, R.J. (1989). *Saxifrages of Europe.* London: Christopher Helm.

14

Propagation not collection

M.I. READ[1] and B.A. THOMAS[2]

Fauna and Flora Preservation Society, 1 Kensington Gore, London, SW7 2AR [1] *, and Department of Botany, National Museum of Wales, Cardiff, CF1 3NP* [2]

ABSTRACT. The annual international trade in wild collected bulbs exceeds 50 million specimens and the effects on wild populations have been severe. At least one bulb species is now extinct in the wild as a result of over-exploitation and the genetic bases for crop improvements of some species have been damaged. The Fauna and Flora Preservation Society's Indigenous Propagation Project seeks to remove pressure from wild populations, provide alternative incomes for those involved in collecting and produce good quality plants for trade. The first field project began in Turkey in December 1991 and with the co-operation of existing trade organizations in Turkey and the Netherlands is promoting the transition to artificial propagation on a local basis. The Fauna and Flora Preservation Society is investigating the need for similar field projects involving plants in other countries.

Introduction

Much of this conference has centred upon the inter-relationships between foreign plants and members of our native flora and their effects upon each other. Some of the other papers make much of the need to conserve our own native flora or our rare garden plants. Here we are exploring the extra dimension of the effects of British horticulture on foreign floras.

Most people are now aware of the need to conserve wild plants and to save natural habitats from destruction. There is much published on the matter and large numbers of people belong to one conservation society or another. Unfortunately not enough people understand how many of their actions can inadvertently work against the general conservation principles. The international trade in wild-collected plants has existed for hundreds of years, but until recently this has not had too much of an impact on wild populations. However, in the last 40 years or so, gardening has become a major pastime in Britain as it has in other parts of Europe and North America. The British horticultural trade alone is now estimated at being worth about £600 million per annum. Much of this revolves around cultivated plants and seed packets involving some 55,000 species, but a significant part of the trade is in wild-collected plants. The types of plants most affected by such trading, termed loosely as bulbs, are easily transportable in their dormant stage and can be packaged attractively.

This paper aims to highlight the dangers of such trade to wild populations and suggests some ways in which alternative actions can prevent their continued wholesale destruction.

The Fauna and Flora Preservation Society, based in London, is the world's oldest international conservation organization. Recently, alongside many other conservation and education activities, the Society's botanists have developed an interest and expertise in issues related to international trade in plants. Research by the Society into the trade in wild-collected bulbs began in 1987, following the work carried out in Turkey by Professor Ekim (Ekim *et al.* 1984) and by Minouk van der Plas-Haarsma (1987) of TRAFFIC Netherlands.

After undertaking research in a number of countries, it became clear that a substantial international trade existed in wild-collected bulb species of many different genera and from countries including Canada, India, Japan, Portugal and the United States. It was apparent that this trade was not being conducted on a sustainable basis, indeed it was also obvious that there had been little or no research carried out by the trade or conservationists to indicate with any accuracy what sustainable levels might be. Many wild populations of plants had already been severely reduced in size and a number of authorities considered that at least one species, *Tecophilaea cyanocrocus* Leyb., had become extinct in the wild as a result of over-collecting.

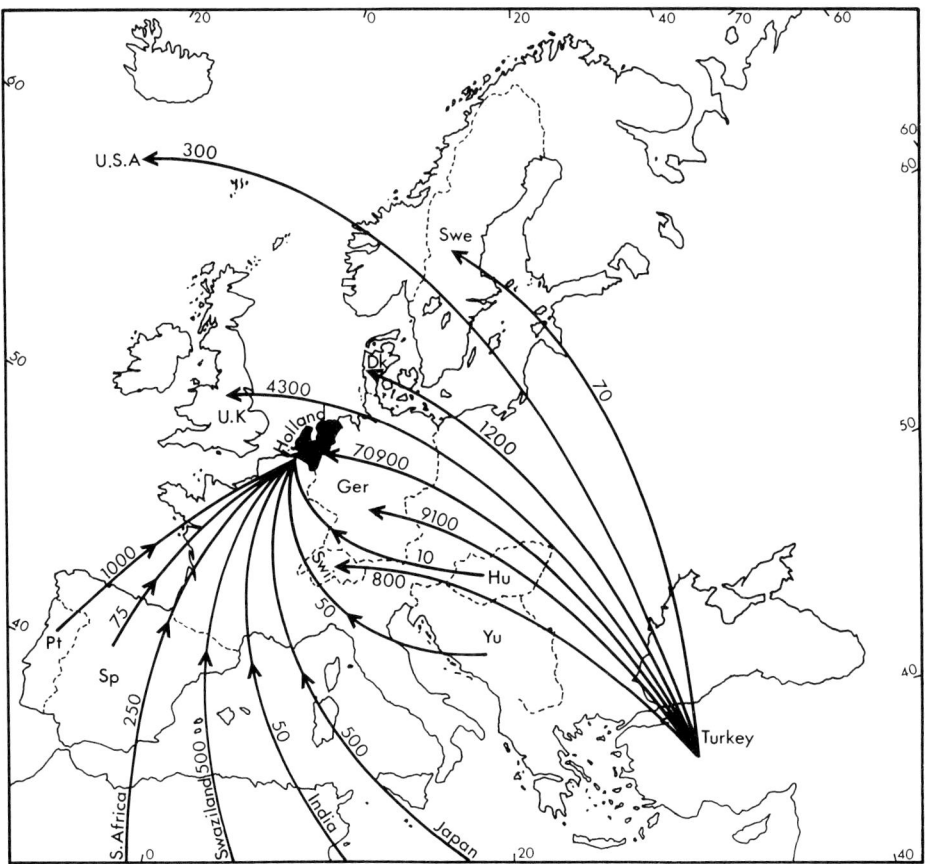

Fig. 1. Numbers of wild bulbs from countries of origin (approximate numbers × 1,000) (from Read, 1989).

The most startling volumes in trade were coming from Turkey, which is a country with an extraordinarily diverse bulbous plant flora (Fig. 1). Fifty to sixty million bulbs were being exported annually with the main species being *Galanthus elwesii*, *Anemone blanda*, *Eranthis hyemalis* and *Cyclamen hederifolium* although the full list probably runs to several dozen species. It was also apparent that, although much of the damage was caused by amateur collectors or small dealers, a large percentage of the total volume of wild bulbs in trade passed through the major bulb houses of the Netherlands (Fig. 2).

But before the problems of international trade are pursued further, the situation in Turkey should be looked at in a historical context. Bulbs such as onion, garlic, leek and saffron, have been grown in the region for 4,000 years, but it was not until the time of the Ottoman Empire that bulbous plants became so sought after for decorative purposes. In the 1630s there were approximately 80 flower shops and 300 florists in Istanbul (Baytop and Mathew 1984).There is also documentary evidence to show that large numbers of bulbs, such as *Hyacinthus*, were being collected from the highlands and taken to the palace gardens in Istanbul. Then in the 17th century trade started to develop with Europe and bulbs, especially those of *Tulipa*, began to be exported in large numbers. From these beginnings, the trade with Europe

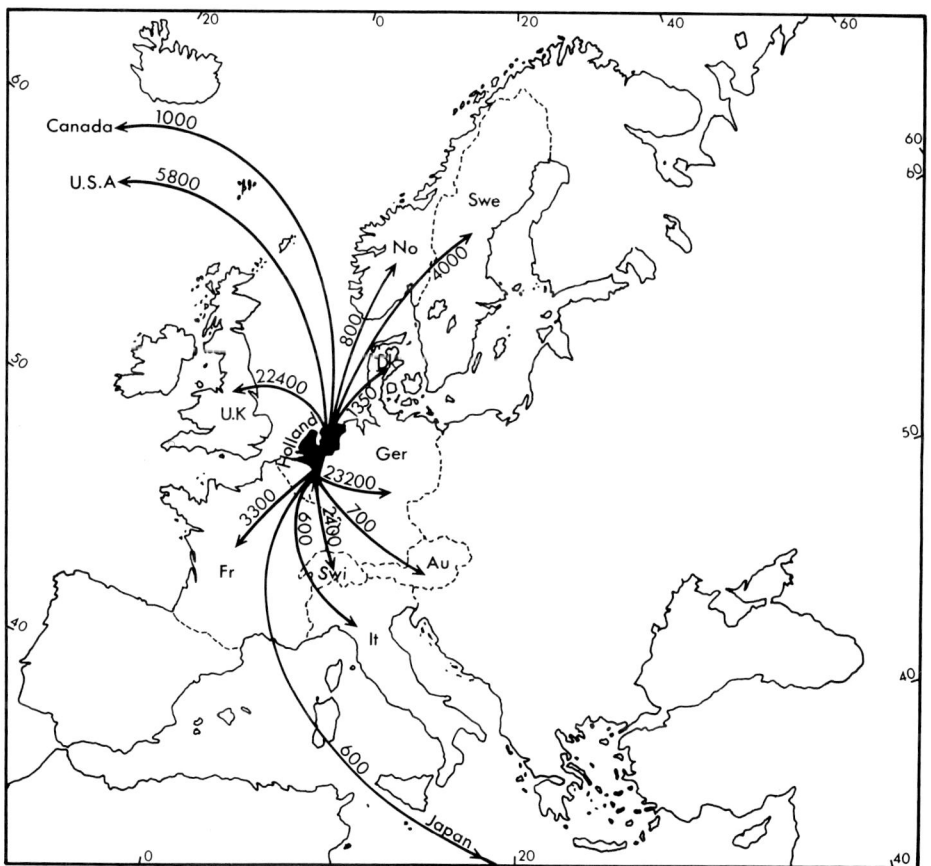

Fig. 2. Numbers of wild bulbs trafficked through Holland (approximate numbers × 1,000) (from Read, 1989).

expanded to such large proportions that thousands of people are now involved. Unfortunately a substantial number of these are directly concerned with the continued collection from the wild.

Early Steps

The Fauna and Flora Preservation Society prides itself in its scientific approach to conservation issues. It is not a vociferous pressure group, but the committee, staff and members of the Society do not shy away from confrontational issues if and when they arise. The first public step in the Society's work to protect wild bulb populations came through publicizing information about the extent of the trade in the press and media. Ultimately the Society published the booklet *Grown in Holland?* in the Spring of 1989 (Read 1989) and the poster 'Cyclamen in Peril' later that year. Until then, the vast majority of bulb buyers in Britain had been totally unaware of the wild origins of some of the bulbs they were buying. Many were not only shocked that they might inadvertently be contributing to the decline of their favourite plants but also by the fact that they had been misled by Dutch companies they had previously held in high esteem. The Society has been joined in much of this work by three organizations in the United States, namely the Natural Resources Defense Council (NRDC), TRAFFIC USA and the Garden Club of America. Valuable work has also been carried out by the Wildlife Trade Monitoring Unit in Cambridge and the Royal Botanic Gardens, Kew. Discussions with many gardeners and research carried out in partnership with the Cyclamen Society in the UK also confirmed early suspicions that the quality of wild bulbs sold to the public was lamentable. Failure rates of 30-50% were found to be not uncommon.

In addition to the valuable element of public education, the publication of *Grown in Holland?* led to us having much more productive discussions with the trade than had previously appeared possible. In particular the Dutch Bulb Exporters Association in Hillegom, and the Bulb Research Centre in Lisse were prepared to take the situation seriously and the progress which the Society has made today would not have been possible without the close co-operation of these organizations. While it is true that the vast majority of bulbs exported by the Netherlands are artificially propagated and contribute much to beautifying gardens and homes world-wide it is now recognized to be unacceptable that the trade should simultaneously be responsible for the extinction of wild bulb populations in other countries. This is quite apart from the economic fact that local extinctions reduce the genetic resources available to the trade for future crop development.

Legislation and labelling

Before turning to the long-term solutions which the Society has been developing, there are two other areas where progress has been made. These are labelling and legislation.

The flagrantly deceitful labelling 'Grown in Holland' appearing on packages of wild-collected bulbs in British shops provided a strong spur to the Society's work. It also underlined the value of providing gardeners with reliable information, which they may then use to select plants which have not been plundered from wild populations. Lengthy discussions involving many organizations have led to the Dutch Bulb Exporters' Association (DBEA) deciding that all species or 'botanical' bulbs exported from Holland must be properly labelled as to their origin. The agreement, which is statutory, is that after the end of July 1992 all wild-collected bulbs leaving Holland must be labelled:

'BULBS FROM WILD SOURCE.'

All other species or 'botanical' bulbs must be labelled:

'BULBS GROWN FROM CULTIVATED STOCK.'

The detailed definitions of what is considered wild and what is considered cultivated have been agreed between the Society and the DBEA and follow exactly the wording used by the Convention on International Trade in Endangered Species of Wild Fauna and Flora (CITES). The definition of 'artificially propagated' which is used is that established by CITES. Details of this are available from the Plants Officer, CITES Secretariat, 6 Rue du Maupas, Case Postale 78, CH-1000 Lausanne 9, Switzerland. Specimens which do not conform to this definition are considered wild.

To warrant the label 'Bulbs grown from cultivated stock' the parental stock must have been 'established and be maintained in a manner not detrimental to the survival of the species in the wild and managed in such a way that the long-term survival of the parent stock is guaranteed'. In other words the propagation must be 'closed', i.e. demonstrably requiring no replenishment from the wild. Growing-on from plants too small for export is therefore not cultivation.

The Society is now actively continuing its efforts to widen this agreement to include dealers in the United Kingdom and the NRDC and TRAFFIC USA are likewise working with dealers in the United States to forge a similar agreement. UK bulb dealers are not yet bound to any legislation. However, as the Society is advising gardeners to look for the correct labelling and does not wish responsible UK-based traders to be unfairly disadvantaged, we are now advising all UK bulb distributors, wholesalers and retailers to adhere to exactly the same wording. In this way gardeners can choose bulbs free from any suspicion that they may be contributing to what has on occasion been a very damaging trade. While there are positive signs of a voluntary agreement by most large retailers in the U.K., the Society still considers legislation to be important and that an EC Regulation on bulb labelling would be a particularly worthwhile measure.

CITES is the major piece of legislation covering bulb species in international trade and is now ratified by well over 100 countries. While much remains to be achieved in the full and correct implementation of CITES it has already become a significant factor controlling trade in plant groups such as cacti and orchids. There are now three bulb genera covered by CITES, namely *Cyclamen*, *Galanthus* and *Sternbergia*. All species of these genera are listed on Appendix II of CITES which means that export is illegal without a permit. A permit should not be issued where the collection and export are considered detrimental to the survival of the species in the wild.

Each party to CITES has to enact legislation to implement CITES and the European Community (EC) is currently drafting a revision of the implementing Regulation that all its member states are obliged to use. This new Regulation finally recognizes the critical state of many populations of bulb species and when adopted (scheduled for 1 January 1993) will be

Fig. 3 (above). *Cyclamen* collector in Trabzon on the Black Sea coast.
Fig. 4 (below). Trial beds at the Ataturk Arboretum, near Istanbul.

notably stricter than the provisions required by CITES. If adopted in its current draft form, member states will be required to ban trade in wild specimens of *Allium grosii, Asphodelus bento-rainhae, Bellevalia hackelii, Colchicum corsicum, C. cousturieri, Crocus etruscus*, five species of *Fritillaria, Hyacinthoides vicentina, Iris boissieri, I. marsica, Leucojum nicaeense, Muscari gussonei*, eleven species of *Narcissus, Ornithogalum reverchonii* and three species of *Scilla*.

In addition to the three genera listed by CITES (*Cyclamen, Galanthus* and *Sternbergia*), EC states will also be required to control trade in *Iris lusitanica, Lilium rubrum* (=*L. pomponium*), *Narcissus bulbocodium*, and *N. juncifolius*. Furthermore, simple monitoring of imports will be required for all other species of *Fritillaria, Leucojum, Lilium, Muscari, Narcissus* and *Scilla* as well as *Iris* Section *Oncocyclus* and all species of *Arisaema, Biarum, Erythronium, Trillium* and *Tulipa*.

The Indigenous Propagation Project

The Fauna and Flora Preservation Society considers the best long-term solution to the problems of plant species threatened by collection to be neither trade restrictions nor consumer boycotts. Rather, the Society wishes to promote artificial propagation of the threatened species within the countries of origin and ideally involve those currently involved in their collection and trade. Consequently the Society has developed and launched the Indigenous Propagation Project (IPP).

The IPP intends to work wherever possible in close co-operation with local governments and traders to:

- improve the prospects of plant populations threatened by collection;
- provide stable and satisfactory alternative incomes to those people involved in collection; and
- provide good quality plants to the trade.

The Society believes the IPP could potentially benefit wild populations of a wide variety of plant groups including cacti, cycads, succulents, tree-ferns, medicinal and herbal plants as well as bulbs.

The Society is in favour of the concept of sustainable exploitation of wild populations as a conservation measure. However, given the difficulty of establishing what levels of collection are sustainable and ensuring that collection conforms to hypothetical models, a transition to artificial propagation and relinquishing of all trade from wild populations is considered to be the best approach in the short term.

The Indigenous Propagation Project and bulbs in Turkey

The IPP in Turkey is the first field project of its kind for the Society and began in earnest in December 1991 (Read & Thomas in press). In Turkey all work is being carried out in co-ordination with existing government and trade bodies and in full partnership with Dogal Hayati Koruma Dernegi (DHKD, the Turkish Society for the Protection of Nature). It is the

Society's intention that in due course the IPP in Turkey will be left under the control of DHKD. This is in line with the Society's policy that benefits of the project should accrue to local people and that as far as possible, stewardship of the project should be in the hands of local people as well.

The IPP in Turkey is beginning to demonstrate the benefits of artificial propagation and to provide education on propagation techniques suitable for small-scale, rural production. In addition, the IPP in Turkey will be carrying out research into the best propagation techniques for a range of species which are currently threatened by collection. Sites have been chosen for these activities. One is the Ataturk Arboretum near Istanbul, two are close to Trabzan on the Black Sea coast (one at 50 m, one at 2100 m altitude) and the fourth near Izmir. A fifth site in the Toros mountains in the south of Turkey will be chosen in due course. It is intended that in the first instance there will be commercial demonstration plots with up to five species, probably *Anemone blanda*, *Cyclamen cilicium*, *C. coum*, *Eranthis hyemalis* and *Sternbergia lutea* all of which have been exported from the wild in very considerable quantities from Turkey in recent years.

Believing that there is considerable potential for increasing the number of species grown in, and exported from, Turkey and that promoting this will ultimately remove the incentive for private collectors to come to Turkey to collect from wild populations, research is also starting in the field of low technology, artificial propagation of a much wider range of species. The list of candidate species for this research is currently *Allium roseum*, *Arum creticum*, *Biarum davisii*, *Crocus abantensis*, *C. baytopiorum*, *C. biflorus* ssp. *crewei*, *C. biflorus* ssp. *pulchricolor*, *C. gargaricus*, *C. olivieri* ssp. *istanbulensis*, *Cyclamen mirabile*, *C. persicum*, *C. pseudibericum*, *C. trochopteranthum*, *Fritillaria alburyana*, *F. aurea*, *F. forbesii*, *F. sibthorpiana*, *F. viridiflora*, *Galanthus fosteri*, *G. gracilis*, *Hyacinthus orientalis*, *Iris pamphylica*, *Lilium martagon*, *Muscari macrocarpum*, *Sternbergia candida*, *S. clusiana* and *S. fischeriana*. A number of the very rarest and geographically most restricted of Turkey's wild bulb species will also be grown with the intention of offering quantities of seed to the trade in due course.

The final two components of the IPP in Turkey in its first year are to be the development of a Code of Conduct for introducing new species to cultivation and the initiation of monitoring of a number of wild populations.

Other countries

Turkey is clearly not the only country where trade is threatening populations of wild plants and the Society is anxious to extend its philosophy of Indigenous Propagation wherever there is a need. Current work by the Society is centring on Portugal, Central and South America and Eastern Europe and has contracted survey studies on the trade in wild-collected plants occurring in these countries.

Conclusions

The Fauna and Flora Preservation Society believes that conservation of plant populations threatened by collection can be achieved in many situations by promoting trade based on

Fig. 5 (above). Trial plots under hazel trees, near Trabzon on the Black Sea coast.
Fig. 6 (below). Modern storage facilities for bulbs awaiting shipment, near Trabzon.

sustainable use and artificial propagation. This is the objective and philosophy of the Indigenous Propagation Project. Within the inevitable constraints of limited resources the Society would like to offer help to any individuals and organizations who believe that their country's bulb flora is suffering from the depredations of the wild bulb trade. The Society is also appealing for information on which species are being collected and from where and would urge that any information, however fragmentary, be sent to its head office in London.

Acknowledgements

Funding for the first two years of the IPP in Turkey has been generously provided by the World Wide Fund for Nature International. Many thanks are also due to Mrs Marjorie Arundel for her support and to Dr Faith Campbell of NRDC, Amanda Hillier of FFPS and Nina Marshall of TRAFFIC USA.

Addendum

The IPP in Turkey is now in its third year. Technical workshops are being held, wild populations are being monitored and research is continuing into the best and most appropriate methods for propagating those species currently being threatened by collection. Most importantly of all, those propagation methods suitable for small-scale, rural production are now being demonstrated to locals as an incentive to start production.

Cultivation at the Ataturk Arboretum is now well under way. At least 60 different plants which are under threat from specialist collectors, afforestation or habitat loss, are in cultivation. It is hoped that releasing large numbers of these species, either as bulbs or seeds, to the trade will remove the incentive for collectors to uproot them from the wild. *Fritillaria michailovskyi* is an excellent example of such an approach. Until about 10 years ago enthusiasts were prepared to pay up to £10 for a single bulb of this rare Turkish species. Now it is propagated in the Netherlands and available at £3 for five bulbs.

Since the Flora and Fauna Society has been involved in Turkey we have been delighted to witness an increasing interest in all quarters regarding plant conservation and especially the development of Turkey's bulb trade. Particularly welcome have been the other initiatives in indigenous propagation involving the Bulb Research Centre at Lisse in the Netherlands, the Turkish Natural Flower Bulb Association based at Yalova, near Istanbul, and the Turkish Government's Horticultural Institute which is also at Yalova.

References

Baytop, T. & Mathew, B. (1984). *The bulbous plants of Turkey*. London: Batsford.
Ekim, T. *et al.* (1984). Taxonomic research on Turkey's geophytes of economic value. Project TBAG/490-A. (In Turkish: English translation available from TRAFFIC Netherlands.)
Plas-Haarsma, M. van der (1987). *Cyclamen* in trade. TRAFFIC Report No.5, TRAFFIC Netherlands.
Read, M. (1989). *Grown in Holland?* London: Fauna and Flora Preservation Society.
Read, M. & Thomas, B.A. (in press). The Indigenous Propagation Project – conservation, development and the wild bulb trade. VI International Symposium on Flower Bulbs. Skierniewice, Poland.

15

The introduction of trees to Roman and Medieval Britain

J.H. DICKSON

Botany Department, The University, Glasgow, G12 8QQ

ABSTRACT: The criteria for confirming the introduction of trees to ancient Britain were discussed. There is documentary evidence and there are fossils, both pollen and coarser remains such as wood, charcoal and propagules. Well represented in Roman and Medieval deposits, Fig (*Ficus carica*) is a particularly telling case. Other trees considered included Sweet Chestnut (*Castanea sativa*) and Walnut (*Juglans regia*).

[Prof. Dickson's paper will be published elsewhere.]

16

The possible effects of cultivated introductions on native willows (*Salix*) in Britain

J.E.J. WHITE

Forestry Authority Dendrologist, Westonbirt Arboretum, Tetbury, Glos., GL8 8QS

ABSTRACT. The threat to willows from commercial exploitation is not so much a threat of extinction, but rather a blurring of taxonomic integrity which can in turn lead to a loss of identity. Most at risk after 1,000 years or more of basket willow cultivation have been the British native osiers. Their survival was certainly never considered important, especially during the boom years for the industry in the 19th century. The present day resurgence of interest in willows does not pose new threats to the genus, but there is concern about the consequences of planting introduced cultivars.

Introduction

There are several ways that *in situ* willow populations can be damaged by invasive commercial planting. Firstly, by deliberately clearing existing vegetation during site preparation and for plant hygiene; this in fact is seldom a serious threat to willows on a large scale. Secondly, there may be competition from vigorous escaped cultivated species and hybrids which could suppress slower growing native stands; this too is a localized problem. Thirdly, cultivated escapes may look so similar to the natural matrix that it is impossible to distinguish between them morphologically; this occurs with White Willow, the wild form of which in the past 100 years has been almost completely disguised. Finally, hybrids can occur between cultivated and wild plants; these hybrids eventually produce entirely different populations and obliterate distinctions between species. In theory it should be possible to contrive a single hybrid made up of all the species occurring in Britain (Meikle 1984) and this is the size of the problem. However, some taxa resist the pressure to crossbreed: European Crack Willow for instance, is usually sterile. The distinction between some species is maintained in nature only by differences of flowering period. A badly chosen cultivated plant grown within insect range of such populations could eventually break down this tenuous barrier. The differences in time of flowering between two distinct species can be 'bridged' in this way. This paper sets out the present degree of risk to taxonomic integrity faced by each of the native tree willows (subgenus *Salix*) and six species and several major hybrids in subgenus *Vetrix* (sallows and osiers). It could be argued that so much damage has already been done to the British native willows that it does not matter what else happens to them. This is in fact only true for one or two species; others remain intact, or more or less so, and the whole subgenus *Chamaetia*, remains unaffected. Loss of distinct native species may simply be a headache for taxonomists, and of no real concern to the commercial producer, but for a lot of wildlife it could mean the difference between life and death (Kennedy & Southwood 1984).

The status of native willows

Salix pentandra, Bay Willow, is a low elevation tree of northern Britain and Europe which is not indigenous south of a line from mid Wales to Great Yarmouth. The catkins appear late in the season with the leaves, so that under normal circumstances hybridization is limited to only late flowering species within the subgenus *Salix*, and seldom occurs at all in nature. In addition, the species has been almost free of artificial contamination because of only limited historical use and commercial potential. There are one or two quite obscure basket willow cultivars such as 'Patent Lumley'. There are also named hybrid osiers with *S. fragilis* and *S. alba* parentage, but these are uncommon. The primary risk to the integrity of the wild species in Britain is continued introduction outside the natural range, which tends not to take any account of slight local variants. The species is morphologically very distinct and strong features can be identified even when they occur in interspecific hybrid progeny.

Salix fragilis s.l., Crack Willow, includes four varieties in Britain. The true European Crack Willow is var. *decipiens*, which has shining yellowish-brown twigs and glabrous glossy leaves seldom over 9 cm long, but up to 3 cm wide. *Salix fragilis* var. *fragilis,* the British Crack Willow, has brittle olive-brown lustrous twigs and long lanceolate leaves (9-15 cm). It has features reminiscent of *S. alba* such as initially slightly pubescent shoots and young leaves. It may indeed be a hybrid between var. *decipiens* and *S. alba*. Apparent sterility may also indicate interspecific hybrid origin. Populations are usually clonal and female. The occasional occurrence of three stamens, a feature of *Salix triandra*, is curious. The third variety of Crack Willow is var. *russelliana*, the Bedford Willow, which is probably the most common and largest growing tree in the *fragilis* aggregate. The foliage is somewhat similar to var. *fragilis*. The shoots are brown, although red and orange variants occur, and side shoots are less inclined to 'crack' off. Meikle (1984) stated that there can be little doubt that this variety is a straightforward sport of *S. fragilis*, which is usually found as a female clone. *Salix fragilis* var. *furcata* is a broad-leaved (3-5 cm) male clone of similar origin and general appearance.

Despite the complexity of this aggregate species, natural populations in Britain appear to be resistant to changes usually associated with cultivated introductions. Status is maintained in part by near sterility and clonal populations, although *fragilis* features in a range of hybrids in the subgenus *Salix*. The integrity of wild British populations does not appear to be under any threat at the present time because there are virtually no commercial uses for the species and it is regarded as unsafe for amenity use and arboriculture.

Salix × meyeriana, the Shiny-leaved Willow, is a hybrid between *S. fragilis* and *S. pentandra* which is widespread in lowland Britain and has often been planted (Meikle 1975). This plant is fertile and provides one outlet for the genetic characteristics of *S. fragilis*. The natural status of *S. × meyeriana* is unknown.

Salix × alopecuroides (= *S. × speciosa*). This hybrid between *S. fragilis* and *S. triandra* is fortunately (from a taxonomic point of view) quite rare. Only male plants have been noted (Meikle 1975). Through the *S. triandra* element of this cross it is theoretically possible for subgeneric hybrids to occur between *Salix* and *Vetrix*, the sallows. Increased commercial use of taxa with this potential would be most unfortunate but this does seem unlikely at present.

Salix × pendulina (*S. babylonica* × *S. fragilis*) is a weeping form of Crack Willow produced artificially in Europe in the 19th century. There are at least three variants, all of them highly ornamental and potentially large trees. These are increasingly planted in landscaping schemes as an alternative to the often rather sickly golden weeping willow. Clonal cultivation is usual; most trees are female but they can be androgynous. Back-crosses have not been noted in wild *S. fragilis* and escapes out of cultivation appear not to have occurred. Unintentional vegetative propagation usually results in weak prostrate shoots which are quickly suppressed by ground vegetation.

Salix × rubens (*S. alba* × *S. fragilis*) is a common and widespread large tree across lowland Britain and Europe. Published descriptions tend to favour *S. alba* characteristics but natural hybrids occur with a complete range of intermediate features. Most British trees appear to have been planted or are derived from artificial plantations. This hybrid has largely replaced natural *S. alba* in many areas and makes confident morphological identification of some varieties of *S. fragilis* very difficult. As this tree was widely planted in the 18th and 19th centuries, it seems unlikely that further use will cause any additional damage to the natural populations of the parent species. There may even be some merit in using local clones in future planting to foster historical interest.

Salix × rubens nothovar. *basfordiana* is known for its brilliant yellow and orange coppice shoots, f. *basfordiana* Meikle and f. *sanguinea* Meikle respectively. These are planted in gardens and amenity landscaped areas. Escapes and deliberate plantings occasionally occur in the countryside. These only impinge on existing hedgerow willows very slightly for space but their visual impact is dramatic during the winter months.

Salix alba, White Willow (strictly *S. alba* var. *alba*), has been replaced over most of Britain by cultivated forms of the hybrid *S. × rubens*. The limits of its natural distribution are probably no longer ascertainable (Meikle 1984). Identification is also seriously confused by the large number of foreign selections now widely planted such as 'Polsdonk' and 'Vries'. The ornamental silver willow 'Sericea' occurs from time to time but it is fairly distinctive. Cricket-bat Willow *S. alba* var. *caerulea* is also easy to recognize but close to Cricket-bat Willow plantations all other species may be grubbed out on the grounds of plant health. In addition to *S. × rubens*, hybrids between *S. alba* and *S. triandra*, and even *S. purpurea*, have been recorded in Europe; they are very rare or have often been incorrectly identified (Meikle 1975).

Salix alba var. *vitellina*, the Golden Willow, is a distinctive tree or coppice. All kinds of clonal selections have been named, and often re-named. Various shades of orange can be seen as escapes from cultivation during the winter months. These do not pose a threat to the identification or existence of wild taxa. This variety was artificially crossed with *S. babylonica* at an early date to produce the common golden Weeping Willow *S.* × *sepulcralis* nothovar. *chrysocoma*. The stock originally used for this is uncertain and the resulting trees are generally poor and prone to disease. It is possible that leaf cast and canker have been carried back to wild populations from such sources.

Salix × *ehrhartiana* (*S. alba* × *S. pentandra*) is a fairly small ornamental osier that has been widely used all over the country but does not appear to pose any threat to natural populations of any other species or their proper identity.

Salix triandra, Almond Willow, is native and widespread in Britain, south of a line from the Humber to Gloucester, but never common. It has traditionally been used to produce heavy rods for the basket industry. Escapes from cultivation of numerous named clones have almost entirely obscured any remnants of a wild population. Many are of continental or unknown origin, some named '*triandra*' are not even that species. 'Black Hollander', for instance, is probably a *S. pentandra* cross. Although *S. triandra* is unlikely to be spoiled further by new planting this remains a threat to the taxonomic integrity of many other species. *Salix triandra* has the ability to hybridize with all of the tree willows, subgenus *Salix*, and also many of the sallows, subgenus *Vetrix*. The most significant bi-subgeneric hybrid is *S.* × *mollissima* (*S. triandra* × *S. viminalis*) which, because it combines the qualities of the two best heavy basket osiers, has been extensively cultivated. Relics of *S.* × *mollissima* withy beds survive to this day as hedgerow plants and riparian scrub. Other *S. triandra* hybrids have been cultivated but less widely. Flowering tends to be late in *S. triandra* and one exceptional male clone, 'Semperflorens', has an extended flowering period from May to October. Bees picking up pollen from this plant have the potential to bridge critical flowering periods in a whole range of other species.

Salix purpurea, Purple Osier (Purple Willow), is a widespread small tree or large bush in the subgenus *Vetrix* which has some helpful morphological aids to identification that occur on no other willow. These include opposite buds (in part), black dead leaves and bitter tasting young bark. They are to some extent carried forward into *S. purpurea* hybrids. Wild native populations survive still in remote and upland regions, but in lowland areas most plants are of cultivated origin. This is the most widely cultivated lightweight osier. It has been artificially crossed with *S. viminalis* (*S.* × *rubra*) for extra weight and *S. repens* (*S.* × *doniana*) to produce very fine material. *S.* × *taylorii* and *S.* × *forbyana* Sm. are multiple hybrids involving *S. caprea* and *S. cinerea* which are difficult to positively identify. A straight cross between *S. purpurea* and *S. cinerea* (*S.* × *pontederiana* (= *S.* × *sordida*)) also occurs in Britain and Europe. The most valuable populations of native *S. purpurea* are in remote areas which are unlikely to be substantially altered by modern planting.

Salix viminalis, the Common Osier, is still grown commercially as a basket willow and it features in the parentage of several hybrids used for biomass production. Apparently good authentic populations can still be found in Britain but it is often impossible to tell if they are derived from local or even British material. The native type makes an excellent basket rod and has been extensively used in the cultivation of osiers. In the search for quality and productivity, growers have imported clones mainly from Europe including Italy and the South, and from the east into Asia and Russia. *Salix viminalis* seeds freely and quickly covers damp waste ground; it grows equally well from cuttings and clonal populations occur along muddy river banks. Although hybridization as a result of dioecism is inevitable, populations appear to remain remarkably pure. Truly native plants can not be guaranteed in lowland areas, however, so there is probably little risk from the expansion of biomass plantings. There are Scottish variants which appear to be natural, and these should be protected if at all possible.

Salix caprea, Goat Willow, is replaced in much of England by hybrid forms and back-crosses. The Scottish variant, however, remains fairly constant, and in remote areas whole populations consist of this ssp. *sphacelata*. True Goat Willow reproduces almost entirely by seed and dioecism obligates out-crossing. Even without human intervention the species in the south has become totally adulterated by crossing with *S. cinerea* subspecies and *S. viminalis*. Hybrids with all the species in subgenus *Vetrix* are known. It is difficult to predict the effects of increased cultivation of related taxa on natural and semi-natural hybrid swarms of sallows which have in the main occurred spontaneously. Hybrids occur with *S. daphnoides* (*S.* × *hungarica* Kerner), which is occasionally planted in upland areas. The inherited bloomed stems show up clearly on the progeny which can be easily identified.

Salix × *sericans* (*S. caprea* × *S. viminalis*) occurs widely as a wild plant and a relic of clonal cultivation. Clonal selections are the subject of experimental work on biomass production. Artificial hybrids have chromosome numbers that are different from those of the natural hybrid (Meikle 1975). Even in nature, seedling progeny of this hybrid cannot remain constant so any threat from newly cultivated clones can only amount to a slight taxonomic irritation.

Salix × *calodendron* (*S. caprea* × *S. cinerea* × *S. viminalis*). The origin of this hybrid is dubious, some authors regarding it as a nothomorph of *S. caprea* × *S. viminalis* while others add *S. aurita* to the equation. The British population appears to be all female and sterile, even in close proximity to compatible male plants. Imported clones pose more of a threat: *S.* × *dasyclados* from Europe, and of approximately the same parentage, is becoming widely used for biomass production. Male and female clones occur, and they could contaminate local subgenus *Vetrix* species and hybrids. Sites within 'bee range' of authentic sallow populations should preferably be avoided.

Salix cinerea ssp. *oleifolia*, the Common Sallow (Rusty Willow), and ssp. *cinerea* the Grey Sallow (Grey Willow), survive as identifiable taxa. Grey Sallow is a continental species that extends naturally only into eastern England. Common Sallow occurs throughout Britain. These subspecies have probably never been cultivated or artificially moved around the country and retain a well-defined natural distribution. Natural hybrid swarms occur, especially with the smaller upland representatives of the subgenus *Vetrix*, *S. phylicifolia* and *S. myrsinifolia*. These would be particularly vulnerable to alteration by introductions of vigorous foreign cultivars anywhere near them.

Salix × *smithiana* (*S. cinerea* × *S. viminalis*) can be distinguished from *S.* × *sericans* only with difficulty; furthermore, in theory the characteristics of *S.* × *calodendron* and

S. ×dasyclados should lie somewhere between the two of them. This tree occurs naturally in Britain and Europe and cultivated populations remain as relics usually identified by clonal evenness. Either of the *S. cinerea* subspecies may be a parent of the hybrid and this is reflected in the vigour of the progeny. Several clones are in current use for biomass production. These should not be planted near to wild *S. cinerea* woodlands or natural hybrid swarms.

Salix aurita, the Eared Sallow (Eared Willow), is widespread all over Britain but probably only authentic in upland areas. The species hybridizes with several other sallows which in some areas completely replace it. *Salix × multinervis* (*S. aurita × S. cinerea*), *S. × fruticosa* (*S. aurita × S. viminalis*) and the very vigorous osier *S. ×stipularis* (*S. aurita × S. viminalis × S. caprea*) can all do this. The remaining natural Eared Sallow populations should be protected from artificial contamination by introduced species and hybrids. The species has the ability to cross into the largely unspoiled subgenus *Chamaetia*, the alpine and dwarf willows.

Conclusions

Present-day willow planting consists of a fairly stable and small Cricket-bat Willow market which is unlikely to move far out of its present 'ideal' geographical range, and a still declining basket willow sector. Should the latter revive at all, new strains are unlikely to be used. Wild willows growing adjacent to osier beds have long since come to terms with their cultivated neighbours. The growth in the market for short rotation coppice willow for biomass does appear likely to continue and there is obviously a need for growers to be aware of vulnerable natural populations nearby. The development of clones likely to produce inter sub-generic hybrids would be most damaging to the British native *Salix*.

Acknowledgements

The author is indebted to Desmond Meikle and Ken Stott for amiable perambulation and conference over many years, and to Paul Tabbush for kindly reading the text.

References

Kennedy, C.E.J. & Southwood, T.R.E. (1984). The number of species of insects associated with British trees. *Journal of Animal Ecology* **53**: 455-478.

Meikle, R.D. (1975). *Salix* L. *In* C.A. Stace (ed), *Hybridization and the Flora of the British Isles*. London: Academic Press.

Meikle, R.D. (1984). *Willows and Poplars of Great Britain and Ireland.* London: Botanical Society of the British Isles.

17

Conservation of rare temperate rainforest conifer tree species: a fast-growing role for arboreta in Britain and Ireland

CHRISTOPHER N. PAGE and MARTIN F. GARDNER

Royal Botanic Garden, Edinburgh, EH3 5LR

ABSTRACT. Conifer germ-plasm can be most assuredly and practically preserved as living trees. Set in permanent, scientifically selected, safe-haven sites, arboreta, if appropriately genetically structured, can provide a vitally important *ex situ* means of long-term conservation of rare tree species, closely integrated with the *in situ* conservation of the same species in the wild.

Building on this potential of arboreta, the Royal Botanic Garden, Edinburgh has, for several years, been developing an extensive Conifer Research and Tree Conservation Programme.

The aims of this Programme are to help to realistically conserve the many conifer tree species which are becoming rapidly reduced or genetically depleted, and which are currently listed as vulnerable, threatened, or endangered in the wild, by constructing scientifically structured long term breeding groups of the more vulnerable material in a network of dedicated arboretum sites across Britain and Ireland.

The National and International Perspective

Our England is a garden that is full of stately views,
Of borders, beds and shrubberies and lawns and avenues,
With statues on the terraces and peacocks strutting by;
But the Glory of the Garden lies in more than meets the eye.
          Rudyard Kipling, 1865-1936

Conifer germ-plasm can be most assuredly and practically preserved as living trees. Set in permanent, scientifically selected, safe-haven sites, arboreta, if appropriately genetically structured, can provide a vitally important *ex situ* means of long term conservation of rare tree species, closely integrated with the *in situ* conservation of the same species in the wild.

In Britain, we have an extremely depauperate indigenous conifer flora, amounting to three species, Scots Pine, Juniper and Yew. The situation becomes especially apparent when we compare our conifer flora with that of other similar-sized islands of the world: Japan, for example, has 46 species of native conifers and New Caledonia, an island about the size of Ireland, 45. The European Tertiary flora was especially rich in conifer species, and included several genera now confined to the Far East of Asia. Most of these failed to survive the Quaternary glaciations of Europe, and being a migrationally highly sedentary group of plants (Page & Clifford 1981), most surviving European species (notably all species of *Picea, Larix* and *Abies* and other *Pinus* species) failed to immigrate again into post-glacial Britain before the Channel rose. Indeed, it is our view that two of our three native conifer species only succeeded in returning because they were bird-dispersed. The third, Scots Pine, is so geographically and morphologically anomalous (e.g. Steven & Carlisle 1959; Ruby & Wright 1976) as well as chemically and genetically distinct within these islands (Thielges 1972; Szmidt & Wang 1993), and has such an unusual post-glacial history, including early radiation from multiple centres (e.g. H.H. Birks 1975; Bennett 1984; Bradshaw & Browne 1987; H.J.B. Birks 1989), that it is our suspicion that it may never have left these shores at all.

How different might the modern vegetation of Britain and Ireland have been if these islands had not been both insular and glaciated! For, if it can be assumed that without the glaciations the indigenous conifer flora would have been about as rich and diverse today as that of Japan (and the hypothesis is probably not far wrong), then in Britain and Ireland the glaciations can be blamed for reducing the species content of the potential post-glacial conifer flora by, we estimate, an amazing 94%; it is interesting to view the overall vegetational (and especially forest!) imbalance in our post-glacial flora in this light.

Conservation and the growing tradition

But we do have a climate that is remarkably appropriate to cultivation of conifer tree species introduced from a great and diverse range of parts of the world. We estimate that at least half of the trees of one third of the Earth's total forests, are hardy somewhere in these islands[1]. This estimate is based on some 100-150 years and more of accumulated experience of growing trees (not *all* of it ours!). As a result, from Scotland to Cornwall and Kerry, there has developed in Britain and Ireland one of the greatest traditions in the world of introducing and growing unusual conifers in gardens and arboreta. For more than two centuries a great diversity of species of conifers have thus played a very significant role in estate policy planting throughout Scotland, England, Wales and Ireland (e.g. Hunter 1883; Hyde 1931; Balfour 1932; Headfort 1932; Miles 1967; James 1981; Hunt & Pett 1991), and have contributed much to the diversity and park-like greening of these areas of our landscape.

Further, the long and important custom of recording the disparate details of the introduction, wild origins, arboretum location, morphology, growth rate, size, longevity, reproduction and husbandry of cultivated coniferous trees, has led to documentation of these introductions with a unique degree of accuracy and completeness within these islands (e.g. Gordon 1880; Dunn 1892a & b; Masters 1892; Veitch 1892, 1906; Webster 1896; Bretschneider 1898; Thurston 1930; Chittenden 1932; Dallimore 1932; Stirling-Maxwell 1932; Orr 1933;

---

[1] One third of the world's forests are conifer-dominated, and at least half of these species can be grown here.

Matthews 1955; Macdonald *et al.* 1957; Hunt 1972; Mitchell 1972a, b & c; Forrest 1985; Mitchell *et al.* 1990), through accumulated statistical data now held as the National Tree Register (Hallett, 1987). Using this superb information base, second to none, it is now possible to plan future planting with a high degree of scientific accuracy and detail, creating forward projections of over a century for rare species. Carefully chosen and sited, and established as genetically structured populations, conifers have an important role to play in the future multiple use of our landscape. They certainly have a special role to play in adding to potential agro-forestry developments a biological diversity and genuine conservational purpose, in the important overall perspective of managing global genetic resources.

Yet currently in the wild, within the world's 600+ species of conifers, at least 334 taxa are listed as vulnerable, threatened, endangered or already extinct (Farjon, Page & Schellvis 1993 – the exact numbers changing almost month by month). This large contingent of some of the world's most important trees also includes many geologically ancient genera which today are relictual, often monotypic and regionally restricted, especially to the Temperate

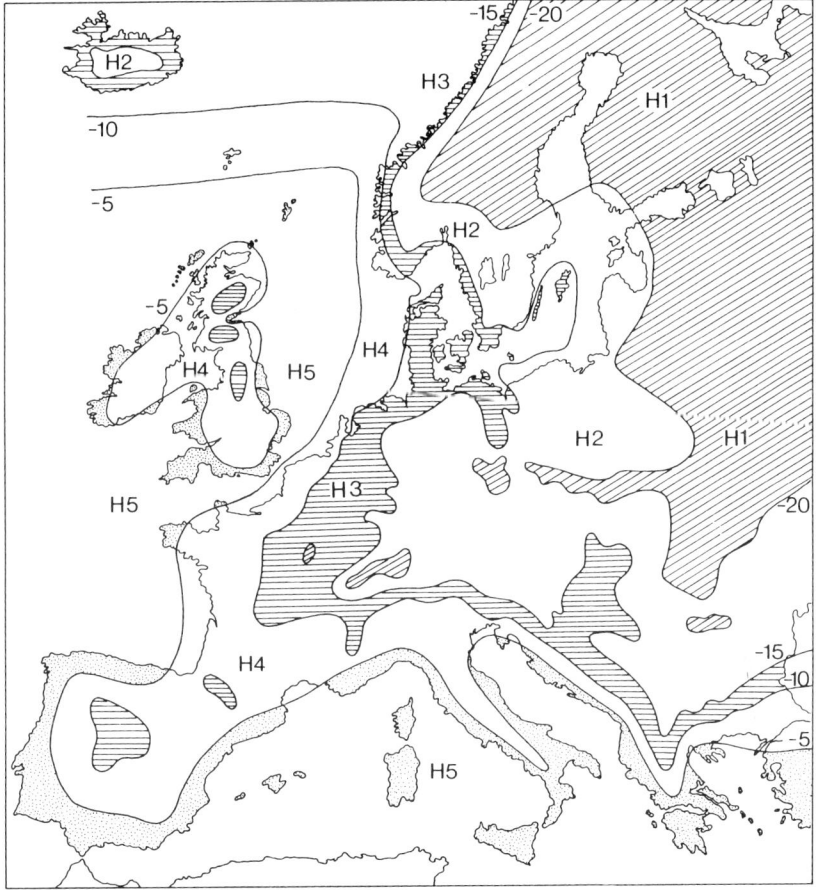

Fig. 1. Hardiness zones for Europe, based on mean minimum January temperature isotherms (after Walters *et al.* 1986). Note the high diversity of available planting zones within the small area of Britain and Ireland.

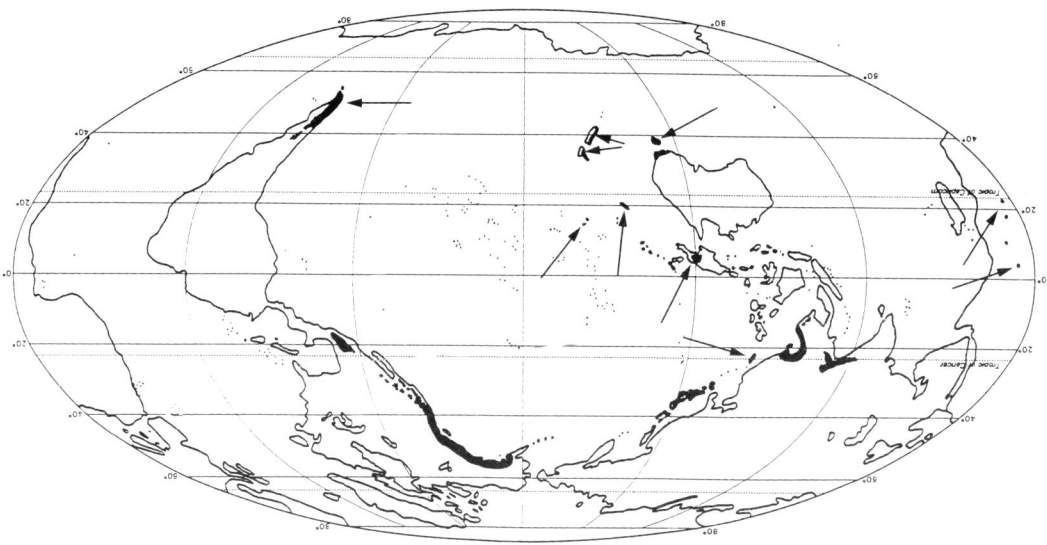

Fig. 2. Approximate world range of the main conifer-dominated Temperate Rainforest communities. Note the small geographic areas of these forests and their close affinity today with the oceanic climates of the Pacific rim. These are the conifer forests with the highest endemism and in which many of the most ancient and relictual conifers exist. Many of these forests are disappearing at the same rate as are the much larger Tropical Rainforests, and most of the more local conifer species within them are especially threatened. Many of these species can, however, be grown successfully in the western and south-western climates of Britain and Ireland.

Rainforests. These important forests occur mainly high on cloud-wrapped tropical mountains, in a wide scatter of locations around the Pacific fringe, and on the southern ends of the southern continents. To the natural rarity of the dominant conifers of these rainforests, man has usually added further range fragmentation and genotype depletion through selective logging of important timber types, grazing of regenerants, sometimes forest burning and, increasingly around the Pacific fringe, wholesale forest removal (Page 1991; Ray & Gardner 1993; Page 1994 & in press).

The early introductions into Britain and Ireland especially, of these Temperate Rainforest conifer species, together with the information base now accumulated from them, show the potential of the climate of these islands for achieving long term *ex situ* conservation of such rare tree species. These include not only the hardier northern hemisphere material such as the Sino-Himalayan introductions of George Forrest, E.H. Wilson and Joseph Rock and those from North America of Archibald Menzies, but most especially much more tender material

Fig. 3. Wholesale clearance and burning of the Chilean conifer *Fitzroya cupressoides* leaves its indelible mark on a wild Temperate Rainforest despite it being a Red Data Book species and also protected from international trade under CITES regulations. (*Photo MG*).

---

from the southern hemisphere, such as the introductions of William Lobb, Richard Pearce and Harold Comber from Chile (e.g. Graham 1980; Magor 1988). Large trees of most of these early introductions still survive and remain healthy in an array of sites around the milder fringes of western Scotland, west Wales, south-west England and southern Ireland (Page 1991, 1992a & b). But it is a sobering thought that most of these are from wild forests that are already extinct and are thus probably the only surviving individuals of their genotypes. Now, 100 to 150 years on from planting in a few sites as multi-individual groups, almost all are steadily displaying their seed-yielding potential: an example is the Chilean *Podocarpus salignus* in west Cornwall (see figs. 6-7), now almost extinct as large trees in the wild (Gardner & Knees 1990; Gardner & Page 1992). The success of these Temperate Rainforest conifers in the milder corners of these islands contrasts with the present sensitivity of some existing natural high-latitude boreal forests to changing climatic parameters (see Kauppi & Posch 1985; Pastor & Post 1988; Bonan *et al.* 1990) and makes such species of considerable wider future interest in these islands if climatic warming takes place (Rook & Page in press). As indicator species of areas of suitable climate for plantings of such species in these islands, the ferns, both native and introduced, as well as other Temperate Rainforest woody species, provide further, particularly sensitive, climatic guides (cf. Page 1982, 1986a & b, 1988a).

Fig. 4. What it's all about (1). *ABOVE*: Pacific north-western Temperate Rainforest dominated by *Thuja plicata* (Western Red-cedar), *Tsuga heterophylla* (Western Hemlock-spruce) and *Pseudotsuga menziesii* (Douglas Fir), Olympic Peninsula, Washington State. *BELOW*: Temperate Rainforest under construction, dominated by the same species, west Argyll. (*Photos CNP*).

Fig. 5. In the right climate all the tree species regenerate freely. Naturally regenerating seedling of *Thuja plicata* in the forest floor of the Temperate Rainforest under construction, west Argyll, August. (*Photo CNP*).

The importance of the conservation of conifers (especially Temperate Rainforest species) in arboreta, in contrast to other methods of germ-plasm storage, is thus the longevity and low cost of the result achieved, and the production of seed as a renewable resource; whereas conifer pollen may remain viable in storage for only about a decade and seed, even in the best controlled conditions only up to 20 years, there is a high expectation of survival of appropriately sited cultivated trees for at least one to two centuries. Once established, most conifers become remarkably self-maintaining, with an enviable longevity and long reproductive life. Thus scientifically structured arboreta are the prime choice for long term *ex situ* conifer germ-plasm conservation. This contrasts with preferred techniques for many flowering plants (especially crop species: e.g. Hawkes 1990) and for many pteridophytes (Page *et al.* 1992). The conifers, being a long-lived woody plant group, are ideal for such a conservation strategy, while their universal dependence on wind for pollination, rather than on a specialized animal group, makes them especially suitable for *ex situ* breeding purposes.

Fig. 6. What it's all about (2). *ABOVE*: Podocarp-dominated Temperate Rainforest, southern Chile, south of Valdivia, February. *BELOW*: Podocarp-dominated Temperate Rainforest, near Truro, west Cornwall, August. These cultivated but self-maintaining *Podocarpus salignus* originate from seed collected by Richard Pearce in Chile in 1845. Planted in a west Cornwall estate, they are now setting seed and, a century and a half later, are surrounded by an abundant supply of naturally self-regenerant seedlings. (*Photos CNP*).

Conservation of rare temperate rainforest conifer tree species 127

Fig. 7 (detail of Fig. 6, lower). The abundant seed set (left) and natural seedling establishment (right) of the Chilean Podocarp *Podocarpus salignus* in west Cornwall, August. (*Photo CNP*).

Fig. 8. What it's all about (3). *ABOVE*: existing planted Temperate Rainforest of Tasmanian type in south-western England: a grove of mature tree ferns *Dicksonia antarctica* thrive in several west Cornwall locations, and will certainly grow in many more. This group is nearly a a hundred years old and speak of the long term suitability of the climate over this time. Nearby are several others now 14 m tall and 200 years old (planted 1792), and each is larger and may be older than most in the wild. All produce spores under these cultivated conditions, and, *BELOW*: regenerate freely and naturally on streamside erosion slopes. (*Photos CNP*).

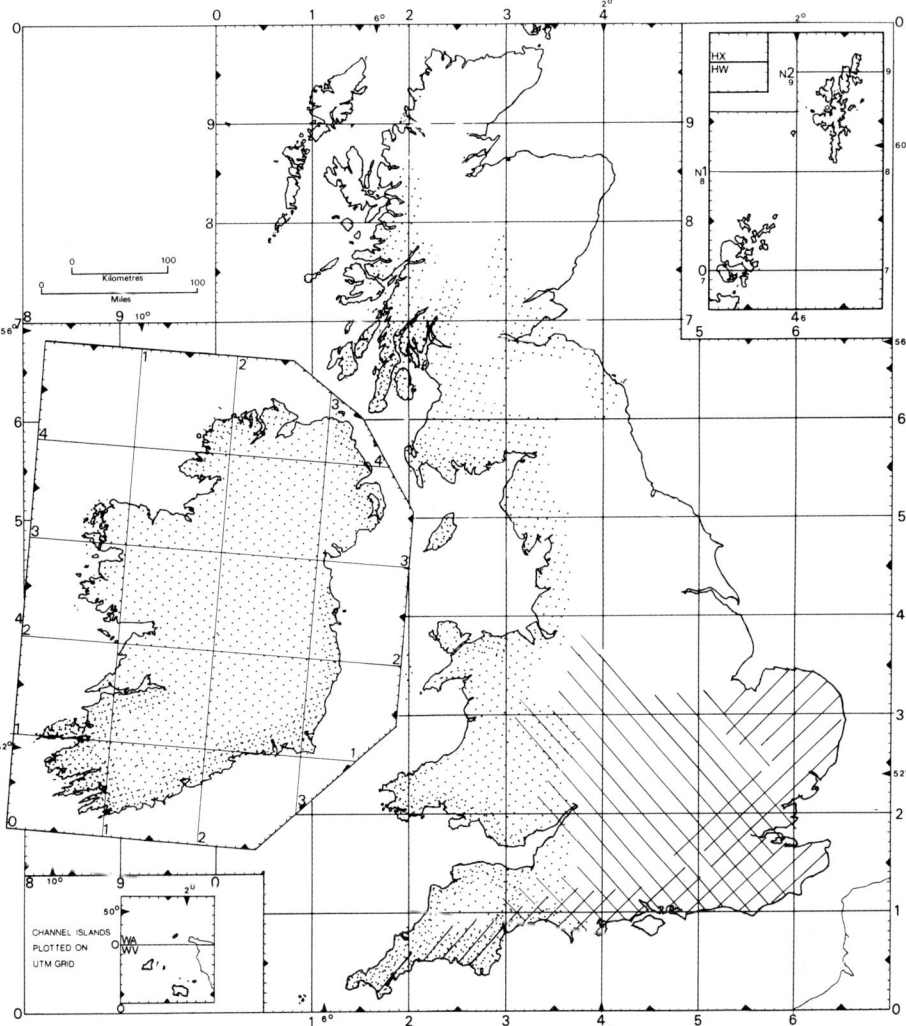

Fig. 9. Main generalized climatic regions in Britain and Ireland being developed in this programme for *ex situ* rare conifer planting. DENSE DOTS (most winter-warm, extreme oceanic, mainly south-western regions with cool, moist summers and the longest growing season), used for planting very tender mainly southern hemisphere Temperate Rainforest species [indicated chiefly by best growth of existing old *Podocarpus salignus, Fitzroya cupressoides, Saxegothaea conspicua, Sequoia sempervirens, Lagarostrobos franklynii* and the tree fern *Dicksonia antarctica*, and of the natural distribution of the native ferns *Hymenophyllum wilsonii* and *Dryopteris aemula*]; SPARSE DOTS (high rainfall, mild Atlantic climate regions), used for planting mainly northern hemisphere Temperate Rainforest species [indicated chiefly by best growth of existing planted *Tsuga heterophylla, Thuja plicata, Calocedrus decurrens* and *Cryptomeria japonica* and of the natural distribution of the native ferns *Polypodium interjectum, Blechnum spicant* and *Oreopteris limbosperma*]; DESCENDING SLASHES (English Shire Counties 'Cedar triangle'), used for planting montane Mediterranean species [indicated chiefly by best growth of existing *Cedrus atlantica* and *Cedrus libani*]; ASCENDING SLASHES (regions of highest summer temperature including East Anglia, south coast and south-east Ireland), used for continental south-eastern species [indicated chiefly by best growth of existing *Ginkgo biloba, Taxodium distichum, Metasequoia glyptostroboides* and *Pseudolarix amabilis*]; UNSHADED AREAS (regions of coldest winter temperatures and shortest growing season) used mainly for continental eastern and high montane species (indicated by best existing growth of *Picea omorika, Hesperopeuce mertensiana* and *Pinus aristata* and the native and planted distribution of *Pinus sylvestris*]. Black diamonds are the location of the main tree growing centres used in this Programme.

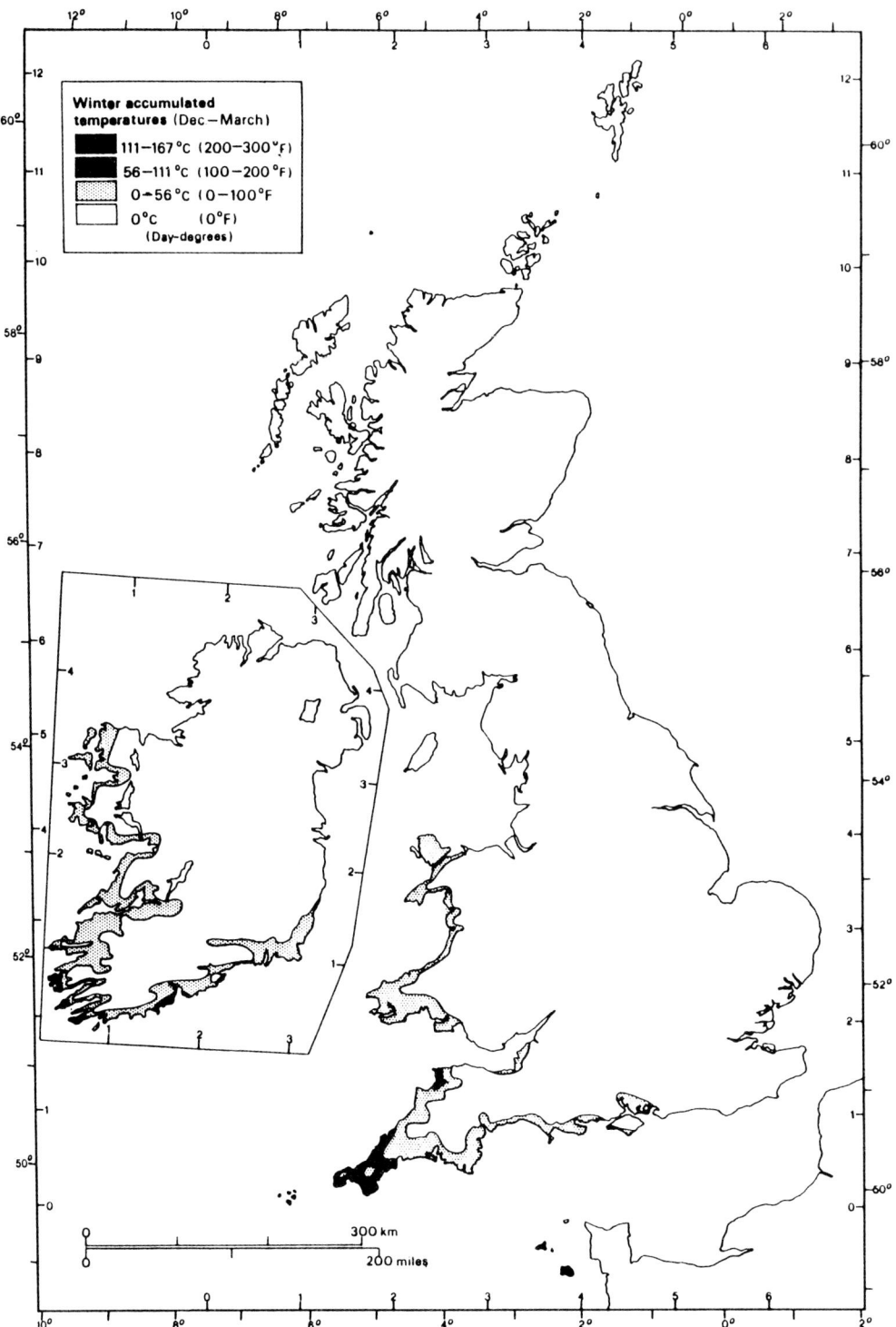

Fig. 10. Winter accumulated temperatures in Britain and Ireland indicate particularly well those areas appropriate for cultivation of some of the most tender Temperate Rainforest species (after Page 1982).

Building on this enormous conservation potential of arboreta for conifers, the tree growing tradition of these islands, and the critical need to prevent further genotype erosion of so many species, the Royal Botanic Garden, Edinburgh has for twenty years been developing an extensive *Conifer Taxonomic Research and Tree Conservation Programme*. The aim of this Programme is to help to conserve the many conifer species which are endangered in the wild, by engineering an array of scientifically structured long term tree breeding groups set in a network of permanent planting sites within these islands.

This Programme, although initiated independently, now has the potential to link through the IUCN/SSC Conifer Specialist Group (Page 1988b, Page & Farjon 1991) to the developing IUCN/WWF Plant Conservation Programme (see Hamann 1990 for details of that programme), as well as to networks elsewhere (such as that already being developed within the Mediterranean region through the Botanic Gardens Conservation Secretariat (Heywood 1990)) and to conservation programmes for other forms of genetic resources (e.g. Burley & Nankoong 1980). Becoming widely recognized as valuable adjunct methods of genotype conservation, especially when integrated with local *in situ* conservation programmes in the wild, are germ-plasm banks associated with the necessary centres of taxonomic and botanic garden expertise in structuring them (e.g. Hawkes 1978, 1990; Heywood 1987; IUCN/WWF 1989).

In parallel with the practical processes of collecting, quarantining, propagating and planting the rare conifer species, are important opportunities to structure taxonomic and biological research. These include the survey, recording and induction of fertility in reclusive conifer material, which enables seed production to be experimentally generated at will by using the methodology developed by Longman *et al.* (1982). Other projects currently developing relate to the natural ecology, genotypic diversity, environmental tolerance and reproductive strategies of the species. The results of such research will form an important information base for use in future forest restoration programmes as part of an integrated conservation approach (Maunder 1992, Maunder & Ramsay 1994), the research and the undertaking of such programmes being pursued jointly with the countries concerned.

Genetic structure of populations and networked distribution

To achieve these unique objectives, two biologically paramount aspects of the programme are currently progressing.

1. The first of these is the planting of whole tree groups forming genetically structured populations.

> *Our plantings are engineered to include as wide a range of genotypic diversity within each population as is available, by initially sampling each from all wild sites. This includes sampling each threatened species systematically from across whole populations, along clinal gradients, and from isolated tree stands, taking into account what is known, in general terms, about population size and gene flow in conifers (e.g. Koski 1970; Muona & Harju 1989; Helgason & Ennos 1991; Di-Giovanni & Kevan 1991), their intrinsically high genetic variability within populations (e.g. Saki* et al.

1971; Lines 1987; Hamrick & Godt 1989; Muona 1989; Kuittinen et al. 1991) and field sampling methodology appropriate to embrace this genetic base (e.g. Faulkener 1976; Frankel & Soule 1981; Soule 1986, 1987; Center for Plant Conservation 1991). Planting numbers are influenced by the morphological and ecological diversity of trees in the field, and generally aim to follow at least the base numbers of 20 to 50 individuals that have commonly been used in commercial conifer seed orchard planting practice (e.g. Faulkener 1975; Giertych 1975).

We are planting in each ex situ *population all the different structural ranges of individuals that can be seen in wild populations. These populations thus differ from those of seed orchards in their high included genetic diversity and their long life expectancy. They are also being initiated to become dynamic in age structure in time, as well as forming a framework for the steady inclusion of other associated threatened flora.*

2. The second of these objectives is the establishment of a *networked distribution* of the sites adopted for these populations. This has especially significant strategic and biological advantages:

a. it makes available more area of land for planting populations than is available at most botanic gardens, and may be expandable as required;

b. it gives greatest opportunity to set differing populations in differing climatic and edaphic conditions;

c. it achieves multi-site insurance through wide site-scatter against potential future disasters striking individual sites;

d. it provides for the necessary biological isolation of potentially cross-breeding congeneric individuals, thus maximizing conservation of high genetic purity; and

e. it establishes a wide scatter of sites around which educational programmes can be constructed on a more local and regional basis.

Genera of conifers likely to succeed in cultivated sites from the west of Scotland (Table 1) to south-western England (Table 2) are presented in the accompanying tables.

Table 1. Schedule of Conifer Genera containing species showing survival rates and reproductive potential appropriate for *ex situ* conservation planting in the west of Scotland.

Key to geographic areas: WNA = Western North America (Pacific Northwest); F= Florida; CM = California to Mexico; TNZ = Tasmania and New Zealand; SIN = Sino-Himalayan region; JAP = Japan; TAI = Taiwan; PAP = Papua New Guinea and adjacent SW Pacific islands; AFR = Africa; AND = Andino-Patagonia; MED = Mediterranean region. *Those prefixed with an asterisk * are the genera whose species are most likely to be established from cuttings.*

	WNA	F	CM	TNZ	SIN	JAP	TAI	PAP	AFR	AND	MED
GINKGOACEAE											
Ginkgo L.					X						
TAXACEAE											
Pseudotaxus Chen					X						
Taxus L.	X		X		X	X	X				
Torreya Arnott	X	X			X	X					
Austrotaxus Compton											
CEPHALOTAXACEAE											
Amentotaxus Pilger											
Cephalotaxus Sieb.& Zucc. ex Endl.					X	X					
PHYLLOCLADACEAE											
Phyllocladus L.C. Rich ex Mirb.											
PODOCARPACEAE											
Saxegothaea Lindb.										X	
Prumnopitys Philippi										X	
Sundacarpus C.N. Page											
Retrophyllum C.N. Page											
Nageia Gaertner											
Afrocarpus C.N. Page									X		
Podocarpus L'Herit.					X	X					
Parasitaxus de Laub.											
Acmopyle Pilger											
Dacrycarpus (Endl.) de Laub.											
Falcatifolium de Laub.											
Dacrydium Solander ex Lamb.											
Halocarpus C.J. Quinn											
Lepidothamnus Philippi											
Lagarostrobos C.J.Quinn				X							
Microcachrys Hook. f.				X							
Microstrobos Garden & Johnson				X							
ARAUCARIACEAE											
Agathis Salisb.											
Araucaria Jussieu										X	

	WNA	F	CM	TNZ	SIN	JAP	TAI	PAP	AFR	AND	MED
SCIADOPITYACEAE											
Sciadopitys Sieb. & Zucc.						X					
TAXODIACEAE											
Athrotaxis D. Don				X							
Cunninghamia R. Br.						X					
Taiwania Hayata					X		X				
Cryptomeria D. Don					X	X					
Sequoiadendron Buchh.	X										
Sequoia Endl.	X										
Metasequoia Miki											
Glyptostrobus Endl.											
Taxodium L.C. Rich.		X									
CUPRESSACEAE											
Neocallitropsis Florin											
Callitris Vent.											
Actinostrobus											
Widdringtonia Endl.											
Tetraclinis Masters											
Platycladus Spach					X						
Microbiota Komaro					X						
Thuja L.	X				X	X					
Pilgerodendron Florin											
Austrocedrus Florin										X	
Libocedrus Endl.											
Papuacedrus Li											
Calocedrus Kurz	X										
Fokienia Henry & Thomas					X						
Fitzroya Hook.f. ex Lindl.										X	
Diselma Hook.f.				X							
Thujopus Sieb. & Zucc.						X					
Chamaecyparis Spach	X	X			X	X	X				
Cupressus L.	X				X						
Juniperus L.	X		X		X	X	X				
PINACEAE											
Pseudolarix G. Gord.					X						
Cedrus Trew					X						X
Larix Mill.	X				X	X					
Keteleeia Carr.											
Pseudotsuga Carr.	X		X								
Cathaya Chun & Kuang					X						
Hesperopeuce Lemmon	X										
Tsuga Carr.	X				X	X	X				
Nothotsuga											
Picea A. Dietr.	X		X		X	X	X				X
Abies Mill.	X		X		X	X	X				X
Ducampopinus A. Chevalier											
Pinus L.	X	X	X		X	X	X				X

Table 2. Schedule of conifer genera containing species showing survival rates and reproductive potential appropriate for *ex situ* conservation planting in south-western England.

Key to geographic areas: WNA = Western North America (Pacific Northwest); F= Florida; CM = California to Mexico; TNZ = Tasmania and New Zealand; SIN = Sino-Himalayan region; JAP = Japan; TAI = Taiwan; PAP = Papua New Guinea and adjacent SW Pacific islands; AFR = Africa; AND = Andino-Patagonia; MED = Mediterranean region. *Those prefixed with an asterisk* * *are the genera whose species are most likely to be established from cuttings.*

	WNA	F	CM	TNZ	SIN	JAP	TAI	PAP	AFR	AND	MED
GINKGOACEAE											
Ginkgo L.					X						
TAXACEAE											
Pseudotaxus Cheng					X						
Taxus L.	X		X		X						
Torreya Arnott	X	X			X						
Austrotaxus Compton								X			
CEPHALOTAXACEAE											
Amentotaxus Pilger					X	X					
Cephalotaxus Sieb.& Zucc. ex Endl.					X	X					
PHYLLOCLADACEAE											
Phyllocladus L.C. Rich ex Mirb.				X							
PODOCARPACEAE											
Saxegothaea Lindb.										X	
Prumnopitys Philippi										X	
Sundacarpus C.N. Page											
Retrophyllum C.N. Page								X			
Nageia Gaertner					X						
Afrocarpus C.N. Page									X		
Podocarpus L'Herit.					X	X	X		X		
Parasitaxus de Laub.											
Acmopyle Pilger											
Dacrycarpus (Endl.) de Laub.				X							
Falcatifolium de Laub.											
Dacrydium Solander ex Lamb.				X							
Halocarpus C.J. Quinn				X							
Lepidothamnus Philippi				X						X	
Lagarostrobos C.J.Quinn				X							
Microcachrys Hook. f.				X							
Microstrobos Garden & Johnson				X							
ARAUCARIACEAE											
Agathis Salisb.				X							
Araucaria Jussieu				X						X	

	WNA	F	CM	TNZ	SIN	JAP	TAI	PAP	AFR	AND	MED
SCIADOPITYACEAE											
Sciadopitys Sieb. & Zucc.						X					
TAXODIACEAE											
Athrotaxis D. Don				X							
Cunninghamia R. Br.					X		X				
Taiwania Hayata					X		X				
Cryptomeria D. Don					X	X					
Sequoiadendron Buchh.											
Sequoia Endl.	X										
Metasequoia Miki					X						
Glyptostrobus Endl.					X						
Taxodium L.C. Rich.		X	X								
CUPRESSACEAE											
Neocallitropsis Florin											
Callitris Vent.				X							
Actinostrobus											
Widdringtonia Endl.									X		
Tetraclinis Masters									X		X
Platycladus Spach											
Microbiota Komarov											
Thuja L.	X										
Pilgerodendron Florin										X	
Austrocedrus Florin										X	
Libocedrus Endl.				X			X				
Papuacedrus Li							X				
Calocedrus Kurz	X				X		X				
Fokienia Henry & Thomas					X						
Fitzroya Hook.f. ex Lindl.			X							X	
Diselma Hook.f.				X							
Thujopus Sieb. & Zucc.						X					
Chamaecyparis Spach		X				X					
Cupressus L.					X						X
Juniperus L.			X			X	X		X		X
PINACEAE											
Pseudolarix G. Gord.					X						
Cedrus Trew					X						X
Larix Mill.					X						
Keteleeia Carr.					X		X				
Pseudotsuga Carr.			X		X	X	X				
Cathaya Chun & Kuang					X						
Hesperopeuce Lemmon											
Tsuga Carr.					X		X				
Nothotsuga					X						
Picea A. Dietr.	X		X				X				
Abies Mill.			X		X						X
Ducampopinus A. Chevalier					X						
Pinus L.		X	X		X		X				X

Conclusions and ultimate conservation purposes

This Programme thus has five main pillars of purpose which it fulfils. These are:

1. scientific taxonomic research;

2. tree species and genotype conservation;

3. education;

4. amenity;

5. enrichment of the managed landscape, with a considerable potential in 'greening' of derelict areas and bringing a genuine conservation purpose to other agencies involved in such activities.

The aims of the Programme are conservational ones that are strictly scientifically structured, but it will achieve its objectives through sound future development of horticultural methods and from the tree planting tradition of these islands. The three main long term practical aims of the Programme structured in this way will be the achievement of:

1. a high degree of clonal longevity, to act as an important buffer against future wild genotype depletion;

2. a unique biological materials resource around which research programmes on rare and little-known materials can be pursued; and

3. soundly structured breeding populations to form future renewable germ-plasm resources for ultimate reintroduction potential.

Fig. 11. Reproductive success in a Temperate Rainforest conifer in cultivation: *Athrotaxis selaginoides* at pollination, originating from Tasmania. (*Photo CNP*).

Fig. 12 (left). Representative coniferous tree species endemic to American Pacific North-western Temperate Rainforest thriving in Britain: *Calocedrus decurrens* (Incense Cedar) self-maintaining in west Cornwall. This species grows well over almost the whole of these islands. (*Photo CNP*).

Fig. 13 (right). Representative coniferous tree species endemic to South American Temperate Rainforest of a few parts of Chile and adjacent Argentina thriving in Britain: a grove of *Araucaria araucana* (Chile Pine, Monkey Puzzle) self-maintaining in west Cornwall. This species grows well when planted over much of the higher rainfall parts of these islands. (*Photo CNP*).

Fig. 14. Representative coniferous tree species endemic to Tasmanian Temperate Rainforest in river valleys (its main headquarters recently flooded by a hydroelectric scheme) thriving in Britain: *Lagarostrobos franklynii* (Huon Pine) self-maintaining (and abundantly fertile) in west Cornwall. This species grows well mainly in the milder western fringes of these islands. (*Photo CNP*).

The *ex situ* populations achieved are integrated with, and support (but do not replace), the aims of *in situ* conservation of wild tree species and natural forest (Page 1989, 1991, 1992a, 1992b, and in press). But in many cases, such cultivated populations established now in well-selected safe-haven sites may considerably outlast those of many of their wild progenitors.

Many trees, and especially long-lived conifers, have already been conserved by being introduced to our gardens where they have flourished and persisted very well, long after removal of the forests from whence they came. Indeed, some such as *Ginkgo* have survived only by being regarded as cultivation-worthy trees – by introductions into Chinese monastery gardens through ancient times. If conservation has been achieved so successfully, almost by accident, in the past, how much more can we achieve today if we add scientific and conservational skills to our popular tree planting and tree growing traditions in these islands?

Acknowledgments

The Edinburgh Conifer Conservation Programme is resourced by the Royal Botanic Garden, Edinburgh and Sainsbury Family Charitable Trusts. Grateful acknowledgement is also made for support to more than 27 corporate and private landowners across these islands, with whom joint dedicated planting programmes are currently progressing. We are grateful to many colleagues for continued stimulating running discussions on the Programme.

References

Balfour, F.R.S. (1932). The history of conifers in Scotland and their discovery by Scotsmen. *In: Conifers in Cultivation. Report of the 1931 Conifer Conference*, pp. 177-211. London: Royal Horticultural Society.
Bennett, K.D. (1984). The post-glacial history of *Pinus sylvestris* in the British Isles. *Quarternary Science Reviews* **3**: 133-155.
Birks, H.H. (1975). Studies in the vegetational history of Scotland. IV. Pine stumps in Scottish blanket peats. *Phil. Trans. Roy. Soc. Lond.* **B270**: 181-226.
Birks, H.J.B. (1989). Holocene isochrone maps and patterns of tree spreading in the British Isles. *J. Biogeography* **16**: 503-540.
Bonan, G.B., Shugart, H.H. & Urban, D.L. (1990). The sensitivity of some high-latitude boreal forests to climatic parameters. *Climatic Change* **16**: 9-29.
Bradshaw, R.H.W. & Browne, P. (1987). Changing patterns in the post-glacial distribution of *Pinus sylvestris* in Ireland. *J. Biogeography* **14**: 237-248.
Bretschneider, E. (1898). *History of Botanical Discoveries in China*. Vol. 1. London: Sampson-Low.
Burley, J. & Nankoong, G. (1980). Conservation of forest genetic resources. Eleventh Commonwealth Forestry Conference, Trinidad, 1980.
Center for Plant Conservation (1991). Genetic sampling guidelines for conservation collections of rare plants. Appendix. *In* D.A. Falk, & K.E. Holsinger, *Genetics and Conservation of Rare Plants*. Oxford & New York: Oxford University Press.
Chittenden, F.J. (ed) (1932). Statistical returns on conifers. *In: Conifers in Cultivation. Report of the 1931 Conifer Conference*, pp. 316-596. London: Royal Horticultural Society.
Dallimore, W. (1932). Reference list of conifers grown out of doors in the British Isles. *In: Conifers in Cultivation. Report of the 1931 Conifer Conference*, pp. 6-40. London: Royal Horticultural Society.
Di-Giovanni, F. & Kevan, P.G. (1991). Factors affecting pollen dynamics and its importance to pollen contamination: a review. *Canad. J. For. Res.* **21**: 1155-1170.
Dunn, M. (1892a). The value in the British Islands of introduced conifers. *Journal of the Royal Horticultural Society* **14**: 73-102.
Dunn, M. (1892b). Statistics of conifers in the British Islands. *Journal of the Royal Horticultural Society* **14**: 481-574.
Farjon, A., Page, C.N. & Schellvis, N. (1993). A world list of conifer taxa of greatest conservation concern. *Biodiversity and Conservation*.
Faulkener, R. (ed) (1975). *Seed Orchards*. London: HMSO (Forestry Commission Bulletin no. 54).

Faulkener, R. (1976). The gene pool of Caledonian Scots Pine – its conservation and uses. *In* R.G.H. Bunce & J.N.R. Jeffers (eds), *Native Pinewoods of Scotland*, pp. 96-99. Cambridge: Institute of Terrestrial Ecology.

Forrest, M. (1985). *Trees and Shrubs Cultivated in Ireland* (ed E.C. Nelson). Dublin: Boethius Press.

Frankel, O.H. & Soule, M.E. (1981). *Conservation and Evolution*. Cambridge: Cambridge University Press.

Gardner, M.F. & Knees, S.G. (1990). Forests without a future. *Country Life* **184**(7): 64-66.

Gardner, M.F. & Page, C.N. (1992). Trees in search of sanctuary. *Country Life* **185**(23): 80-81.

Giertych, M. (1975). Seed Orchard designs. *In* R. Faulkener (ed), *Seed Orchards*, pp. 25-37. London: HMSO (Forestry Commission Bulletin no. 54).

Gordon, G. (1880). *The Pinetum*. London: Bohn.

Graham, B. (1980). Concerning William Lobb. *The Cornish Garden* **23**: 11-15.

Hallett, V. (1987). The National Tree Register. *The Garden* **112**: 28-29.

Hamann, O. (1990). The botanic gardens as part of the IUCN/WWF plants conservation programme. *In* J.E.H. Bermejo, M. Clemente & V. Heywood, *Conservation Techniques in Botanic Gardens*, pp. 39-47. Koenigstein: Koeltz.

Hamrick, J.L. & Godt, M.J.W. (1989). Allozyme diversity in plant species. *In* A.H.D. Brown, M.T. Clegg, A.L. Kahler & B.S. Weir, *Plant Population Genetics, Breeding and Genetic Resources*, pp. 43-63. Sunderland, Mass.: Sinauer Associates.

Hawkes, J.G. (1978). The taxonomist's role in the conservation of genetic diversity. *In* H.W. Street (ed), *Essays in Plant Taxonomy*. London: Academic Press.

Hawkes, J.G. (1990). Germplasm banks: a method for endangered plant conservation. *In* J.E.H. Bermejo, M. Clemente & V. Heywood, *Conservation Techniques in Botanic Gardens*, pp. 49-56. Koenigstein: Koeltz.

Headfort, The Marquess of (1932). Conifers in the gardens and parks of Ireland. *In: Conifers in Cultivation. Report of the 1931 Conifer Conference*, pp. 212-222. London: Royal Horticultural Society.

Helgason, T. & Ennos, R.A. (1991). The outcrossing rate and gene frequencies of a native Scots pinewood population determined using isozyme markers. *Scottish Forestry* **45**: 111-119.

Heywood, V.H. (1987). The changing role of the botanic garden. *In* D. Bramwell, V.H. Heywood & H. Synge (eds), *Botanic Gardens and the World Conservation Strategy*. London: Academic Press.

Heywood, V.H. (1990). Objectives and strategies for a network of Mediterranean botanic gardens. *In* J.E.H. Bermejo, M. Clemente & V. Heywood, *Conservation Techniques in Botanic Gardens*, pp. 57-61. Koenigstein: Koeltz.

Hunt, D. (1972). Reference list of conifers and conifer allies grown out of door in the British Isles. *In: Conifers in the British Isles: Proceedings of the 1970 Conifer Conference*, pp. 109-122. London: Royal Horticultural Society.

Hunt, D. & Pett, D.E. (1991). *Historic Gardens in Cornwall*. Truro: Cornwall Gardens Society and Cornwall Gardens Trust.

Hunter, T. (1883). *Woods, Forests and Estates in Perthshire*. Perth: Henderson, Robertson & Hunter.

Hyde, H.A. (1931). *Welsh Timber Trees, Native and Introduced*. Cardiff: National Museum of Wales.

IUCN/WWF (1989). The Botanic Gardens Conservation Strategy. Gland: IUCN, WWF.

James, N.D.J. (1981). *A History of English Forestry*. Oxford: Blackwell.

Kauppi, P. & Posch, M. (1985). Sensitivity of boreal forests to possible climatic warming. *Climatic Change* **7**: 45-54.

Koski, V. (1970). A study on pollen dispersal as a mechanism of gene flow in conifers. *Comm. Inst. For. Fenn.* **80**: 1-78.

Kuittinen, H., Murna, O., Karkkainen, K. & Bozzan, Z. (1991). Serbian Spruce, a narrow endemic with much genetic variation. *Canad. J. Forestry Res.* **211**: 363-367.

Lines, R., (1987). Seed origin variation in Sitka Spruce. *Proc. Roy. Soc. Edinburgh* **93B**: 25-39.

Longman, K.A., Dick, J. & Page, C.N. (1982). Cone induction with gibberellin for taxonomic studies in Cupressaceae and Taxodiaceae. *Biologia Plantarum* **24**: 195-201.

Macdonald, J., Wood, R.F., Edwards, M.V. & Aldhous, J.R. (1957). *Exotic Forest Trees in Great Britain*. London: HMSO (Forestry Commission Bulletin no. 30).

Magor, E.W.M. (1988). Origins of conifers grown in Cornwall. *The Cornish Garden* **31**: 65-74.

Masters, M.T. (1892). List of conifers and taxads in cultivation in the open air in Great Britain and Ireland. *Journal of the Royal Horticultural Society* **14**: 179-256.

Matthews, J.D. (1955). Production of seed by forest trees in Britain. *Rep. For. Res. For. Comm.* **1953-4**: 64-78.

Maunder, M. (1992). Plant reintroduction: an overview. *Biodiversity and Conservation* **1**: 51-61.

Maunder, M. & Ramsay, M. (1994). The reintroduction of plants into the wild: an integrated approach to the conservation of native plants. *In* A.R. Perry & R.G. Ellis (eds), *The Common Ground of Wild and Cultivated Plants*, pp. 81-88. Cardiff: National Museum of Wales.

Miles, R. (1967). *Forestry in the English Landscape*. London: Faber & Faber.

Mitchell, A.F. (1972a). Conifer Statistics. *In: Conifers in the British Isles: Proceedings of the 1970 Conifer Conference,* pp. 123-285. London: Royal Horticultural Society.

Mitchell, A.F. (1972b). Noteworthy specimens of conifers in the British Isles. *In: Conifers in the British Isles: Proceedings of the 1970 Conifer Conference,* pp. 286-293. London: Royal Horticultural Society

Mitchell, A. (1972c). *Conifers in the British Isles*. London: HMSO.

Mitchell, A.F., Hallett, V.E. & White, J.E.J. (1990). *Champion Trees in the British Isles*. London: HMSO (Forestry Commission Field Book 10).

Muona, O. (1989). Population genetics and forest tree improvement. *In* A.H.D. Brown, M.T. Clegg, A.L. Kahler & B.S. Weir, *Plant Population Genetics, Breeding and Genetic Resources,* pp. 43-63. Sunderland, Mass.: Sinauer Associates.

Muona, O. & Harju, A. (1989). Effective population sizes, genetic variability and mating systems in natural stands and seed orchards of *Pinus sylvestris*. *Silvae Genetica* **38**: 221-228.

Orr, M.Y. (1933). Plantae Chinenses Forrestianae: Coniferae. *Notes Royal Bot. Gard. Edinb.* **18**: 119-158.

Page, C.N. (1982). *The Ferns of Britain and Ireland*. Cambridge: Cambridge University Press.

Page, C.N. (1986a). Pteridophyta. *In* S.M. Walters *et al.* (eds), *The European Garden Flora*. Cambridge: Cambridge University Press.

Page, C.N. (1986b). Coniferae. *In* S.M. Walters *et al.* (eds), *The European Garden Flora*. Cambridge: Cambridge University Press.

Page, C.N. (1988a). *Ferns. Their Habitats in the Landscape of Britain and Ireland*. London: Collins New Naturalist Series.

Page, C.N. (1988b). The new conifer group. *Species – Newsletter of the Species Survival Commission, I.U.C.N.* **10**: 21.

Page, C.N. (1989). The role of Edinburgh Royal Botanic Garden in the international conservation of conifers. *International Dendrology Society Yearbook* 1989: 112-115.

Page, C.N. (1991). Cornish gardens as green banks for the survival of temperate rainforest trees. *The Cornish Garden* **34**: 5-9.

Page, C.N. (1992a). Many benefits of tree 'green banks'. *The West Briton*, 16 Jan 1992.

Page, C.N. (1992b). A Cornwall-linked botanic garden and temperate rainforest tree conservation strategy into the 21st century. *The Cornish Garden* **35**: 69-75.

Page, C.N. (1994). The ex-situ conservation of temperate rainforest conifer tree species: a British-based programme. *Biodiversity and Conservation* **2**.

Page, C.N. & Clifford, H.T. (1981). Ecological biogeography of Australian conifers and ferns. *In* A. Keast (ed), *Ecological Biogeography of Australia*, pp. 472.-498. The Hague: W. Junk.

Page, C.N., Dyer, A.F., Lindsay, S. & Mann, D.G. (1992). Conservation of pteridophytes: the *ex situ* approach. *In* J.M. Ide, A.C. Jermy & A.M. Paul (eds), *Fern Horticulture: Past, Present and Future Perspectives*. Proceedings of the International Symposium on the Cultivation and Propagation of Pteridophytes, London, 7-11 July 1991, pp. 269-278. Andover: Intercept.

Page, C.N. & Farjon, A. (1991). The Conifer Specialist Group. *Species – Newsletter of the Species Survival Commission* **17**: 70-71.

Pastor, J. & Pos, W.M. (1988). Response of northern forests to CO_2-induced climatic change. *Nature* **334** (6177): 55-58.

Ray, D. & Gardner, M.F. (1993). The conifer conservation programme – an overview. *Scottish Forestry* **47**: 175-182.

Rook, D. & Page, C.N. (in press). The importance of arboreta in the context of global warming. *Symposium Arboretum Musilla, Finland, August 1992*.

Ruby, J.L. & Wright, J.W. (1976). A revised classification of geographic varieties of Scots Pine. *Silvae Genetica* **25**: 169-175.

Saki. K.-I., Miyazaki, Y. & Matsura, T. (1971). Genetic studies in natural populations of forest trees. 1. Genetic variability on the enzymatic level in natural forests of *Thujopsis dolabrata*. *Silvae Genetica* **20**: 168-173.

Soule, M.E. (ed) (1986). *Conservation Biology. The Science of Scarcity and Diversity*. Sunderland, Mass.

Soule, M.E. (ed) (1987). *Viable Populations for Conservation*. Cambridge: Cambridge University Press.

Steven, H.M. & Carlisle, A. (1959). *The Native Pinewoods of Scotland*. Edinburgh: Oliver & Boyd.

Stirling-Maxwell, Sir J. (1932). The influence of exotic conifers on silviculture in the British Isles. *In: Conifers in Cultivation. Report of the 1931 Conifer Conference,* pp. 43-54. London: Royal Horticultural Society.

Szmidt, A.E. & Wang, X.-R. (1993). Molecular systematics and genetic differentiation on *Pinus sylvestris* (L.) and *P. densiflora* (Sieb. & Zucc.). *Theor. Appl. Genet.* **86**: 159-165.

Thielges, B.A. (1972). Intraspecific variation in foliage polyphenols of *Pinus* (subsection *Sylvestres*). *Silvae Genetica* **21**: 114-119.

Thurston, E. (1930). *British and Foreign Trees and Shrubs in Cornwall.* Cambridge: Cambridge University Press.

Veitch, J.T. (1892). The coniferae of Japan. *Journal of the Royal Horticultural Society* **14**: 19-38.

Veitch, J.T. (1906). *Hortus Veitchii.* London: James Veitch and Son.

Walters, S. M., Brady, A., Brickell, C.D., Cullen, J., Green, P.S., Lewis, J., Matthews, V.A., Webb, D.A., Yeo, P.F. & Alexander, J.C.M. (eds) (1986). *The European Garden Flora.* Vol. 1. Cambridge: Cambridge University Press.

Webster, A.D. (1896). *Hardy Coniferous Trees.* London: Hutchinson.

18

Hedera

H.A. MCALLISTER

Botanic Gardens, University of Liverpool, Ness, Neston, South Wirral, Cheshire, L64 4AY

Introduction

This study of ivies was initiated by Alison Rutherford in 1974 in connection with the Irish Ivy Survey (Rutherford 1979), the Irish Ivy, *Hedera hibernica* (Fig. 1a), being a commonly cultivated and very frequently naturalized ivy. Jacobsen (1954) had reported that *Hedera helix* was diploid with 2n=48 while *H. hibernica* was tetraploid with 2n=96. If this could be confirmed, it suggested an absolute way of distinguishing between the two species.

Results

These counts were confirmed and collections of wild local ivies made. By chance, one of the first of these to be counted was from Cnicht in Snowdonia and proved to be tetraploid though it looked like *H. helix* and not at all like the commonly cultivated Irish Ivy. This led to the discovery that the native ivies of the south-west, from the Isle of Wight, West Country, Ireland, Isle of Man and West Wales to the Solway coast of Scotland, were tetraploid, while the native ivies to the east and north were diploid. A comparison of the two cytotypes showed that they could be distinguished by the form of their trichomes, often referred to as 'scale-hairs' in ivies. In the diploids (*H. helix*) the rays of the trichomes stand out at an angle to the surface of the leaf (Fig. 1b) while in the tetraploids (*H. hibernica*) the rays of the trichomes on the leaf lamina lie parallel and often appressed to the leaf surface (Fig. 1c) (McAllister & Rutherford 1990). The Irish Ivy appears to be a distinctive clone, or more probably a group of clones, of *H. hibernica* whose leaves are of a uniform, semi-adult form.

Thanks to members of the BSBI and other friends, wild ivies kept arriving from the Atlantic islands, N. Africa, Europe and Asia. Chromosome counts were made of these, and their morphology, especially of the trichomes, studied. It was noticed that the native ivies from north of the ancient Tethys sea (its remains are the Mediterranean, Black and Caspian Seas and the Himalayas) have few, long, white hairs on each trichome (Fig. 1d) while species from south of the Tethys have numerous, short, orange-red hairs which are more or less laterally fused on each trichome. Only in the Caucasus do species of the two groups meet, with polyploid species of presumed hybrid derivation, with species of both groups in their

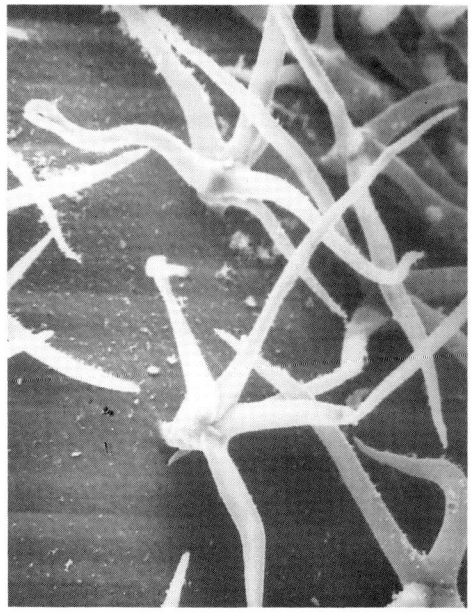
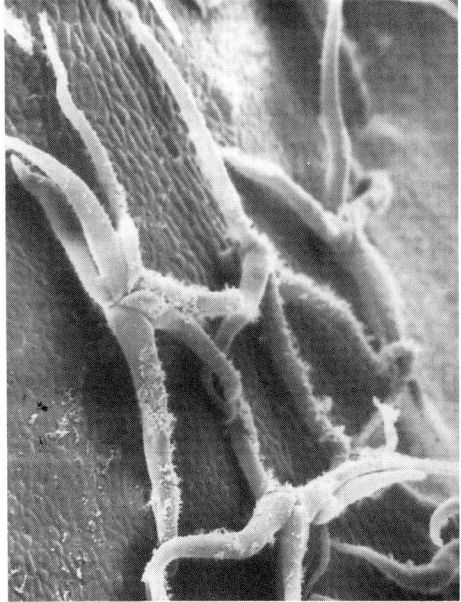
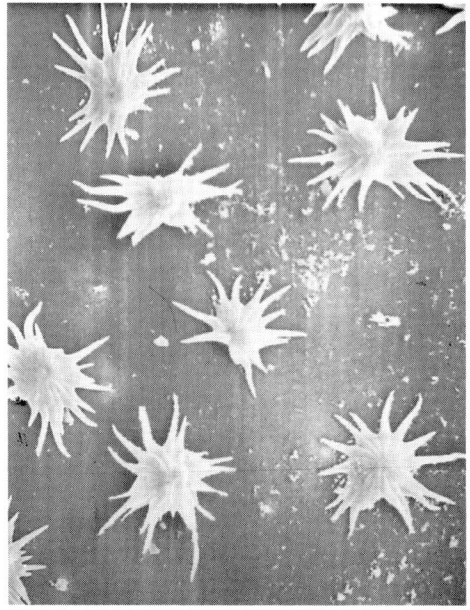

Fig. 1. *a* (top left), *Hedera hibernica* (2n=96), Wirral, Cheshire, herbarium specimen; b (top right), *H. helix* (2n =48), Helensburgh, Dunbartonshire, SEM of trichomes; *c* (bottom left), *H. hibernica* (2n=96), Cranstal, Isle of Man, SEM of trichomes; *d* (bottom right), *H. maderensis*, near Funchal, Madeira, SEM of trichomes.

ancestry, on the Atlantic coast of Europe. Thus *H. azorica* and *H. helix* are diploids of the first type, and presumably closely related, while the tetraploid *H. hibernica* and the hexaploid *H. maderensis* are probably of inter-group hybrid derivation.

Hedera canariensis in the Canaries, *H. algeriensis* and *H. maroccana* in north Africa, *H. cypria* from Cyprus, *H. colchica* and *H. pastuchovii* in the Caucasus, *H. nepalensis* in the Himalayas and China, *H. shensiensis* (if this is a distinct species) in China and *H. rhombea* in Japan, Korea and Taiwan, all belong to the southern group (Rutherford, McAllister & Mill 1993).

Macaronesian and North African ivies

This gives a very different picture from that usually described for North Africa and Macaronesia (the Atlantic islands), from where all ivies are often referred to *H. canariensis*. The diploid *H. azorica* is quite distinct from all other species in this area in its trichomes which give the shoot tip a furry appearance. Its large, matt leaves are variable in the degree of lobing, ranging from almost unlobed in a plant from the island of Pico to having five or seven deep and almost equal lobes as in the early introduction from São Miguel.

The diploid *H. canariensis* from the Canary Islands also has large, matt leaves (but very different trichomes such that the shoot tip appears almost glabrous to the naked eye) but is very uniform throughout the three islands in which it occurs, its leaves having three, shallow, forward-pointing lobes and very rounded cordate bases.

The hexaploid *H. maderensis* is also very uniform throughout Madeira with typical, five-lobed, ivy-shaped leaves, very similar in general appearance but slightly smaller than those of the Irish Ivy.

In North Africa, the glossy-leaved tetraploid *H. algeriensis* (often wrongly labelled *H. canariensis* and usually seen in its variegated form, 'Gloire de Marengo') has a characteristic triangular-shaped, three-lobed leaf with the lateral lobes pointing obliquely forwards (McAllister 1988). It is probably the most distinctive ivy, being easily recognizable by both the shape and glossiness of its leaves.

The diploid, large-leaved species from the Atlas Mountains (*H. maroccana*) has typically ivy-shaped, 5-lobed leaves similar to those of *H. maderensis* but larger and with the veins standing proud on the upper surface near the junction with the petiole. What may be a single clone of this species with very attractive, neat, thin leaves is very commonly cultivated throughout southern Spain and Portugal. Despite its fresh, green appearance, it is very drought tolerant, sometimes being found in association with cacti.

Ivies in the Iberian Peninsula

As is so often the case, the situation in the Iberian Peninsula is very interesting. In the north and the Pyrenees the tetraploid *H. hibernica* is found; in the Sierra Nevada and other mountain ranges in the south unusual variants of the diploid *H. helix* occur. One type, with rather shallowly lobed leaves is similar to *H. helix* and is found throughout the

Mediterranean region. The other has small glossy leaves, very white veins, and shoots which become positively geotropic (i.e. grow downwards) during the winter and is differentiated as *H. helix* ssp. *rhizomatifera*. Hexaploids, found around Algeciaras in southern Spain, Monchique in south-west Portugal and in the Sierra do Sintra west of Lisbon, are referred to *H. maderensis* ssp. *iberica*. Their large leaves are variable in shape, but are often long-lobed with trichomes similar to those of the Madeiran ivy.

Sicilian Ivy

All wild ivies collected on Sicily, though morphologically *H. helix* on trichome characteristics, are tetraploid and rather larger-leaved than is usual for *H. helix*. Though the plants grown at Ness are now over ten years old and have flowered, no fruit has been set and it is not known whether they are black- or yellow-fruited. The yellow-fruited ivies so far examined have been diploid and morphologically typical of Mediterranean *H. helix* and are probably merely a yellow-berried form of the Common Ivy. However, an ivy with large yellow fruits has been described from the Caucasus but no material of this has been seen.

The Cyprus Ivy

The native ivy on Cyprus has always been described as *H. helix*, but a glance at the excellent non-flowering specimens collected on Cyprus by R.D. Meikle (**K**) showed trichomes of the southern type. Through the kindness of Mr Meikle and his contacts in Cyprus, seed was obtained in the spring of 1978, making the first known introduction to this country of the endemic Cyprus Ivy. It proved to be hexaploid (2n=144) with triangular leaves which were much longer than broad and with attractive whitish markings over the veins. Its closest relationships seem to be with *H. pastuchovii* of the Caucasus, but that species lacks the whitish markings and has the very characteristic smell of *H. colchica* which is absent from the Cyprus plant. The Cyprus Ivy has therefore been described as a new species *H. cypria*.

Key to species of *Hedera*

The young shoot and leaves of all species of *Hedera* bear hairs, which consist of a more or less obvious stalk, a multicellular central boss, and 3-20 unicellular radiating rays which are fused for varying proportions of their length. The size of these hairs, the number of rays, and the degree of fusion of these rays are useful taxonomic characters most easily seen in slightly expanded young leaves on the underside of which the individual hairs can be clearly distinguished.

1. Hairs white in living state, the majority on any plant with 4-8(-10) rays and 0.5-0.9 mm in diameter; rays appressed or not (Figs. 1b, 1c) 2
   Hairs orange (living ray cells with orange-brown sap), the majority on any plant with 8-18 rays 0.2-0.7 mm in diameter; rays always appressed to leaf surface (Fig. 1d) . 6

2. Rays of hairs on underside of lamina lying in one plane parallel to the leaf surface (Fig. 1c) (occasional hairs on veins may have rays standing out at an angle to the leaf surface), leaf lobes often as broad as long (Fig. 1a) .... *hibernica* (2n=96)
   Rays of hairs on underside of lamina sticking out at all angles to the leaf surface (Fig. 1b), leaf lobes usually longer than broad 3

3. Leaves usually small, 5-8(-10) cm across, more or less glossy, terminal lobe longer and broader than lateral lobes making leaves 'ivy-shaped' .... (*helix s.l.*) 4
   Leaves usually larger (more than 8 cm across), matt, lobes more or less equal so that leaf is not 'ivy-shaped' .... *azorica* (2n=48)

4. Fruit black, leaves usually dark green, 3-5 lobed 5
   Fruit yellow or orange, leaves barely lobed, heart-shaped, paler grass green ....
   .... *helix* ssp. *poetarum* (2n=48)

5. Plant not rhizomatous, very variable .... *helix* ssp. *helix* (2n=48, (96))
   Plant producing long (to over 30 cm) white downward growing rhizomes in winter, leaves small (up to 5.5 × 4.5 cm) with conspicuous white markings over the veins on the upper surface (S. Spain, Sierra Nevada, Huelva) ....
   .... *helix* ssp. *rhizomatifera* (2n=48)

6. Leaves basically triangular in outline, obviously longer than broad, not very large, mostly less than 10 cm long 7
   Leaves usually broader, ivy-shaped (3-5 lobed) or broadly cordate, small or large 9

7. Plant slender (petioles less than 1 mm in diameter), leaves thin, greyish yellow green with greyish markings over veins, fruit yellow-orange .... *nepalensis* (2n=48)
   Plant parts thicker (petioles more than 1 mm in diameter), leaves thick, leathery with or without whitish markings over veins, fruit black 8

8. Distinct white markings over veins in young leaves (Cyprus) .... *cypria* (2n=144)
   Markings over veins less distinct (Caucasus, Elburz Mts.) .... *pastuchovii* (2n=144)

9. Plant similar in stature to *H. helix* or smaller (leaves mostly less than 8 cm across), leaves 'ivy-shaped', 3-5 lobed 10
   Plant more robust than *H. helix* with many leaves more than 8 cm across, leaves not normally 'ivy-shaped' except *H. maderensis* 11

10. Fruit orange-yellow (Himalayas, China) .... *nepalensis* (2n=48)
    Fruit black (Japan, Korea, Taiwan) .... *rhombea* (2n=48)

11. Leaves very large when well grown (at least in cultivated clones of *H. colchica*), usually more than 10 cm across, 0-3 lobed 12
    Leaves smaller, usually less than 10 cm across when well grown, 3-5 lobed 14

12. Leaves very glossy, mostly truncate at base .... *algeriensis* (2n=96)
    Leaves matt, cordate at base 13

13. Leaves very thick and leathery, bases somewhat cordate, lobes acute . *colchica* (2n=192)
    Leaves thinner, cordate 'lobes' large, hemispherical, leaf lobes blunt *canariensis* (2n=48)

14. Leaves 'ivy-shaped', (3-)5 lobed, cordate at base 15
    Leaves 3- or variably lobed and toothed, truncate or somewhat cuneate at base and so not typically ivy-shaped .... *maderensis* ssp. *iberica* ined. (2n=144)

15. Well developed leaves less than 10 cm across, young stems and petioles green .... *maderensis* (2n=144)
    Well developed leaves more than 10 cm across, young stems and petioles red .... *maroccana* (2n=48)

References

Jacobsen, P. (1954). Chromosome numbers in the genus *Hedera* L. *Hereditas* **40**: 252-254.
McAllister, H.A. (1988). Canary and Algerian ivies. *Plantsman* **10**(1): 27-29.
McAllister, H.A. & Rutherford, A. (1990). *Hedera helix* L. and *H. hibernica* (Kirchner) Bean (Araliaceae) in the British Isles. *Watsonia* **18**(1): 7-15.
Rutherford, A. (1979). The BSBI Irish Ivy Survey. *BSBI News* **22**: 8-9.
Rutherford, A., McAllister, H.A. & Mill, R.R. (1993). New ivies from the Mediterranean area and Macaronesia. *Plantsman* **15**(2): 115-128.

19

The genus *Cotoneaster* in the British Isles

JEANETTE FRYER

Cornhill Cottage, Honeycritch Lane, Froxfield, Petersfield, Hampshire, GU32 1BE

ABSTRACT. Identification of the estimated 300 taxa within the genus *Cotoneaster*, many of which are apomicts, can be difficult but Professor Karl-Evert Flinck and Dr Bertil Hylmö devised a workable method of separating the genus into sections, subsections and series and succeeded in making order out of chaos (Flinck & Hylmö 1966). Forty-five taxa of *Cotoneaster* are included in Professor Clive Stace's *New Flora of the British Isles* (Stace 1991). The following is a classification of these 45 taxa using Flinck & Hylmö's scheme.

Classification

Section 1 ***Cotoneaster***
   Petals erect, obovate. Flowering of inflorescence extended.

  Subsection A ***Adpressi***
   Flowers usually with 10-15 stamens.

   Series ***Distichi***
      Species mostly from Yunnan and Burma. Stipules persistent. Flowers nodding.

      1. C. nitidus Jacq.
         Flowers single, leaves and fruits shiny.

   Series ***Horizontales***
      Species mostly from the Hubei Province in China. Branches frequently herring-bone in habit, not rooting at the nodes. Flowers erect.

2. *C. horizontalis* Decne.
This species is well known. Height to 1 m.

3. *C. atropurpureus* Flinck & Hylmö
More prostrate than *C. horizontalis*. Leaves dark green, apex rounded, often truncate. Petals with a black spot at the base.

4. *C. hjelmqvistii* Flinck & Hylmö
Much taller, to 4 m if supported. Leaves shiny clear green, saucer-shaped, up to 2 cm long.

5. *C. divaricatus* Rehd. & Wils.
Forms a dense bush. Fruits oblong ellipsoid, becoming dark red-maroon.

Series *Adpressi*
Species from India, Pakistan and China. Branching irregular. Rooting at the nodes. Leaves undulate.

6. *C. adpressus* Bois
An outbreeding diploid, variable. Low-growing shrub.

7. *C. nanshan* A. Vilm. ex Mottet
Flowers much larger than in *C. adpressus*. Petals with ragged edges. Larger in all parts.

Series *Nitentes*
Most species from the Sichuan Province in China. Fruits purple-black.

8. *C. nitens* Rehd. & Wils.
Prettier in flower than in fruit. Forms a rounded shrub up to 3 m high.

Subsection B *Cotoneaster*
Flowers with 20 stamens.

Series *Lucidi*
Mainly from China, the exception being *C. lucidus* from Siberia. Shrubs to 4 m. Fruits shiny-black. Early leaf fall.

9. *C. lucidus* Schlect.
One of the earliest to flower. Fruit with three nutlets. Leaves with good autumn colour.

10. *C. laetevirens* (Rehd. & Wils.) Klotz
Taller than *C. lucidus* and with narrower leaves. Nutlets usually two per fruit.

11. C. villosulus (Rehd. & Wils.) Flinck & Hylmö
Leaves somewhat bullate. Fruits obovoid, nutlets two.

Series ***Cotoneaster***
Species from Europe. Young branches and lower surface of leaves tomentose.

12. C. cambricus Fryer & Hylmö ined.
(Syn.: *C. integerrimus* Medic. var. *anglicus* Hrab.-Uhr.)
Endemic to the Great Orme, North Wales. Differs from *C. integerrimus* of continental Europe in the colour of its wood, leaf shape and colour, indentation of the veins, number of flowers in the inflorescence, pedicel and peduncle length, petal : sepal length ratio, fruit size and nutlet number per fruit.

Series ***Mucronati***
Species from Sikkim, Assam and Yunnan. Stiff, erect shrubs. Calyx ± campanulate.

13. C. mucronatus Franch.
Diploid. Can be very variable. Fruits red, to 12 mm, nutlets 2-3.

14. C. simonsii Baker
Leaves and flowers smaller than in *C. mucronatus*. Fruits to 8-10 mm, orange-red, nutlets 3-4.

Series ***Bullati***
Majority of species from Sichuan. Height to 5 m. Leaves bullate. Up to 5 nutlets per fruit.

15. C. bullatus Bois
Leaves to 7 cm. Fruits red.

16. C. rehderi Pojark.
Larger in all parts that *C. bullatus*. Leaves to 15 cm, more bullate. Fruits darker red when fully ripe.

17. C. moupinensis Franch.
Similar to *C. rehderi* but has black fruits.

Series ***Dielsiani***
Central China, mostly Sichuan. Deciduous shrubs. Leaves yellowish-grey, villose-tomentose on lower surface.

18. C. dielsianus E. Pritzel ex Diels
Up to about 3 m high. Flowers 3-7 per inflorescence. Fruits rich red.

19. C. splendens Flinck & Hylmö
Height 1-1.5 m. Flowers 1-3 per inflorescence. Fruits more orange-red.

Series *Franchetioides*
Majority of species from Yunnan. Evergreen shrubs. Leaves silvery-white tomentose on lower surface.

20. C. franchetii Bois
An elegant, arching shrub. Anthers pale mauve. Fruits obovoid, orange-red.

21. C. amoenus Wils.
Pedicels and peduncles shorter. Anther pale mauve. Smaller and more densely branched than *C. franchetii*.

22. C. sternianus (Turrill) Boom
A stiffly erect shrub with thick, rugose leaves. Anthers white. Fruits suborbicular, red.

23. C. insculptus Diels
Anthers white. Leaves very shiny, veins deeply impressed. Lowers and more spreading than *C. sternianus*.

Series *Zabelioides*
Mainly from the northern provinces of China. Pedicels and peduncles thin and long. Flowers more open. Petals pale pink.

24. C. zabelii Schneid.
Shrub to 2 m in height.

Section 2 *Chaenopetalum*
Petals spreading, suborbicular. Flowers opening all together.

Subsection C *Chaenopetalum*
Leaves over 2 cm.

Series *Multiflori*
Species from Central Asia. Nutlets 2, stuck together as one.

25. C. multiflorus Bunge
A beautiful shrub. Flowers white, 8-10 mm across. Fruits shiny carmine, 10-12 mm.

Series *Insignes*
Species from middle-eastern countries. Leaves pubescent on lower surface. Fruits black with apex slightly open.

26. *C. ellipticus* (Lindl.) Loud.
A wide-spreading shrub with dull black fruits.

27. *C. hissaricus* Pojark.
Smaller in all parts than *C. ellipticus*. Fruits more purple-black.

Series *Cooperi*
Majority of species from Bhutan. Leaves with lower surface soon becoming glabrous. Apex of fruit closed.

28. *C. transens* Klotz
Large shrub or small tree. Fruits remaining red until late in the season, finally turning purple-black.

Series *Bacillares*
Species from the Himalayas. Fruits black, apex wide open.

29. *C. bacillaris* Wall. ex Lindl.
A tall many-stemmed shrub. Leaves obovate. Fruits to 8 mm.

30. *C. ignotus* Klotz
A tree, usually with single trunk. Leaves larger and more rounded than those of *C. bacillaris*.

31. *C. obtusus* Wall. ex Lindl.
Smaller, a medium-sized shrub. Leaves yellowish-green when young. Pedicels and peduncles very red in fruit.

32. *C. affinis* Lindl.
Similar to *C. bacillaris* but with longer leaves which are more pubescent on both surfaces. Fruits larger than in *C. bacillaris*.

Series *Chaenopetalum*
Species also from the Himalayas. Deciduous large shrubs and small- to medium-sized trees. Fruits red or yellow.

33. *C. frigidus* Wall. ex Lindl.
A very variable outbreeding diploid with many cultivars.

34. *C.* × *watereri* Exell
Hybrids between *C. frigidus* and *C. salicifolius*, which sometimes occur.

Series *Salicifolii*
Species from the Sichuan and Hubei Provinces. Evergreen. Leaves narrow, often somewhat rugose. Fruits red or yellow.

35. *C. salicifolius* Franch.
Outbreeding diploid, very variable. There are also many cultivars of this species.

Series *Pannosi*
Species from Yunnan. Evergreen. Leaves leathery, tomentose on lower surface.

36. *C. pannosus* Franch.
Pendulous shrub, 2-3 m high.

37. *C. lacteus* W. Smith
Larger and more erect than *C. pannosus*. Young growth covered with tawny hairs. Leaves with veins more deeply impressed.

Series *Radicantes*
Species from central China. Prostrate or trailing shrubs. Nutlets more than 2 per fruit.

38. *C. dammeri* Schneid.
An outbreeding diploid. A carpeting species.

39. *C.* × *suecicus* Klotz
Hybrids between *C. dammeri* and *C. conspicuus* sometimes occur.

Subsection D *Microphylli*
Leaves 2 cm or less.

Series *Microphylli*
From China, Tibet, India, Nepal and Burma. Low evergreen shrubs with leathery leaves. Nutlets 2 per fruit.

40. *C. conspicuus* Marq.
A very variable outbreeding diploid. Fruits orange. Many cultivars.

41. *C. integrifolius* (Roxb.) Klotz
Leaves dark, shiny, fruits large, carmine. This species is frequently confused with *C. microphyllus* which has more rounded leaves and cherry-red fruits.

42. *C. linearifolius* (Klotz) Klotz
Smaller in all parts than *C. integrifolius*. Fruits minute, bead-like.

43. C. congestus Baker
Leaves dull, paler green. Fruits red. Very shy fruiting.

44. C. buxifolius Wall. ex Lindl.
Not a common species. Fruits very dark red-maroon.

45. C. cashmiriensis Klotz
A ground-hugging species with bright red fruits.

All taxa mentioned have been found bird-sown, naturalized, or in the case of *C. cambricus*, wild, in the British Isles.

References

Flinck, K.E. and Hylmö, B. (1966). A list of Series and Species in the genus Cotoneaster. *Botaniska Notiser* **119**(3).

Stace, C. (1991). *New Flora of the British Isles*. Cambridge: Cambridge University Press.

Index

Common English names and a few other non-scientific references are printed in Roman type. Scientific names are in *italics*. Figures in Roman type refer to text entries, those in *italics* refer to distribution maps, and those in **bold** refer to illustrations.

Abies, 120, 134, 136
Acacia, 23
   cyclops, 23
Acaena, 23
Acer campestris, 9
   pseudoplatanus, 20, 42
Achillea millefolium, 36
   ptarmica, 14
      'The Pearl', 14
Acmopyle, 133, 135
Actinostrobus, 134, 136
Adiantum capillus-veneris, 56
Aesculus hippocastanum, 20, 32
Aethusa cynapium ssp. *agrestis*, 36
Afrocarpus, 133, 135
Agathis, 133, 135
Ajuga reptans, 10, 11
   'Argentea', 10
   'Burgundy Glow', 11
   'Pink Elf', 13
   'Rainbow', 10
Alchemilla alpina, 9
Alder, 11, 13
   Cut-leaved, **12**
Allium, 92
   ampeloprasum ssp. *babingtonii*, 20
   grosii, 107
   roseum, 108
Alnus glutinosa, 11, **12**, 13
Alopecurus pratensis
   'Aureo-variegatus', 10
Amentotaxus, 133, 135
Amomyrtus luma, 23
Amphipod, 19
Andromeda polifolia
   Macrophylla', 13
Anemone, Wood, 14
Anemone blanda, 14, 93, 103, 108
   nemorosa, 14

   'Robinsoniana', 14
   'Virescens', 14
Anthoxanthum odoratum, 8
Anthyllis vulneraria, 13
   ssp. *carpatica* var.
      pseudovulneraria, 35
   ssp. *polyphylla*, 35
Aphanes arvensis, 8
Apple, 20
Aquilegia vulgaris, 14
   'Nora Barlow', 14
Araucaria, 133, 135
   araucana, **138**
Arbutus unedo, 27
Archangel, Yellow, 10, 48
   Variegated, 48
Arcitalitrus dorrieni, 19
Arctostaphylos uva-ursi, 9
Arisaema, 107
Armadillidium nasatum, 19
Arrhenatherum elatius, 10
Artemisia, 92
   absinthium, 11
      'Lambrook Silver', 11
Arum creticum, 108
Ash, 9, 12
   One-leaved, **12**
Aspen, 12, 13
Asphodelus bento-rainhae, 107
Asplenium ruta-muraria, 51
   scolopendrium
      'Bolton's Nobile', 54
      'Crispum', 51, 54
      'Fred Jackson', 54
      'Multifidum', 51
      (Ramosum group), 54
   septentrionale, 55
   trichomanes, 54
      (Incisum group), 54
      'Percy Greenfield', 54
      ssp. *pachyrachis*, 55

   var. *subequale*, 55
Athrotaxis, 134, 136
   selaginoides, **137**
Athyrium filix-femina
   'Kalothrix', 51
   'Victoriae', 54
Austrocedrus, 134, 136
Austrotaxus, 133, 135
Avens, Mountain, 9
   Water, 78
      (double-flowered), 78
      (hose-in-hose), 78
Azolla filiculoides, 56
Balsam, Indian, 42
Bearberry, 9
Beech, 9-12
   Cut-leaved, **12**
Begonia, 92
Bellevalia hackelii, 107
Bellflower, Adria, 48
   Clustered, 9
   Trailing, 48
Bellis perennis, 7, 14, 15
   'Dresden China', 14
   'Hen and Chicks', **Plate 1**
   'Tuberosa Monstrosa', 14
Betula, 92
   nana, 9
   pendula, 9, 11-13
      'Tristis', 12
      'Youngii', 12
Biarum, 107
   davisii, 108
Bindweed, Hairy, 7
Birch, Dwarf, 9
   Silver, 9, 11-13
Bird's-foot-trefoil, 35
   Alien, 35, 36
Bistort, Common, 12
Blackthorn, 11
Blechnum spicant, 129
Blue-sow-thistle, Common, 31

Bluebell, 33, 98
   Bracteate, 78, **79**
Bog-rosemary, 13
Box, 9, 12
Brassica napus ssp. *oleifera*, 46
Broom, 13
Buddleja asiatica, 23
   davidii, 21, 23
   variabilis, 21
Bugle, 10, 11, 13
Burnet, Fodder, 37
   Salad, 37
Butcher's-broom, 15
Buttercup, Creeping, 14
     (double-flowered), 78
   Hairy, 37
   Meadow, 14
   St Martin's, 37
Butterfly-bush, 21
Buxus sempervirens, 9, 12
Callicoma serratifolia, 19
Callitris, 134, 136
Calluna vulgaris, 10, 11, 14
   'H.E. Beale', 14
   'Silver Queen', 11
   'Sister Anne', 11
Calocedrus, 134, 136
   decurrens, 129, **138**
Caltha palustris, 8, 14
Calystegia pulchra, 7
Campanula glomerata, 9
   gracilis, 19
   portenschlagiana, 48
   poscharskyana, *44*, 48
   rotundifolia, 98
Campion, Red, 13, 14
   Sea, 14
Canary-grass, Reed, 10
Cannabis, 90
Cannabis sativa, 46
Capsella bursa-pastoris, 7
Cardamine pratensis, 14
Carex bigelowii, 10
   ornithopoda, 10
   pendula, 8
Carpinus betulus, 9, 11-13
Carpobrotus, 22
Carrot, Wild, 36
Castanea sativa, 111
Castanospermum australe, 90
Catchfly, Alpine, 9
   Sticky, 14
Catharanthus roseus, 90
Cathaya, 134, 136
Cedar, Incense, **138**
Cedrus, 134, 136

atlantica, 129
libani, 129
Celandine, Lesser, 11, 13, 14
Centaurea cyanus, 36
   nigra, 38
Cephalotaxus, 133, 135
Chamaecyparis, 134, 136
Chasmanthe, 22
Cheiranthus cheiri, 20
Cherry, Bird, 11
   Wild, 9, 12
Chestnut, Sweet, 111
Chrysanthemum segetum, 36
Cicerbita macrophylla, 31
Cirsium oleraceum, 46
   vulgare, 23
Cleavers, 7
Clover, 32, 35
   Red, 11, 37
   White, 11
Cock's-foot, 10
Cocos nucifera, 23
Colchicum corsicum, 107
   cousturieri, 107
Columbine, 14
Comfrey, 65-70
   Caucasian, 70
   Creeping, 69
   Prickly, 68
   Russian, 68
   White, 70
Convallaria majalis, 9
Cornflower, 36
Cortaderia selloana, 23
Corylus avellana, 11, 12, 15
Cotoneaster, 20, 45, 151-157
   adpressus, 152
   affinis, 155
   amoenus, 154
   atropurpureus, 152
   bacillaris, 155
   bullatus, 153
   buxifolius, 157
   cambricus, 153, 157
   cashmiriensis, 157
   congestus, 157
   conspicuus, 156
   dammeri, 156
   ×*C. conspicuus*, 156
   dielsianus, 153
   divaricatus, 152
   ellipticus, 155
   franchetii, 154
   frigidus, 155
   ×*C. salicifolius*, 155
   hissaricus, 155

hjelmqvistii, 152
horizontalis, 152
ignotus, 155
insculptus, 154
integerrimus, 153
   var. *anglicus*, 153
integrifolius, 156
lacteus, 156
laetevirens, 152
linearifolius, 156
lucidus, 152
microphyllus, 156
moupinensis, 153
mucronatus, 153
multiflorus, 154
nanshan, 152
nitens, 152
nitidus, 151
obtusus, 155
pannosus, 156
rehderi, 153
salicifolius, 156
Section *Chaenopetalum*, 154
Section *Cotoneaster*, 151
Series *Adpressi*, 152
Series *Bacillares*, 155
Series *Bullati*, 153
Series *Chaenopetalum*, 155
Series *Cooperi*, 155
Series *Cotoneaster*, 153
Series *Dielsiani*, 153
Series *Distichi*, 151
Series *Franchetioides*, 154
Series *Horizontales*, 151
Series *Insignes*, 155
Series *Lucidi*, 152
Series *Microphylli*, 156
Series *Mucronati*, 153
Series *Multiflori*, 154
Series *Nitentes*, 152
Series *Pannosi*, 156
Series *Radicantes*, 156
Series *Salicifolii*, 156
Series *Zabelioides*, 154
simonsii, 153
splendens, 154
sternianus, 154
Subsection *Adpressi*, 151
Subsection *Chaenopetalum*, 154
Subsection *Cotoneaster*, 152
Subsection *Microphylli*, 156
×*suecicus*, 156
transens, 155
villosulus, 153
×*watereri*, 155

zabelii, 154
Couch, Common, 7
Cowslip, 9, 13, 33, 38
Crane's-bill, Meadow, 14
Crassula helmsii, 27, **44**, 47
Crataegus, 98
    laevigata, 14
        'Paul's Scarlet', 14
    monogyna, 12, 13, 15
        'Pendula Rosea', 12
Creeping-Jenny, 9, 10
Crepis setosa, 37
Crocosmia, 22, 26, 90
    aurea, 24, **25**
    ×*crocosmiiflora*, 24, **25**, 26
    pottsii, 24, **25**, 26
Crocus abantensis, 108
    baytopiorum, 108
    biflorus ssp. *crewei*, 108
        ssp. *pulchricolor*, 108
    etruscus, 107
    gargaricus, 108
    olivieri ssp. *istanbulensis*, 108
Cryptomeria, 134, 136
    japonica, 129
Cuckooflower, 14
Cunninghamia, 134, 136
Cupressus, 134, 136
Cyclamen, 105, 107
    cilicium, 108
    collector, **106**
    coum, 108
    hederifolium, 103
    mirabile, 108
    persicum, 108
    pseudibericum, 108
    trochopteranthum, 108
Cypripedium calceolus, 86
Cyrtomium falcatum, 56
Cytisus scoparius, 13
Dacrycarpus, 133, 135
Dacrydium, 133, 135
Dactylis glomerata, 10
Dactylorhiza praetermissa, 86
Daisy, 7, 14
    Hen and Chicks, 15
    Oxeye, 37
    Shasta, 37
Daucus carota, 36
Deschampsia cespitosa, 15
Dianthus caryophyllus, 20
    deltoides, 9
Dicksonia antarctica, **128**, 129
Digitalis, 90
    purpurea, 77
Diplacus aurantiacus, 62

glutinosus, 62
puniceus, 62
Diselma, 134, 136
Dock, Blood-veined, 11
    Fiddle, 35
Draba aizoides, 9
Dropwort, 14
Dryas octopetala, 9
Dryopteris aemula, 55, 129
    ×*brathaica*, 55
    filix-mas, 55
        'Bollandiae', 55
    ×*D. aemula*, 55
    ×*D. carthusiana*, 55
Ducampopinus, 134, 136
Echium pininana, 22
Elder, 10, 11
Elodea canadensis, 22
    nuttallii, 22
Elymus repens, 7
Epilobium brunnescens, 23
    montanum, 8
    nummulariaefolium, 23
Epipactis helleborine, 7
Eranthis hyemalis, 103, 108
Erica erigena, 27
    tetralix, 11
        'Alba Mollis', 11
    vagans, 10, 27
        'Valerie Proudley', 10
Erysimum cheiri, 20
Erythronium, 107
Euphorbia amygdaloides, 11
Fagus sylvatica, 9-11, **12**
    'Cockleshell', 11
    'Pendula', 12
    'Purpurea Pendula', 12
    'Zlatia', 10
Falcatifolium, 133, 135
Fallopia japonica, 21, 31, **44**, 47
Fascicularia bicolor, 23
    pitcairniifolia, 23
Fern, Killarney, 56
Feverfew, 77
Ficus carica, 111
Field-rose, 7
Fig, 111
Figwort, Water, 10
Filipendula ulmaria, 10
    vulgaris, 14
Fir, Douglas, **124**
Fitzroya, 134, 136
    cupressoides, **123**, 129
Flax, 46
    New Zealand, 23
    Perennial, 9

Fokienia, 134, 136
Foxglove, 77
Foxglove virus, 77
Foxtail, Meadow, 10
Fraxinus excelsior, 9, **12**
    f. *diversifolia*, 11
    'One-leaved Ash', 11
Fritillaria, 107
    alburyana, 108
    aurea, 108
    forbesii, 108
    meleagris, 9
    michailovskyi, 110
    sibthorpiana, 108
    viridiflora, 108
Fritillary, Snake's-head, 9
Fuchsia, 24, 26
    gracilis, **25**
    magellanica, 24, **25**
        var. *gracilis*, 24
    cf. *magellanica s.s.*, 24
    'Riccartonii', 24, **25**
Fumaria, 36
Fumitory, 36
Galanthus, 105, 107
    elwesii, 103
    fosteri, 108
    gracilis, 108
    nivalis, 20
Galium aparine, 7
    odoratum, 9
Galligaskins, 14
'Gardeners Garters', 10
Garlic, 103
Gaultheria mucronata, 23
    shallon, 22
Genista pilosa, 9
Gentian, Spring, 9
Gentiana verna, 9
Geranium pratense, 14
Geum rivale, 78
Ginkgo, 133, 135, 139
    biloba, 129
Glastonbury Thorn, 15
Glyptostrobus, 134, 136
Gorse, 13
Greenweed, Hairy, 9
Griselinia, 23
Groundsel, 78
Guelder-rose, 9, 10
Gunnera, 22
    manicata, 22
    tinctoria, 22, 27
Hair-grass, Tufted, 15
Hakea, 23
Halocarpus, 133, 135

Haloragis micrantha, 27
Haplophthalmus danicus, 19
Harebell, 98
Hart's-tongue, 9, 54, 78
Hawk's-beard, Bristly, 37
Hawthorn, 12, 13, 15
　Midland, 14
Hazel, 11, 12, 15, **107**
Heath, Cornish, 10
　Cross-leaved, 11
Heather, 10, 11, 14
Hebe, 23
Hedera, 145-150
　algeriensis, 147, 149
　　'Gloire de Marengo', 147
　azorica, 147, 149
　canariensis, 147, 149
　colchica, 147, 148, 149
　cypria, 147, 148, 149
　helix, 10, 145, **146**, 147, 148, 149
　　ssp. *helix*, 149
　　ssp. *poetarum*, 149
　　ssp. *rhizomatifera*, 148, 149
　　'Buttercup', 10
　helix s.l., 9, 11, 149
　hibernica, 145, **146**, 147, 148
　maderensis, **146**,147, 149, 150
　　ssp. *iberica*, 148, 150
　maroccana, 147, 150
　pastuchovii, 147-149
　nepalensis, 147, 149
　rhombea, 147, 149
　shensiensis, 147
Helianthus annuus, 46
Helictotrichon pratense, 37
Helleborine, Broad-leaved, 7
Hemlock-spruce, Western, **124**
Hemp, 46
Heracleum mantegazzianum, 18, 21
Hesperopeuce, 134, 136
　mertensiana, 129
Hippophae rhamnoides, 9
Holcus mollis, 10
Holly, 9, 11, 12
　Hedgehog, 11
Honeysuckle, 7
　Dutch, 15
Hop, 10
Hopper, Woodland, 19
Hornbeam, 9, 11-13
Horse-chestnut, 20, 32
Humulus lupulus, 10
Hyacinthoides non-scripta, **79**, 98
　vicentina, 107

Hyacinthus, 103
　orientalis, 108
Hymenophyllum wilsonii, 129
Hypericum, 71-75
　addingtonii, 75
　androsaemum, 72
　aegypticum, 74
　androsaemum, 71, 72
　×*arnoldianum*, 73
　barbatum, 72
　beanii, 74
　calycinum, 71, 74, 75, **Plate 4**
　canadense, 22
　× *cyathiflorum* 'Gold Cup', 75, **Plate 4**
　densiflorum, 73
　　×*H. kalmianum*, 73
　　×*H. lobocarpum*, 73
　× *dummeri*, 75, **Plate 4**
　elatum, 72
　elodes, 72
　forrestii, 74, **Plate 3**
　fragile, 73
　frondosum, 73
　× 'Gold Cup', 75
　× 'Hidcote', 74, 75, **Plate 4**
　'Hidcote Variety', 74
　hircinum, 71, 72
　　ssp. *albimontanum*, 72
　　ssp. *cambessedesii*, 72
　　ssp. *hircinum*, 72
　　ssp. *majus*, 72, **Plate 3**
　hookerianum, 75
　humifusum, 71, 72
　×*inodorum*, 72, **Plate 3**
　　'Elstead', 72
　kalmianum, 73
　× 'Lawrence Johnson', 75
　lobocarpum, 73
　nummularium, 72
　olympicum, 71, 72, 73
　　'Citrinum', 73
　　'Sulphureum', 73
　　f. *minus*, 73
　　f. *olympicum*, 73
　　f. *uniflorum*, 73
　patulum, 71, 74
　　var. *henryi*, 74
　　var. *forrestii*, 74
　polyphyllum, 73
　prolificum, 73
　　×*H. densiflorum*, 73
　pseudohenryi, 72
　pulchrum, 73
　richeri ssp. *richeri*, 72
　Section *Drosocarpium*, 72

Section *Myriandra*, 71, 73
　undulatum, 72
　×*vanfleetii*, 73
　xylosteifolium, 72, **Plate 3**
Ilex aquifolium, 9, 11, 12
　'Ferox', 11
　'J.C. van Tol', 11
Impatiens glandulifera, 21, 42
Inula salicina, 19
Iris, Yellow, 9, 10
Iris boissieri, 107
　lusitanica, 107
　marsica, 107
　pamphylica, 108
　pseudacorus, 9, 10
　Section *Oncocyclus*, 107
Ivy, 9-11, 145-150
　Cyprus, 148
　Irish, 145, 147
　Sicilian, 148
Jack-a-napes-on-horseback, 14
Jacob's-ladder, 9
Juglans regia, 111
Juncus effusus, 15
　planifolius, 19, 27
Juniper, 12, 120
　Irish, 13
Juniperus, 134, 136
　communis, 12
　　'Hibernica', 13
　　'Hornibrookii', 12
　　'Repanda', 12
Kalmia, 22
Keteleeria, 134, 136
Knapweed, Common, 38
Knotgrass, Cornfield, 36
Knotweed, Japanese, 31, 47
Lactuca sativa, **79**
Lady-ferns, Crested, 55
Lady's-mantle, Alpine, 9
Lagarostrobos, 133, 135
　franklynii, 129, **139**
Lamiastrum galeobdolon, 10
　'Silver Carpet', 10
　ssp. *argentatum*, 10, *44*, 48
Lampranthus, 22
Larix, 120, 134, 136
Lavatera olbia, 78
Leek, 103
Lepidothamnus, 133, 135
Lettuce, **79**
Leucanthemum maximum, 37
　×*superbum*, 37
　vulgare, 37
Leucojum, 107
　aestivum, 27

nicaeense, 107
Leycesteria crocothyrsos, 90
Libocedrus, 134, 136
Lilium, 107
   martagon, 108
   pomponium, 107
   rubrum, 107
Lily-of-the-valley, 9
Linaria vulgaris (peloric form), **79**
Linum perenne, 9
   usitatissimum, 46
Lobelia erinus, 22
Lonicera, 93
   periclymenum, 7, 15
Lotus corniculatus, 35
Lucerne, 34
Lychnis alpina, 9
   flos-cuculi, 8, 13
     'dwarf form', 13
   viscaria, 14
Lysimachia nummularia, 9, 10
Lythrum salicaria, 8
   'Firecandle', 13
Magnolia, 22
Maize, 46
Mallow, Musk, 9
Malus domestica, 20
Malva moschata, 9
Maple, Field, 9
Margyricarpus pinnatus, 23
   setosus, 23
Marigold, Corn, 36
Marjoram, 10, 11
Marsh-marigold, 8, 14
Matricaria discoidea, 42, *44*
Matteuccia struthiopteris, 56
Meadowsweet, 10
Medicago lupulina, 37
   sativa
     ssp. *falcata*, 34
     ssp. *sativa*, 34
Medick, Black, 37
   Sickle, 34
Medicks, 35
Melica uniflora, 10
Melick, Wood, 10
Mentha suaveolens, 10
Mertensia paniculata, 99
Metasequoia, 134, 136
   glyptostroboides, 129
Microbiota, 134, 136
Microcachrys, 133, 135
Microstrobos, 133, 135
Milium effusum, 10
   'Bowles Golden Grass', 10
Millet, Wood, 10

Mimulus, 32, 59-64
   aurantiacus, 62
   caespitosus, 62
   ×*caledonicus*, 64
   cardinalis, 62
   cupreus, 61, 63, 64
     'Andean Nymph', 63
     'Inca Sunset', 63
     'Roter Kaiser', 63
     'Whitecroft Scarlet', 63
     ×*M. nummularius*, 63
     ×*M. variegatus*, 63
   guttatus, 59, 60, 63, 64
   ×*hybridus*, 63, 64
   langsdorffii, 60, 63
   langsdorfii, 59
   lewisii, 62
   luteus, 59-61, 63, 64
     complex, 62-64
     var. *rivularis*, 60, 63
     ×*M. guttatus*, 60
   ×*maculosus*, 63, 64
   moschatus, 62
   nummularius, 60, *61*, 61, 63, 64
     ×*M. variegatus*, 63
   ×*polymaculus*, 64
   punctatus, 59
   puniceus, 62
   primuloides, 62
   ringens, 62
   rivularis, 60
   ×*robertsii*, 60, 64
   Section *Diplacus*, 62
   Section *Erythranthe*, 62
   Section *Eumimulus*, 62
   Section *Mimulus*, 62
   Section *Paradanthus*, 62
   Section *Simiolus*, 62
   ×*smithii*, *61*, 61, 63, 64
   Subgenus *Mimulus*, 62
   Subgenus *Schizoplacus*, 62
   Subgenus *Synplacus*, 62
   tilingii, 62
     var. *caespitosus*, 62
   variegatus, 61, 63, 64
     ×*M. nummularius*, 61
Mint, Round-leaved, 10
Misopates orontium, 7
Molinia caerulea, 10
Monkeyflowers, 32
Monkey Puzzle, **138**
Montbretia, 24, 26
Moor-grass, Purple, 10
Muscari, 107
   gussonei, 107
   macrocarpum, 108

Mycelis muralis, 27
Myrtus apiculata, 23
Nagara nana, 19
Nageia, 133, 135
Narcissus, 92, 107
   bulbocodium, 107
   juncifolius, 107
Neocallitropsis, 134, 136
Nostoc punctiforme, 22
Nothotsuga, 134, 136
Oak, Pedunculate, 9, 11-13
   Sessile, 11
Oat-grass, False, 10
   Meadow, 37
Onion, 103
Onoclea sensibilis, 56
Orchid, Loose-flowered, 86
Orchis laxiflora, 86
   militaris, 86
Oreopteris limbosperma, 129
Origanum vulgare, 10, 11
Ornithogalum reverchonii, 107
Osier, 116
   Common, 116
   Purple, 116
Osmunda regalis, 53
Oxlip, 9
Oxtongue, Hawkweed, 37
Paeonia mascula, 84
Palm, Coconut, 23
Papuacedrus, 134, 136
Parasitaxus, 133, 135
Parsley, Fool's, 36
Parsley-piert, 8
Pasqueflower, 9
Pear, Plymouth, 32, 87
Pelargoniums, 92
Penny-cress, Perfoliate, 4
Periwinkle, Madagascar, 90
Pernettya mucronata, 23
Persicaria bistorta 'Superbum', 13
Phacelia campanularia, 99
Phalaris arundinacea, 10
Philadelphus, 78
Phormium tenax, 23
Phygelius capensis, 22
Phyllitis crispa, 51
   scolopendrium, 9, 78
Phyllocladus, 133, 135
Picea, 120, 134, 136
   omorika, 129
Picris hieracioides ssp.
   grandiflora, 37
Pigmyweed, New Zealand, 47
Pilgerodendron, 134, 136
Pine, Chile, **138**

Huon, **139**
Scots, **3**, 9, 120
Pineappleweed, 42, 43
Pink, Clove, 20
  Maiden, 9
Pinus, 120, 134, 136
  aristata, 129
  sylvestris, **3**, 9, 129
Plantago lanceolata, **15**
  ('Rose-flowered' form), **15**
  major, 11, 15
  media, 15
Plantain, Greater, 11, 15
  Hoary, 15
  Rose, 14
Platycladus, 134, 136
Podocarpus, 133, 135
  salignus, 123, **126**, **127**, 129
Polemonium caeruleum, 9
Polygonum cuspidatum, 21
  rurivagum, 36
Polypodium cambricum, 55
  'Cambricum', 51, **52**
  interjectum, 129
  vulgare s.l., 51
Polystichum setiferum
  'Divisilobum Wollaston', 53
  'Gracillimum', 54
  'Plumosum Bevis', 54
  'Plumosum Drueryi', 54
  'Plumosum Green', 54
Populus tremula, 12, 13
Primrose, 9
  Scottish, 9
Primula, 14
  elatior, 9, 14
    'Hose in Hose', **Plate 2**
  'Galligaskins', 14
  'Hose in Hose', 14
  'Jack-a-napes-on-horseback', 14
  'Jack in the Green', 14
  scotica, 9
  veris, 9, 13, 38
  vulgaris, 9, 14
    'Alba Plena', 14, **Plate 2**
    'Jack in the Green', **Plate 2**
    'Lilacina Plena', 14
Prumnopitys, 133, 135
Prunella grandiflora, 37
  ssp. *pyrenaica*, 37
  vulgaris, 37
Prunus avium, 9, 12
  'Plena-Pendula', 12
  padus, 11
  spinosa, 11
Pseudolarix, 134, 136

amabilis, 129
Pseudotaxus, 133, 135
Pseudotsuga, 134, 136
  menziesii, **124**
Pteris cretica, 56
Pulsatilla vulgaris, 9
Purple-loosestrife, 8, 13
Pyrus cordata, 32, 87
Quercus petraea, 11
  robur, 9, 11-13
Ragged-robin, 8, 13
Ranunculus acris, 14
  ficaria, 11, 13, 14
    'Brazen Hussy', 11
    (double-flowered), **Plate 1**
    'Colarette', 14
    'Green Petal', 14
    var. *major*, 13
  marginatus var. *marginatus*, 37
  repens, 14, 78
  sardous, 37
    var. *trachycarpus*, 37
Rape, Oil-seed, 46
Red-cedar, Western, **124**
Retrophyllum, 133, 135
Rhododendron, 18, 42, 43
  Wild, 26
Rhododendron, 21, 26, 96
  blandfordiiflorum, 96
  bullatum, 96
  catawbiense, 26
  cinnabarinum, 96
    ssp. *cinnabarinum*, 96
    ssp. *tamaense*, 96
    ssp. *xanthocodon*, 96
    var. *pallidum*, 96
    var. *purpurellum*, 96
  concatenans, 96
  edgeworthii, 96
  keysii, 96
  maximum, 26
  ponticum, 21, 26, 42, *44*
    ssp. *ponticum*, 26
  roylei, 96
  sciaphilum, 96
  Subsection *Cinnabarina*, 96
  Subsection *Edgeworthia*, 96
Rhubarb, 91
Ribes sanguineum, 22, 27
Ribwort, **15**
Rosa arvensis, 7
Rowan, 11, 13
Rubus spectabilis, 22
Rumex pulcher, 35
  sanguineus, 11
Ruscus aculeatus, 15

Rye-grass, 33
Saffron, 103
Sage, Wood, 11
Salix, 90, 113-118
  alba, 11, 114, 115
    'Polsdonk', 115
    'Sericea', 115
    'Splendens', 11
    var. *alba*, 115
    var. *caerulea*, 115
    var. *vitellina*, 116
      ×*S. babylonica*, 116
    'Vries', 115
    ×*S. fragilis*, 115
      var. *decipiens*, 114
    ×*S. pentandra*, 116
    ×*S. purpurea*, 115
    ×*S. triandra*, 115
  ×*alopecuroides*, 115
  aurita, 118
    ×*S. cinerea*, 118
    ×*S. viminalis*, 118
    ×*S. caprea*, 118
  babylonica, 116
    ×*S. fragilis*, 115
  ×*calodendron*, 117
  caprea, 12, 116, 117
    'Kilmarnock', 12
    ssp. *sphacelata*, 117
    ×*S. cinerea* sspp., 117
    ×*S. cinerea*×*S. viminalis*, 117
    ×*S. daphnoides*, 117
    ×*S. viminalis*, 117
    nothomorph, 117
    ×*S. aurita*, 117
  cinerea, 116, 118
    ssp. *cinerea*, 117
    ssp. *oleifolia*, 117
    ×*S. myrsinifolia*, 117
    ×*S. phylicifolia*, 117
    ×*S. viminalis*, 117
  ×*dasyclados*, 117, 118
  ×*doniana*, 116
  ×*ehrhartiana*, 116
  ×*forbyana*, 116
  fragilis, 114, 115
    var. *decipiens*, 114
    var. *fragilis*, 114
    var. *furcata*, 114
    var. *russelliana*, 114
    ×*S. pentandra*, 115
    ×*S. triandra*, 115
  ×*fruticosa*, 118
  ×*hungarica*, 117
  lanata, 9

×*meyeriana*, 115
×*mollissima*, 116
×*mutinervis*, 118
×*pendulina*, 115
pentandra, 114
  'Patent Lumley', 114
×*pontederiana*, 116
purpurea, 12, 116
  ×*S. cinerea*, 116
  ×*S. repens*, 116
  ×*S. viminalis*, 116
repens, 11, 12
  'Boyd's Pendulous', 12
  var. *argentea*, 11
×*rubens*, 115
  nothovar. *basfordiana*, 115
    f. *basfordiana*, 115
    f. *sanguinea*, 115
×*rubra*, 116
×*sepulcralis* nothovar.
    chrysocoma, 116
×*sericans*, 117
×*smithiana*, 117
×*sordida*, 116
×*stipularis*, 118
Subgenus *Salix*, 113-116
Subgenus *Vetrix*, 114-117
Subgenus *Chamaetia*, 114, 118
×*taylorii*, 116
triandra, 114-116
  'Black Hollander', 116
  'Semperflorens', 116
  ×*S. viminalis*, 116
viminalis, 116, 117
Sallow, Common, 117
  Eared, 118
  Grey, 117
Sambucus nigra, 10, 11
Sanguisorba minor
  ssp. *minor*, 37
  ssp. *muricatus*, 37
Sarracenia purpurea, 18, 22
Saxegothaea, 133, 135
  conspicua, 129
Saxifraga, 97
Sciadopitys, 134, 136
Scilla, 107
Scrophularia auriculata, 10
Scrophulariaceae, 78
Sea-buckthorn, 9
Sedge, Bird's-foot, 10
  Pendulous, 8
  Stiff, 10
Selfheal, 37
  Greater, 37
Senecio vulgaris, 78

Sequoia, 134, 136
  sempervirens, 129
Sequoiadendron, 134, 136
Shepherd's-purse, 7
Silene dioica, 13, 14
  (flore plena), **Plate 1**
  uniflora, 14
Sisyrinchium angustifolium, 19, 22
  bermudiana, 22
Skimmia, 92
Snapdragon, Lesser, 7
Sneezewort, 14
Snowdrop, 20, 33
Soft-grass, Creeping, 10
Soft-rush, 15
Sollya heterophylla, 99
Sorbus aria, 11
  'Lutescens', 11
  aucuparia, 11, 13
  'Sheerwater Seedling', 13
Spiraea douglasii, 22
Spotted-orchids, 32
Spurge, Wood, 11
Sternbergia, 105, 107
  candida, 108
  clusiana, 108
  fischeriana, 108
  lutea, 108
Sundacarpus, 133, 135
Sunflower, 46
Sycamore, 20, 42
Symphytum, 65-70
  asperum, 65-70
    ×*S. officinale*, 66-68
  bulbosum, 65
  caucasicum, 65, 70
  grandiflorum, 65, 68-70
    ×*S.* ×*uplandicum*, 70
  × 'Hidcote Blue', 65, **68**, 70
  officinale, 65-68
    complex, 65
    'Peppermint Stripe', 66, 67
  orientale, 65, 70
  tauricum, 65
  tuberosum, 65
  × *uplandicum*, 65-70
    complex, 67
    'variegatum', 67
    ×*S. tuberosum*, 65
    'variegatum', 68
    ×*S. officinale*, 67
Taiwania, 134, 136
Tanacetum parthenium, 77
  vulgare, 11
  var. *crispum*, 11
Tansy, 11

Taxodium, 134, 136
  distichum, 129
Taxus, 133, 135
  baccata, 9, 12, 13
  'Dovastoniana', 12
  'Fastigiata', 13
  'Repandens', 12
Tecophilaea cyanocrocus, 102
Tetraclinis, 134, 136
Teucrium scorodonia, 11
  'Crispum', 11
Thistle, Cabbage, 46
Thorn, Glastonbury, 15
Thuja, 134, 136
  plicata, **124**, **125**, 129
Thujopsis, 134, 136
Thyme, Wild, 11
Thymus polytrichus, 11
  var. *languinosus*, 11
Tmesipteris, 19
Toadflax, Common (peloric form),
    79
Toadflax family, 78
Torreya, 133, 135
Tree-fern, 19
Trichomanes speciosum, 56
Trichoniscus stebbingi, 19
Trichorina tomentosa, 19
Trifolium pratense, 11, 37
  'Susan Smith', 11
  repens, 11, 23
  'Calvary Clover', 11
Trillium, 107
Tsuga, 134, 136
  heterophylla, 124, 129
Tulipa, 103, 107
Ulex europaeus, 13, 23
Vernal-grass, Sweet, 8
Veronica filiformis, 21
Vetch, 32
  Kidney, 13, 35
  [Alien], 35
Viburnum opulus, 9, 10
Viola odorata, 9
Violet, Sweet, 9
Wahlenbergia, 19
  polytrichifolia, 99
Wallflower, 20
Walnut, 111
Whitebeam, Common, 11
Whitlowgrass, Yellow, 9
Widdringtonia, 134, 136
Willow, Almond, 116
  Bay, 114
  Bedford, 114

Crack, 113-115
  (British), 114
  (European), 114
Creeping, 11, 12
Cricket-bat, 115
Eared, 118
Goat, 12, 117
Golden, 116
Grey, 117
Purple, 12, 116
Rusty, 117
Shiny-leaved, 115
Weeping, 115
  (Golden), 116
White, 11, 113, 115
Woolly, 9
Willowherb, Broad-leaved, 8
  New Zealand, 23
Wood-lice, Exotic, 19
Woodruff, 9

Woodsia, 53
  alpina, 55
  ilvensis, 53
Wormwood, 11
Yarrow, 36
Yellow-rattle, 32
Yew, 9, 12, 13, 120